Proceedings of the
Second International Conference
on the
Behaviour of Off-Shore Structures

HELD AT
IMPERIAL COLLEGE, LONDON, ENGLAND
AUGUST 28th - 31st, 1979

VOLUME 2

Sponsored by:
Norwegian Institute of Technology
Delft University of Technology
Massachusetts Institute of Technology
University of London

Secretariat and Publishers:
BHRA Fluid Engineering
Cranfield
Bedford MK43 0AJ
UK

Editors: H.S.Stephens
Mrs. S.M.Knight

Volume 2 contains 30 of the invited papers and 9 of the contributions presented at the 2nd International Conference on the Behaviour of Off-Shore Structures. Volume 1 contains 26 of the invited papers and 13 of the contributions, as well as the 3 general lectures. The remaining papers and contributions together with an edited record of the discussion will appear in Volume 3.

The Organisers are not repsonsible for statements or opinions made in either the papers, the contributions or the discussion.

These papers have been reproduced by offset printing from the author's orginal typescript to minimise delay.

CITATION

When citing papers from this volume, the following reference should be used:-

Title, Author, Proc. 2nd International Conference on the Behaviour of Off-shore Structures, BHRA Fluid Engineering, Cranfield, Bedford, England.
Volume 2, Paper No., Pages, (August, 1979).

 British Library Cataloguing in Publication Data

```
BOSS '79 (Conference), London, 1979
  BOSS '79.
  1. Offshore structures - Congresses
  I. Stephens, Herbert Simon
  II. Knight, S  III. British Hydromechanics
  Research Association. Fluid Engineering
  627'.98          TC1505

ISBN 0-906085-34-9
```

Published by

BHRA Fluid Engineering
Cranfield, Bedford, MK43 0AJ, England.

Printed by the Cotswold Press Ltd., England

© copyright 1979. BHRA Fluid Engineering

Set of 3 vols: ISBN 0 - 906085 - 34 - 9
Vol. 1   ISBN 0 - 906085 - 35 - 7
Vol. 2   ISBN 0 - 906085 - 36 - 5
Vol. 3   ISBN 0 - 906085 - 37 - 3

ORGANISING COMMITTEE

Professor B.G. Neal (Chairman)	Dept. of Civil Engineering, Imperial College, London
Prof Dr Ir E.W. Bijker THE NETHERLANDS	Dept. of Civil Engineering, Delft University of Technology, The Netherlands.
Professor A.W. Bishop	Dept. of Civil Engineering, Imperial College, London
Professor R.E.D. Bishop	Dept. of Mechanical Engineering, University College London
Prof. C. Chryssostomidis U.S.A.	Dept. of Ocean Engineering, Massachusetts Institute of Technology, U.S.A.
Dr R. Eatock Taylor	Dept. of Mechanical Engineering, University College London
Prof. N. Janbu NORWAY	Divn of Soil Mechanics and Foundation Eng., Norwegian Institute of Technology, Norway
Professor K.O. Kemp	Dept. of Civil Engineering, University College London
Dr. P.R. Vaughan	Dept. of Civil Engineering, Imperial College, London
G.A.J. Young	Assistant Director BHRA Fluid Engineering
H.S. Stephens (Secretary-General)	Assistant Director, BHRA Fluid Engineering

INTERNATIONAL COMMITTEE

The Netherlands

Prof Dr Ir E.W. Bijker	Dept. of Civil Engineering, Delft University of Technology
Mr. F.K. Ligtenberg	Director, Institute T.N.O. for Building Materials and Structures
Mr. A. Paape	Assistant Director, Delft Hydraulics Laboratory

Norway

Prof. O.M. Faltinsen	Division of Ship Hydrodynamics, The Norwegian Institute of Technology
Prof. N. Janbu	Division of Soil Mechanics and Foundation Engineering, The Norwegian Institute of Technology
Prof. T. Moan	Division of Ship Structures, The Norwegian Institute of Technology

U.S.A.

Dr. I. Dyer	Department of Ocean Engineering, Massachusetts Institute of Technology
Dr. A.A.H. Keil	School of Engineering, Massachusetts Institute of Technology
Prof. J.J. Connor	Department of Civil Engineering, Massachusetts Institute of Technology

U.K.

Prof. A.W. Bishop	Dep. of Civil Engineering, Imperial College, London
Prof. R.E.D. Bishop	Dept. of Mechanical Engineering, University College London
Prof. K.O. Kemp	Dept. of Civil Engineering, University College London
Prof. B.G. Neal	Dept. of Civil Engineering, Imperial College, London

2nd INTERNATIONAL CONFERENCE ON THE BEHAVIOUR OF OFF-SHORE STRUCTURES

CONTENTS
(Volume 2)

The following papers were presented at the conference:

Paper		Page
	Foundation/Structure Interaction and Fluid/Structure Interaction	
48	A Probabilistic Model for Stiffness Degradation of Steel Jacket Structures. D.C.Angelides and J.J.Connor, Massachusetts Institute of Technology, U.S.A.	1
49	Modal Superposition v Direct Solution Techniques in the Dynamic Analysis of Offshore Structures. J.H.Vugts and I.M.Hines, Shell Internationale Petroleum Maatschappij BV, The Netherlands. R.Nataraja and W.Schumm, Lloyd's Register of Shipping, U.K.	23
50	The Importance of Structural Motion in the Calculation of Wave Loads on an Offshore Structure. P.Fish and R.Rainey, Atkins Research and Development, U.K.	43
51	A Linear Analysis of Interaction Problems in Offshore Platforms. R.Eatock Taylor, University College London, U.K.	61
52	An Initial Study of the Dynamic Behaviour of the New Christchurch Bay Tower. B.R.Ellis and A.P.Jeary, Building Research Establishment, U.K.	87
53	Use of Generalized Coordinates in the Non-Linear Time-Domain Analysis of Steel Jackets. A.K.Basu and R.P.Singh, Indian Institute of Technology, India.	97
*54	A Review of Hydrodynamic Loads on Off-Shore Structures and Their Formulation. J.H.Vugts, Shell Internationale Petroleum Maatschappij BV, The Netherlands.	
	Wave and Current Induced Motions	
55	Hydrodynamic Coefficients of Rectangular Barges in Shallow Water. J.A.Keuning, Delft Hydraulics Laboratory, The Netherlands. W.Beukelman, Delft University of Technology, The Netherlands.	105
56	Reciprocity Relations for Offshore Structures Vibrating at Acoustic Frequencies. C.C.Mei, Massachusetts Institute of Technology, U.S.A.	125
57	Dynamic Response of Tethered Production Platform in a Random Sea State. C.L.Kirk and E.U.Etok, Cranfield Institute of Technology, U.K.	139
58	Hydrodynamic Model for a Dynamically Positioned Vessel. M.S.Triantafyllou, Massachusetts Institute of Technology, U.S.A.	165
59	Hydrodynamic Analysis of Tankers at Single-Point-Mooring Systems. O.M.Faltinsen, O.Kjaerland, N.Liapis and H.Walderhaug, Norwegian Institute of Technology, Norway.	177

* These papers will be published in Volume 3.

60	Production Riser Analysis. M.P.Harper, Koninklijke/Shell Exploratie en Produktie Laboratorium, Rijswijk, Netherlands.	207
61	On the Hydrodynamic Analysis of Arbitrarily shaped and Multibodied Marine Structures - Application to Wave-Energy Devices and Ship/Dock Interaction. M.Katory, Hong Kong Polytechnic, Hong Kong. A.A.Lacey, Institute of Mathematics, Oxford, U.K.	213
62	Floating Breakwaters and Topologically Generated Circulation. B.MacMahon, Imperial College, U.K.	223
63	Wave Induced Motions of Marine Deck Cargo Barges. W.P.Stewart, Atkins Research and Development, U.K.	229

Geotechnical Performance of Off-Shore Structures

64	Three Gravity Platform Foundations. I.Foss and J.Warming, Det Norske Veritas, Norway.	239
*65	How Successful have Performance Monitoring Programs been for Gravity Base Structure? E.Dibiagio, F.Myrvoll and S.Borg Hansen, Norwegian Geotechnical Institute, Norway.	
66	Large Model Tests for the Oosterschelde Storm Surge Barrier. L. de Quelerij and J.K.Nieuwenhuis, Rijkswaterstaat, The Netherlands. M.A.Koenders, Delft Soil Mechanics Laboratory, The Netherlands.	257
67	Developments in Piling for Offshore Structures. W.J.Rigden, BP Trading Ltd., U.K. J.J.Pettit, BSP International Foundations Ltd., U.K. H.D. St. John, Building Research Establishment, U.K. T.J.Poskitt, Queen Mary College, U.K.	279
*68	Performance of Dynamically Loaded Pile Foundations. R.G.Bea, J.M.E.Audibert, J.B.Stevens and M.R.Akky, Woodward-Clyde Consultants, U.S.A.	
69	The Influence of Hydrostatic Pressure on the Process of Underwater Excavation with Special Reference to the Underwater Bulldozer. V.A.Lobanov, Odessa Insitute of Civil Engineering, U.S.S.R.	297

Ultimate Limit State

70	Buckling Strength and Post-Collapse Behaviour of Tubular Bracing Members including Damage Effects. C.S.Smith, W.Kirkwood and J.W.Swan, Admiralty Marine Technology Establishment, U.K.	303
71	Current Research into the Strength of Cylindrical Shells used in Steel Jacket Construction. P.J.Dowling and J.E.Harding, Imperial College, U.K.	327
72	Buckling of Compressed, Longitudinally Stiffened Cylindrical Shells. A.C.Walker and S.Sridharan, University College London, U.K.	341
73	Transverse Impact on Beams and Slabs. I.C.Brown and S.H. Perry, Imperial College, U.K.	357

* These papers will be published in Volume 3.

74	Strategy for Monitoring, Inspection and Repair for Fixed Offshore Structures. P.W.Marshall, Shell Oil Company, U.S.A.	369
75	Offshore Oil Production and Drilling Platforms. Design Against Accidental Loads. S.Fjeld, Det Norske Veritas, Norway.	391
76	Some Aspects of Double-Skin Composite Construction for Sub-Sea Pressure Chambers. P.Montague and C.D.Goode, University of Manchester, U.K.	415

Interpretation of Full Scale Data

77	Environmental and Structural Instrumentation of Platforms on the Norwegian Continental Shelf. I.Holand, The Norwegian Institute of Technology, Norway. S.Berg and G.Beck, Offshore Technology Testing and Research Group (OTTER), Norway.	425
78	Observed Foundation Behaviour of Concrete Gravity Platforms Installed in the North Sea 1973-1978. O.Eide, K.H.Andersen and T.Lunne, Norwegian Geotechnical Institute, Norway.	435
*79	Full-Scale Wave Loading on Cylinders. P.Holmes and R.G.Tickell, University of Liverpool, U.K.	
80	Recent Signal Processing Advances in Spectral and Frequency Wave-number Function Estimation and their Application to Offshore Structures. A.B.Baggeroer, Massachusetts Institute of Technology, USA.	457
81	Preliminary Results of the Geotechnical Instrumentation Installed Below the TCP 2 Gravity Structure (Frigg Field) During its First Winter (1977-78) Period. J.P.Mizikos, SNEA(P) - France.	477
*82	Presentation of the Brent B Instrumentation Project. I.Foss, Det Norske Veritas, Norway.	
83	Measuring Equipment for Field Investigations on Near Surface Wave Forces. F.Busching, E.Martini and U.Sparboom, Tech, Univ. Braunschweig, Federal Republic of Germany.	493

Safety

84	Methods of Reliability Analysis for Jacket Platforms. M.J.Baker and T.A.Wyatt, Imperial College of Science and Technology, U.K.	499
85	Probabilistic Reliability Analysis. A.L.Bouma, University of Technology, Delft, The Netherlands. Th. Monnier and A.Vrouwenvelder, Institute TNO for Building Materials and Building Structures, The Netherlands.	521
86	The Safe Design and Construction of Steel Spheres and End Closures of Submersibles, Habitats and Other Pressurised Vessels. D.Faulkner, University of Glasgow, U.K.	543

* These papers will be published in Volume 3.

87	Future Safety Considerations. A matter of Scarcity and Probabilistic Approach. Ch.J.Vos and B.J.G. Van der Pot, Delta Marine Consultants, Netherlands. J.K.Vrijling, Volker Stevin Group, Netherlands.	557

Future Problems

*88	A General Review of Future Problems and their Solution. D.Thornton, The British Petroleum Co. Ltd., U.K.	
89	Adjustment of Structural Concrete Techniques to Offshore Conditions. A.J.Harris and B.M.Fox, Harris & Sutherland, U.K.	571
90	Seabed Containment Structures for Hydrocarbon Production. J.A.Derrington, M.J.Collard and J.M.Skillman, Sir Robert McAlpine & Sons Limited, U.K.	577
91	Netherlands Marine Technological Research. G.A.Heyning, Netherlands Industrial Council for Oceanology, The Netherlands.	593
92	Marine Fouling on Platforms in the Northern North Sea R.Ralph, Aberdeen University, U.K.	605

* These papers will be published in Volume 3.

BOSS'79

PAPER 48

Second International Conference
on Behaviour of Off-Shore Structures

Held at: Imperial College, London, England
28 to 31 August 1979

A PROBABILISTIC MODEL FOR STIFFNESS DEGRADATION OF STEEL JACKET STRUCTURES

D.C. Angelides and J.J. Connor

Massachusetts Institute of Technology, U.S.A.

Summary

This paper is concerned with the development and application of an integrated model for the short and long term dynamic response of an offshore structure to random wave excitation. A discrete model of the upper structure is coupled with a finite element model for the pile-soil medium, and the system is subjected to "probabilistic" storms characterized by three components: duration, intensity, and discretized significant wave height variation over the storm duration.

Soil (clay) degradation, due to cyclic excitation, is followed during the passage of a storm. The soil deformation is determined with a frequency domain finite element non-linear iterative model based on hyperbolic type constitutive relations with parameters specified from existing experimental information (Andersen, BOSS 1976). Variation of the pile stiffness with the level of excitation at the top of the pile and number of cycles is determined by repeated application of the model.

The upper structure is solved separately, in the frequency domain, and the foundation stiffness, obtained from the soil-pile model, are adapted iteratively. Sarpkaya's 1977 experimental information is used to specify the values of C_M and C_D corresponding to the flow and response parameters. The end result is an expression for soil-pile stiffness as a function of the significant wave height H_S and the number of cycles N of the excitation.

A Markov model is developed for soil degradation and the transition probability propagation through a storm is derived with a simulation technique using either "historical" or "synthetic" random storms. Continuous soil degradation is considered during each storm. Between successive storms, the two limiting cases of no recovery and total recovery of the soil stiffness are examined. Finally a reliability analysis (first passage problem) is developed for the soil-pile stiffness.

Sponsored by: Delft University of Technology, The Netherlands
Massachusetts Institute of Technology, U.S.A.
The Norwegian Institute of Technology, Norway
University of London, England

Secretariat provided by: BHRA Fluid Engineering

Copyright: ©   BHRA Fluid Engineering
Cranfield, Bedford, England

INTRODUCTION

An assessment of the behavior of an offshore structure requires a combination of hydrodynamical, structural, goetechnical, and statistical expertise. Uncertainties associated with the environmental conditions and system properties influence the response of the structure. The primary sources are: (i) statistical representation of the sea state, (ii) hydrodynamic loading parameters, and (iii) dynamic soil properties and their degradation under cyclic loading.

Reduction in foundation stiffness due to soil degradation (clays subjected to cyclic loading suffer degradation when the cyclic stress is above a lower limit, (NGI, 1975; Andersen, 1976)) increases the fundamental period of the structure and the dynamic amplification of the structural response. Deep water structures, such as the HONDO and COGNAC, have periods in the region of 5 seconds. Although the dominant wave excitation is in the region of 10-15 seconds, a shift in the structural period magnifies the low frequency excitation and may have a significant long term effect.

Fatigue and other time-dependent phenomena are usually analyzed with methods that do not consider the order in which the individual cycles of the excitation are applied. One case where the loading history is important is soil fatigue. Fatigue induced by cyclic loading on a soil is a consequence of nonlinear behavior and therefore models based on superposition of the accumulative effect of cycles of different intensities without distinguishing the order in which the individual cycles occur (Miner's rule) have been found to be inaccurate (Van Eekelen, 1977). In this paper an integrated methodology is developed for the analysis of an offshore structure in a nonstationary wave environment. Deterministic and stochastic concepts are applied in parallel to obtain an improved representation of: (1) forcing, (2) short and long-term behavior of the soil (clay) under cyclic loading and (3) the response of the upper structure. For the purpose of investigating the sensitivity of the response of an offshore structure to force, wave environment, and soil uncertainties as well as degradation of the foundation stiffness with the number of cycles of excitation, the following models are developed: (1) a simplified frequency domain model for the upper structure, (2) a frequency domain model for the soil-pile interaction including degradation with the number of cycles, (3) a probabilistic model for the representation of the sea state, (4) a stochastic model for the degradation of the foundation stiffness.

MODELLING STRATEGY

The upper structure is modelled as an "equivalent planar beam" with lumped masses for the dynamic response. Hydrodynamic forcing is evaluated on the original three-dimensional structure and Sarpkaya et al's information (1977) is used to select values for C_M and C_D (coefficients in Morison's equation) which are consistent with the statistical average measures of relative motion for each member of the structure. Iteration is required but the process converges rapidly.

Soil-pile interaction is analyzed with an iterative linear quasi-three dimensional finite element program which can treat either end bearing or floating piles subjected to harmonic forcing (horizontal, vertical, moment) applied to the top of the pile. The program solves for the displacements due to a representative range of values of the foundation force and moment, and computes the equivalent foundation stiffnesses. These stiffness values are input in the upper structure model, which then adapts the foundation stiffness to the particular wave excitation by iterating on the response.

Parameters of the sea state model are: the mean rate of arrival of storms, the joint probability distribution of storm duration and average intensity, and the random process that describes the variation of a statistical wave height measure (significant wave height, H_s, or visual wave height, H_v) during each storm. The model is nonstationary at a micro-scale (within each storm duration), but is stationary at a macro-scale (storm arrivals, storm duration and intensity parameters have probability characteristics that do not depend on time). The evolution of H_s or H_v within each storm is represented through a nonstationary random process with parameters that depend on storm duration and intensity. This feature allows one to model the initial buildup and the terminal decay of storm intensity. All parameters are inferable from historical records and numerical estimates are obtained in this study using visual wave-height data from the North Atlantic. The model is well-suited for simulation, i.e., for producing "synthetic storms" which can be used as realistic inputs to the structure-foundation system.

In the stochastic model for the degradation of the foundation stiffness, the indicator of soil degradation is taken to be the value of the horizontal foundation stiffness k_{xx}. Degradation of stiffness is modeled by a homogeneous Markov process, discrete in state and continuous in time. Occurrence of storms is defined as a Poisson process. A state transition probability matrix $\underline{P}$ is generated by repeated numerical simulation of storms followed by analysis of stiffness degradation, starting from different initial states. Two extreme cases of recovery of stiffness are analyzed: (1) total recovery following each storm and (2) continuous degradation with no recovery. In each case, it is easy to calculate the probability that critical states will be reached during an arbitrary time interval and for any initial state distribution.

A more detailed description of the various models is presented in the following sections. Results are included for a typical North Sea structure.

UPPER STRUCTURE MODEL

The structural geometry and pile configuration are assumed to be doubly symmetric with one of the symmetry axes coinciding with the direction of wave propagation. An approximate 2-dimensional model is generated by lumping the masses and hydrodynamic forces at the various panel nodes and simulating the stiffness with an equivalent planar beam. Fig. 1 illustrates this discretization. There are 2 displacement measures per node, the horizontal translation and the rotation of the equivalent beam cross-section. Rotatory inertia is neglected. The vertical variation of quantities such as translation, velocity, is assumed to be step-wise, i.e., they are taken as constant over the tributary zone for a node, defined here as the region halfway above and below the node. Phase differences between the hydrodynamic forcing for members located in different vertical planes is included in the evaluation of the total nodal forces.

Rigidity coefficients for the equivalent beam are generated with a complementary energy argument. One applies self-equilibrating force systems to a typical panel, evaluates the complementary energy and then by differentiating with respect to the shear force and bending moment, obtains the flexural and shear rigidities. Considering the X bracing shown in Fig. 2, the approach leads to

$$\frac{1}{(EI)_{eq}} = \frac{2}{h^2 A_c E_c} \tag{1}$$

$$\frac{1}{(GA_s)_{eq}} = \frac{L^3}{\ell h^2} \frac{1}{2 A_d E_d} \tag{2}$$

for an individual truss parallel to the flow direction.

FOUNDATION MODEL

The strategy followed here is to evaluate the response of a single pile subjected to periodic forcing, allowing for nonlinear soil behavior, and generate quasi-linear pile stiffness coefficients which are combined to simulate the pile foundation system. A finite element pile model developed by Blaney et al (1976) is extended to evaluate vertical stiffness of piles, simulate floating piles, and treat nonlinear soil behavior in an approximate way (Angelides, 1978).

The soil is assumed to be linear viscoelastic with hysteretic type damping reproduced through a complex modulus $G(1+2iD)$, where G is the dynamic secant shear modulus and D is the damping ratio. The relation between the damping ratio and the shear modulus is taken as (Hardin and Drnevich, 1972)

$$D = D_{max} (1 - \frac{G}{G_{max}}) \tag{3}$$

$$D_{max} = 0.3 - 0.0065 \log N \tag{4}$$

where G_{max} is the dynamic shear modulus for small strain amplitudes and N is the number of cycles. Frequency and confining pressure effects are disregarded for simplicity.

An expression for the secant shear modulus is obtained by modifying a hyperbolic type stress-strain relation to incorporate the effect of cyclic loading. It's form is:

$$G = \frac{G_{max}}{1+a\gamma[1+b\frac{\log N}{\log(c+dT)}]} \tag{5}$$

where γ is the shear strain amplitude, T is the period of the excitation, and a,b,c,d are constants determined from experimental data. Results published by NGI (1975), Andersen (1976), and Fisher et al (1976) provide the basis for this relation.

Fig. 3 illustrates the finite element discretization. Toroidal elements are employed for the core region and the far field is represented with a semi-analytical energy transmitting boundary which simulates radiation damping, Kausel (1974). Lateral radiation occurs only for frequencies above the fundamental frequency of the soil stratum.

The nonlinear soil behavior is treated by iterating on the shear modulus at the element level within the core region. Elastic behavior is ensured in the far field by adjusting the location of the transmitting boundary. Poisson's ratio and the damping coefficient are adjusted along with the shear modulus. However, the bulk modulus is assumed to be constant. The corrections are applied over the total volume of each torus to maintain the axi-symmetric stiffness characteristics of the system. Results presented here are based on using the maximum shearing strain in the r-z plane evaluated at $\psi=0$ and at the centroid of each element.

$$\bar{\gamma} = \left|\sqrt{\gamma_{rz}^2 + (\varepsilon_{rr} - \varepsilon_{zz})^2}\right|_{\psi=0} \tag{6}$$

This choice of strain measure is approximate. Other measures such as $\gamma_{r\theta}$ at $\psi=0$ are presently being examined.

Because of the nonlinear soil behavior, the total set of forces has to be applied to the pile. Fig. 4 defines the notation employed here. One specifies the magnitudes F_x, F_z, M, the frequency of excitation , the number of cycles N, and solves for the complex displacement measures v_x, v_z, ϕ in terms of the real force measures.

$$\begin{bmatrix} v_x \\ \phi \\ v_z \end{bmatrix} = \begin{bmatrix} f_{xx} & f_{x\phi} & 0 \\ f_{x\phi} & f_{\phi\phi} & 0 \\ 0 & 0 & f_{zz} \end{bmatrix} \begin{bmatrix} F_x \\ M \\ F_z \end{bmatrix} \tag{7}$$

Inverting the coefficient matrix results in the complex stiffness coefficients for a particular periodic excitation.

$$\begin{bmatrix} F_x \\ M \end{bmatrix} = \begin{bmatrix} k^*_{xx} & k^*_{x\phi} \\ k^*_{x\phi} & k^*_{\phi\phi} \end{bmatrix} \begin{bmatrix} v_x \\ \phi \end{bmatrix} \tag{8}$$

$$F_z = k^*_z v_z$$

where
$$k^* = \bar{k}e^{j\xi} = k+jc \qquad j=\sqrt{-1} \tag{9}$$

The pile configuration selected for the numerical study consisted of two rows of piles connected with a rigid cap (see Fig. 5). Noting the kinematic constraints, $v_z=d\theta$, and $\phi=\theta$, and applying simple statics, one obtains the corresponding foundation stiffness terms,

$$\begin{bmatrix} F_{tot} \\ M_{tot} \end{bmatrix} = \bar{n} \begin{bmatrix} k^*_{xx} & k^*_{x\theta} \\ k^*_{x\theta} & k^*_{\theta\theta}+d^2 k^*_{zz} \end{bmatrix} \begin{bmatrix} v_x \\ \theta \end{bmatrix} \tag{10}$$

Since the displacements depend on the foundation stiffness, iteration is required for the evaluation of the system response due to wave excitation. This problem is discussed in the following section.

SYSTEM RESPONSE TO WAVE EXCITATION

The sea state is not a stationary process. However, for short time intervals, of the order of one to several hours, the free surface fluctuation, $\eta(x,t)$, from the still water level (S.W.L.) can be approximated as a stationary process (Kinsman, 1965; Nordenstrøm, 1971; Cartwright, 1974). This allows one to use a spectral density function for the short term description. The well known model which takes $\eta(x,t)$ as a zero mean, Gaussian, stationary process that may be represented by a linear summation of an infinite number of sinusoids with phase angles randomly distributed between $0-2\pi$ is used in this study.

Returning to the upper structure discretization, shown in Fig. 1, the total hydrodynamic force at a node is obtained by summing the contributions of the members contained in the nodal tributary zone, defined here as the region halfway above and below the node. Introducing the double subscript ij to denote the jth member in zone i, and applying a modified form of Morison's equation, which attempts to account for motion of the structure (see Berge and Penzien, 1974; Moan, Haver and Vinje, 1975), the total force is written as

$$P_i(t) = \sum_j \{\rho(C_{M_{ij}} -1)\frac{\pi}{4} D_{ij}^2 \ell_{ij} (\ddot{u}_{ij}-\ddot{v}_i) + \rho\frac{\pi}{4} D_{ij}^2 \ell_{ij} \ddot{u}_{ij} + \frac{1}{2}\rho C_{D_{ij}} \ell_{ij} |\dot{u}_{ij}-\dot{v}_i|(\dot{u}_{ij}-\dot{v}_i)\} \quad (11)$$

where $C_{M_{ij}}$, $C_{D_{ij}}$ are the inertia and drag coefficients consistent with the flow and structural response characteristics for member (ij); D_{ij}, ℓ_{ij} are the diameter and projected length; $\ddot{u}_{ij}$, $\dot{u}_{ij}$ are the water particle acceleration and velocity; and $\ddot{v}_i$, $\dot{v}_i$ are the structural acceleration and velocity at node level i.

Since a spectral analysis approach is to be used the nonlinear term in Eq. (11) is linearized with an equivalent linearization technique (Krylov and Bogoliubov, 1947; Lin, 1967). Defining $\dot{r}_{ij}$ as the relative velocity for member ij,

$$\dot{r}_{ij} = \dot{u}_{ij} - v_i \quad (12)$$

one writes the quadratic term as

$$|\dot{r}_{ij}|\dot{r}_{ij} = a_{ij} \dot{r}_{ij} \quad (13)$$

Noting that the excitation is a zero mean Gaussian process and the system is linear (iteratively linear soil), the expression for a_{ij} reduces to

$$a_{ij} = \sqrt{\frac{8}{\pi}} \sigma_{\dot{r}_{ij}} \quad (14)$$

where $\sigma_{\dot{r}_{ij}}$ is the root mean square value of the relative velocity. Letting $S_{\dot{r}_{ij}\dot{r}_{ij}}$ represent the spectral density function (one sided) for $\dot{r}_{ij}$, $\sigma_{\dot{r}_{ij}}$ is determined from

$$\sigma_{\dot{r}_{ij}}^2 = (m_o)_{ij} \quad (15)$$

where $(m_\lambda)_{ij} = \int_0^\infty \omega^\lambda S_{\dot{r}_{ij}\dot{r}_{ij}}(\omega) d\omega \quad (16)$

One of the uncertainties influencing the response is associated with the values for C_M and C_D. Sarpkaya et al (1977) present data which exhibit a significant variation with Reynolds number (Re), Keulegan-Carpenter number (K) and relative roughness (k/D). In this work C_M and C_D are selected from Sarpkaya's information. In particular since the sea state is a random process, the following definitions for Reynolds number and Keulegan-Carpenter number are introduced:

$$(Re)_{ij} = \frac{\sigma_{\dot{r}_{ij}} D_{ij}}{\nu}, \quad (K)_{ij} = \frac{\sigma_{\dot{r}_{ij}} (T_o)_{ij}}{D_{ij}} \quad (17)$$

where ν is the kinematic viscosity and $(T_o)_{ij}$ is defined as

$$(T_o)_{ij} = 2\pi \sqrt{\frac{(m_o)_{ij}}{(m_2)_{ij}}} \quad (18)$$

Sarpkaya's data is fitted and explicit analytical relations for C_D, C_M are derived. These expressions, coupled with iteration on $\sigma_{\dot{r}_{ij}}$ allow one to identify the "consistent" values for the coefficients and a_{ij}.

It should be noted that Sarpkaya's experiments were conducted for single harmonic rectilinear flow past a fixed cylinder whereas the real conditions involve orbital motion, random excitation, and flexible cylinders. The procedure outlined above represents an attempt to model the hydrodynamic forcing more realistically than is now performed. However, more experimental research is needed.

Defining $S_{\eta\eta}(\omega_n)$ as the ordinate of a one sided wave spectrum, at the discrete frequency ω_n, the water particle velocity $\dot{u}_{ij}(t)$ can be expressed as

$$\dot{u}_{ij}(t) = \sum_{n=1}^{M} \sqrt{2S_{\eta\eta}(\omega_n)\Delta\omega}\, \omega_n\, G_{in} \cos(k_n x_{ij} - \omega_n t + \psi_n) \qquad (19)$$

where ψ_n is a random phase angle uniformly distributed between 0 and 2π; G_{in} and k_n are given by

$$\omega_n^2 = gk_n \tanh k_n h$$

$$G_{in} = \frac{\cosh k_n(z_i + h)}{\sinh k_n h} \qquad (20)$$

g = gravitational acceleration

A two parameter wave spectrum recommended by ISSC (1967) is adopted in this study.

$$S_{\eta\eta}(\omega) = \frac{H_s^2 T_z}{8\pi^2} \left(T_z \frac{\omega}{2\pi}\right)^{-5} \exp\left[-\frac{1}{\pi}\left(T_z \frac{\omega}{2\pi}\right)^{-4}\right] \qquad (21)$$

where H_s is the significant wave height and T_z is the zero crossing period. An additional simplification was introduced by using Wiegel's 1978 relation between H_s and T_z. An analytical approximation for this relation is

$$T_z = 1.72\, H_s^{0.559} \qquad (H_s \text{ in feet}) \qquad (22)$$

Substituting for $\dot{u}_{ij}$ and $\ddot{u}_{ij}$ in Eq. (11) and separating the fluid and structural response terms, leads to the desired form for P_i,

$$P_i = -M_{ai}\ddot{v}_i - C_{di}\dot{v}_i + F_i$$

$$F_i = \text{Im}\left\{\sum_{n=1}^{M} A_n F_{in}\, e^{j(-\omega_n t + \psi_n + \gamma_{in})}\right\} \qquad (23)$$

where

$$A_n^2 = 2S_{\eta\eta}(\omega_n)\Delta\omega \qquad \omega_n = n\Delta\omega$$

$$F_{in}^2 = B_{in}^2 + C_{in}^2 \qquad \tan\gamma_{in} = \frac{C_{in}}{B_{in}}$$

$$B_{in} = \sum_j R_{ijn}^{(1)} \cos(k_n x_{ij}) - R_{ijn}^{(2)} \sin(k_n x_{ij})$$

$$C_{in} = \sum_j R_{ijn}^{(1)} \sin(k_n x_{ij}) + R_{ijn}^{(2)} \cos(k_n x_{ij})$$

$$R_{ijn}^{(1)} = \omega_n^2\, G_{in}\, \rho\, \frac{\pi}{4}\, C_{M_{ij}} D_{ij}^2 \ell_{ij}$$

$$R_{ijn}^{(2)} = \frac{1}{2} \rho \omega_n G_{in} C_{D_{ij}} D_{ij} \ell_{ij} a_{ij}$$

$$M_{ai} = \rho \frac{\pi}{4} \sum_j (C_{M_{ij}} - 1) D_{ij}^2 \ell_{ij}$$

$$C_i = \frac{1}{2} \rho \sum_j C_{D_{ij}} D_{ij} \ell_{ij} a_{ij} \qquad (24)$$

Phase shift due to finite width of the structure is accounted for in the $k_n x_{ij}$ terms.

Assemblage of the system equations is straight forward. Superimposing the upper-structure and foundation stiffness at the base node, the equations for periodic excitation are written as:

$$(\underline{M} + \underline{M}_a)\ddot{\underline{V}} + (\underline{C} + \underline{C}_d)\dot{\underline{V}} + (\underline{K} + \underline{K}_F)\underline{V} = \underline{F} \qquad (25)$$

where $\underline{v} = \{v_1, \theta_1, v_2, \theta_2, \ldots\}$; $\underline{F} = \{F_1, 0, F_2, \ldots\}$. $\underline{M}, \underline{C}, \underline{K}$ are the conventional mass, viscous damping and complex stiffness, $\underline{M}_a$ is the added mass, $\underline{C}_d$ defines the hydrodynamic drag damping, and $\underline{K}_F$ represents the complex foundation stiffness matrix contained in Eq. 10.

Equation (25) is solved in the frequency domain for the complex nodal displacement measures, and the final result is expressed as a superposition,

$$v_i = I_m(v'_i) = I_m \left\{ \sum_{n=1}^{M} A_n v_{in} e^d (-\omega_n t + \psi_n + \beta_{in}) \right\}$$

$$\theta_i = I_m(\theta'_i) = I_m \left\{ \sum_{n=1}^{M} A_n \theta_{in} e^j (-\omega_n t + \psi_n + \delta_{in}) \right\} \quad (26)$$

Iteration on the hydrodynamic coefficients and linearized drag term requires the spectral density function for the relative velocity. Combining Eq. (19) and Eq. (26), one obtains:

$$\dot{r}_{ij} = \dot{u}_{ij} - \dot{v}_i = \sum_{n=1}^{M} A_n \omega_n H_{ijn} \cos(-\omega_n t + \psi_n + \zeta_{in})$$

where

$$H_{ijn}^2 = G_{in}^2 + v_{in}^2 + 2G_{in} v_{in} \cos(k_n x_{ij} - \beta_{in})$$

$$\tan \zeta_{in} = \frac{G_{in} \sin k_n x_{ij} + v_{in} \sin \beta_{in}}{G_{in} \cos k_n x_{ij} + v_{in} \cos \beta_{in}} \quad (27)$$

Noting that ψ_n are independent random phase angles, the autocorrelation function of $\dot{r}_{ij}(t)$ is

$$R_{\dot{r}_{ij}\dot{r}_{ij}}(\tau) = \frac{1}{2} \sum_{n=1}^{M} A_n^2 \omega_n^2 H_{ijn}^2 \cos(\omega_n \tau) \quad (28)$$

Then:

$$\sigma_{\dot{r}_{ij}}^2 = R_{\dot{r}_{ij}\dot{r}_{ij}}(0) = \frac{1}{2} \sum_{n=1}^{M} A_n^2 \omega_n^2 H_{ijn}^2 \quad (29)$$

The spectral density function has the following form:

$$S_{\dot{r}_{ij}\dot{r}_{ij}}(\omega_n) = \omega_n^2 H_{ijn}^2 S_{\eta\eta}(\omega_n) \quad (30)$$

The foundation force and moment are evaluated with (10) and (26):

$$F_{tot} = Im\{\bar{n}(k_{xx}^* v_1 + k_{x\theta}^* \theta_1)\}$$

$$M_{tot} = Im\{\bar{n}(k_{x\theta}^* v_1 + (k_{\theta\theta}^* + d^2 k_{zz}^*)\theta_1)\} \quad (31)$$

where Im denotes the imaginary part. A typical expression has the form:

$$F = Im\{\bar{n} \sum_{n=1}^{M} A_n [\bar{k}_x v_{1n} e^{j(\beta_{1n}+\xi_x)} + \bar{k}_{x\theta} \theta_{1n} e^{j(\delta_{1n}+\xi_{x\theta})}] e^{j(-\omega_n t + \psi_n)}\} = \bar{n} \sum_{n=1}^{M} A_n F_n \sin(-\omega_n t + \psi_n + \mu_n) \quad (32)$$

Comparing (32) with (27) and noting the development of the autocorrelation function for $\dot{r}_{ij}$, it follows that

$$R_{FF}(\tau) = \frac{\bar{n}^2}{2} \sum_{n=1}^{M} A_n^2 F_n^2 \cos(\omega_n \tau) \quad (33)$$

The measures required for iteration are:

$$\sigma_F^2 = R_{FF}(0) \quad (34)$$

$$S_{FF}(\omega_n) = \bar{n}^2 F_n^2 S_{\eta\eta}(\omega_n) \quad (35)$$

Similar expressions apply for the foundation moment.

When the foundation is fixed, this approach has to be modified. The foundation force and moment are now equal to the shear and moment at node 1 of the equivalent beam. One can write

$$F = \text{Im}\left[(1+2D_s i) F_1(t)\right]$$
$$M = \text{Im}\left[(1+2D_s i) M_1(t)\right] \qquad (36)$$

where D_s is the hysteretic damping coefficient for the upper structure and F_1, M_1 involves the equivalent beam real stiffness coefficients and the complex displacements for nodes 1 and 2. A typical expression is

$$F = \text{Im}\left[(S_1(v_1-v_2)+S_2\theta_1+S_3\theta_2)(1+2D_s j)\right] = \sum_{n=1}^{M} A_n F_{Bn} \sin(-\omega_n t + \psi_n + \nu_n) \qquad (37)$$

where S_j denotes the beam stiffness coefficients.

Iteration on the foundation stiffness coefficients is based on r.m.s. values σ_F, σ_M and the associated average zero crossing periods T_F, T_M. To simplify the computations, an arithmetic average, T_{FM}, of the zero crossing periods is used. Also in the analysis of the particular structure used in this study, it was observed that the iteration is accelerated if one takes σ_F as the primary measure and establishes relations for σ_M and zero crossing periods in terms of σ_F.

Of central interest is the variation in foundation stiffness with significant wave height and number of cycles of wave excitation. The response of the upperstructure on a rigid foundation is evaluated in order to establish relations between σ_M, T_{FM} and σ_F. Stiffness coefficients are generated for a range of σ_F and number of cycles, N, and are input in the upper structure program. The hydrodynamic coefficients $C_{M_{ij}}$, $C_{D_{ij}}$ and hydrodynamic drag damping C_d are modified according to the procedure described earlier. A zero response is assumed initially. Iteration proceeds until the following convergence measures are satisfied.

$$\varepsilon_F = \left|\frac{(\sigma_F')^2 - (\sigma_F)^2}{(\sigma_F')^2}\right| < \bar{\varepsilon}_F$$

$$\varepsilon_r = \max_{\substack{\text{all} \\ i,j}} \left|\frac{(\sigma_{\dot{r}_{ij}})^2 - (\sigma_{\dot{r}_{ij}})^2}{(\sigma_{\dot{r}_{ij}}')^2}\right| < \bar{\varepsilon}_r \qquad (38)$$

where the prime denotes the current value.

SENSITIVITY APPLICATIONS

With the analytical tool developed one can assess the sensitivity of the response to various assumptions for the hydrodynamic coefficients and foundation stiffnesses. Extensive comparison studies were carried out by Angelides (1978) and some typical results are included here.

Fig. 6 compares constant versus variable hydrodynamic coefficients for a hypothetical structure. The r.m.s. values are 0.493 inches for variable and 0.426 inches for constant C_D, C_M, indicating that the variable coefficient procedure is conservative for this structure. The usual upper bound design values for the hydrodynamic coefficients, i.e., $C_M=2$ and $C_D=1.4$, were used for the constant coefficients case. Drag dominates for this structure; the opposite trend was observed for inertia dominant structures.

Sensitivity to surface roughness is illustrated in Fig. 7. Results indicate that that the sensitivity is more significant for large H_s which is to be expected since the ratio of drag force to inertia force increases as the ratio of wave height to member diameter increases and Sarpkaya's data shows that relative roughness has more effect on C_D than on C_M.

The effect of foundation flexibility is shown in Fig. 8. Including the foundation lowers the natural frequency and, for this case, excites the fundamental mode and increases the r.m.s. values of displacements.

The last example illustrates foundation stiffness degradation. A typical structure with exaggerated (weak) soil properties was analyzed following the iterative procedure described earlier. Fig. 9 shows the influence of loading intensity and number of cycles on the horizontal foundation stiffness components and the r.m.s. displacement. The effect is pronounced at high wave height. These results can be used to track the variation in stiffness during a storm.

STOCHASTIC STORM MODELLING

As mentioned previously, wave excitation can be considered to be a stationary process over a limited time interval, of the order of one to few hours, whereas a storm can last many hours. One can account for non-stationarity by varying the wave spectrum over the duration of the storm. The approach adopted here is based on discretizing the storm duration, S, with hourly intervals and specifying the significant wave height for each interval as input to eq. (21). In what follows, a strategy for generating statistical estimates from historical information and its application to visual wave data for the North Atlantic is described.

Storms, their duration, S, and their inter-arrival time, I, are defined in Fig. 10 in terms of a statistical wave height measure, H (visual observation, H_v, or significant wave height, H_s). The storm duration is taken as the time that H exceeds a specified minimum value H'. A low value of H' may be considered for fatigue and accumulation of damage, while a high value would be specified for extreme response analysis. This study was concerned with the modelling of degradation of foundation stiffness, and the value of 1.6 meters was selected.

The model for the evolution of H in time has some stationary characteristics and some nonstationary features. The parameters for the probability distribution for storm duration, S; storm inter-arrival time, I; and the average intensity, a_o, of each storm are stationary whereas the actual evolution of H within a storm is represented through a nonstationary random function.

The evolution of H along each historical storm is approximated by a Fourier series,

$$\bar{H}(t) = a_o + \sum_{n=1}^{N} [a_n \cos \frac{2\pi nt}{S} + b_n \sin \frac{2\pi nt}{S}] \tag{39}$$

with N adjusted to optimize the fit. Nonstationority of the H(t) process is due only to the evolutionary character of $\bar{H}(t)$, whereas fluctuations about $\bar{H}(t)$, identified as $\Delta H(t)$, are portions of a stationary process. Fig. 11 shows a typical historical storm with $\bar{H}(t)$ approximated with 5 harmonics.

Storms are categorized according to duration and average intensity, and the probability distributions for the parameters of the model are derived from the historical information on H(t). These are: mean storm arrival rate, ρ, taken to be a property of a Poisson process; probability distribution of S; probability distribution of a_o conditional over S; probability distributions for the Fourier coefficients conditional over (a_o, S); and the probability distributions for ΔH conditional over (a_o, S).

Distributions derived by Sunder (1979) based on historical data are used for numerical examples. The steps for generation of "synthetic" storms are the following:

A value $S=S_1$ is generated randomly from $P(S)$. Then $a_o=a_{o1}$ is generated randomly from $P(a_o|S_1)$. Next, the higher order Fourier coefficients are generated randomly from $P(a_{j1}|a_{o1},S_1)$, $P(b_{j1}|a_{o1},S_1)$,.... Using these coefficients, the shape $\bar{H}(t)$ is obtained within the storm and discretized for a one hour interval. Finally ΔH is generated randomly from $P(\Delta H|a_{o1},S_1)$ at each discrete time and added to $\bar{H}$. Fig. 12 shows one of the synthetic storms.

STIFFNESS DEGRADATION SIMULATION

Clay under cyclic loading suffers degradation due to the development of excess pore pressure. Several investigators such as Andersen (1976), Andersen et al (1977), Bonin et al (1976) distinguish between short term and long term behavior. Short-term behavior is defined as the case when there is no significant drainage and therefore no dissipation of excess pore pressure and no recovery of the soil. Long-term behavior refers to the case where drainage with the associated dissipation of excess pore pressure and recovery of the soil occur. The short-term condition of the clay may include one or more storms (a reasonable assumption adopted in this study is that during a storm no drainage occurs).

For the modelling of the stiffness degradation in this work, one starts with curves, such as plotted in Fig. 9, which define the decrease in stiffness with number of cycles for specified significant wave heights. Since a storm is represented by variable H_s or H_v applied for discrete time intervals, a procedure for tracking the stiffness throughout the storm is required.

The strategy adopted here is illustrated in Fig. 13. Three intervals are considered; N_j is the number of cycles of the sea state characterized by a significant wave height $^jH_{sj}$ and a zero crossing period T_{zj} for jth time interval. Assuming the structure starts from the initial undisturbed state, O, the stiffness follows the path O-A-B for the first time interval. Increasing the height to H_{sc} involves first a horizontal translation to point C, the equivalent starting point for the current stiffness and wave values, and then translation along the H_{sc} curve for N_2 cycles to point D. Decreasing the wave height shifts the starting point to E, and degradation occurs along the H_{sz} curve to point Z. The process is repeated for successive intervals until the end of the storm is reached.

The methodology presented above models the degradation of the foundation stiffness k for a given deterministic storm. In the following a homogeneous Markov type model discrete in state (each state is associated with a range of k) and continuous in time is developed for the degradation of the foundation stiffness in a time interval during which a series of storms of stochastic character occurs.

The two required characteristics of the model are: (1) a transition probability matrix and (2) the mean storm arrival rate ρ.

The stiffness domain k is partitioned and discrete stiffness states are identified. Fig. 14 illustrates this discretization; k_i is the representative value for state i and k_n represents the trapping (failure) state.

The transition probability matrix P is derived by simulation as follows: The k-path is followed for each storm of a randomly chosen set of r "historical" or "synthetic" storms, starting from each of the discrete values of stiffness except the one associated with the trapping state. Defining n_{ij} as the number of times the system, which was initially in state i, lands in state j after the set of r storms is passed through the structure, the transition probability for stiffness is determined with

$$P_{ij} = \frac{n_{ij}}{r}, \quad \begin{array}{l} i = 1, 2, \ldots n-1 \\ j = 1, 2, \ldots \ldots n \end{array} \quad (40)$$

The last row corresponds to the trapping state and requires $P_{ni}=0$ for $i \neq n$ and $P_{ni}=1$ for i=n.

Seven states were used for the results presented here and the transition probability matrices obtained with one hundred North Atlantic storms passed through the structure corresponding to Fig. 9a are listed in Table 1. These results are adequate for an initial assessment.

Having obtained the transition probability matrix P, the first passage problem, i.e.: the probability of the first passage time $\tilde{\tau}$ to the trapping state or in other words the probability associated with the occurance of the trapping state in time t_o is obtained. Two extreme cases for the behavior of the foundation stiffness between successive storms are considered: (1) no recovery, (2) total recovery. These assumptions were imposed due to a lack of knowledge of the recovery mechanism between storms. The real case is bounded by these two extremes. Case (1) is presented first.

The notation $q^{(0)}$ is introduced to represent the initial state probability vector and $q^{(n)}$ the state probability vector at the end of the n-th successive storm. Then under the Markovian and homogeneity assumption:

$$q^{(n)} = (P^T)^n \, q^{(0)} \tag{41}$$

Furthermore, if $q^{(t_o)}$ is the state probability vector at time t_o, then,

$$q^{(t_o)} = \left(\sum_{m=0}^{\infty} P[m,t_o] \, (P^T)^m \right) q^{(0)} \tag{42}$$

and the last element of $q^{(t_o)}$ is the probability associated with the occurance of the trapping state in time t_o, starting from $q^{(0)}$. In eq. (42) $P[m,t_o]$ is the probability mass function for the number of storms in a given time interval t_o.

Recalling that $P[m,t_o]$ is assumed to be a Poisson process with rate ρ, an explicit expression is established for the coefficient matrix of eq. (42) by noting that P^T can be written as:

$$P^T = Q \, \alpha \, Q^{-1} \tag{43}$$

where α is a diagonal matrix containing the eigenvalues and Q contains the eigenvectors. Then,

$$(P^T)^n = Q \, \alpha^n \, Q^{-1} \tag{44}$$

and (42) reduces to

$$q^{(t_o)} = (Q \, [\delta_{ij} \, \exp(-\rho t_o(-P_{ii}+1))] Q^{-1}) q^{(0)} \tag{45}$$

For the other limiting case (i.e.: total recovery between storms) the probability of reaching the trapping state in time t_o is given by

$$q_n^{(t_o)} = q_n^{(0)} + \sum_{i=1}^{n-1} q_i^{(0)} \left\{ \sum_{m=0}^{\infty} \cdot P[m,t_o][1-(1-P_{in})^m] \right\} \tag{46}$$

Taking storm occurrence as a Poisson process, (46) reduces to

$$q_n^{(t_o)} = 1 - \sum_{i=1}^{n-1} q_i^{(0)} \exp(-P_{in} \, \rho t_o) \tag{47}$$

Results generated with the North Atlantic wave data and a hypothetical structure are listed in Table 2. A fairly high value was used for the trapping state; a lower value would give smaller probabilities. Comparing the transition probabilities in Table 1, one observes that the synthetic storms generated a more conservative (the off-diagonal elements are higher) transition probability matrix than the historical storms. It appears that the synthetic storms are slightly more intense, and degradation is accelerated. Generation of results using a lower trapping state is presently being carried out.

CONCLUSIONS

The objective of this study was to develop an integrated procedure for investigating the effect of forcing and system uncertainties and degradation of foundation stiffness due to cyclic excitation. Our approach combined improved modelling techniques for the structural system and stochastic models for simulating the sea state variability. To reduce the uncertainty associated with the wave loading, the hydrodynamic coefficients were adjusted according to the actual values of the flow parameters rather than specify constant values.

Sensitivity studies indicate that relative surface roughness of the structural members has a negligible effect on the response of the structure considered in this study. However, it may be important for other structures having smaller diameter members, and the ability to allow for variable relative roughness is useful.

Soil uncertainty, introduced through the dynamic shear modulus, dominated over the other sources in our numerical studies. Significant shifts in the response were observed for reasonable modifications in the soil properties.

The stochastic storm model is well suited for simulation of long term environmental loading. Coupled with the technique developed to follow soil degradation due to cyclic loading, one has a computationally simple, yet analytically rigorous capability to assess long term response. It can readily be generalized to other problems where simple superposition of the accumulative effect of cycles of different intensities is not valid.

Only limited results for the first passage problem are presented here since the objective was to illustrate the approach. Effort in this area is continuing, and we hope that other investigators will be attracted to the procedures described in the paper.

ACKNOWLEDGEMENT

This work was funded by the Instituto Tecnológico Venezolano del Petróleo, Caracas, Venezuela. We are most appreciative of their support.

REFERENCES

Andersen, K.H., Hansteen, O.E., Høeg, K., and Prevost, J.H.: "Soil Deformations Due to Cyclic Loads on Offshore Structures", N.G.I., Report 52412-6, March 1977.

Andersen, K.H.: "Behaviour of Clay Subjected to Undrained Cyclic Loading", BOSS '76, 1976, Vol. 1, pp. 392-403

Angelides, D.C.: "Stochastic Response of Fixed Offshore Structures in Random Sea", Ph.D. Thesis, Dept. of Civil Eng., M.I.T., Oct. 1978.

Berge, B., and Penzien, J.: "Three-Dimensional Stochastic Response of Offshore Towers to Wave Forces", OTC 1974, Paper No. OTC 2050, pp. 173-183.

Bonin, J.P., Deleuil, G., and Zaleski-Zamenhof, L.C.: "Foundation Analysis of Marine Gravity Structures Submitted to Cyclic Loading", OTC 1976, paper No. OTC 2475, pp. 571-584.

Blaney, G.W., Kausel, E., and Roesset, J.M.: "Dynamic Stiffness of Piles", 2nd International Conference on Numerical Methods in Geomechanics, V.P.I., 1976, ASCE, pp. 1001-1012.

Cartwright, D.E.: "Theoretical and Technical Knowledge", Paper 1. The Science of Sea Waves after 25 Years, The Dynamics of Marine Vehicles and Structures in Waves, London, 1974.

Fischer, J.A., Koutsoftas, D.C., and Lu, T.D.: "The Behaviour of Marine Soils Under Cyclic Loading", BOSS '76, 1976, Vol. II, pp. 407-417.

Hardin, B.O. and Drnevich, V.P.: "Shear Modulus and Damping in Soils: Design Equations and Curves", Journal of the Soil Mechanics and Foundations Division, ASCE, SM 7, July 1972, pp. 667-692.

ISSC: Proceedings of the Third International Ship Structures Congress, Oslo, Sept. 1967.

Kausel, E.: "Forced Vibrations of Circular Foundations on Layered Media", Technical Report, R74-11, Jan. 1974, Dept. of Civil Eng., M.I.T., pp. 240.

Kinsman, B.: "Wind Waves, Their Generation and Propagation on the Ocean Surface", Prentice-Hall Inc., New Jersey, 1965, pp. 676.

Krylov, N. and Bogoliubov, N.: "Introduction to Nonlinear Mechanics: Approximate Asymptotic Methods. Vol. 11. Princeton Univ. Press, 1947.

Lin, Y.K.: "Probabilistic Theory of Structural Dynamics", McGraw-Hill, Inc., N.Y., 1967.

Moan, T., Haver, S. and Vinje, T.: "Stochastic Dynamic Response Analysis of Offshore Platforms, with Particular Reference to Gravity-Type Platforms", OTC 1975, Paper No. OTC 2407, pp. 707-716

Nordenstrøm, N.: "Methods for Predicting Long Term Distributions of Wave Loads and Probability of Failure for Ships", Det Norske Veritas, Report No. 71-2-S, 1971.

Norwegian Geotechnical Institute: "Research Project, Repeated Loading on Clay, Summary and Interpretation of Test Results", Oct. 1975, 74037-9.

Sarpkaya, T., Collins, N.J., and Evan, S.R.: "Wave Forces on Rough-Walled Cylinders at High Reynolds Numbers", Offshore Technology Conference 1977, Vol III, paper No. OTC 2901, pp. 175-184.

Shyam Sunder, S.: "Stochastic Modelling of Ocean Storms", SM Thesis, Dept. of Civil Engineering, M.I.T., Feb. 1979.

Van Eekelen, H.A.M.: "Single-Parameter Models For Progressive Weakening of Soils by Cycling Loading", Geotechnique 27, No 3, pp. 357-368, 1977.

Wiegel, R.L.: "Waves and Wave Spectra and Design Estimates", Deep-Sea Oil-Production Structures, January 23-27, 1978, Berkeley.

Discretized Stiffness $k_{xx} * 10^{-7}$	k_1	k_2	k_3	k_4	k_5	k_6	k_7
	1.20	1.10	1.00	0.90	0.80	0.70	0.65

a) Discrete Stiffness States (see Fig. 9a)

				FINAL			
I N I T I A L	0	0.52	0.22	0.16	0.05	0.03	0.02
	0	0.52	0.22	0.16	0.05	0.03	0.02
	0	0	0.74	0.16	0.05	0.03	0.02
	0	0	0	0.88	0.07	0.03	0.02
	0	0	0	0	0.95	0.03	0.02
	0	0	0	0	0	0.97	0.03
	0	0	0	0	0	0	1.00

b) Historical Storms

				FINAL			
I N I T I A L	0	0.37	0.32	0.16	0.08	0.05	0.02
	0	0.37	0.32	0.16	0.08	0.05	0.02
	0	0	0.68	0.17	0.08	0.05	0.02
	0	0	0	0.83	0.1	0.05	0.02
	0	0	0	0	0.93	0.04	0.03
	0	0	0	0	0	0.94	0.06
	0	0	0	0	0	0	1.00

c) Synthetic Storms

Table 1 Transition Probability Matrices

Time Period	Probability of First Passage			
	Historical Storms		Synthetic Storms	
	Recovery	No Recovery	Recovery	No Recovery
170 days			.860	.992
365 days	.986	.997	.986	.999

Data based on Table 1, $\rho = 1/44.83$ hours^{-1}, and $q^{(0)} = \{0.7, 0.2, 0.05, 0.03, 0.01, 0.004, 0.006\}$

Table 2 Probability of First Passage

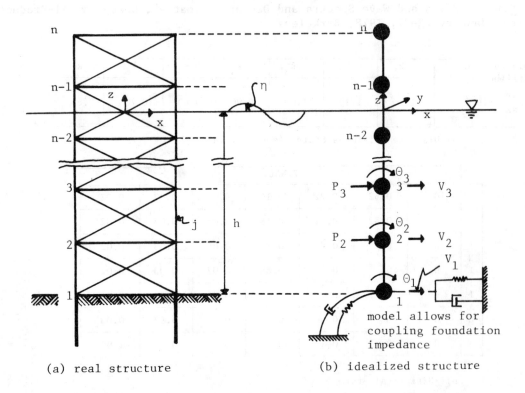

Fig. 1 System Idealization

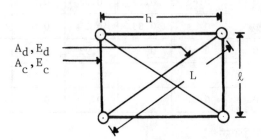

Fig. 2 Individual Truss with X-Bracing

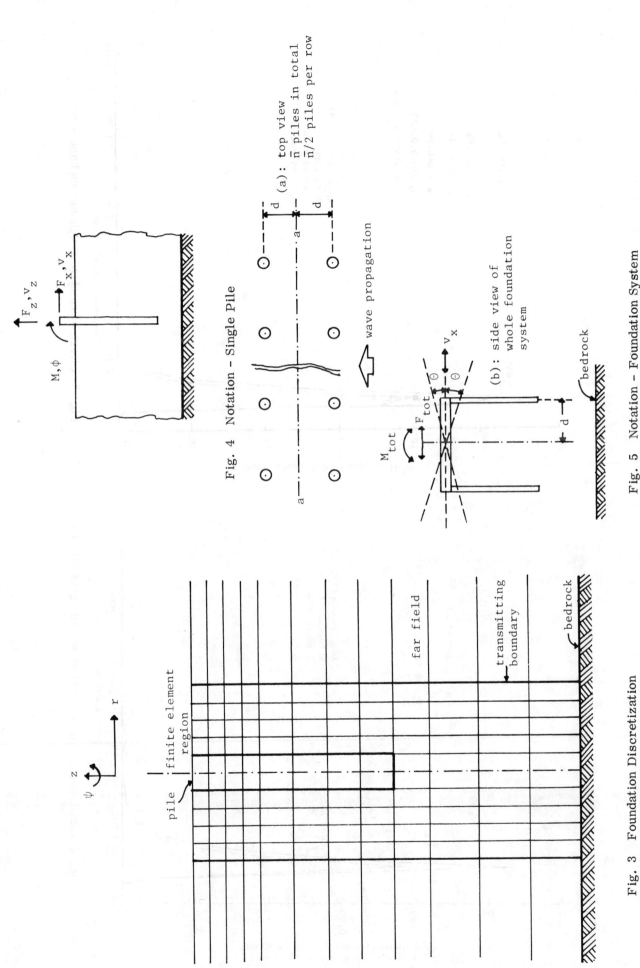

Fig. 4 Notation – Single Pile

Fig. 5 Notation – Foundation System

Fig. 3 Foundation Discretization

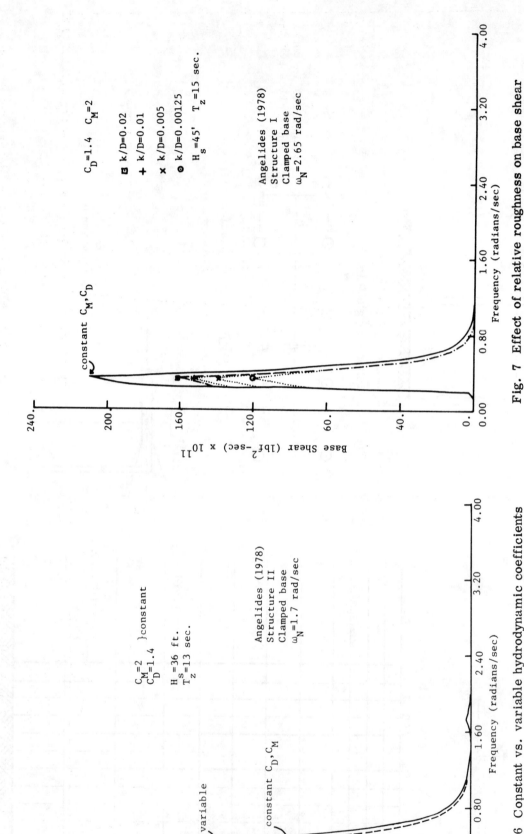

Fig. 7 Effect of relative roughness on base shear

Fig. 6 Constant vs. variable hydrodynamic coefficients

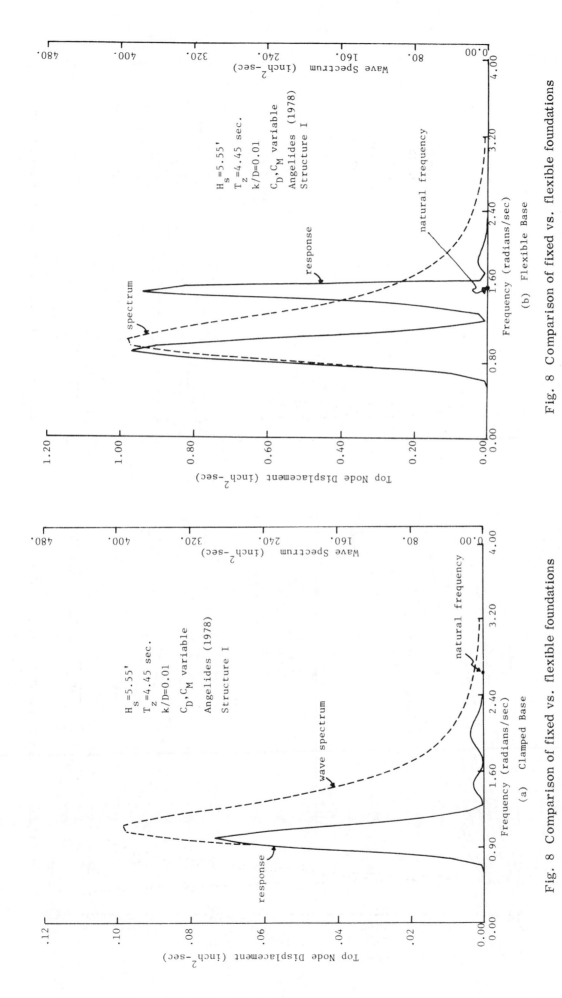

Fig. 8 Comparison of fixed vs. flexible foundations

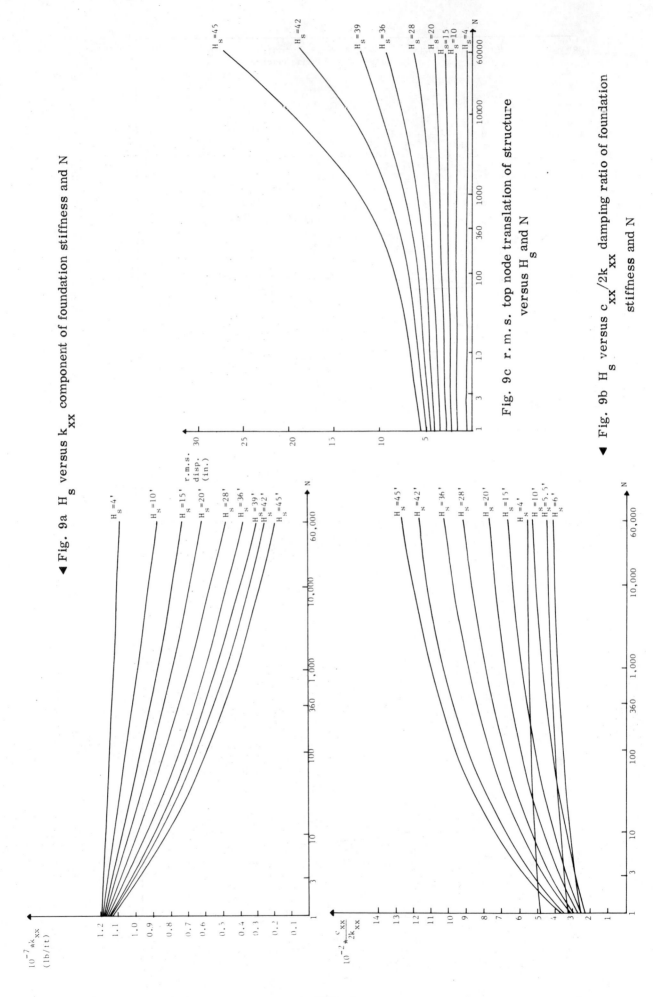

Fig. 9a H_s versus k_{xx} component of foundation stiffness and N

Fig. 9b H_s versus $c_{xx}/2k_{xx}$ damping ratio of foundation stiffness and N

Fig. 9c r.m.s. top node translation of structure versus H_s and N

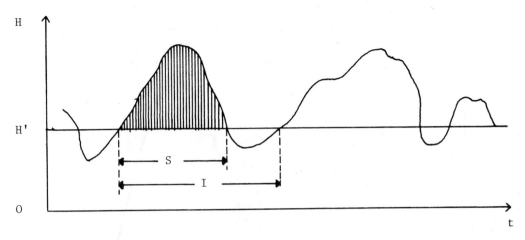

Fig. 10 Definition of Storm Duration, S and Inter-arrival Time between Storms, I

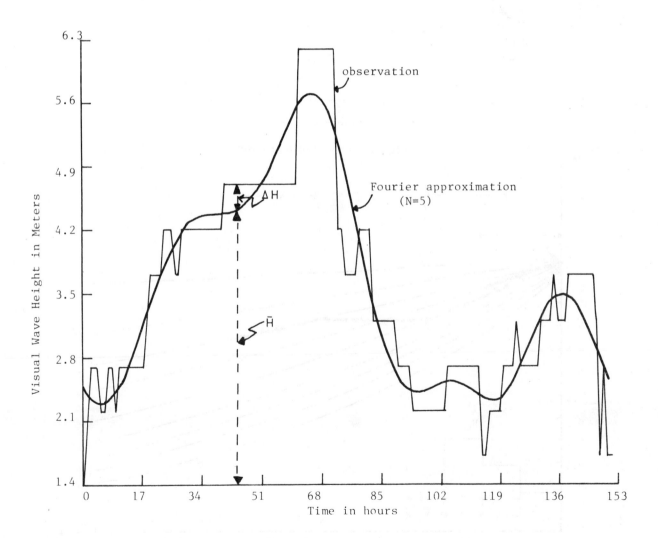

Fig. 11 Fourier Series Approximation for a Storm

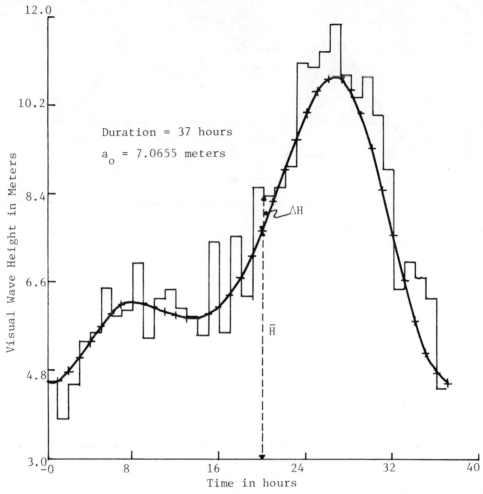

Fig. 12 "Synthetic Storm"

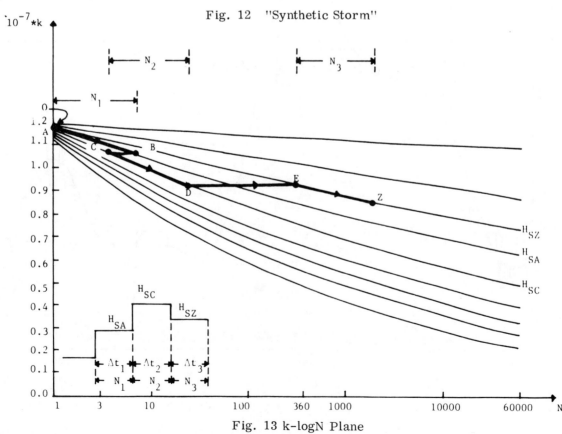

Fig. 13 k-logN Plane

20

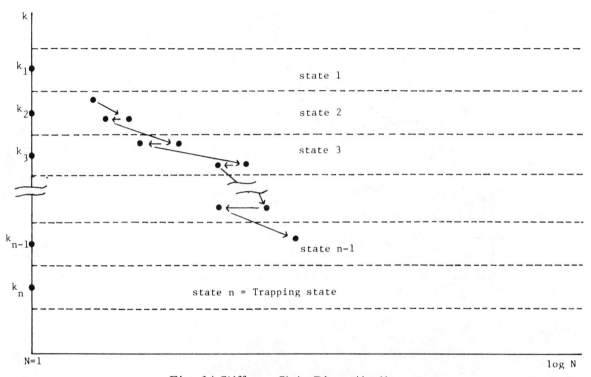

Fig. 14 Stiffness State Discretization

BOSS'79

PAPER 49

**Second International Conference
on Behaviour of Off-Shore Structures**

Held at: Imperial College, London, England
28 to 31 August 1979

MODAL SUPERPOSITION v DIRECT SOLUTION TECHNIQUES IN THE DYNAMIC ANALYSIS OF OFFSHORE STRUCTURES

J.H. Vugts and I.M. Hines

Shell Internationale Petroleum Maatschappij BV, The Netherlands

and

R. Nataraja and W. Schumm

Lloyd's Register of Shipping, U.K.

Summary

In this paper the results of the dynamic analysis of an offshore structure, according to three different methods, are presented and compared. The structure is a realistic steel jacket platform having approximately 1500 degrees of freedom. Loading is the same for all analyses and is provided by a periodic wave of 5 second period propagating in an oblique direction. All calculations are performed in the frequency domain using a NASTRAN based dynamic analysis package.

Two derivatives of the structure are used for the study. The first is made artificially stiff in order to make it dynamically insensitive to the wave loading. The second is made artificially flexible in order to produce significant dynamic response to the prescribed wave excitation. The following calculations are performed and compared from both the technical and relative cost standpoints:

a) Direct solution in the physical coordinates.
b) Modal superposition analysis using varying numbers of modes.
c) An alternative modal superposition procedure using varying numbers of modes to account for dynamic effects, supplemented by the full static solution which replaces the corresponding static component contained in the solution of the truncated modal series.

The results indicate that pure modal superposition analysis is unacceptable if an accurate recovery of member end forces (and hence stresses) is the objective. The enhanced modal solution offers a significant improvement in force recovery. The ultimate reliability of the technique is, however, still subject to a number of reservations. From a cost point of view there is no incentive to revert to modal analysis either. Thus a firm recommendation is made to utilise direct solution dynamic analysis techniques wherever possible.

Sponsored by: Delft University of Technology, The Netherlands
Massachusetts Institute of Technology, U.S.A.
The Norwegian Institute of Technology, Norway
University of London, England

Secretariat provided by: BHRA Fluid Engineering

Copyright: © BHRA Fluid Engineering
Cranfield, Bedford, England

1. INTRODUCTION

1.1 Dynamic Analysis Requirements for Offshore Structures

Offshore exploration and production is moving into deeper and more hostile environments. In the past the design of fixed offshore structures has been based largely upon well established static structural analysis procedures, any dynamic response being accounted for by applying dynamic amplification factors to base shear and overturning moment and/or level shear values at various locations along the structure. The flexibility, and hence natural periods, of the newer deepwater platforms are increasing and thereby approaching values at which there is significant wave energy. Resultant dynamic excitation can then have a significant effect upon member stress levels. In such cases it is essential to incorporate a full and detailed dynamic analysis within the design procedure.

A realistic dynamic analysis of a fixed offshore structure is required not only to predict structural response to short-term extreme events, such as design storms, but also (in many cases even especially so) to calculate response to less severe sea states which have a controlling influence upon the fatigue life of component parts of the structure. The fatigue aspect imposes the most severe demands upon a dynamic analysis tool. It is these demands which will be briefly discussed hereafter.

An offshore structure represents a partially submerged three dimensional spaceframe; an idealized mathematical model of such a platform may have several thousand degrees of freedom. The requirement is for an accurate prediction of member stress levels, at a number of locations around the ends of members, for a wide variety of three dimensional random loading conditions. Fatigue damage is a highly localized phenomenon. A structural model adequate for fatigue analysis must therefore retain the global force patterns throughout the structure, whilst also representing the influence of local force distributions and construction details. These requirements demand a detailed structural model, since the stresses at discrete locations around individual members must all be considered individually. Gross simplifications are in conflict with the nature of the problem. There appears no justification for performing the dynamic analysis using a simplified idealization, either through a reduced model or by using eigenvalue condensation techniques. Simplified idealizations make it difficult to maintain the correct phase relationships between forces and moments acting upon different parts of the structure. Condensation techniques are an aid to the exact solution of static structural analysis problems but inherently involve approximations in a dynamic problem. They require considerable insight and engineering judgement and inevitably reduce the realism of the problem description. By focussing upon these aspects it becomes apparent that a full dynamic analysis using a very detailed structural model is the only acceptable approach if local stress recovery for a realistic fatigue investigation is the ultimate goal.

1.2 Calculation of Dynamic Response

The above illustrates that the practical value of a fatigue analysis of an offshore platform is highly dependent upon an accurate determination of the local stress response of a complex multi-degree-of-freedom system. This poses a formidable analysis problem but solutions can be obtained using either of two basic dynamic analysis techniques. In the first, termed modal superposition, the N coupled equations of motion in the physical coordinate system are transformed into an equivalent set of N uncoupled single degree of freedom equations in modal or generalised normal coordinates. Such a transformation can only be applied when the system damping matrix is of a special form, bearing a particular relationship to a combination of the mass and stiffness matrices of the system (Ref. 1). In such a system the eigenvalues are real and proportional to the square of the natural frequencies whilst the corresponding eigenvectors represent the normal modes of vibration of the system. Solution takes place in the transformed normal coordinates, where each single degree of freedom system is solved independently, after which response in physical coordinates is regained by superimposing the modal solutions and applying the inverse transformation.

Solution of all eigenvalues and eigenvectors is impractical for a large system. If it is possible to show that relatively few modes are sufficient to represent the response of the structure then the method becomes both attractive and economical. It is this basic assumption which has encouraged the use of modal analysis for the dynamic analysis of large structures. The fundamental difficulty arises in deciding how many modes, and which ones in particular, to retain in order to ensure accurate stress recovery. Decisions of this nature are fundamental to the modal analysis procedure and these require experience, engineering judgement and possibly a good deal of faith. It will be shown that trial calculations on a large structure indicate that force, and hence stress recovery, can be poor even if large numbers of modes are included. In addition there appears to be no rational means of estimating the possible errors resulting from truncation of the modal series. Given these reservations the adequacy of modal analysis techniques for the dynamic analysis of offshore structures is questionable if the objective is a detailed, reliable estimation of stresses at particular locations throughout the structure.

The alternative to modal superposition analysis is to perform the solution of the equations of motion directly in the physical coordinates. The advantage of this approach is that no subjective judgements are required. The solution does not rely upon a truncated series and is a complete or exact solution to the mathematical model formulated. If both the loading and the structural description are linear a frequency domain approach can be adopted in which the direct solution is performed by straightforward matrix manipulations at a number of discrete frequencies within the range of the excitation. Provided the computational requirements of the matrix operations are not prohibitive, and the linearization of the system can be justified, the frequency domain direct solution offers a sound "fool proof" approach which is not subject to the uncertainties which cloud modal superposition analysis.

1.3 Present State of the Art

The dynamic analysis of fixed offshore structures has received extensive coverage in published literature over the past few years. Many publications have addressed specific problems associated with a realistic idealization of the structure and its environment, or concentrated upon developing solutions to the resulting complex mathematical models (e.g. Refs. 2-4). A subjective review of some of the broader issues governing the dynamic analysis of offshore structures appears in Vugts and Hayes' paper (Ref. 5). Here an attempt was made to draw attention to uncertainties in the fundamental problem areas, to compare currently available analytical tools which might be of practical value within an integrated design procedure and to provide a review of the current state of the art in this relatively unknown field. Within the scope of this review several uncertainties in the adequacy of modal analysis techniques were highlighted. Of the two basic dynamic analysis techniques, modal analysis has received a more widespread application to offshore problems (Refs. 6, 7, 8). It is generally accepted that modal analysis is more economical, offers adequate accuracy, particularly when supplemented by a static solution which attempts to include the response in the higher modes, and does not impose severe computational burdens which means that modest computer facilities can be used. The direct solution has not been exploited to a large extent by the offshore industry. For large structures the method requires powerful computing facilities, capable of performing the required matrix operations efficiently. It is generally felt to be uncompetitive in terms of cost when compared with modal analysis techniques.

However, the value of an accurate result cannot easily be estimated. A relatively expensive, but reliable and straightforward prediction of the local stress history represents an infinitely better solution than a cheaper result which is subject to technical reservations.

1.4 Objectives of the Present Study

This paper attempts to address the reservations surrounding modal analysis by directly comparing the results of three trial dynamic analyses of a realistic steel jacket structure.

The analyses compared are as follows:

(i)   Direct solution in the physical coordinates.

(ii) Modal analysis using varying numbers of modes.

(iii) Modal analysis using relatively few modes to account for dynamic effects supplemented by the full static solution.

All calculations were performed in the frequency domain using Lloyd's Register's version of the McNeal Schwendler NASTRAN package. The study was intended to effect a bench test comparison of modal analysis and direct dynamic solution techniques in order to try and provide answers to the following fundamental questions:

(i) Is pure modal superposition a reliable means of estimating platform displacements, member end forces and moments, and how are results influenced by the number of modes used?

(ii) What effect does supplementing the modal analysis with the full static solution have upon response parameters, and how do results compare with the direct solution?

(iii) How do the costs of the modal and direct solutions compare for a large offshore structure?

2. REVIEW OF THE CALCULATIONS PERFORMED

In order to provide a realistic comparison of direct and modal dynamic analysis techniques the response of a full scale steel jacket structure has been investigated. Two separate mathematical models were used for the study, both being derived from a Shell North Sea jacket, which is presently under construction. A perspective line sketch of the structure is shown in Fig. 1.

The geometric configuration of both models was identical to the true structure but the stiffness characteristics of the original jacket were modified in order to produce models exhibiting different dynamic response characteristics. The first model was made artificially stiffer by increasing steel modulus and soil spring stiffness values in order to reduce its natural period to 0.8 seconds. The second model was made artificially more flexible by reducing its steel modulus and soil spring stiffness values in order to give a first mode natural period of 4 seconds. Both models were loaded by a linear wave train having a period of 5 seconds, propagating at an oblique angle of 30 degrees through the structure to generate a truly three dimensional loading distribution. The stiff structure is dynamically insensitive to wave excitation at this frequency and its response is therefore almost entirely static. The flexible structure's natural frequency is close to the excitation frequency, however, resulting in significant dynamic response.

The following frequency domain analyses were performed on both structures.

(i) Direct solution in the physical coordinates.

(ii) Modal superposition analysis using 10, 20, 30, 40 and 50 modes.

(iii) As (ii) but supplemented by the static solution which includes the response of the truncated higher modes.

For each structure the results of the second and third analyses have been compared with the response obtained from the direct analysis. The direct analysis is considered to be the reference solution since it represents an exact solution to the prescribed mathematical model.

3. MATHEMATICAL MODELS

3.1 Structural Idealization

Two separate NASTRAN finite element models were used for the dynamic analysis comparisons, both derived from a realistic North Sea structure. The platform measures 110.72 m from the sea bed to the upper deck and has been designed for a water depth of 83 m. A detailed finite element beam idealization of the original platform was used in order to model the characteristics of the structure as completely and realistically as possible. The resulting mathematical model had 251 nodes, 477 members and a total of 1506 degrees of freedom. Line sketches of the finite element model are shown in Figs. 2-4.

The stiffness of the jacket structure was represented using NASTRAN beam elements (CBAR) having six degrees of freedom at each node. Pile-structure interaction was included by specifying a (6 x 6) foundation stiffness matrix at each of the pile attachment points at the base of the platform. The equivalent linear stiffness coefficients of the pile/soil system were determined from a separate analysis prior to the NASTRAN idealization. Since a detailed investigation of deck response was not required, the deck structure was modelled simplistically using equivalent beam elements which maintained the correct load distribution through the structure.

The mass matrix of the model included contributions from the structural steel, pile grout, entrained fluids and the added mass resulting from fluid reaction forces. The structural mass contributions are evaluated automatically within NASTRAN using member cross-sectional dimensions and material density data. The mass of grout and entrained fluid was specified as additional non-structural mass per unit length for the relevant members. A consistent structural mass matrix was formulated for the jacket structure, whereas the additional deck mass, corresponding to an estimated operational deck load, was lumped equally at six discrete locations on the lower deck framing.

Added mass contributions of all submerged members were determined using a pre-processor program which utilises the NASTRAN idealization. Added mass components in two directions normal to each member, together with the associated rotational inertia components, were evaluated in local member axes, and assigned to each node. These were then transformed into the global structure axes. The coupled mass matrices resulting from these transformations were input to the NASTRAN idealization in the form of additional mass matrices associated with each submerged node.

All conductor tubes were retained in the structure idealization in order to include their significant contribution to the total wave loading on the platform, as well as the structural and added mass associated with them. The conductors did not contribute to the stiffness of the structure and degrees of freedom associated with them were not considered in the dynamic response analysis.

Damping was included in all the analyses. All damping contributions together (e.g. soil, hydrodynamic, and structural) were represented using an equivalent viscous damping coefficient corresponding to 2% of critical in all the structure modes.

3.2 Natural Periods

A limited eigenvalue analysis of the original structure showed that the platform had a fundamental natural period of 1.59 seconds. This compared favourably with preliminary hand calculations using Rayleigh's method. The flexible and stiff derivatives of the base case were generated by factoring soil stiffness and steel modulus values. The factors were calculated using the linear relationship between eigenvalue and stiffness. The resulting natural periods and the stiffness factors used are shown in Table 1.

3.3 Wave Loading

Using a Lloyd's Register pre-processor wave loads were determined on all submerged members for a linear wave train of 5 second period, steepness ratio 1/14, propagating at an angle of 30 degrees to the X-axis of the platform. The wave characteristics are defined in Table 2.

The wave force analysis was performed in the frequency domain and all calculations were carried out in complex notation in order to retain both the phase and amplitude relationships throughout the structure. Individual member wave loads were calculated using a linearized form of Morison's equation; the relative velocity postulate was not included in drag force calculations. Wave forces per unit length were determined at a number of discrete locations along each member. The number of points used to subdivide each member was determined on the basis of the member length and the wavelength. Wave particle kinematics were determined at each point according to Airy wave theory. Local wave force distributions were numerically integrated over the submerged length and then converted to equivalent nodal loads in a consistent manner. This is achieved by applying fixed member end restraints and then evaluating the resultant nodal shear forces and bending moments.

4. CALCULATION PROCEDURE

4.1 General Remarks

The Lloyd's Register of Shipping fatigue analysis system for fixed offshore structures (Ref. 9), utilises the NASTRAN finite element package as the central dynamic processor. An extensive, though not exhaustive, survey of commercially available dynamic analysis packages was conducted within SIPM during 1977. This revealed that NASTRAN is one of very few packages which offers a direct frequency domain solution procedure in conjunction with the large problem solving capacity demanded by todays mammoth steel offshore structures. The custom designed Lloyd's package thus offered a suitable tool with which to perform a bench test comparison of modal and direct dynamic analyses.

4.2 Dynamic Analysis Using NASTRAN

The NASTRAN finite element package offers various methods for performing a comprehensive time or frequency domain dynamic analysis and incorporates both modal superposition and direct solution techniques. Recognising that pure modal analyses may result in poor stress recovery, NASTRAN also incorporates an enhanced modal analysis procedure, termed mode acceleration (Refs. 10, 11). This technique includes the additional effect of the almost purely static response in the higher modes, which is neglected in pure modal superposition analysis.

The mode acceleration technique is similar, although not identical to the so-called static back substitution procedure (Ref. 8). In static back substitution the dynamic and static components of the total modal response are separated, after which the static modal components are subtracted from the total response obtained by the superposition analysis. The remaining dynamic modal components are then supplemented by a full static solution.

An analytical comparison of the two techniques revealed that the two methods are fully equivalent apart from the manner in which modal damping forces are treated. In the mode acceleration solution modal damping forces are not included, whilst their influence is duly incorporated in the static back substitution solution. In order to compare the influence of these subtle differences the static back substitution solution algorithm was programmed into NASTRAN, using a high level DMAP alter package (Ref. 14). The dynamic response of the flexible structure was then determined and compared with the results obtained from the mode acceleration solution. No measurable differences in response were obtained. This is probably attributable to the low value of damping used in the model. Any significant differences between these two techniques are expected to be confined to response levels obtained in the resonant peak of any mode where damping forces are of prime importance.

Using Lloyd's Register's offshore structure analysis facilities the following calculations were performed on both the stiff and the flexible structures:

(i)   Eigenvalue extraction of the first 50 modes.

(ii) Evaluation of response to prescribed wave excitation using 5 modal superposition analyses with 10, 20, 30, 40 and 50 modes.

(iii) As for (ii) but using the mode acceleration technique.

(iv) Determination of response to prescribed wave excitation using the direct solution method.

The eigenvalue analyses were performed on the full uncondensed models, using a search technique, the Inverse Power Method with Shifts (Ref. 12), which is one of several algorithms available in NASTRAN. This technique is a derivative of the standard inverse power method which includes an artificial shift point from which the eigenvalue search proceeds. The modification minimizes or eliminates the disadvantages of the standard inverse power method which include difficulty of the procedure in the presence of zero eigenvalues and slow convergence for closely spaced eigenvalues.

Figures 5-8 show the first three primary structural modes (these being bending modes in two orthogonal directions and a torsional mode), and a local conductor framing mode (number 29) by way of example. The results of the modal and direct dynamic response analyses, in terms of nodal displacements and member end forces and moments, were retained on backing storage in order to provide a data base for the post-processing operations which were used during the comparisons.

4.3 Interpretation of Response Data

The detailed analyses performed on both structures generated an enormous amount of data. A logical and consistent comparison of the response parameters thus necessitated a large scale data processing exercise.

The quantity of data to be interpreted was reduced somewhat by considering a limited number of response variables. The parameters selected were:

(i)   two horizontal displacements at each node point

(ii) three rotations in global axes at each node point.

(iii) axial forces at both ends of each member

(iv) two bending moments at both ends of each member.

The data handling was separated into two distinct phases.

Phase 1

In order to compare the response obtained from modal and direct dynamic analyses a post-processor computer program was developed to determine the ratio of any response parameter obtained from a modal solution to the equivalent value obtained from the direct, reference solution, i.e.

$$\text{Ratio} = \frac{\text{Value of parameter from modal solution}}{\text{Value of parameter from direct solution}}$$

Both the magnitude and phase of each parameter were ratioed in this manner.

One of the difficulties inherent in the ratio procedure is the possibility of numerical ill-conditioning if two small values are compared. An attempt was made to reduce this effect. An initial scan of the data was made and the average of the third one highest values of each selected parameter was obtained. If both the numerator and the denominator of the ratio were two orders of magnitude lower than the average value, as defined above, then these cases were excluded from the comparison. Ill-conditioning may however still occur if both values are close to the threshold and the reference value is thus very small.

Phase 2

The ratios resulting from the comparison were assembled into histograms showing the percentage of the total number of occurrences falling within a series of pre-determined ratio class intervals. Class intervals on either side of unity, up to a maximum value of 2.0 and a minimum value of 1/2.0 were chosen in order to incorporate under- and overestimates relative to the direct reference solution. In addition the percentages of occurrences exceeding the limits of the histograms were determined and listed.

A total of six histograms was generated for each modal solution, and each structural model. These are

for all node points:

(a) Two horizontal displacements together

(b) All three rotations together

(c) The five degrees of freedom of (a) and (b) together

and for all member ends:

(d) Axial forces

(e) Bending moments in two planes together

(f) The three forces/moments (d) and (e) together.

Thus a total of 240 histograms were produced, namely 2(structures) x 2(modal solutions) x 5(number of modes included) x 6(number of parameters) x 2(magnitude and phase components). Data in the histograms has not been normalized with respect to class width and therefore the histograms do not represent probability density functions.

5. PRESENTATION OF RESULTS

5.1 General Comments

As outlined in the introduction, the objective of the study was to perform both a technical and cost comparison of modal and direct dynamic analysis techniques. The use of two structural models, exhibiting markedly different dynamic response characteristics, was originally considered necessary in order to fully investigate the reliability of modal analysis techniques. For the stiff model, exhibiting little or no dynamic response, a direct dynamic solution is not strictly necessary. The original intention was to compare a modal analysis of the stiff structure with a relatively inexpensive static solution. However, the stiff model was, in fact, also subjected to a direct dynamic analysis, since it degenerates to the static solution for a dynamically insensitive structure. This also puts all comparisons on a consistent basis.

It was not possible to predict at the outset whether the results of both models would show the same patterns. However, once the analyses were complete a review of the results obtained showed that response comparisons for the stiff and flexible models indicated the same general trends. The response comparisons of the stiff model merely reinforced and duplicated the trends shown by the analyses of the dynamically sensitive structure. In view of this similarity, and in the interest of brevity, the subsequent discussion and example histograms focus upon the magnitudes of the response variables for the flexible model. The arguments presented are, however, equally applicable to either of the structural models and to both the magnitude and phase of the response parameters.

5.2 Comparison of Calculated Response Parameters

5.2.1 Pure Modal Analysis

Figures 9-10 show the histograms of the ratio of horizontal displacement and member end rotations obtained from pure modal analyses using 10 and 50 modes, to those obtained by direct solution. The horizontal displacement histogram for 10 modes shows widespread scatter. However, looking into the differences in greater detail the values of displacements at a number of nodes on the main legs and at deck level have been compared with the results from the direct solution. Good agreement was observed. This tends to confirm the generally accepted opinion that modal analysis, using relatively few modes, provides a reasonable estimate of gross platform behaviour as represented in terms of global horizontal displacements at various levels along the structure. The scatter in the 10 mode displacement histogram is probably attributable to differences in the displacements at a large number of local bracing nodes. The displacement behaviour in these regions is dependent upon the large number of higher, and more local, modes which do not contribute significantly to the overall behaviour of the platform and which have been truncated from the 10 mode series. Increasing the number of modes reduces the scatter in the displacement histogram significantly. When using 50 modes all the lateral displacements lie within $\pm$ 5% of the direct solution values, as can be seen in Figure 9.

The member end rotation histograms for both 10 and 50 modes are shown in Figure 10. Widespread scatter is evident in the results and only a marginal improvement is observed as the number of modes increases. The axial force and bending moment histograms shown in Figures 11-12 show the same broad scatter pattern. Once again increasing the number of modes does not reduce the scatter significantly. From these results it is clear that large numbers of member end rotations, axial forces and bending moments as calculated by modal analysis fall well below an acceptable level of accuracy, with an apparent bias towards underestimation.

5.2.2 Mode Acceleration Analysis

Figures 13-16 show the lateral displacement, rotation, axial force, and bending moment histograms obtained from mode acceleration analysis using 10 and 50 modes, respectively. Comparing this with their counterparts in Figures 9-12 obtained from pure modal analysis, a significant improvement in recovery of member end forces can be observed even when only 10 modes are used. This is evident from the much narrower scatter in the histograms with large numbers of ratios occurring close to the reference value of unity. Similar improvements were reported by Maddox (Ref. 13) who compared modal analysis techniques for a simple cantilever beam model. Increasing the number of modes results in some additional improvement for all variables. This indicates confirmation that the major contribution due to the static response in all variables is included by the mode acceleration (or alternatively static back substitution) technique.

Despite the dramatic improvement in accuracy offered by mode acceleration a detailed investigation of all the histograms has revealed a small, but significant, number of occurrences of high/low ratios which fall outside acceptable levels of accuracy. These occurrences are, by way of example, evident in Figures 14 and 16, for the member end rotation and bending moment histograms when using 10 modes. A more detailed analysis of the results will be necessary in order to investigate whether these arise because of numerical ill-conditioning, or whether they represent an inherent weakness in the mode acceleration (static back substitution) method. If the latter is the case it still needs to be established if these unacceptable cases can be explained in a rational manner.

5.3 Comparison of Relative Cost

The cost of a detailed dynamic analysis is high, but it still represents only a small part of the total design expenditure. Notwithstanding this, a cost comparison of modal and direct dynamic analyses is a useful exercise, not only in producing a guideline to the likely cost of performing a particular analysis, but especially as a means of verifying commonly accepted theories regarding the economic (dis)advantages of a particular approach.

The computer costs of performing modal and direct dynamic analyses for a single wave excitation frequency, as incurred during this study, have been extrapolated in order to estimate the cost variation with the number of excitation frequencies. Care has been taken in the extrapolation to factor only those components of the analysis costs which are dependent upon the number of frequencies. The results of the extrapolation are shown in Figure 17, where the costs are the relative computing charges for Lloyd's Register's NASTRAN based analysis package on their in-house IBM 370/158 machine.

Modal solutions, including mode acceleration, incur an initial cost overhead resulting from the eigenvalue analysis which is a mandatory pre-requisite for superposition analysis. Once the eigenvalues and mode shapes have been determined the cost of the superposition analysis increases linearly with the number of solution frequencies. The same is also true for the mode acceleration technique, which is slightly more expensive, as indicated by its steeper gradient in Figure 17.

An extensive eigenvalue analysis is not necessary for a direct dynamic solution. In fact eigenvalues and eigenvectors are not required at all. However, in practice, a limited eigenvalue analysis is preferable since it gives the engineer a valuable insight into the location of the platform's natural frequencies, relative to wave excitation, and the subsequent susceptibility of the structure to respond dynamically. This information can be obtained from a very limited eigenvalue analysis, which is restricted to the extraction of the first few (e.g. 3-5) primary structural modes. A significant reduction in analysis costs can thus be obtained.

From Fig. 17 it becomes apparent that the cost of a direct dynamic analysis using Lloyd's Register's package is approximately 40% more expensive per frequency than a superposition analysis using 10 modes. If no preliminary eigenvalue analysis is performed the overall cost of the direct solution is marginally lower for up to 11 excitation frequencies and significantly less expensive than a modal analysis using 50 modes. An investigation of the solution techniques within NASTRAN has revealed that the computational costs of eigenvalue extraction, modal superposition and direct solution effectively vary linearly with the number of degrees of freedom. This implies that the relative cost comparisons shown in Figure 17 apply regardless of the number of degrees of freedom. A modal solution using N modes incurs approximately the same cost as a direct solution for N excitation frequencies as shown in Figure 17. These results are in contradiction to the generally accepted theory that direct solutions are prohibitively expensive for large structures. Such comparisons, when coupled with the poor member force recovery discussed earlier make it extremely difficult to justify the use of modal solutions if alternative direct solution facilities exist.

6. CONCLUSIONS

(i) The objective of a detailed dynamic analysis of an offshore structure is to predict, as reliably and realistically as possible, the structural response to the environmental loading. The response parameters of prime interest are the member end forces/moments, and the stress levels at discrete locations around the ends of members. In certain cases nodal displacements may also be of separate interest. The need for high precision follows from the requirement of accurate stress recovery.

(ii) The results of the present study indicate, beyond doubt, that pure modal superposition is unacceptable if a realistic prediction of detailed member end forces and moments (and hence stresses) is the primary objective. The comparisons with exact direct solutions confirm that modal analysis provides reasonably reliable estimates of gross platform behaviour, expressed in terms of global horizontal displacements, even when (relatively) few modes are used. However the member end rotations, axial forces and bending moments, which are dependent upon small differential deformations, may be grossly in error, with subsequent poor stress recovery. Increasing the number of modes does not improve the accuracy significantly unless all or the great majority of the modes are included. For a large structure, having possibly thousands of degrees of freedom, such an approach is both impractical and prohibitively expensive. It is also in contradiction with the fundamental justification for the application of modal superposition in the first instance, which is that relatively few modes are sufficient to represent the behaviour of the structure.

(iii) The enhanced mode acceleration (or static back substitution) technique offers a significant improvement in force recovery. The considerable improvement in accuracy observed for the comparisons made in this study illustrates that the static response of the higher modes, which are truncated in the modal series, has a large influence upon the member force contributions. The mode acceleration/static back substitution technique thus offers far better force recovery than pure modal superposition analysis. Despite this, the test comparisons reported here indicate that the method may still fail to predict member forces and moments accurately for a small, though significant number of member ends. At present it is not possible to explain these occurrences in a rational manner. A more extensive study of the results will be necessary to establish whether these represent an inherent weakness in the technique, or are merely the result of numerical ill-conditioning caused by the magnitude of the response parameters in the ratio procedure. Until these issues are examined critically the ultimate reliability of the technique, and hence justification for its inclusion within an integrated design analysis, must remain subject to doubt.

(iv) In contrast to modal techniques the direct solution offers an exact solution to the prescribed mathematical model and is not subject to intuitive judgements based upon previous experience. Its straightforward fool proof approach makes it easier to use and hence to incorporate within an integrated design procedure.

(v) The relative cost comparisons performed in this study demonstrate that, contrary to commonly accepted theories, frequency domain direct solutions are cost effective, if not less expensive than frequency domain modal solutions. For this reason there appears to be no cost incentive for the use of modal solutions if comparable direct solution facilities exist.

(vi) As a logical consequence of the above results and conclusions it is firmly recommended that, whenever stress recovery is the primary objective of a dynamic analysis of an offshore structure, a direct solution of the equations of motion in the physical coordinates should be performed.

REFERENCES

1. Caughey, T.K.

   "Classical normal modes in damped linear dynamic systems"
   Trans. of ASME, Journal of App. Mech. pp. 269-271 (June 1960)

2. Malhorta, A.K. and Penzien, J.

   "Response of offshore structures to random wave forces"
   Proc. ASCE J. Struct. Div. pp. 2155-2173 (October 1970)

3. Maddox, N.R.

   "Fatigue analysis for deepwater fixed bottom platforms"
   Proc. 6th Offshore Technology Conference Paper 2051
   Organised by ASPE et al. Dallas, Texas (May 6-8 1974)

4. Eatock Taylor, R.

   "Structural dynamics of offshore platforms"
   Proc. Conference on offshore structures. Paper 10
   Organised by Inst. Civil Eng. London (7-8 October 1974)

5. Vugts, J.H. and Hayes, D.J.

   "Dynamic analysis of fixed offshore structures. A review of some basic aspects of the problem". Engineering Structures, Vol. 1, No. 3 (April 1979)

6. Hallam, M.G., Boudreaux, R.H., Millman, D.N., Heaf, N.J.

   "Dynamic fatigue analysis of Murchison tower structure"
   Proc. 10th Offshore Technology Conference. Paper 3163
   Organised by ASPE et al., Houston Texas (8-11 May 1978)

7. Gray, R.M., Berge, B. and Koehler, A.M.

   "Dynamic analysis of the North Sea Forties field platforms"
   Proc. 7th Offshore Technology Conference. Paper 2250
   Organised by ASPE et al., Houston Texas (5-8 May 1975)

8. Marshall, P.W. and Kinra, R.K.

   "Dynamic and fatigue analysis for deepwater fixed platforms."
   Second Annual ASCE-EMD Speciality Conference, Raleigh, N.C. (May 1977)

9. Schumm, W.

   "Dynamic analysis of deepwater platforms".
   Int. Symposium on Integrity of offshore structures. Paper 2
   Organised by Institute of Engineers and Shipbuilders, Glasgow (6-7 April 1978)

10. McNeal Schwendler Corporation

    "Dynamic data recovery". Section 9.4, McNeal Schwendler NASTRAN
    Theoretical manual pp. 9.4.1 - 9.4.3 (1970)

11. Thomson, W.T.

    "Theory of vibration with applications" pp. 318-320, Prentice Hall (1972)

12. <u>Wilkinson, J.H.</u>

    "The algebraic eigenvalue problem" Clarendon Press, Oxford (1965)

13. <u>Maddox, N.R.</u>

    "On the number of modes necessary for accurate response and resulting forces in dynamic analysis" Trans. of ASME, pp. 516-517 (June 1975)

14. <u>McNeal, R.H., McCormick, C.W.</u>

    "The NASTRAN computer program for structural analysis"
    Computers and Structures, Vol. 1 pp. 389-412 (1971)

Model	Natural Period (sec.)	Factor of Stiffness
Stiff	0.8	3.975
Flexible	4.0	0.159

TABLE 1 - NATURAL PERIODS OF STIFFNESS FACTORS

Water Depth	83 m
Wave Period	5 sec.
Wave Length	39.03 m
Wave Steepness	1/14
Wave Height	2.79 m
Approach angle (to X axis)	30 deg.

TABLE 2 - WAVE PARAMETERS

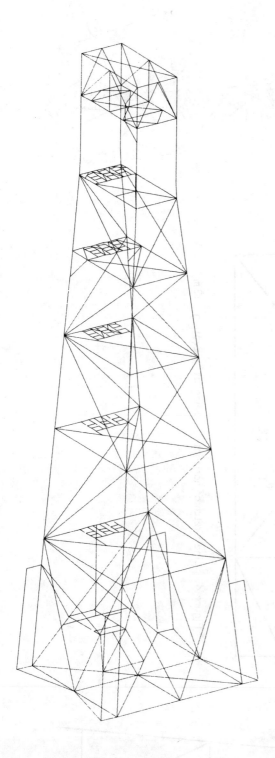

Figure 1. Perspective line sketch of the jacket

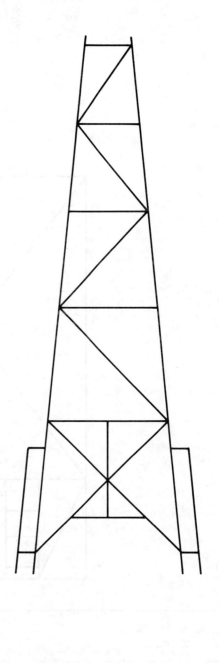

Figure 2. Line sketch of Row 1

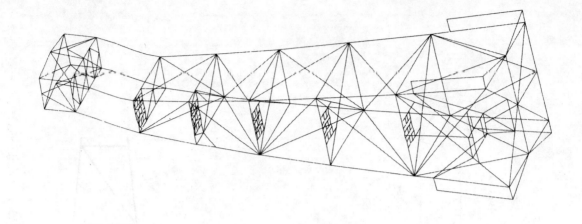

Figure 5. 1st Mode - Y Bending

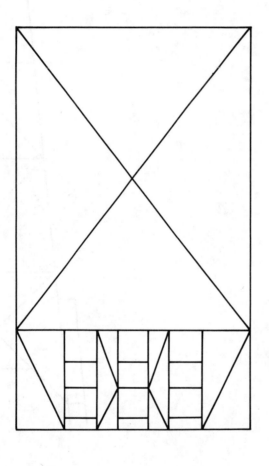

Figure 4. Line sketch of horizontal framing at elevation 88m

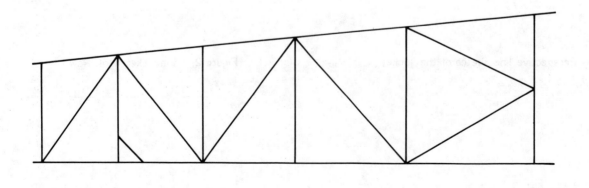

Figure 3. Line sketch of Row A

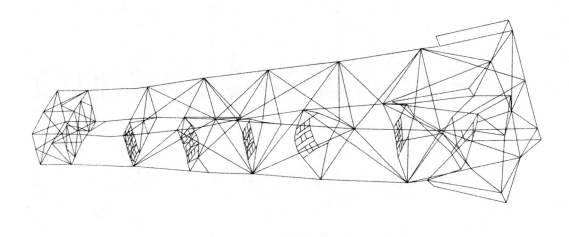

Figure 8. 29th Mode – Local conductor framing vibration

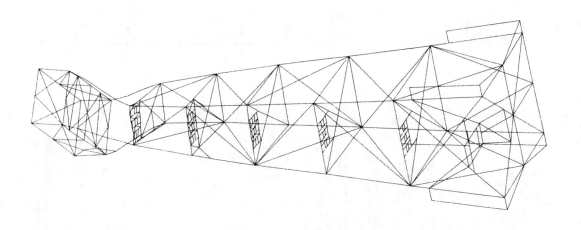

Figure 7. 3rd Mode – Torsion

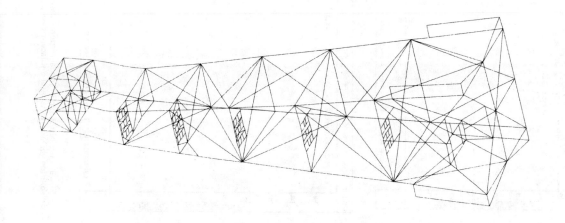

Figure 6. 2nd Mode – X Bending

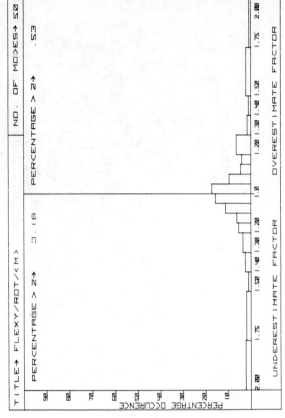

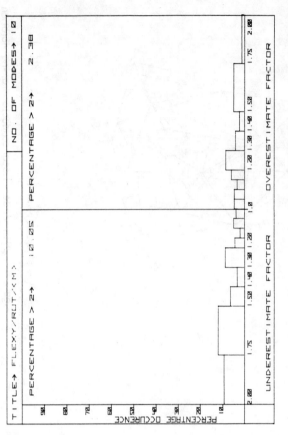

Fig. 9 Ratio histogram of modal response parameters referenced to corresponding direct solution values. Horizontal displacements - pure modal, 10 and 50 modes

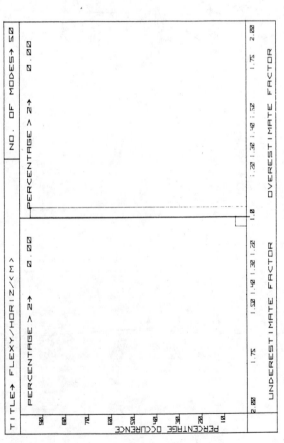

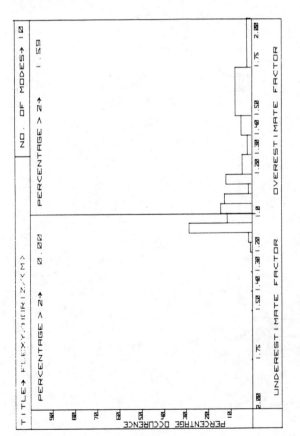

Fig. 10 Ratio histogram of modal response parameters referenced to corresponding direct solution values. Member end rotations - pure modal, 10 and 50 modes

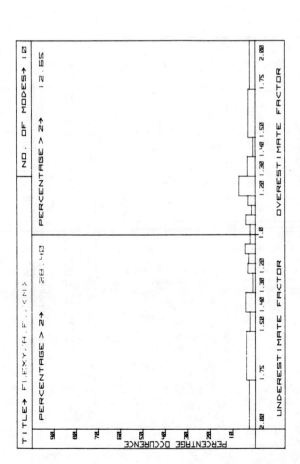

Fig. 11 Ratio histogram of modal response parameters referenced to corresponding direct solution values. Member end axial forces - pure modal, 10 and 50 modes

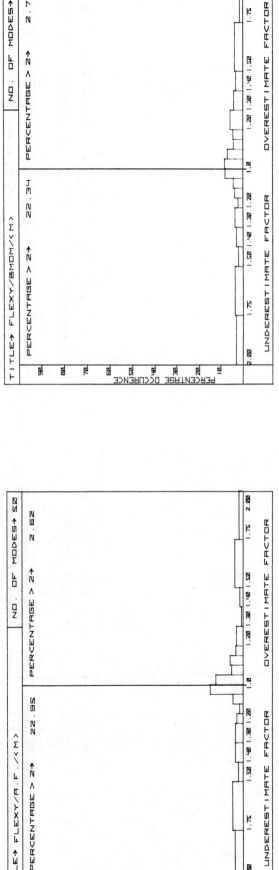

Fig. 12 Ratio histogram of modal response parameters referenced to corresponding direct solution values. Member end bending moments - pure modal, 10 and 50 modes

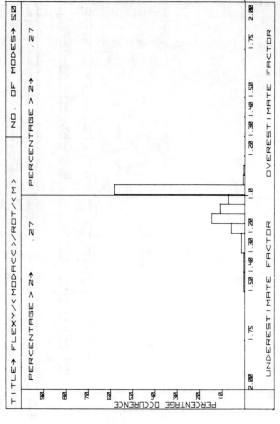

Fig. 14 Ratio histogram of modal response parameters referenced to corresponding direct solution values. Member end rotations - mode accleration, 10 and 50 modes

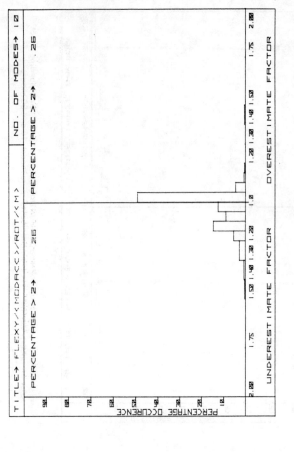

Fig. 13 Ratio histogram of modal response parameters referenced to corresponding direct solution values. Horizontal displacements - mode acceleration, 10 and 50 modes.

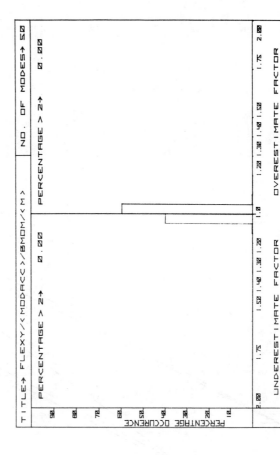

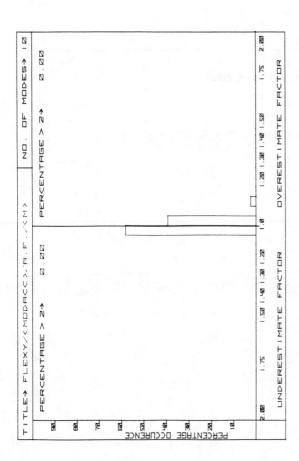

Fig. 15 Ratio histogram of modal response parameters referenced to corresponding direct solution values. Member end axial forces - mode acceleration, 10 and 50 modes

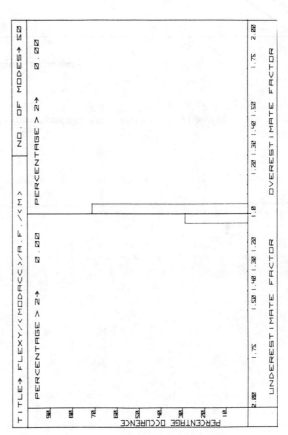

Fig. 16 Ratio histogram of modal response parameters referenced to corresponding direct solution values. Member end bending moments - mode acceleration, 10 and 50 modes

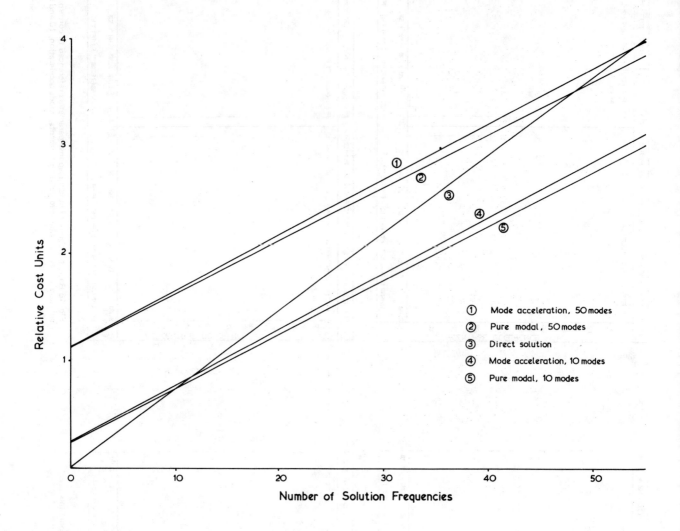

Figure 17   Relative cost comparison of modal and direct dynamic analysis

BOSS'79

PAPER 50

Second International Conference
on Behaviour of Off-Shore Structures

Held at: Imperial College, London, England
28 to 31 August 1979

THE IMPORTANCE OF STRUCTURAL MOTION IN THE CALCULATION OF WAVE LOADS ON AN OFFSHORE STRUCTURE

P. Fish and R. Rainey

Atkins Research and Development, U.K.

Summary

Structural motion is important in the calculation of wave loads on offshore structures because it is known that the mass, damping, stiffness and loading terms of conventional linear dynamic analysis are, in reality, functions of this structural motion. Solving such complex non-linear equations to give answers simple enough to be practically useful, is, however, difficult.

One such non-linear problem resulting from structural motion is the "hydrodynamic damping" effect on the taller steel jacket structures attributable to the drag term in Morison's equation. The practical problem is to tackle the non-linear equations in order to find a suitable damping matrix to use in existing linear analysis programs.

This Paper describes a "fundamentalist" solution to the problem by simplifying a typical tall (156 m water depth) jacket structure and solving both the full non-linear and the linearised equations in the time domain. Suitable damping values can then be deduced by comparing the respective motion time histories. It is, however, a laborious and not entirely satisfactory procedure, because the appropriate damping value clearly varies with sea state and it is hard to form a global picture of this variation. There are also problems of isolating the effects of starting transients and of non-linear phenomena other than damping.

The authors propose that the solution to the problem is to adopt one of the frequency domain non-linear analysis schemes that have recently appeared in the offshore literature. The potential application of one such technique is described in some detail and the controlling damping parameter that it produces is derived. Computationally, the technique would use much of the same software as the "fundamentalist" approach and normally at a reduced cost.

Sponsored by: Delft University of Technology, The Netherlands
Massachusetts Institute of Technology, U.S.A.
The Norwegian Institute of Technology, Norway
University of London, England

Secretariat provided by: BHRA Fluid Engineering

Copyright: © BHRA Fluid Engineering
Cranfield, Bedford, England

NOMENCLATURE

A	Local cross-sectional area of structure
C	Structural and soil damping matrix
C'	Implied damping matrix, steady part
C''	Implied damping matrix, fluctuating part
C_H	Hydrodynamic damping matrix
C_D	Local drag coefficient
C_M	Local mass coefficient
$\underline{F}$	Loading vector
$\underline{F}'$	Implied loading vector, steady part
$\underline{F}''$	Implied loading vector, fluctuating part
dF	Local element of loading on structure
K	Structural and soil stiffness matrix
K'	Implied stiffness matrix, steady part
K''	Implied stiffness matrix, fluctuating part
K_H	Hydrodynamic stiffness matrix
M	Structural mass matrix
M'	Implied mass matrix, steady part
M''	Implied mass matrix, fluctuating part
M_H	Hydrodynamic added mass matrix
dS	Local element of structural length
U	Local water velocity
x	Local structural displacement
$\underline{x}$	Displacement vector
ζ	Damping ratio (% critical)
η	Instantaneous water level
θ	Amplitude of rotation in x-z plane
ρ	Fluid density (sea water)
ω_n	Natural frequency of structure with added mass included

1. BACKGROUND AND APPROACH

To allow for structural motions, it is normal to analyse the wave loading on offshore structures by the following dynamic equations :

$$[M]\underline{\ddot{x}} + [C]\underline{\dot{x}} + [K]\underline{x} = \underline{F}(t) \tag{1.1}$$

where $\underline{x}$ is the vector of displacements in the structure's N degrees of freedom and M, C and K are mass, damping and stiffness matrices. The dependence of the wave loading vector $\underline{F}(t)$ on structural motions is allowed for approximately by adding extra components M_H, C_H and K_H to M, C and K (using, for example, an approximation to Morison's equation as described in Section 2). The great advantage of this "linear theory" approach is that (1.1) remains a set of linear differential equations amenable to discussion and solution in the normal dynamicist's language of "transfer functions", "response spectra" etc.

It is known, however, that the situation is, in reality, a good deal more complicated and that the wave loading can be much more fully analysed by the equations :

$$[M(\underline{x},\underline{\dot{x}},\underline{\ddot{x}},t)]\underline{\ddot{x}} + [C(\underline{x},\underline{\dot{x}},\underline{\ddot{x}},t)]\underline{\dot{x}} + [K(\underline{x},\underline{\dot{x}},\underline{\ddot{x}},t)]\underline{x} = \underline{F}(\underline{x},\underline{\dot{x}},\underline{\ddot{x}},t) \tag{1.2}$$

in which mass, damping and stiffness matrices and also the wave loading vector, are allowed to vary with time and the state of motion of the platform.

Until recently, this extra complication has been of academic interest only, because the mathematical methods used to solve equations like (1.2) in offshore platforms were so cumbersome that it was impractical to assess the significance and even the accuracy of any given solution. In the last year or two, however, various iterative schemes have appeared in the offshore literature (Refs. 1, 2 and 3) which allow the solution to be found by stages in the frequency domain, thus promising answers which are sufficiently clear and simple to be of practical use. In particular, it has been shown (Refs. 1 and 3) by these means and confirmed in model tests, that the time-dependence of [K] in (1.2) leads to an important class of unstable motions in tethered buoyant platforms and that the x-dependence of $\underline{F}$ in (1.2) leads to similar unstable motions and oscillations in floating platforms.

The background to this Paper, therefore, is that methods are becoming available to allow for a much fuller assessment of the importance of structural motion in the calculation of wave loads on offshore structures. It is clear, however, that much detailed work is needed, not least on the choice of suitable non-linear terms $M(\underline{x},\underline{\dot{x}},\underline{\ddot{x}},t)$, $\underline{F}(\underline{x},\underline{\dot{x}},\underline{\ddot{x}},t)$ etc. above, for the new methods to be utilised. Accordingly, this Paper takes an offshore structure for which the fluid loading $\underline{F}(\underline{x},\underline{\dot{x}},\underline{\ddot{x}},t)$ is well known (a steel jacket in the Morison equation régime) and tackles a well known non-linear problem ("hydrodynamic damping") showing first how it has been tackled conventionally.

2. FLUID-STRUCTURE INTERACTION ON LARGE STEEL JACKETS

2.1. The Engineering Problem

Now that the offshore jacket type of platforms are designed for waters of increasing depth, their dynamic characteristics become progressively more important. This is because increasingly tall structures naturally tend to have increasingly long natural periods, so that more and more wave energy is of high enough frequency to give a resonant structural response. In particular, therefore, it becomes progressively more important to have an accurate picture of the damping mechanisms of the structure, since these control this resonant response.

Unfortunately, for these tall jacket structures, it appears that the relatively well understood soil and material damping effects are of less importance than the damping due to the fluid in which the structure is placed. So the fluid-structure interaction problem of hydrodynamic damping comes to the fore : it seems essential, for safety reasons, that damping is not overestimated and wasteful in terms of construction materials if it is under-estimated.

2.2. Analysis by Morison's Equation

The response of a structure with N degrees of freedom may be described by N dynamic equations as follows :

$$[M]\underline{\ddot{x}} + [C]\underline{\dot{x}} + [K]\underline{x} = \underline{F} \tag{2.1}$$

where [M] is the mass matrix of the structure including water used for flooding and marine growth mass, [C] is the matrix due to structural and soil damping and [K] is the matrix due to structural and soil stiffness.

The force vector F accounts for the hydrodynamic forces acting on each of the structure's constituent bodies.* Morison et al (Ref. 4) proposed that the force exerted by a flowing fluid on a rigid cylindrical body of incremental length dS may be represented with reasonable accuracy by the sum of a drag and an inertia force :

$$dF = \tfrac{1}{2} C_D \rho\, DU^2 dS + C_M \rho\, A\dot{U} dS \tag{2.2}$$

then $\underline{F} = \{\int_0^\eta dF\}$ where η is the instantaneous water level at the axis of the body.

The drag and inertia coefficients C_D and C_M are semi-empirical and, strictly, must be considered as varying along the member and with time. However, they are normally specified as an average for a particular member and this value is used at every calculation point within the wave, though it is believed that a drag coefficient relating to steady flows may overestimate the hydrodynamic damping in waves (see Ref. 5).

Therefore, if the structure is responding to water waves, then in a dynamic analysis using Morison's equation, hydrodynamic damping may be included by one of two methods :

Method A

Equation (2.2) is used with relative terms for velocity and acceleration as follows :

$$dF = \tfrac{1}{2} C_D \rho D(U-\dot{x})|U-\dot{x}|dS + \rho A\dot{U}dS + (C_M-1)\rho A(\dot{U}-\ddot{x})dS$$

       DRAG FORCE   FROUDE- INERTIA
                                  KRILOV   FORCE
                                  FORCE (2.3)

which accounts for fluid-structure forces and their interaction as follows :

---

*In this Paper, the term "body" refers equally to a single member of a lattice structure and to a main platform leg. When the body diameter is a large fraction of the wave length (D/L > 0.2), there is a significant variation of the pressure gradient across the width of the body and the incident wave is therefore diffracted. In the present analysis, we are solely concerned with body sizes for which D/L < 0.2.

(i) A drag force dependent on the relative velocity between water particles and each body.

(ii) An inertia force (known as the Froude-Krilov force) due to the undisturbed pressure field around each body.

(iii) An inertia force dependent on the relative acceleration between the water particles and each body. This force is due to the disturbed pressure field around each body and represents the force due to the time varying "added hydrodynamic mass" of the body.

In this method, it is normally assumed that C_M and C_D are equal to the values applicable on a rigid, stationary cylinder.

Method B

The non-interactive form of Morison's equation (2.2) is used to represent the fluid force, while the structural side of equation (2.1) is modified as follows :

$$[M+M_H]\underline{\ddot{x}} + [C+C_H]\underline{\dot{x}} + [K]\underline{x} = \underline{F} \qquad (2.4)$$

The $\ddot{x}$-dependence of the inertia force in (2.3) is here approximated by a constant "added mass" matrix, M_H, and the $\dot{x}$-dependence of the drag force in (2.3) by an additional damping matrix C_H, thus producing the highly convenient set of linear differential equations mentioned in Section 1.

Method A, which employs equations (2.1) and (2.3), is the normal approach adopted for research studies (Refs. 1, 2 and 10), especially when the structure is very responsive. However, Method B, because it is based on the simpler linear equations (2.2) and (2.4), facilitates a much less costly solution procedure and is, as mentioned in Section 1, the method adopted by most commercially available programs for the dynamic analysis of offshore platforms, etc. (Ref. 10).

Effectively, therefore, the analysis of steel jackets by Morison's equation reveals that hydrodynamic damping is an essentially non-linear phenomenon and that, to treat it as a linear effect, as is in practice usually necessary, requires the estimation of a hydrodynamic damping matrix C_H.

2.3. The Mathematical Problem

This task of estimating a term in a linear equation (2.4) so as to make it mimic as closely as possible a non-linear equation (2.1 and 2.3) is a purely mathematical one. The full analysis (1.2) of Section 1 is here needed, not for its own sake, but to demonstrate what is wanted of the linear analysis (1.1), so that one of its terms can be chosen correctly. Since the non-linear terms in (2.3) which we wish to mimic vary with sea state (because U varies with sea state), there is no reason to suppose that there is one single "best approximation" to them. Because of this considerable complication, the problem has, in fact, been waiting for something like the new methods described in Section 1; its history in the offshore literature does not contain an established solution.

Paulling (Ref. 7), for example, approaches the problem by envisaging a hypothetical fluid in which drag forces are not square-law, as in (2.2) and (2.3) but just linearly proportional to velocity, and he derives a linear drag coefficient which gives the correct energy dissipation in the fluid for some given structural motion guessed beforehand. By virtue of the assumed linearity, the total drag force is just the drag force due to the body velocity alone, plus that due to the fluid velocity alone, and thus he can isolate one as a damping term on the LHS of (2.1) and the other as a forcing term on the RHS of (2.1). Clearly, in cases where most of the energy dissipation in the fluid is attributable to body velocity, this "overall energy balance" technique can be expected to predict the damping term accurately and in cases where it is mostly attributable to fluid velocity, the forcing term. Hence the method has been widely accepted for estimating hydrodynamic damping in problems like the resonant pitching of semi-submersible rigs in waves (cited by Paulling), but not in our problem of a stiff structure which moves far less than the water around it.

Malhotra and Penzien (Ref. 6), to take another example, approach the problem of finding a "best linear approximation" more explicitly. They guess a value for the linear damping matrix C_H in (2.4), solve (2.4) to predict the structural motion (which can be done readily, since (2.4) is linear), and then calculate the exact fluid forces (i.e. as given by (2.3)) that would be felt by the structure if it did actually move in that way. By comparing these forces with the forces used in the motion prediction (i.e. used in the guessed equation (2.4)), they derive an "error force" and then repeat the whole process in an iterative manner, so that the mean-square value of this "error force" is minimised. The technique has the advantage that only equations like (2.4) need be solved in the iterative optimisation process, but it is not clear how realistic the "minimum mean-square error force" criterion is. For example, an error force mainly at the structure's resonant frequency might well correspond to a far larger difference between actual and predicted structural motions than an error force of larger mean-square value, but different frequency content.

For these reasons, this Paper takes a fresh look at the problem by starting with a "fundamentalist" analysis in which a simplified, but realistic, structure is analysed, both by Method A above (using a specialised "time domain" computer program to calculate actual motion histories) and by Method B with various values of hydrodynamic damping C_H, so that a suitable value for C_H (suitable, that is, for the sea state in question) can be deduced by comparing answers. This gives a general picture of the difficulty and importance of the problem. The potential analysis of the same structure by the latest iterative mathematical techniques is then described, and the scope of the answers which would be produced outlined, together with the computational problems involved.

First, however, a suitable realistic offshore structure must be chosen for the work, and simplifications made so that the analysis can be done at reasonable cost.

3. STRUCTURE FOR ANALYSIS

3.1. Reduction to Simpler Structure

The structure chosen for analysis is the E.E.C. Phase III Reference Structure (see Fig. 1) which has a suitably long natural period because it stands in 156 m of water and has a flexibility parameter (= deck motion ÷ base motion, in fundamental mode) of 40. Due to the complexity and expense that would be involved in analysing the full structure, it was reduced to a simple two-dimensional model consisting of 33 large cylinders forming a rigid plane frame (see Fig. 2). Each cylinder represents a number of different kinds of real members, at various inclinations, in the same physical zone of the Reference Structure and it must be statically and dynamically equivalent to the combined effect of all the real members in that zone. Further details of the reduction method are explained in a report by one of the authors (Ref. 8).

3.2. Dynamic Similarity

The Reduced Structure will be dynamically equivalent to the full Reference Structure if the drag force, the Froude-Krilov force and the inertia force given by

Morison's equation are equivalent for every zone of both structures. This is achieved by using modified drag and inertia coefficients for the Reduced Structure. It is evident that some simplifications must be made which are not strictly accurate; nevertheless, it is proposed that the model is sufficiently good to serve the purpose of estimating hydrodynamic damping.

3.3. Determination of Stiffness and Damping

The response of the Reduced Structure is restricted to two degrees of freedom at the centre of mass (translation and rotation). A system of horizontal and vertical springs and dashpots (see Fig. 2) is used to model the stiffness and damping characteristics of the Reference Structure. In order to determine [K] and [C] to be used in equations (2.1) and (2.4), a complete dynamic analysis of a two degree of freedom platform on springs was derived for this study (see Fig. 3). A harmonic response was assumed for the fundamental sway mode with natural frequency ω_n. The stiffness matrix is found to contain off-diagonal terms indicating the strong coupling that exists between translation and rotation. The damping matrix (used to describe soil, structural and hydrodynamic damping - see later) is then derived from :

$$[C] = \frac{2\zeta}{\omega_n}[K]$$

where ζ is the "damping ratio" expressed as a percentage of critical damping ($\zeta=1$). Stiffness proportional damping was used, as this approach gives higher values than mass proportional damping (which may or may not include added mass effects).

4. TIME DOMAIN APPROACH

4.1. The General Method

As explained briefly in Section 2, the "fundamentalist" time domain approach to the hydrodynamic damping question is the direct technique of comparing motion (in our case rotation and displacement) time-histories from a full non-linear analysis (Method A of Section 2) with those from linear analyses (Method B of Section 2) incorporating various damping terms.

If fluid-structure interaction is important under any particular set of conditions, then it is to be expected that the amplitudes of rotation and displacement predicted by Methods A and B will be different. As a consequence of the amplitude comparison, it is possible to deduce how much extra damping should be used in Method B in order to simulate the effects of hydrodynamic damping and so obtain the "correct" amplitude as predicted by Method A.

4.2. Computer Program PLATDYN

The computer program PLATDYN (PLATform DYNamics) was available to analyse the dynamic response of fixed, tethered and floating structures in the time domain. PLATDYN calculates the motion history of a flexible fixed structure by simulating it as a rigid body mounted on a number of springs and dashpots whose stiffness and damping are chosen to match the actual stiffness and damping characteristics of the entire soil-structure system.

These motion histories of the structure, in any or all of its degrees of freedom, are computed by PLATDYN using a modified Runge-Kutta method applied to equations (2.1) and (2.3). They allow for all the externally applied forces arising from hydrostatic and hydrodynamic considerations and from any mooring cables, springs and dashpots.

The hydrodynamic forces are derived by application of Morison's equation, using linear wave theory, to elemental parts of each constituent body in the structure. Each body is described by its external dimensions and by three drag and mass coefficients applicable to Morison's equation. The true immersed volume and draft of each body element is calculated at each time step when evaluating the fluid-induced forces. The wave-induced components are found from the particle velocity and acceleration of an undisturbed wave at the centre of the submerged volume and allow for the inclination of the body axis to the vertical.

4.3. Modifications to PLATDYN

Two methods of applying Morison's equation were given in Section 2; equations (2.1) and (2.3) were the fully interactive method including the non-linear $\dot{x}^2$ term and it is assumed that the solution to this equation gives the "correct" time history analysis, defining "truth values" of displacement. The version of PLATDYN using Morison's equation in this form will be called PLATDYN "A". The other method, given by equations (2.2) and (2.4), ignores $\dot{x}^2$ in the drag term and the added hydrodynamic inertia is included in the mass matrix instead of there being (correctly) a time-dependent inertia force on the right-hand side governed by mass coefficients C_M. Version PLATDYN "B" incorporates the latter method. Using the one program PLATDYN, with minor changes for both analyses, provided a consistent method in which the differences in answers could be directly attributed to the effects of fluid-structure interaction.

4.4. Hydrodynamic Damping in Still Water

Soil and structural damping were eliminated in order to isolate the contribution due to the hydrodynamic $\dot{x}^2$ drag force term in Morison's equation. As expected, the version B run predicted oscillations of constant amplitude, while the result of the version A run revealed decaying vibrations. The ratio of successive amplitudes in each period of computed response showed the still water hydrodynamic damping for this structure amounts to 0.5% critical. This is due solely to the motion of the structure through the (still) water. However, when the water particles themselves are in motion due to the presence of currents and waves, then hydrodynamic damping may become significantly more important and "implied" values are used to account for the relative motion between the fluid particles and the dynamically responsive structure.

4.5. Implied Hydrodynamic Damping in Waves

At least four runs of PLATDYN are necessary in order to obtain one estimate of implied hydrodynamic damping in waves, since the solution of Morison's equation with and without fluid-structure interaction is needed for at least two soil damping values.

There are some difficulties in applying the method, in particular :

(i) The implied hydrodynamic damping will depend on sea state and structural dynamic properties (as mentioned in Section 2).

(ii) The amplitude difference varies during the response time history in regular waves and would need to be time-averaged for random waves.

As only a limited range of sea states and structural properties have been tested (see Table 1), it is not possible to define all the conditions under which a significant difference in amplitude occurs. Point (i) means that a further study should therefore include careful variations in these quantities. Point (ii) is important because of the implication that picking off peak values of the response amplitude and using these to determine "implied hydrodynamic damping" is not an entirely satisfactory procedure. A comparison of the r.m.s. of the displacements obtained from

PLATDYN versions A and B would be more appropriate for both regular and random wave loads, but to do this would necessitate long runs on the computer. It was therefore decided to use only the peak displacements for comparison in this analysis, but the time history of response would be studied carefully in each run performed.

(i) Regular Waves

Two cases were considered in which the response of the structure was computed in regular waves. In the first instance, the wave frequency was twice the natural frequency of the structure and differences in amplitudes of horizontal displacement and rotation, as predicted by versions A and B of PLATDYN were negligible. However, if the structure is excited at its resonant frequency in 3 m waves, significant differences in amplitudes occur : with 1% total damping, version B methods overestimate the motion amplitude by 10% - in practice, this might noticeably reduce the predicted fatigue life of a proposed structure. It was found that between 1% and 1½% implied hydrodynamic damping was required in order that version B methods should give the same response amplitudes as version A methods. (Underestimating the fatigue life may not be as serious as the use of too much hydrodynamic damping, which would clearly result in the amplitude of response being underestimated by version B by roughly 10% for each 1% of extra damping. In any event, values of hydrodynamic damping in still water have been shown to be of no relevance in the context of a dynamic analysis in waves).

(ii) Design Wave

To represent a typical design wave, the response of the structure to a single 24 m wave of 12.25 seconds period was computed. The wave length was four times the overall dimensions of the structure, so that only one quadrant of the wave was acting on the structure at any one time. Although this represents a non-resonant condition (the natural frequency of the structure was three times the wave frequency), the importance of structural motion in modifying the wave force is highly significant.

Fig. 4 shows the time histories for the first 10 seconds of response as computed by versions A and B of PLATDYN. The response is entirely transient and it would have been desirable to have continued the time history. Nevertheless, it is clear that version B predicts displacements which may be 100% greater than the "correct" values. Fig. 5 indicates how values of implied hydrodynamic damping are obtained : the values are 12% at the first response peak and 6% at the second peak. (These values are not strongly dependent on the amount of input damping ζ since the gradient of both the A and B lines is small). Therefore, calculation methods which do not allow for fluid-structure interaction are not satisfactory for a design wave dynamic analysis and the input damping may be increased substantially without the risk of underestimating the displacement.

It should be noted at this stage that it is not possible to identify which physical mechanism is giving rise to such large values of hydrodynamic damping. (In Section 2, it was shown that Method A not only includes $\dot{x}^2$ in the drag force, but also allows for time-varying added mass.) The frequency domain approach discussed in the next Section will provide a simple answer to this question.

(iii) Random Waves

A number of runs were conducted using both versions of PLATDYN for various values of input damping ζ. However, all runs gave virtually identical results for 10 seconds of integration, indicating that, in this particular random sea, structural motion is not important in the calculation of wave loads.

4.6. Discussion

From the limited range of structural and wave conditions studied in this investigation, it is possible to reach a tentative conclusion that A-type calculation methods (based on Morison's equation using relative velocity and acceleration) will usually predict smaller displacements of steel lattice structures in the fundamental sway mode than the B-type methods (which assume that the incident wave force is not modified by the structural response).

The actual importance of this with regard to the design of offshore structures is that an underestimate of the hydrodynamic damping may lead to an excessively conservative design or, alternatively, an overestimate of the hydrodynamic damping may result in an overestimate of fatigue life. Thus, it is likely that under certain conditions, structural motion is important in the calculation of wave loads. One of the authors (Ref. 9) has, therefore, proposed a simple design procedure in which hydrodynamic damping values are obtained for the Reduced Structure (as described above) and which are then used in the full B-type calculation of wave loads on the complete jacket structure.

However, a great deal of computing effort and expense is required in order to determine just one value of hydrodynamic damping for a particular jacket structure in a given sea state. The need, therefore, is for a cheap method which will give hydrodynamic damping values under any conditions, as well as provide general insight into the problem of non-linear motions. The next Section explains how the frequency domain approach is capable of an elegant solution to some of the drawbacks of the more traditional time domain methods.

5. THE FREQUENCY DOMAIN APPROACH

5.1. The General Method

The general idea of the "frequency domain" approach to the solution of non-linear equations like (1.2) in Section 1, is to reach the complex equation by a sequence of simple ones which are solved in turn to produce a "first iteration" solution, a "second iteration" solution, etc. The fundamental point is that the structural motion histories are approximated by a converging sequence of functions - in general, these are functions of time (i.e. motion histories), but for our application of oscillatory wave loading, the simplicity of the "simple equations" allows one to think in the highly convenient terms of equivalent functions of frequency (i.e. response spectra). This all means that the natural setting for the subject is the branch of abstract mathematics known as Functional Analysis, in which entire functions are thought of as points in a general infinite-dimensional space - indeed, the origin of the method lies in that mathematical literature.

In our application, however, the complexity is in the applications to real structures, rather than in the abstract ideas involved, so it is necessary to keep the discussion in the engineering notation best suited to the real problem. The fullest notation for a general problem in the dynamics of offshore structures is that of equation (1.2) and the general argument can be developed for that equation as in Ref. 1. For the case of a stiff structure, however, the structural mass, damping and stiffness matrices are always thought of as constants and this allows the same ideas to be developed with less clutter. We start, therefore, by writing the exact structural equations (2.1) and (2.3) in the form :

$$M\underline{\ddot{x}} + C\underline{\dot{x}} + K\underline{x} = F(\underline{x},\underline{\dot{x}},\underline{\ddot{x}},t) \qquad (5.1)$$

This is now solved by the iterative scheme :

$$M\underline{\ddot{x}}_{n+1} + C\underline{\dot{x}}_{n+1} + K\underline{x}_{n+1} = \underline{F}(\underline{x}_n,\underline{\dot{x}}_n,\underline{\ddot{x}}_n,t) \qquad (5.2)$$

$$+ \frac{d\underline{F}}{d\underline{x}}(\underline{x}_n,\underline{\dot{x}}_n,\underline{\ddot{x}}_n,t)\{\underline{x}_{n+1} - \underline{x}_n\}$$

$$+ \frac{d\underline{F}}{d\underline{\dot{x}}}(\underline{x}_n,\underline{\dot{x}}_n,\underline{\ddot{x}}_n,t)\{\underline{\dot{x}}_{n+1} - \underline{\dot{x}}_n\}$$

$$+ \frac{d\underline{F}}{d\underline{\ddot{x}}}(\underline{x}_n,\underline{\dot{x}}_n,\underline{\ddot{x}}_n,t)\{\underline{\ddot{x}}_{n+1} - \underline{\ddot{x}}_n\}$$

which produces a converging sequence $\underline{x}_1(t)$, $\underline{x}_2(t)$, $\underline{x}_3(t)$ .... of successive approximations to the solution $x(t)^2$.

This method remains in the time domain at this stage. It is alternatively possible (Ref. 2) to devise an all-frequency domain approach, although the physical identity of the terms is then lost. All that has been done is to expand the RHS as a truncated Taylor Series about the function $\underline{x}_n(t)$ where its value was last established, and thus produce a (time-varying) linear equation for $\underline{x}_{n+1}(t)$. From the perspective of Functional Analysis, the method appears as no more than a generalisation to functions of the very well known Newton-Raphson method for solving algebraic equations.

We now emphasise the structure of (5.2) by rearranging it into the form :

$$[M+M'+M''(t)]\underline{\ddot{x}}_{n+1} + [C+C'+C''(t)]\underline{\dot{x}}_{n+1} + [K+K'+K''(t)]\underline{x}_{n+1} = \underline{F} + \underline{F}' + \underline{F}''(t) \quad (5.3)$$

in which the derivative terms on the RHS of (5.2) have led to extra "implied" mass, damping and stiffness matrices which come respectively from the rate of change of loading with acceleration $[d\underline{F}/d\underline{\ddot{x}}]$ with velocity $[d\underline{F}/d\underline{\dot{x}}]$ and with displacement $[d\underline{F}/d\underline{x}]$. (The parts of the derivative terms which depend on $\underline{x}_n$, the given previous motion estimated, have led just to extra "implied" loading terms on the RHS). Moreover, these extra terms have all been split into their steady components (M', C', etc.) and their (zero-mean) fluctuating components (M''(t), C''(t) etc.) which are periodic functions of time. This emphasises that the core of the method is the use of periodic differential equations - these are more difficult to solve than the ordinary constant differential equations of (1.1) but give the iterative scheme a very rapid convergence which is, in general, "quadratic", implying a doubling of the number of correct decimal digits at each stage. Indeed, for stiff structures, it is rarely necessary to complete more than one iteration for good accuracy.

For each iteration, then, there are a number of new parameters to be calculated by evaluating particular values or particular rates-of-change of the existing expression (2.3) for hydrodynamic loading. The problem then becomes one of solving the periodic differential equation (5.3). The idea here is to rewrite the equation as :

$$[M+M']\underline{\ddot{x}}_{n+1} + [C+C']\underline{\dot{x}}_{n+1} + [K+K']\underline{x}_{n+1} = F+F'+F''(t) - \{M''(t)\underline{\ddot{x}}_{n+1} + C''(t)\underline{\dot{x}}_{n+1} + K''(t)\underline{x}_{n+1}\} \qquad (5.4)$$

so that the final bracketed term on the right-hand side of (5.4) can be thought of as a "feedback loop" around the "linear system" formed by the rest of the equation. See Fig. 6.

A solution is then sought by "iterating round the feedback loop", i.e. cutting the loop, calculating the output and feeding back a correction based on this output, repeatedly. Because the terms in the "feedback loop" are periodic, their action on the periodic output estimates is just to produce sum-and-difference frequency and so the whole solution can be visualised readily enough as a "network" in the frequency domain.

Figure 7, taken from Reference 3, shows the "network" for the surge motion of a tethered buoyant platform in regular waves, where both the forcing term and the feedback coefficient K''(t) on the RHS on (5.4) have the same frequency as the waves. Calculation of actual numerical values, can, of course, be done by computer. (The computing effort involved is, in fact, trivial compared with the work of evaluating the time-varying coefficients in (5.3)), but the graphical approach gives the kind of global picture of "what is happening" (i.e. which are the crucial variables, the worst-case sea states, etc.) that one normally associates only with the simple linear theory of equation (1.1) in Section 1. The method has been used successfully in the past to predict non-linear motions of floating and tethered buoyant structures (especially dynamically unstable motions, which can be analysed very naturally as instabilities in the feedback loop described above).

5.2. Application to the Hydrodynamic Damping Problem

It is at once clear that a considerable advantage of this frequency domain approach to the problem of hydrodynamic damping is that one does not have to contend with the phenomena of transients, since network diagrams like Fig. 7 deal explicitly with periodic solutions. By contrast, it is obvious from the computed time history shown in Fig. 4, that the time domain approach gives answers which may be dominated by the computer's starting transient - this may give a quite disproportionate picture of the importance of hydrodynamic damping because, in real life, "design waves" are not solitary shocks to the structure but part of a slow build-up storm with typically a very narrow sea spectrum.

The details of the problem can be made apparent by continuing the mathematical development already started. Referring to equation (5.2) and noting that $F(\underline{x},\underline{\dot{x}},\underline{\ddot{x}},t)$ is given by equation (2.3), we see first that $dF/d\underline{\ddot{x}}$ will be what is known normally as "added mass" of the structure. It will produce terms M' and M''(t) in (5.3) which are respectively the added mass of the structure in still water and the fluctuating component of added mass attributable to the parts of the structure in the splash zone which gain and lose added mass as a wave passes. Next, the term $dF/d\underline{x}$ will reflect the change in hydrodynamic loading with platform position. Since the platform needs to move of the order of the water depth to affect the hydrostatic loading drastically and of the order of a wave length to affect the hydrodynamic loading drastically, this term leads to implied stiffnesses K' and K''(t) in (5.3) which are quite negligible on a steel jacket when compared with the structural stiffness K.

Finally, the term $dF/d\underline{\dot{x}}$ leads to the critical implied damping matrices C' and C''(t) in (5.3). They are defined simply as the steady and fluctuating components of the rate-of-change of platform loading with velocity - a parameter which clearly derives from the $\underline{\dot{x}}$ term in (2.3) and which could be computed readily enough by evaluating (2.3) for the previous estimate of platform motion ($\underline{x}_n$) and for that plus a small incremental velocity. It is clear that this rate-of-change will hardly alter as the "previous estimate of platform motion" is revised, because the structure is known to move very little in comparison with the water, and therefore the hydrodynamic loading calculation and thus these derivatives C' and C''(t), as well as M' and M''(t), will hardly change from iteration to iteration.

The conclusion is that the first iteration (taking $\underline{x}_o \equiv 0$) to the platform motion should give a very good approximation to the platform motion from its equation:

$$[M+M'+M''(t)]\underline{\ddot{x}}_1 + [C+C'+C''(t)]\underline{\dot{x}}_1 + K\underline{x}_1 = \underline{F}(0,t) \qquad (5.5)$$

where M', M''(t), C', C''(t) are the rates-of-change-of-loading terms calculated for $\underline{x} = 0$, i.e. at equilibrium. Hence the whole problem of hydrodynamic damping boils down to choosing a damping matrix C_H, which will enable a constant-coefficient equation like (2.4) to mimic the variable-coefficient (5.3). Clearly, a good start would be to put $C_H = C'$ (M' being already added in as a conventional "added mass"); the errors involved are then simply described by the lower network arrows in a diagram like Fig. 7, for these describe the "feedback action" of the time varying matrices $M''(t)$ and $C''(t)$.

The "global picture of what is happening" promised above is therefore that the best choice of hydrodynamic damping on a jacket structure in waves is closely connected to the average rate-of-change of hydrodynamic loading with velocity (and is thus clearly a function of sea state). The errors associated with making it simply equal to this rate-of-change can be gauged by drawing a network diagram like Fig. 7, from which it will be clear what the worst-case sea states are from the point of view of these errors. One obvious case is when the waves are at half one of the platform's natural frequencies, for then the time-varying added mass matrix produces a second-harmonic response (precisely as in Fig. 7 itself) which will be magnified by platform resonance.

5.3. Computational Aspects

It was mentioned in 5.1 above that the computational effort needed to solve the frequency domain approach's sequence of "simple equations" is relatively trivial. The effort comes in the calculation of the various coefficients ("implied damping", "implied stiffness", etc.) used by these "simple equations". This latter task, it was also mentioned above, is a matter of calculating particular values of existing formulae like equation (2.3) in Section 2.

From a computational point of view, therefore, the implementation of the frequency domain approach described here would merely involve the rearrangement of the subroutines in an existing time domain program like PLATDYN to calculate certain new variables, plus the addition of some post-processing software of relatively modest needs in computing time.

To be specific, for each iteration, the loading term $\underline{F}(x,\dot{x},\ddot{x},t)$ in (5.1) needs to be calculated for series of points in time and associated previously calculated structural displacements and velocities. In addition, its rates-of-change with respect to structural velocity and acceleration also need to be calculated, implying two extra loading calculations per degree of freedom (at an incremental acceleration and an incremental velocity in that degree of freedom) at each of the points in time. If the aim was to resolve motions up to, say, the third harmonic of the frequency of a regular wave, the number of points in time needed would be eight (an eight-point Fast Fourier Transform gives the magnitude and phase of the fundamental and three harmonics, plus the steady component). Hence the total number of "instantaneous state" runs per iteration would be 8 + 2 x 8 x (number of structural degrees of freedom).

This compares with a time domain program like PLATDYN which for, say, a run of 100 time steps using a fourth order integration routine, needs 400 "instantaneous state" calculations. The implication is that one iteration of the frequency domain method described above would need less computing time than a time domain calculation if the structure had less than 25 degrees of freedom, and more computing time if more than 25.

For the particular problem of "hydrodynamic damping" considered in this Paper, this figure of 25 degrees of freedom should strictly be doubled to 50 because the added mass matrix would be already calculated by other means and so the rate-of-change-with-acceleration calculation can be omitted. In any event, non-linear offshore structures with sufficient degrees of freedom to make the frequency domain method uneconomic would probably not be analysed in the time domain either, because, as mentioned in Section 1, the information produced would be just too complex to be of practical use. Certainly, a frequency domain analysis of the two degree of freedom structure devised in Section 3 presents no problems in terms of computer time.

6. CONCLUSIONS

A suitable value of linearised "hydrodynamic damping" for a given degree of freedom of a tall steel jacket structure in a given sea state can be found by simplifying the structure and computing motion time histories, firstly with full non-linear equations, and then with various linearised damping terms.

However, the procedure is expensive on computer time, especially if very long time histories are computed in order to remove the confusing effects of starting transients. Also, it is difficult to separate the effects of "hydrodynamic damping" from those of, say, the varying hydrodynamic added mass associated with varying platform immersion. Finally, there is no guide as to which sea states give the most critical values of linearised "hydrodynamic damping", or the most significant non-linear effects.

The frequency domain methods recently introduced for the analysis of floating and tethered structures would be significantly cheaper on computing time, provided the structure is again simplified down to less than about fifty degrees of freedom. More importantly, the methods offer comprehensive answers to the "which sea state is critical" problem, by means of diagrams showing "global" behaviour. In particular, it seems that a time-averaged rate-of-change-of-loading parameter introduced by these techniques is the controlling variable for determining damping.

REFERENCES

(1) Rainey, R.C.T. : "Parasitic motions of offshore structures". Paper accepted for publication by R.I.N.A., (November 1978).

(2) Taudin, P. : "Dynamic response of flexible offshore structures to regular waves". Offshore Technology Conference Paper OTC3160 (1978).

(3) Rainey, R.C.T. : "The dynamics of tethered platforms". Journal of the R.I.N.A. (March 1978).

(4) Morison, J.R., O'Brien, M.P., Johnson, J.W. and Schaaf, S.A. : "The force exerted by surface waves on piles". Petroleum Transactions, AIME, Vol. 189, pp149-154. (1950).

(5) Verley, R.L.P. : "An experimental investigation into hydrodynamic damping in waves". BHRA Project No. RP13104. (1978).

(6) Malhotra, A.K. and Penzien, J. : "Analysis of tall open structures subjected to stochastic excitation". Conference on Dynamic Waves in Civil Engineering, University of Wales, Swansea. (1970).

(7) Paulling, J.R. : "Wave induced forces and motions of tubular structures". Symposium on Naval Hydrodynamics, Pasadena. (1970).

(8) Dean, R.B., Fish, P.R. and Heaf, N.J. : "Analysis of hydrodynamic damping in offshore structures". Report no. 5 for L.E.A. Offshore Management Limited. Atkins Research and Development. (January 1978).

(9) Fish, P.R., Dean, R.B. and Heaf, N.J. : "Fluid-structure interaction in Morison's equation for the design of offshore structures". Journal of Engineering Structures (in preparation). (1979).

(10) Hallam, M.G., Heaf, N.J. and Wootton, L.R. : "Dynamics of marine structures : Methods of calculating the dynamic response of fixed structures subject to wave and current action". CIRIA Report UR8.

Run No.	Sea State	Wave Length L m	Wave Height H m	Wave Period Tw s	Nat. Period of struc. Ts s	Nat. Freq. of struc. Ns Hz	Soil damping % critical	Computed Hydrodynamic Damping C_H % crit.
1	Still water	-	-	-	2.165	0.462	0	0.5
2	Still Water	-	-	-	4.330	0.231	0	0.5
3	Regular	30	3.0	4.33	4.330	0.231	1	-
4	Regular	7.5	0.75	2.165	4.330	0.231	1	0
5	Regular	30	3.0	4.33	4.330	0.231	1	1.1 - 1.4
6	Regular	30	3.0	4.33	4.330	0.231	3	1.1 - 1.4
7	Regular	30	3.0	4.33	4.330	0.231	5	1.1 - 1.4
8	Design	240	24	12.25	4.330	0.231	1	6 - 12
9	Design	240	24	12.25	4.330	0.231	3	6 - 12
10	Design	240	24	12.25	4.330	0.231	5	6 - 12
11	Random	-	3.0 (Significant)	4.33 (Zero-crossing)	4.330	0.231	1	0
12	Random	-	3.0 (Significant)	4.33 (Zero-crossing)	4.330	0.231	5	0

Table 1 Production runs using program PLATDYN versions A and B

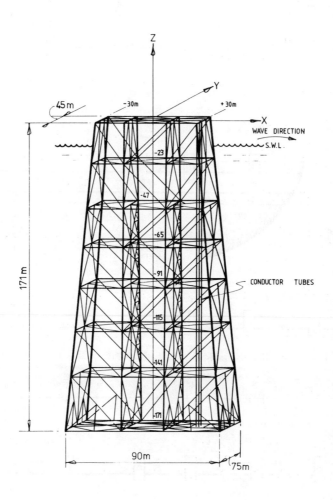

Fig. 1   E.E.C. Phase III reference structure

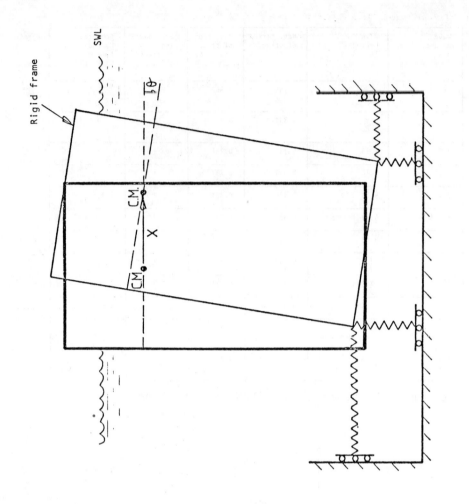

Fig. 3 Mode shape of displaced structure showing location of centre of mass

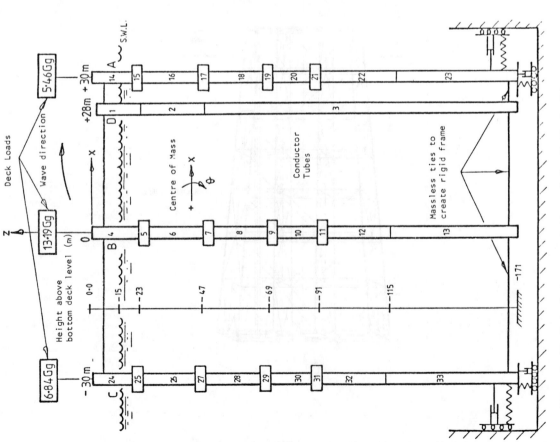

Fig. 2 Reduced representation of the E.E.C. Phase III reference structure (not to scale)

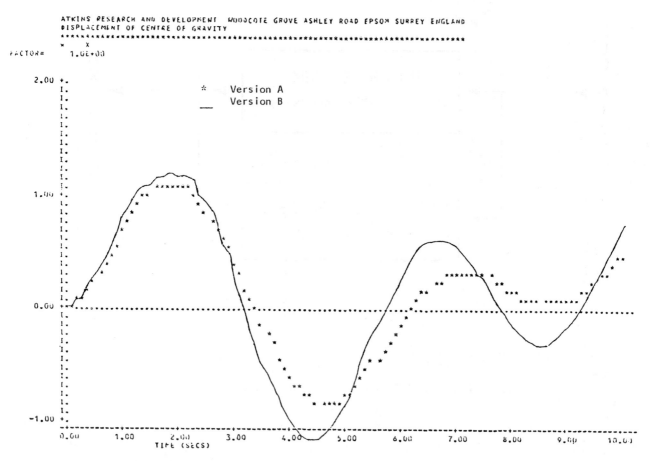

Fig. 4 Response time-history of the reduced structure in a typical design wave

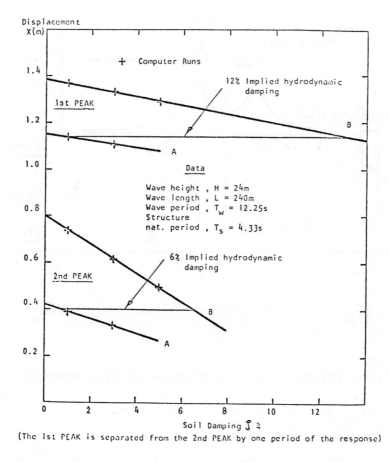

Fig. 5 Determination of hydrodynamic damping in a typical design wave

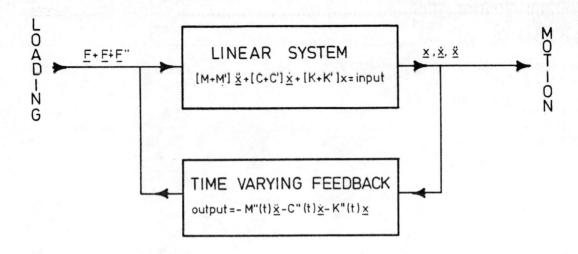

Fig. 6 Periodic differential equations as feedback systems

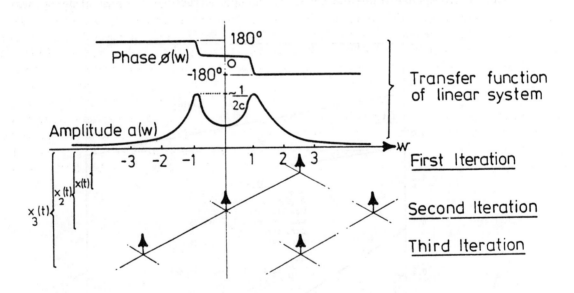

Fig. 7 Illustrative example of the mathematical technique

BOSS'79

PAPER 51

Second International Conference
on Behaviour of Off-Shore Structures
Held at: Imperial College, London, England
28 to 31 August 1979

A LINEAR ANALYSIS OF INTERACTION PROBLEMS IN OFFSHORE PLATFORMS

R. Eatock Taylor

University College London, U.K.

Summary

This paper examines the problems of fluid-structure and soil-structure interaction for bottom supported platforms. The structures may be 'stiff' or 'compliant', but it is assumed that their behaviour may be described by linear dynamics theory. A modal analysis with a frequency domain computation is used. Because of the significant spatial variation of damping effects, arising from hydrodynamic damping near the free surface, damping in the structure, and foundation damping it is inappropriate to assume proportional damping for the fluid-structure-soil system. This problem, and the difficulty of frequency dependent coefficient matrices in the equations of motion (added mass, radiation damping and foundation stiffness matrices), is overcome using the 'dry' modes of the 'fixed base' structure. A rational basis is thereby developed for obtaining a reduced dynamic model, which accurately represents the highly complex interaction problem. Typical results are given for gravity platforms, including the effects of three dimensional added mass and damping and a layered soil. The rationale for these techniques is discussed in the context of interpretation of full scale data from North Sea platforms.

Sponsored by: Delft University of Technology, The Netherlands
Massachusetts Institute of Technology, U.S.A.
The Norwegian Institute of Technology, Norway
University of London, England

Secretariat provided by: BHRA Fluid Engineering
Copyright: © BHRA Fluid Engineering
Cranfield, Bedford, England

NOMENCLATURE

Symbols
- $\underset{\sim}{A}$ — Column matrix of generalised actions
- $\underset{\sim}{B}$ — Damping matrix
- $\underset{\sim}{D}$ — Column matrix of generalised displacements
- $\underset{\sim}{M}$ — Inertia matrix
- $\underset{\sim}{n}$ — Normal to submerged body surface
- n — Number of generalised coordinates
- $\underset{\sim}{P}$ — Column matrix of principal coordinates of fixed base structure
- p — Hydrodynamic pressure
- $\underset{\sim}{S}$ — Stiffness matrix
- $\underset{\sim}{T}$ — Transformation matrix relating generalised displacements
- t — Time
- $\underset{\sim}{X}$ — Matrix of mode shapes arranged columnwise
- x, y, z — Cartesian coordinates
- η — Structural damping ratio
- ρ — Fluid density
- Σ — Submerged surface of structure
- Φ — Velocity potential
- ϕ — Spatial potential
- $\underset{\sim}{\chi}$ — Rigid body mode
- $\underset{\sim}{\psi}$ — Fixed base distortion mode (nodal coordinates)
- ψ — Fixed base distortion mode (continuum form)
- ω — Frequency

Subscripts
- A — Superstructure degrees of freedom
- B — Degrees of freedom connecting superstructure to rigid base
- D — Diffracted wave potential
- I — Incident wave potential
- r, s — Index defining generalised coordinate number
- R — Rigid body coordinates
- S — Fixed base superstructure distortion coordinates

Dressings
- * — Modified or transformed matrix
- − — Complex amplitude of time varying quantity
- + — Matrix of "added" hydrodynamic quantities
- W — Matrix of wave forces
- T — Transpose of a matrix

1. INTRODUCTION

In the study of the structural dynamics of offshore platforms it is highly desirable to employ linear analyses if at all possible. For wave excitation of certain types of structure, such as massive concrete gravity platforms, the hypothesis of linearity seems especially appropriate. For other types of platform, for example framed structures, the assumption of a linear relationship between wave height and structural response is not generally tenable, although a linearised theory may be possible at the higher wave frequencies and smaller wave heights associated with response at a structural resonance. The range of validity of the linearity assumption has yet to be clarified on the basis of data from different types of instrumented platforms.

It seems reasonable to suppose, however, that linear dynamics theory will continue to fulfil a vital rôle in the analysis of certain types of offshore platforms, both in design assessment and in interpretation of structural responses of instrumented platforms. Although linear formulations in the time domain have been used for offshore platforms, a frequency domain approach is more appropriate. The random nature of ocean waves points to the techniques of spectral analysis; and it is also found that the interactions between fluid, structure and foundation are considerably more amenable to description in the frequency domain than in the time domain. This paper, therefore, employs a linear frequency domain formulation. In the interests of economy a modal approach is adopted.

Numerous invesitgators have of course employed conventional modal techniques in analysing the dynamics of offshore structures. Their approach has been based either on the use of structural dynamics computer program packages, such as NASTRAN and STARDYNE; or on the development of special purpose programs such as those described by Penzien et al (Ref.1), Marshall (Ref.2) and Hallam et al (Ref.3) for offshore framed structures; and by Moan et al (Ref.4) and Nataraja and Kirk (Ref.5) for concrete gravity platforms. A feature common to these analyses is the method of dealing with soil-structure and fluid structure interactions. The structure is assumed to be supported on springs and dampers (whose properties may be obtained from analytical half space solutions, finite element idealisations, etc.), and undamped natural modes and frequencies are defined for the total structure plus foundation system, including added mass effects due to the surrounding fluid. These modes may be termed "wet system modes".

Although this technique appears simple to implement, it suffers several disadvantages. Many of these have been previously discussed in the context of the earthquake response of land-based structures (e.g. Vaish and Chopra (Ref.6)). One problem leading to both theoretical and practical difficulties, is that the soil springs are inevitably frequency dependent, so that natural frequencies and modes must be defined for a system with variable coefficient matrices. Even if the underlying soil is assumed to be a linearly elastic continuum, the system is non-conservative. The difficulty is compounded by the frequency dependence of hydrodynamic added mass terms, although this problem has commonly been ignored up until now.

A related difficulty concerns the distribution of damping in the total system. Generally the foundation is much more heavily damped than the structure, and it can be misleading to make the conventional assumption of proportional damping. An artifice is required (e.g. Roesset et al (Ref.7)) to ensure the appropriate weighting of damping contributions over the total extent of the structure and foundation. But again the difficulty is compounded by hydrodynamic effects, in this case associated with damping localised near the free surface. This has been discussed in the review by Vugts and Hayes (Ref.8).

An alternative to using wet system modes is to base the analysis on the undamped modes of the fixed base structure. The generalised coordinates become the principal coordinates associated with the fixed base structure, together with the coordinates describing the displacements of the interface between structure and soil. This substructuring approach has been described by Gutierrez and Chopra (Ref.9) for the case of a general land-based flexible foundation on an arbitrary soil, and the extension to a system including rigid elements (for example an idealised caisson of a gravity platform) has been made by Eatock Taylor (Ref.10). Such a formulation has several advantages for studying the effects of soil-structure interaction. It is shown subsequently in this paper that this is also a particularly appropriate technique to resolve the fluid-structure interaction problem of offshore structures, provided that the fixed base modes are those associated with undamped vibration of the structure in vacuo. These may be called "dry fixed base modes".

The purpose of this paper, then, is to show how the conventional modal approach may be adapted to resolve efficiently the highly complex interaction problems associated with the dynamics of offshore platforms. The adaptations required to solve the soil-structure interaction problem are described in §2, and the fluid-structure interaction is formulated and solved in §3. This approach is then applied to a gravity platform analysis in §4.

2. THE USE OF 'FIXED BASE' MODES FOR THE SOIL-STRUCTURE INTERACTION PROBLEM

2.1 Review of some advantages

The references cited above have listed many of the advantages of a formulation based on fixed base modes, in the context of dynamically sensitive land-based structures. It is appropriate to emphasize two of these here. Firstly, for a variety of reasons, and particularly in the case of offshore structures, it is often necessary to perform several analyses involving one structure on different soil profiles. Whereas a solution based on system modes would require for each soil profile a completely new analysis for structure and soil, the alternative using fixed base modes requires only trivial calculations of the interaction effects after the basic analysis of the structure has been completed once and for all. In a similar manner, finite element calculations for the soil strata may be performed in the absence of details of the structure above.

A second feature of this approach is the economy with which accurate solutions may be obtained, using a very small number of modes. This is related to the manner in which damping is represented. In the system mode approach, an effect of the heavy soil damping is to exert strong coupling between the modes. Thus the transformed damping matrix formed by the change from generalised to principal coordinates has strong off-diagonal elements. A possible approach is to ignore these terms, but where greater accuracy is required it is not unusual to derive results from an exact solution of the coupled equations for the first few system modes (e.g. Clough and Mojtahedi (Ref.11), Warburton (Ref.12)). The aim is to retain the economy of the mode superposition procedure assuming significant contributions from only the first few modes - while incorporating the effects of damping coupling between these modes. Unfortunately, however, this assumption may lead to important errors. It has been shown (Ref.13) that higher modes (of natural frequency say ten times that of the lowest natural frequency) may indeed make a significant contribution to response in the region of the fundamental resonance, even though the excitation of these higher modes by the externally applied forces may be small. Furthermore, it has been found (Ref.12) that numerical difficulties can arise from the system mode approach in the case of relatively soft soils, when the lowest mode is close to a rigid body mode.

An analysis using the undamped fixed base modes overcomes those difficulties. Since the fixed base structure may generally be assumed to be proportionally damped, its behaviour may be expressed accurately in terms of a reduced set of uncoupled equations, and an economic solution for the completed system may then be obtained by assembling and solving exactly the equations of motion written in terms of the foundation degrees of freedom and the first few principal coordinates of the fixed base structure.

A simple example illustrating this feature has been discussed in Ref.14. It is a two-dimensional idealisation of a wide range of structures represented by five lateral degrees of freedom, resting on a half-space foundation where the additional freedoms of swaying and rocking are introduced. It is found that the highest system mode (the seventh) can make a significant contribution to the foundation shear, when the structure is excited harmonically at the top at the funamental resonant frequency of the structure-foundation system: the use of six system modes rather than the complete set of seven leads to a 16% error in the foundation shear, in one of the cases analysed. This is in spite of the fact that the natural frequency associated with the seventh mode is approximately fifteen times the excitation frequency, and that the generalised force in this mode is very small. The importance of this system mode is associated with the high degree of damping coupling between the first and seventh modes. If, however, fixed base modes are used, the results converge very rapidly: if only the first fixed base mode is included, the foundation reactions are found to be within 1% of the exact solution.

2.2 The equations of motion

The approach using fixed base modes may be employed to advantage for any type of foundation. It has been found to be a particularly valuable tool in achieving economy and accuracy in the dynamic analysis of a class of offshore structures: those that are supported on a foundation which behaves essentially as a rigid body, for example concrete gravity platforms having very stiff caissons or rafts. Thus Ref.10 and Ref.15 have used this procedure to clarify the soil-structure interaction problem for gravity platforms. In § 3 of this paper we show how the use of fixed-base modes is also particularly suited to solution of the fluid-structure interaction problem. But before we embark on that development, it is necessary first to review the equations of motion in the absence of fluid effects, and to illustrate their use.

Let the degrees of freedom above the level of the rigid body base be D_A (corresponding to some finite element idealisation of the superstructure); let the degrees of freedom which couple the superstructure to the rigid base be D_B; and let the rigid body degrees of freedom of the base be D_R. The equations of motion of the structure undergoing harmonic excitation may therefore be written

$$\begin{bmatrix} M_{AA} & M_{AB} \\ M_{AB}^T & M_{BB} \end{bmatrix} \begin{bmatrix} \ddot{D}_A \\ \ddot{D}_B \end{bmatrix} + (1 + i\eta_S) \begin{bmatrix} S_{AA} & S_{AB} \\ S_{AB}^T & S_{BB} \end{bmatrix} \begin{bmatrix} D_A \\ D_B \end{bmatrix} = \begin{bmatrix} A_A \\ A_B \end{bmatrix} \qquad (1)$$

A structural damping ratio η_S has been assumed, and the appropriate mass and stiffness submatrices for the superstructure are $M_{\alpha\beta}$ and $S_{\alpha\beta}$ respectively. A_A corresponds to the external loads on the superstructure at degrees of freedom A. A_B corresponds to the external loads at degrees of freedom B, plus the loads on the superstructure transmitted from the rigid base. Thus A_B includes the effects of soil-structure interaction, inertial effects of the base, and external (e.g. hydrodynamic) loads on the base.

The superstructure motions are given by the base rigid body motions, plus the distortions of the superstructure relative to a fixed base. D_A may therefore be expressed in terms of D_R and the principal coordinates P_S associated with the characteristic modes of the fixed base superstructure. D_B, connected to the rigid base, may be written directly in terms of D_R. Thus, as in Ref.10 and Ref.15 :

$$\begin{bmatrix} D_A \\ D_B \end{bmatrix} = \begin{bmatrix} X_{AS} & T_{AR} \\ 0 & T_{BR} \end{bmatrix} \begin{bmatrix} P_S \\ D_R \end{bmatrix} \qquad (2)$$

X_{AS} is a matrix containing the fixed base modes Ψ_{S_r} arranged columnwise. The transformation of Eq.(2) is exact if all n_A modes corresponding to the degrees of freedom A are included in X_{AS} (i.e. $r = 1, \ldots n_A$); it is approximate if only the first n_S modes are retained in the solution (i.e. $r = 1, \ldots n_S$). T_{BR} is a simple geometric transformation matrix and

$$T_{AR} = -S_{AA}^{-1} S_{AB} T_{BR} \qquad (3)$$

It may be noted that the matrix

$$T_R = \begin{bmatrix} T_{AR} \\ T_{BR} \end{bmatrix} \qquad (4)$$

may be thought of as a matrix containing the n_R rigid body modes of the complete structure, when the base undergoes rigid body displacements in each of its degrees of freedom D_R.

Through the use of Eq.(2) it is possible to transform Eq.(1) to the form

$$\begin{bmatrix} M_{SS} & M_{SR} \\ M_{RS} & M_{RR} \end{bmatrix} \begin{bmatrix} \ddot{P}_S \\ \ddot{D}_R \end{bmatrix} + \begin{bmatrix} S_{SS}(1+in_S) & 0 \\ 0 & 0 \end{bmatrix} \begin{bmatrix} P_S \\ D_R \end{bmatrix} = \begin{bmatrix} A_S^+ \\ A_R^* \end{bmatrix} \qquad (5)$$

The transformed matrices are defined in the Appendix, and it may be noted that because of the orthogonality of the modes Ψ_{S_r} the matrices S_{SS} and M_{SS} are in fact diagonal. The transformed actions are given by

$$A_S^+ = X_{AS}^T A_A$$
$$A_R^* = T_{AR}^T A_A + T_{BR}^T A_B \qquad (6)$$

A_S^+ represents the generalised hydrodynamic actions exciting the superstructure in its fixed base distortion modes. A_R^* is the set of actions associated with responses of the structure in its rigid body modes: it includes hydrodynamic loads due to fluid pressures on superstructure and base, together with base inertias and the effects of soil-structure interaction. Typically the latter effects are expressible by terms proportional to $\dot{D}_R$ and $\ddot{D}_R$. Hence A_R^* may be expanded in the form:

$$A_R^* = T_{AR}^T A_A + A_R - M_{RR}^* \ddot{D}_R - B_{RR}^* \dot{D}_R - S_{RR}^* D_R \qquad (7)$$

where A_R is the component due to external (hydrodynamic) loads on the base. M_{RR}^* is the inertia matrix for the base. The matrices S_{RR}^* and B_{RR}^* may be found from elastic half space theory, finite element solutions etc.

The transformed Eq.(5) is a coupled set of equations in $(n_S + n_R)$ unknowns. Compared with the original number of generalised coordinates in the finite element idealisation of the structure, this is a very small number of unknowns. The reduction has been achieved by discarding the contributions of the higher fixed base modes. It has been found in fact that even the extreme case of $n_s = 1$ can give quite acceptable results for a gravity platform undergoing motions parallel to a vertical plane (i.e. ignoring torsion). This was foreshadowed in results described above, and has been convincingly demonstrated in some analyses of gravity platforms presented by Duncan (Ref.16).

2.3 Some results

These results (Ref.16) were obtained using the 'fixed base' mode approach described herein. They provide a useful illustration of the effects of soil-structure interaction, and for this reason some of the results are reviewed here. They relate to the typical gravity platform design shown in Fig 1, which is similar to the Brent B CONDEEP, and is designated platform 1 in this paper. Transfer functions are given for deck deflection and base shear, corresponding to excitation by sinusoidal waves. They have been calculated according to Eq.(5) and Eq.(7), on the basis of a finite element idealisation of the fixed base structure. The results are plotted as solid lines in Figs 2 and 3. (For the purposes of this illustration, fluid interaction effects have been approximated through the use of a simple two dimensional constant added mass coefficient and zero hydrodynamic damping. The method of improving upon this representation is discussed subsequently).

The transfer functions have been non-dimensionalised by the equivalent quasistatic results, so that they are plotted as magnification factors. Four soil conditions are considered, having the specifications given in Fig 1. The frequency dependent impedance functions for the uniform elastic half-space were approximated by the equations of Veletsos and Verbic (Ref.17), while those for the layered soils were obtained from polynomial fits to the analytical solutions of Luco (Ref.18). Figs 2 and 3 show the influence of soil type on deck deflection and total shear at the base of the platform. As the soil is softened, the natural frequency of the fundamental "system" resonance of course decreases, and the magnification factor at this frequency decreases. On the other hand, the introduction of the thin layer of soft soil beneath the platform increases the magnification factor at the second system resonance, particularly for the base shear (the second system resonances for soil conditions 2 and 3 are approximately 3.3 rads/sec and 2.8 rads/sec respectively). This is because the second system mode involves relatively large horizontal displacements at the foundation: for the half space this mode is highly damped, but the effect of the overlying soft layer is to decrease the damping associated with swaying at the foundation (Ref.18). This characteristic is discussed in further detail by Duncan (Ref.16), who examines the behaviour of several stress resultants for two different platforms. A related effect is the influence of caisson diameter: in general, the larger the diameter, the closer will be the first and second system resonances. For certain platforms on soft soils, the two humps corresponding to these two resonances merge into one another, forming a rather broad and ill-defined region where dynamic magnification occurs.

The transfer functions plotted as solid lines in Figs 2 and 3 were calculated using 6 "fixed base" structure modes in Eq.(5). Results have also been obtained using only one "fixed base" mode, and these are indicated by plotting symbols in the figures. The differences are scarcely perceptible up to a frequency of about 3 radians/sec. The influence of higher "fixed base" modes is only evident above this frequency, giving rise to the peaks and troughs characteristic of phase differences between the wave forces on the three towers of the platform. This conclusion, backed by more detailed studies (Refs.19 and 20), suggests that the significant features of the overall dynamics of gravity platforms in two dimensions may be characterised by a simple three degree of freedom spring-mass-damper model. Distortion of the superstructure is represented by one degree of freedom, and the other two freedoms correspond to swaying and rocking of the foundation. The stiffness and damping terms associated with the latter are frequency dependent, but all parameters may easily be found following the procedure of (Ref.16). With the addition of a fourth degree of freedom, vertical motions (heave) may be included, and it would be relatively straightforward to extend these ideas to a three dimensional representation.

The accuracy of these simple models, clearly demonstrated in Figs 2 and 3, is of
considerable importance in the interpretation of data from instrumented platforms.
Methods are being developed whereby reliable estimates may be made of the stiffness
and damping associated with structure and underlying seabed. Often these must be
based on poor quality data from a small number of transducers. Parameter evaluation
may be optimised by reduction of the basic structural idealisation to a model with
only a few degrees of freedom. Duncan (Ref.21) has demonstrated the advantages of
employing fewer degrees of freedom in the structural model than input parameters from
the instrumentation, in obtaining best fits for characteristic stiffness and damping
parameters. The model described herein is a most satisfactory basis for this approach,
providing an efficient theoretical representation of the effects of soil-structure
interaction. It requires, however, an accurate and efficient method of describing
the fluid-structure interaction phenomenon, and this is described in the following
section.

3. THE CHOICE OF "DRY" MODES FOR EVALUATING THE FLUID LOADING

3.1 Definition of the generalised fluid forces

For the purposes of this analysis, we assume that the flow may be represented by the
approximation of an ideal fluid. Viscous effects are therefore neglected, but the
methodology of this section is applicable to a wide range of structures including
gravity platforms, large diameter articulated columns, and many types of floating
structure. The fluid forces are those induced by waves and those due to the motions
of the structure. We are concerned here principally with the transverse motions of
platforms, for which the motion induced loads are proportional to the horizontal
accelerations and velocities of the structure. They may therefore be described
loosely in terms of added mass and damping effects. Our problem is how to evaluate
these effects and to incorporate them into the analysis of the previous section.

Let us consider again Eq.(5), with the understanding that the wave loads and motion
induced loads are included on the right hand side. The matrices $\underset{\sim}{M}_{\alpha\beta}$ thus refer only
to the structural mass (including any enclosed liquids in storage tanks etc.) By
retaining this form of the equations of motion, we effectively uncouple the structural
and the fluid problems. The most time-consuming part of the structural analysis is
evaluation of the lowest fixed base modes and natural frequencies of the complex
finite element structural model. But this may be performed independently of the
surrounding fluid, without any complications due to frequency dependent added masses
etc. The modes thereby obtained may be classified as "dry" modes.

We have still however to specify the generalised fluid forces on the right hand side
of Eq.(5), corresponding to the coordinates $\underset{\sim}{P}_S$ and $\underset{\sim}{D}_R$. Let Σ be the submerged surface of the structure, comprising the submerged surface Σ_A of the superstructure and
the surface Σ_B of the rigid base. Let the total dynamic pressure at a point (x,y,z)
on Σ be $p(x,y,z,t)$. This is a continuous function of position, whereas the fluid
loads associated with terms $\underset{\sim}{A}_A$ or $\underset{\sim}{A}_B$ in Eq.(6) are consistent nodal loads, obtained
using the finite element shape functions of the structural model. The generalised
fluid loads in Eq.(5) may be expressed in terms of the pressures as follows. Let

$$\underset{\sim}{\psi}_{S_r} = \begin{bmatrix} \psi_{S_{rx}}(x,y,z) \\ \psi_{S_{ry}}(x,y,z) \\ \psi_{S_{rz}}(x,y,z) \end{bmatrix} \qquad (r = 1, \ldots n_S)$$

be the continuum form of the rth fixed base mode, analogous to the finite degree of freedom representation $\underset{\sim}{\psi}_{S_r}$. Thus $\underset{\sim}{\psi}_{S_r}$ is obtained from $\underset{\sim}{\Psi}_{S_r}$ through substitution of the shape functions. Furthermore, let

$$\underset{\sim}{n} = \begin{bmatrix} n_x(x,y,z) \\ n_y(x,y,z) \\ n_z(x,y,z) \end{bmatrix}$$

be the normal from the surface into the fluid. The generalised fluid forces associated with the principal coordinates of the fixed base structure may then be written (Eq.(6)):

$$A_S^+ = X_{AS}^T A_A = -\int_{\Sigma_A} p \, \chi_S^T \, \underset{\sim}{n} \, d\Sigma \tag{8}$$

where χ_S is a matrix containing the fixed base modes $\underset{\sim}{\psi}_{S_r}$ arranged columnwise. Eq.(8) may be verified by consideration of the virtual work associated with a virtual displacement in each of the fixed-base modes in turn.

The generalised fluid forces associated with the rigid body cordinates may be obtained in a similar manner. Let

$$\psi_{R_r} = \begin{bmatrix} \psi_{R_{rx}}(x,y,z) \\ \psi_{R_{ry}}(x,y,z) \\ \psi_{R_{rz}}(x,y,z) \end{bmatrix} \qquad (r = 1, \ldots n_R)$$

be the continuum form of the rth rigid body mode, and let these modes be arranged columnwise in the matrix χ_R. (This is the continuum analogue of the finite degree of freedom matrix $\underset{\sim}{T}_R$ given in Eq.(4)).

The generalised fluid forces corresponding to the coordinates $\underset{\sim}{D}_R$ in Eq.(7) are then

$$A_R^+ = T_{AR}^T A_A + A_R = -\int_{\Sigma_A} p \, \chi_R^T \, \underset{\sim}{n} \, d\Sigma - \int_{\Sigma_B} p \, \chi_R^T \, \underset{\sim}{n} \, d\Sigma \tag{9}$$

3.2 The fluid boundary value problem

The next stage in formulating the equations of motion using the dry fixed base modes is to define the fluid pressure p in Eq.(8) and (9). Since the fluid (of density ρ) is assumed to be ideal, the boundary value problem may be specified in terms of a velocity potential Φ. The dynamic pressure is thus given by

$$p = \rho \frac{\partial \Phi}{\partial t} \tag{10}$$

(The dynamic buoyancy and second order effects in Bernoulli's equation are ignored). Since our objective is to obtain linear transfer functions corresponding to sinusoidal waves, it is useful to note that the total velocity potential itself then varies harmonically, at frequency ω. We take

$$\Phi = \operatorname{Re}(\phi \, e^{i\omega t}) \tag{11}$$

The total flow field results from the incident wave (with associated potential Φ_I), a wave diffracted by the body (ϕ_D), and fluid motions induced by response of the structure in its modes of vibration. The latter response is defined in Eq.(2) in terms of $\underset{\sim}{P}_S$ and $\underset{\sim}{D}_R$ which may be written

$$\underset{\sim}{P}_S = \operatorname{Re}[\underset{\sim}{\bar{P}}_S \, e^{i\omega t}]$$

$$\underset{\sim}{D}_R = \operatorname{Re}[\underset{\sim}{\bar{D}}_R \, e^{i\omega t}] \tag{12}$$

Since the problem is linear, the potential due to the body motions may also be expressed as a superposition of potentials associated with the fixed-base modes (ϕ_{S_r}) and with the rigid body modes (ϕ_{R_r}). Thus

$$\phi = \phi_I + \phi_D + \sum_{r=1}^{n_S} \bar{P}_{S_r} \phi_{S_r} + \sum_{r=1}^{n_S} \bar{D}_{R_r} \phi_{R_r} \tag{13}$$

The incident wave potential ϕ_I may be assumed known. The unknown potentials satisfy Laplace's equation, a free surface boundary condition, a rigid wall boundary condition at the seabed, and a radiation condition at infinity. (Further details of these aspects, in the context of a related problem, are described in Ref.22.) Furthermore, the following boundary conditions must be satisfied on the submerged surface of the structure:

$$\frac{\partial \phi_D}{\partial n} = -\frac{\partial \phi_I}{\partial n} \quad \text{on } \Sigma \tag{14a}$$

$$\frac{\partial \phi_{S_r}}{\partial n} = -i\omega \, \underset{\sim}{\psi}_{S_r}^T \underset{\sim}{n} \quad \text{on } \Sigma_A, \quad r = 1, \ldots n_S \tag{14b}$$

$$\frac{\partial \phi_{S_r}}{\partial n} = 0 \quad \text{on } \Sigma_B, \quad r = 1, \ldots n_S \tag{14c}$$

$$\frac{\partial \phi_{R_r}}{\partial n} = -i\omega \, \underset{\sim}{\psi}_{R_r}^T \underset{\sim}{n} \quad \text{on } \Sigma \quad r = 1, \ldots n_R \tag{14d}$$

The boundary value problems for ϕ_D, ϕ_{S_r} and ϕ_{R_r} are well-posed, subject to certain continuity requirements on Σ, and the solutions may be approximated by some numerical technique such as the boundary integral method (Refs.22, 23). The analysis is uncoupled from the structural dynamics problem, except through the specification of the modes $\underset{\sim}{\psi}_{S_r}$ and $\underset{\sim}{\psi}_{R_r}$ for the boundary conditions in Eq.(14).

3.3 Evaluation of the fluid loads

We may now express the generalised fluid loads of Eq.(8) and Eq.(9) in terms of the known velocity potentials. We make use of Eq.(10) - (14). In Eq.(8) we have

$$\underset{\sim}{A}^+_{S_r} = - \int_{\Sigma_A} p \, \underset{\sim}{\psi}^T_{S_r} \underset{\sim}{n} \, d\Sigma$$

$$= \text{Re} \left\{ \rho \left[\int_{\Sigma_A} (\phi_I + \phi_D) \frac{\partial \phi_{S_r}}{\partial n} d\Sigma + \sum_{s=1}^{n_S} \bar{P}_{S_s} \int_{\Sigma_A} \phi_{S_s} \frac{\partial \phi_{S_r}}{\partial n} d\Sigma \right. \right.$$

$$\left. \left. + \sum_{s=1}^{n_R} \bar{D}_{R_s} \int_{\Sigma_A} \phi_{R_s} \frac{\partial \phi_{S_r}}{\partial n} d\Sigma \right] e^{i\omega t} \right\} \quad (15)$$

In Eq.(9) we have

$$\underset{\sim}{A}^+_{R_r} = - \int_{\Sigma_A + \Sigma_B} p \, \underset{\sim}{\psi}^T_{R_r} \underset{\sim}{n} \, d\Sigma$$

$$= \text{Re} \left\{ \rho \left[\int_{\Sigma} (\phi_I + \phi_D) \frac{\partial \phi_{R_r}}{\partial n} d\Sigma + \sum_{s=1}^{n_S} \bar{P}_{S_s} \int_{\Sigma} \phi_{S_s} \frac{\partial \phi_{R_r}}{\partial n} d\Sigma \right. \right.$$

$$\left. \left. + \sum_{s=1}^{n_R} \bar{D}_{R_s} \int_{\Sigma} \phi_{R_s} \frac{\partial \phi_{R_r}}{\partial n} d\Sigma \right] e^{i\omega t} \right\} \quad (16)$$

These may be written in terms of generalised wave forces $\underset{\sim}{A}^W_S$, $\underset{\sim}{A}^W_R$, and added mass and hydrodynamic damping matrices, $\underset{\sim}{M}^+_{SS}$, $\underset{\sim}{M}^+_{SR}$, $\underset{\sim}{M}^+_{RR}$ and $\underset{\sim}{B}^+_{SS}$, $\underset{\sim}{B}^+_{SR}$, $\underset{\sim}{B}^+_{RR}$ respectively, as defined in the Appendix. Thus we obtain

$$\underset{\sim}{A}^+_S = \underset{\sim}{A}^W_S - \text{Re} \left\{ \left[(-\omega^2 \underset{\sim}{M}^+_{SS} + i\omega \underset{\sim}{B}^+_{SS}) \bar{P}_S + (-\omega^2 \underset{\sim}{M}^+_{SR} + i\omega \underset{\sim}{B}^+_{SR}) \bar{D}_R \right] e^{i\omega t} \right\} \quad (15a)$$

$$\underset{\sim}{A}^+_R = \underset{\sim}{A}^W_R - \text{Re} \left\{ \left[(-\omega^2 \underset{\sim}{M}^{+T}_{SR} + i\omega \underset{\sim}{B}^{+T}_{SR}) \bar{P}_S + (-\omega^2 \underset{\sim}{M}^+_{RR} + i\omega \underset{\sim}{B}^+_{RR}) \bar{D}_R \right] e^{i\omega t} \right\} \quad (16a)$$

If these expressions are substituted in Eq.(5), written for harmonic motion, we finally obtain equations of motion of the form

$$\begin{bmatrix} \{-\omega^2 (M_{SS} + M_{SS}^+) + i(\eta_S S_{SS} + \omega B_{SS}^+) + S_{SS}\} & \{-\omega^2 (M_{SR} + M_{SR}^+) + i\omega B_{SR}^+\} \\ \{-\omega^2 (M_{RS} + M_{RS}^+) + i\omega B_{RS}^+\} & \{-\omega^2 (M_{RR} + M_{RR}^* + M_{RR}^+) + i(B_{RR}^* + B_{RR}^+) + S_{RR}^*\} \end{bmatrix} \begin{bmatrix} \bar{P}_S \\ \bar{D}_R \end{bmatrix}$$

$$= \begin{bmatrix} A_S^W \\ A_R^W \end{bmatrix} \quad (17)$$

This is a small set of equations which may be readily solved at each frequency of interest. The number of unknowns is $(n_S + n_R)$ and the number of fluid boundary value problems to be solved is one more than this (since ϕ_D is to be obtained, see Eq.(13)). This is an order of magnitude fewer than the number of unknowns required by a direct analysis of the fluid-structure interaction problem, for example using fluid finite elements.

4. APPLICATION AND CONCLUSIONS

4.1 Numerical study

In order to illustrate application of the preceding theory, results for a gravity platform are presented and discussed. A comparison is made between results obtained using the full three dimensional frequency dependent hydrodynamic analysis, developed in §3 (and designated 3D) and results from a simpler strip theory approximation to the hydrodynamic loads (designated 2D). Details of the structure (platform 2) are given in Fig 4. It consists of a circular cylindrical base, of radius 60 m, resting on an elastic seabed and supporting a single circular cylindrical tower of radius 15 m. The structure has been deliberately chosen to have a very simple geometry, to facilitate comparison between the hydrodynamic analyses. Typical wave forces and generalised added masses are illustrated. Results are also presented for typical response transfer functions over the frequency range $0 < \omega < 4$ rad s^{-1}. These correspond to excitation by a sinusoidal wave of unit height, from which response statistics may be obtained corresponding to any chosen wave spectrum.

The responses were evaluated from Eq.(17) using the lowest two fixed base modes of the tower (designated modes S1 and S2 respectively), and the rigid body modes swaying and rocking of the complete structure (designated R1 and R2 respectively). Because of the symmetry of this structure, only modes in a single vertical plane need be included in an analysis of response to a long-crested sea. These modes are shown in Fig 5. The distortion modes are those of a tower having uniform cross-sectional properties, with a point mass and rotational inertia at deck level. Although in this special case analytical expressions for the modes could have been employed, they were in fact calculated from an idealisation with eight finite elements as an illustration of the general procedure that has been developed. The natural frequencies of the lowest two fixed base tower modes are $\omega_{S1} = 3.90$ rad s^{-1} $\omega_{S2} = 20.9$ rad s^{-1}.

The wave forces in the 2D strip theory analysis were applied as consistent loads to the finite element nodes shown in Fig 6, assuming a linear variation of force along a single element. The value of wave force per unit length at each node was calculated using the expression given by MacCamy and Fuchs (Ref.24): this is the exact wave diffraction solution for a surface piercing circular cylinder which extends to the seabed. Thus except at very low frequencies, where the influence of the base is significant, the 2D expressions for wave force per unit length at the nodes are essentially exact for the example structure. The corresponding consistent loads applied to the finite element nodes are, however, approximate, because of the assumption of a linear variation between nodes. This may be expected to give increasingly inaccurate results at higher frequencies. (The source of error may be illustrated by considering an extreme short wave case when the topmost submerged element carries a load varying approximately exponentially from a maximum at the upper node to near zero at some point above the lower node: the assumption of a linear variation between nodes would seriously overestimate the consistent load terms). It should be emphasized that this error would arise from an unduly coarse structural idealisation, and it is not related to the hydrodynamic analysis. It can be simply circumvented by using a sufficient number of finite elements or by incorporating exact analytical expressions for the consistent load terms.

The three dimensional hydrodynamic analysis was effected using the boundary integral method (Refs 22, 23). This involved approximate solution of the integral equations for the velocity potentials, evaluated at the panel points shown in Fig 6. For the example structure it was appropriate to employ an axisymmetric variant of the theory, which has been validated for this type of structure by comparison with analytical and experimental results (Ref 23). Fig 7 shows the resulting total wave overturning moment on the structure (equivalent to the generalised force corresponding to the rocking mode, R2), plotted against frequency. This is the moment due to a sinusoidal wave of unit height. The results from the 3D analysis are compared in Fig 7 with the overturning moment calculated in the 2D analysis program using the consistent loads on the coarse finite element mesh of Fig 6. As anticipated, the values from the 2D program are seriously in error at higher frequencies, lying well above the values from the 3D program using the panel points shown in Fig 6. It should be observed, however, that Fig 7 is particularly intended to highlight the discrepancy. The values of total horizontal forces predicted by the two programs are very much closer than these values of moments. And the discrepancy could be eliminated for this structure by using a similar number of finite element nodes to the number of panel points employed in the hydrodynamic analysis.

As noted earlier, the geometry of platform 2 has been selected to enable this comparison between 2D and 3D programs. Generally the tower or towers of such a platform would taper. In a calculation of the wave forces on a non-cylindrical tower the 3D hydrodynamic analysis based on the boundary integral procedure would be inherently more accurate than the 2D analysis, regardless of the number of finite elements used in the structural idealisation.

The comparison of added mass as calculated for platform 2 by the two analyses is given in Fig 8. The generalised added masses $M^+_{RR_{11}}$, $M^+_{SS_{11}}$ and $M^+_{RS_{11}}$, defined in the Appendix and calculated using the three dimensional hydrodynamic analyses, are plotted against frequency. These terms correspond respectively to the added mass for rigid body swaying of the structure, the generalised added mass associated with distortion in the first fixed base mode, and the cross added mass term coupling these two modes. The two dimensional strip theory terms corresponding to the first two of these are $M^*_{RR_{11}}$ and $M^*_{SS_{11}}$, calculated assuming a constant added mass coefficient of unity for translation of a circular cross-section perpendicular to the axis of the cylinder. Fig 8 presents the ratios $M^+_{RR_{11}}/M^*_{RR_{11}}$ and $M^+_{SS_{11}}/M^*_{SS_{11}}$.

These are an indication of the three dimensional characteristics of the flow, and the frequency dependence of the added mass terms for this structure. In the sway mode 3D effects are paramount, flow over the top of the base causing a significant reduction in added mass. In the first fixed-base distortion mode the added mass is found to be strongly frequency dependent, particularly between 0.5 and 1.0 rad s^{-1}. Also plotted in Fig 8 is the ratio $M^+_{RS_{11}}/(M^+_{RR_{11}} M^+_{SS_{11}})^{\frac{1}{2}}$, indicating the order of magnitude of cross-coupling terms.

Corresponding hydrodynamic damping terms for platform 2 are shown in Fig 9, calculated by the 3D analysis. These are a measure of the radiation damping caused by oscillation of the submerged body near the free surface, an effect ignored by the simple strip theory. The damping terms have therefore been non-dimensionalised using the strip theory added masses described above. The dimensionless damping associated with the first fixed base distortion mode is seen to be much more significant than hydrodynamic damping in the sway mode. This is to be expected because of the concentration of activity near the free surface in the former case. It is found that the maximum hydrodynamic damping is in fact larger than the structural damping in the towers, for which the value $\eta_s = 0.05$ was used in Eq.(5). But high values of hydrodynamic damping are restricted to a rather narrow frequency range.

The responses of platform 2 are shown in Figs 10-12, associated with the two soil conditions specified in Fig 4. These illustrate the differences between results from the two dimensional and three dimensional analyses. The deck displacement magnitude per unit wave height is given in Fig 10 for soil condition 5. This is a firm soil overlain by a soft layer. The first peak in the transfer function, near $\omega = 0.6$ rad s^{-1}, corresponds to the peak in the horizontal wave force (analogous to that in the wave overturning moment, Fig 7). All of the transfer functions studied here are dominated by this peak, which is characteristic of wave diffraction around a vertical cylinder. A second peak is however clearly evident, and this is associated with response at the fundamental "resonance" of the coupled fluid-structure-soil system. The 2D analysis indicates that this resonance occurs at a frequency of about 1.5 rad s^{-1}; this analysis also suggests a second resonance near $\omega = 2.5$ rad s^{-1} (which may be compared with the behaviour of platform 1 on layered soils 2 and 3). In the vicinity of the first and second resonances the results from the 3D analysis are considerably lower, but this is partly due to the reduced wave forces predicted by the 3D analysis. This point is discussed further in connection with the total overturning moment.

Fig 11 shows the deck displacement transfer function associated with soil condition 6, a uniformly soft elastic half space. Again there is evidence of the fundamental resonance, but the peak is much less pronounced because of the greater soil damping associated with the softer seabed. As in the previous case, the results of the 3D analysis lie well below the 2D line in the region of this resonance.

The total dynamic overturning moment at the seabed for the soft soil (condition 6) is plotted in Fig 12. In form it is similar to the deck displacement (Fig 11). It may however be compared directly with the wave overturning moment shown in Fig 7. Indeed the latter may be defined as the total quasistatic overturning moment at the seabed. Hence the ratios of the values plotted as solid lines in Fig 12 and Fig 7 give a curve of dynamic magnification factor for the 2D analysis. This curve, with the corresponding magnification factors associated with the 3D results, is also included in Fig 12. It highlights the fundamental resonance and clarifies some differences between the two hydrodynamic analyses.

By examining magnification factors, we can eliminate differences due to the different wave forces, predicted by the computer programs based on two dimensional and three dimensional hydrodynamic theories. Thus the remaining discrepancies arise from differences in added mass, and the neglect of hydrodynamic damping in the 2D analysis. Fig 12 shows that the resonant peak predicted by the 2D analysis is at a lower frequency than suggested by the 3D results. This is associated with the overestimated added mass in the 2D strip theory (c.f. Fig 8). It also appears that the magnification factor at resonance is slightly smaller in the 3D results (although more points would be required to clarify this feature). This would be consistent with a contribution from hydrodynamic (radiation) damping. It may be seen in Fig 9, however, that for this structure the generalised hydrodynamic damping parameters have their maxima at frequencies well below the fundamental resonance. Furthermore, in this case of a soft soil, tower distortion (mode S1) makes a relatively smaller contribution than the rigid body modes (R1 and R2). The effects of hydrodynamic damping would therefore be expected to be quite small under this particular combination of circumstances. Even so, the results of Figs 10-12 provide evidence of potentially important modifications to the results if use is made of the three dimensional frequency dependent hydrodynamic analysis.

4.2 Conclusions

This paper has been concerned both with soil-structure interaction and fluid-structure interaction. A linear theory has been developed, based on the techniques of modal analysis. By using "dry" modes of the structure vibrating in vacuo it has been possible to uncouple the fluid boundary value problem from the structural dynamics. For platforms supported on a compliant seabed, the dry modes may be those of the fixed base structure. It has been shown that by this device the boundary value problem for the soil may also be solved independently. The result is an efficient technique for solution in the frequency domain, leading to transfer functions for displacements or stresses. These may be combined with wave spectra, to obtain response spectra and response statistics for random wave excitation.

It has been shown in §2 that widely differing degrees of damping in structure and foundation may be accommodated by the use of fixed base modes. It is unnecessary to introduce inaccuracy through the assumption of proportional damping in the combined soil-structure system. Even the use of a non-proportional damping matrix is found to be problematical if the equations of motion are formulated in terms of a limited number of system modes: damping coupling with higher modes can introduce significant contributions to behaviour at the fundamental resonance. It is shown, however, that accurate results may be obtained with a small number of fixed base modes. This paper does not discuss specifically the relative accuracies of results for displacements and stresses. If local stress resultants are evaluated by a direct superposition of a limited number of modal contributions, they will generally be less accurate than the corresponding displacements. It has been shown elsewhere, however, that this problem may be circumvented: the stresses are calculated quasi-statically, by loading the structure with its external actions and the inertia forces associated with dynamic response in its lowest modes. In this way dynamic stresses are obtained to the same degree of accuracy as displacements, while the advantages of the modal superposition procedure are retained.

The platform examined in §4 has illustrated the phenomenon of fluid-structure interaction, and provided comparison of results computed by a strip theory approach with corresponding results from the theory of §3. The influence of three dimensional effects on the added mass terms was found to be of considerable significance: for this structure undergoing swaying, for example, an added mass coefficient of 0.6 would be more appropriate than the conventional value of unity. Some of the generalised added mass parameters were also found to vary strongly with frequency, particularly those associated with the distortion modes. Even more strongly dependent on frequency was the hydrodynamic (radiation) damping. High values of damping were found in the distortion modes, but only over a narrow range of frequencies (0.5 - 1.1 rad s^{-1}). For this particular structure the hydrodynamic damping decayed to rather low values at the fundamental resonant frequency, so that the influence on resonant magnification factors was small. On the other hand the influence of added mass on the value of the fundamental "wet system" natural frequency highlighted the potential importance of using the accurate three dimensional analysis.

The theory described in this paper leads to a linear analysis for a class of fluid-structure-soil interaction problems, which exploits the economy of the mode superposition approach. A three dimensional hydrodynamic analysis is generally required for the wave diffraction problem, and the additional cost of computing the radiation potentials for the lowest dry modes is relatively minor. For the example structure the influence of three dimensional and free surface effects was relatively modest, and the simpler theory gave conservative results. But it is not difficult to visualise geometries for which these effects would be more significant.

The validity of the approach is related to the validity of potential flow theory for wave diffraction analysis. Experiments are in hand using articulated models to simulate structural distortions, in order to investigate this range of validity.

Another difficult area for bottom supported platforms is the use of linear soil impedances (which may be based on finite element analysis). It would appear that an important way of clarifying these uncertainties would be a careful investigation of the responses of instrumented platforms, using the foregoing theory.

ACKNOWLEDGEMENTS

I am most grateful for the assistance of Dr. P.E. Duncan in the preparation of this paper. The work was sponsored by the Science Research Council Grant GR/A/4685.9 to the London Marine Technology Centre.

REFERENCES

1. Penzien, J., Kaul, M.K. and Berge, B.: "Stochastic response of offshore towers to random sea waves and strong motion earthquakes". Computers and Structures, 2, pp 733-756 (1972).

2. Marshall, P.W.: "Dynamic and fatigue analysis using directional spectra". Proc. Eight Offshore Technology Conf., Houston, Paper 2537, II, pp 143-157 (1976).

3. Hallam, M.G., Heaf, N.J., Boudreaux, R.H. and Millman, D.N.: "Dynamic and fatigue analysis of Murchison tower structure". Proc. Tenth Offshore Technology Conf., Houston, Paper 3163, II, pp 1001-1010 (1978).

4. Moan, T., Syversten, K. and Haver, S.: "Dynamic analysis of gravity platforms subjected to random wave excitation". Presented to Society of Naval Architects and Marine Engineers Spring Meeting, San Francisco (May 1977).

5. Nataraja, R. and Kirk, C.L.: "Dynamic response of a gravity platform under random wave forces". Proc. Ninth Ann. Offshore Technology Conf., Houston, Paper 2904, III, pp 199-208 (1977).

6. Vaish, A.K. and Chopra, A.K.: "Earthquake finite element analysis of structure-foundation systems". J. Eng. Mech. Div. ASCE, 100, pp 1101-1116 (1974).

7. Roesset, J.M., Whitman, R.V. and Dobry, R.: "Modal analysis for structures with foundation interaction". J. Struct. Div., ASCE, 99, pp 399-415 (1973).

8. Vugts, J.H. and Hayes, D.J.: "Dynamic analysis of fixed offshore structures". Eng. Struct. 1, pp 114-120 (1979).

9. Gutierrez, J.A. and Chopra, A.K.: "A substructure method for earthquake analysis of structures including structure-soil interaction". Earthq. Eng. Struct. Dyn., 6, pp 51-69 (1978).

10. Eatock Taylor, R.: "Structural dynamics of offshore platforms". Proc. Conf. Offshore Structures, pp 125-132, Institution of Civil Engineers (1975).

11. Clough, R.W. and Mojtahedi, S.: "Earthquake response analysis considering non-proportional damping". Earthq. Eng. Struct. Dyn., 4, pp 489-496 (1976).

12. Warburton, G.B.: "Soil-structure interaction for tower structures". Earthq. Eng. Struct. Dyn., 6, pp 535-556 (1978).

13. Duncan, P.E. and Eatock Taylor, R.: "A note on the dynamic analysis of non-proportionally damped systems". Earthq. Eng. Struct. Dyn., 7, pp 99-105 (1979).

14. Duncan, P.E. and Eatock Taylor, R.: "Inaccuracies in the dynamic analysis of non-proportionally damped systems". Report No. OEG/78/5, Department of Mechanical Engineering, University College London (1978).

15. Eatock Taylor, R.: "A preliminary study of the structural dynamics of gravity platforms". Proc. Seventh Ann. Offshore Technology Conf., Houston, Paper 2406, III, pp 695-706 (1975).

16. Duncan, P.E.: "Simple models for the dynamics of deepwater gravity platforms". Eng. Struct. 1, pp 65-72 (1979).

17. Veletsos, A.S. and Verbic, B.: "Basic response functions for elastic foundations". J. Engng. Mech. Div. ASCE, 100, pp 189-202 (1974).

18. Luco, J.E.: "Impedance functions for a rigid foundation on a layered medium". Nucl. Eng. and Design, 31, pp 204-217 (1974).

19. Eatock Taylor, R.: "A two degree of freedom model for the dynamics of offshore structures". Earthq. Eng. Struct. Dyn., 6, pp 331-346 (1978).

20. Eatock Taylor, R. and Duncan, P.E.: "The dynamics of offshore gravity platforms: some insights afforded by a two degree of freedom model". Earthq. Eng. Struct. Dyn., 6, pp 455-472 (1978).

21. Duncan, P.E.: "A scheme for the interpretation of data from instrumented offshore platforms". J. Sound Vib., 64, (1979).

22. Eatock Taylor, R. and Waite, J.B.: "The dynamics of offshore structures evaluated by boundary integral techniques". Int. J. Num. Meth. Engng. 13, pp 73-92 (1978).

23. Eatock Taylor, R. and Dolla, J.P.: "Hydrodynamic loads on vertical bodies of revolution". Report No. OEG/78/6, Department of Mechanical Engineering, University College London (1978).

24. MacCamy, R.C. and Fuchs, R.A.: "Wave forces on piles: a diffraction theory". Beach Erosion Board Tech. Mem. No. 69 (1954).

APPENDIX

DEFINITION OF MATRICES APPEARING IN EQS (5), (15) and (16).

The transformed matrices of Eq.(5) are given by

$$\underset{\sim}{S}_{SS} = X_{AS}^T \, S_{AA} \, X_{AS}$$

$$\underset{\sim}{M}_{SS} = X_{AS}^T \, M_{AA} \, X_{AS}$$

$$\underset{\sim}{M}_{SR} = X_{AS}^T \, M_{AA} \, T_{AR} + X_{AS}^T \, M_{AB} \, T_{BR} = M_{RS}^T$$

$$\underset{\sim}{M}_{RR} = T_{AR}^T \, M_{AA} \, T_{AR} + T_{AR}^T \, M_{AB} \, T_{BR} + T_{BR}^T \, M_{BA} \, T_{AR} + T_{BR}^T \, M_{BB} \, T_{BR}$$

The generalised wave forces in Eq.(15a) and Eq.(16a) are defined by the terms

$$A_{S_r}^W = \mathrm{Re}\left\{ \left[\rho \int_{\Sigma_A} (\phi_I + \phi_D) \frac{\partial \phi_{S_r}}{\partial n} d\Sigma \right] e^{i\omega t} \right\}$$

$$A_{R_r}^W = \mathrm{Re}\left\{ \left[\rho \int_{\Sigma} (\phi_I + \phi_D) \frac{\partial \phi_{R_r}}{\partial n} d\Sigma \right] e^{i\omega t} \right\}$$

The generalised added mass and damping matrices in Eq.(15a) and Eq.(16a) are defined by the terms:

$$M_{SS_{rs}}^+ = \mathrm{Re}\left\{ \frac{\rho}{\omega^2} \int_{\Sigma_A} \phi_{S_s} \frac{\partial \phi_{S_r}}{\partial n} d\Sigma \right\}$$

$$M^+_{SR_{rs}} = \text{Re}\left\{\frac{\rho}{\omega^2} \int_{\Sigma_A} \phi_{R_s} \frac{\partial \phi_{S_r}}{\partial n} d\Sigma\right\}$$

$$M^+_{RS_{rs}} = \text{Re}\left\{\frac{\rho}{\omega^2} \int_{\Sigma} \phi_{S_s} \frac{\partial \phi_{R_r}}{\partial n} d\Sigma\right\}$$

$$M^+_{RR_{rs}} = \text{Re}\left\{\frac{\rho}{\omega^2} \int_{\Sigma} \phi_{R_s} \frac{\partial \phi_{R_r}}{\partial n} d\Sigma\right\}$$

$$B^+_{SS_{rs}} = \text{Im}\left\{\frac{-\rho}{\omega} \int_{\Sigma_A} \phi_{S_s} \frac{\partial \phi_{S_r}}{\partial n} d\Sigma\right\}$$

$$B^+_{SR_{rs}} = \text{Im}\left\{\frac{-\rho}{\omega} \int_{\Sigma_A} \phi_{R_s} \frac{\partial \phi_{S_r}}{\partial n} d\Sigma\right\}$$

$$B^+_{RS_{rs}} = \text{Im}\left\{\frac{-\rho}{\omega} \int_{\Sigma} \phi_{S_s} \frac{\partial \phi_{R_r}}{\partial n} d\Sigma\right\}$$

$$B^+_{RR_{rs}} = \text{Im}\left\{\frac{-\rho}{\omega} \int_{\Sigma} \phi_{R_s} \frac{\partial \phi_{R_r}}{\partial n} d\Sigma\right\}$$

It may be noted that the following symmetry relationships should be satisfied:

$$\underset{\sim}{M}^+_{RS} = \underset{\sim}{M}^{+^T}_{SR}$$

$$\underset{\sim}{B}^+_{RS} = \underset{\sim}{B}^{+^T}_{SR}$$

These may be proved through application of Green's theorem and the boundary conditions satisfied by the radiation potentials, including Eq.(14c). In practice, the numerical approximations employed in obtaining the potentials ϕ_R and ϕ_S may lead to a very slight departure from symmetry in these terms (involving discrepancies of the order of 1% for the cases studied here).

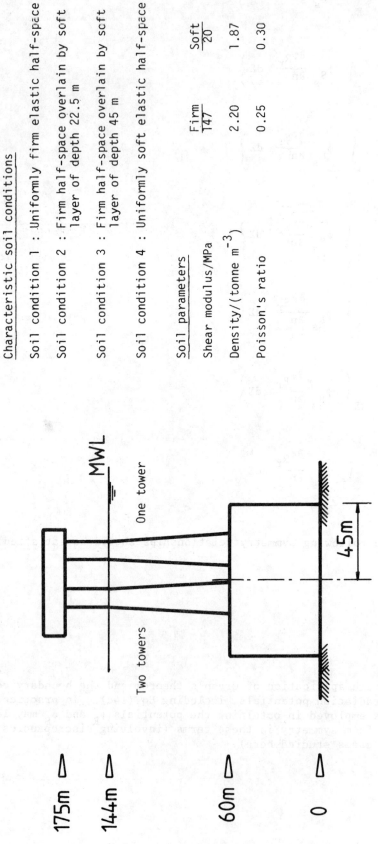

Fig. 1 Principal dimensions and characteristic soil data for platform 1.

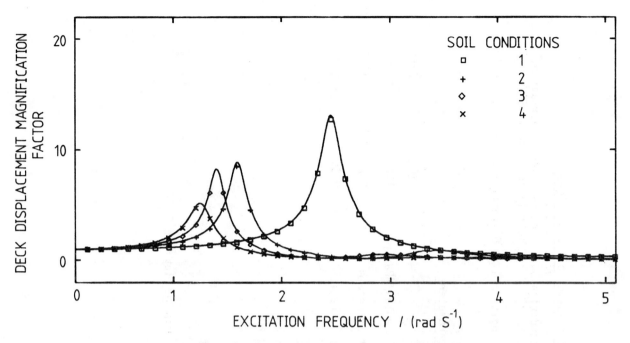

Fig. 2 Magnification factor for deck displacement of platform 1 on soil conditions 1 to 4

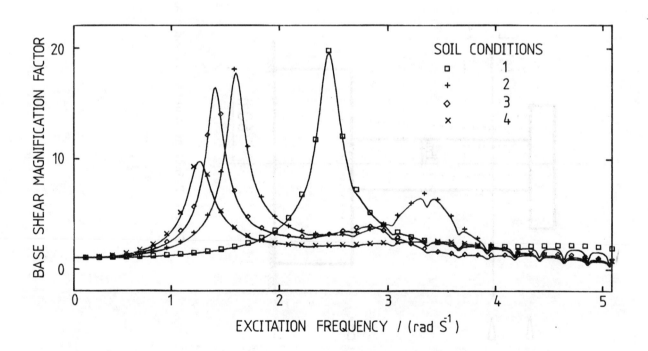

Fig. 3 Magnification factor for base shear of platform 1 on soil conditions 1 to 4

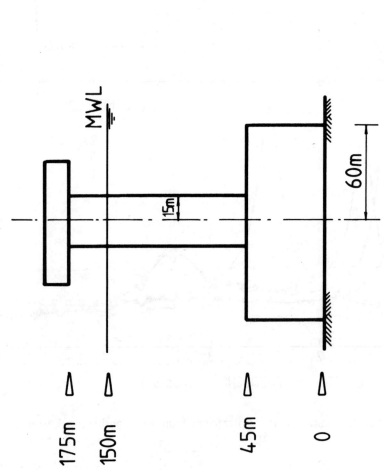

Fig. 4 Principal dimensions and characteristic soil data for platform 2.

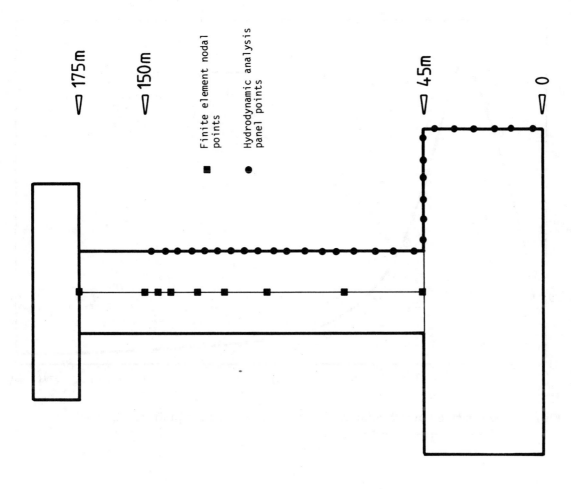

Fig. 6 Finite element nodal points and panel points for axisymmetric hydrodynamic analysis of platform 2.

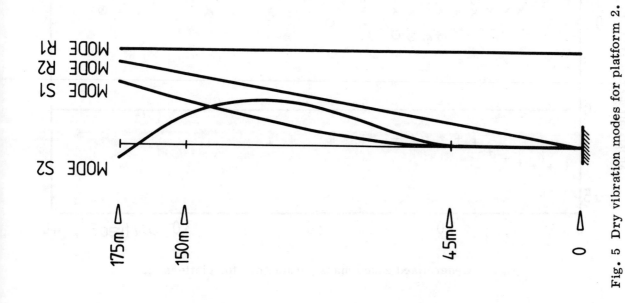

Fig. 5 Dry vibration modes for platform 2.

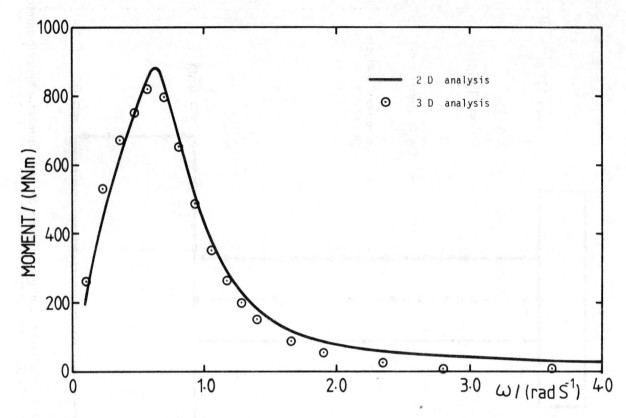

Fig. 7 Wave induced overturning moment about sea bed for platform 2, due to a sinusoidal wave of unit height.

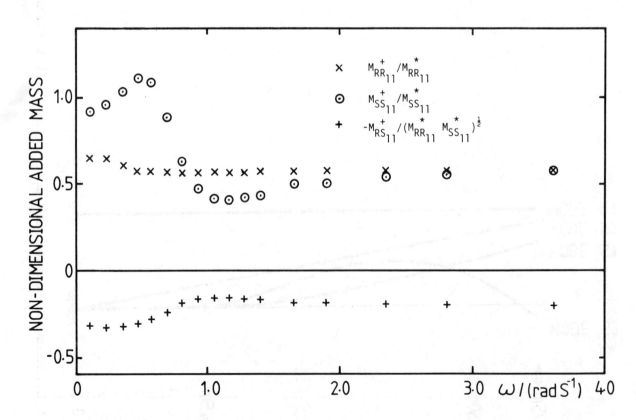

Fig. 8 Generalised added mass parameters for platform 2.

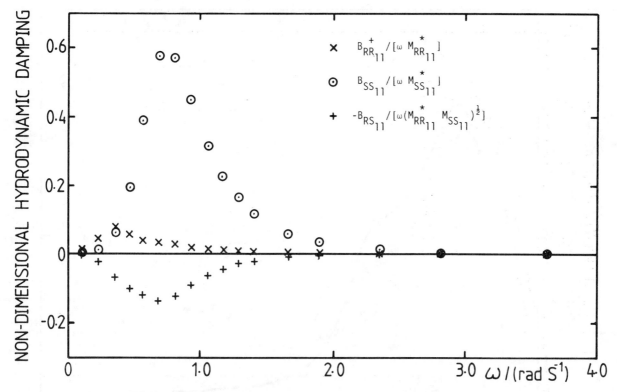

Fig. 9 Generalised hydrodynamic damping parameters for platform 2.

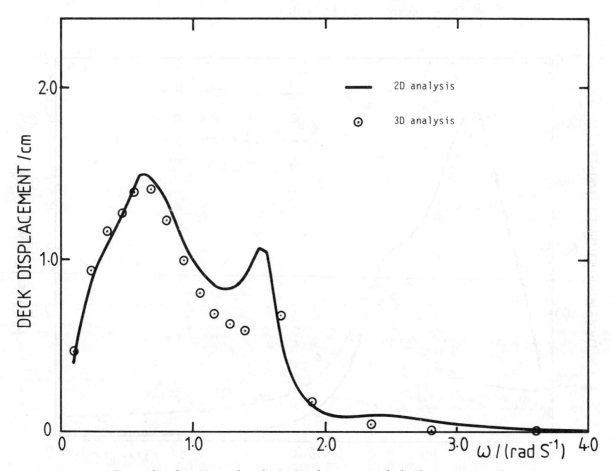

Fig. 10 Transfer functions for deck displacement of platform 2 on soil condition 5, due to a sinusoidal wave of unit height

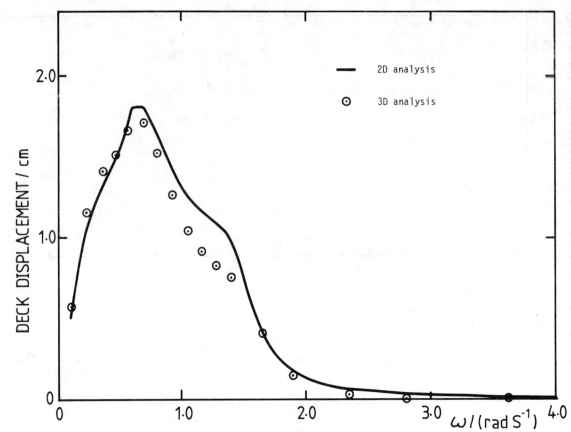

Fig. 11 Transfer functions for deck displacement of platform 2 on soil condition 6, due to a sinusoidal wave of unit height

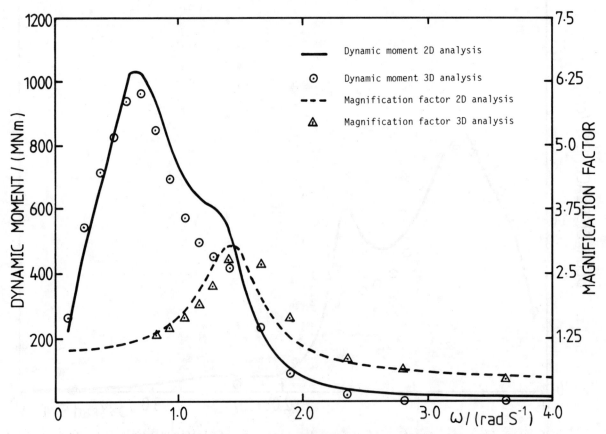

Fig. 12 Transfer functions and magnification factors for total overturning moment about sea bed for platform 2 on soil condition 6, due to a sinusoidal wave of unit height

BOSS'79

PAPER 52

Second International Conference on Behaviour of Off-Shore Structures
Held at: Imperial College, London, England
28 to 31 August 1979

AN INITIAL STUDY OF THE DYNAMIC BEHAVIOUR OF THE NEW CHRISTCHURCH BAY TOWER

B.R. Ellis and A.P. Jeary

Building Research Establishment, U.K.

Summary

The National Maritime Institute's Christchurch Bay tower was constructed at Calshot Spit and deployed in Christchurch Bay in Spring 1978. As part of their study of the dynamic behaviour of structures, the Building Research Establishment was invited to monitor the behaviour of the new tower. The dynamic characteristics of the tower were measured at both Calshot and Christchurch Bay, and it was found that neither soil-structure interaction nor water-structure interaction had a significant effect on the tower's behaviour. Some of the many problems involved with monitoring structural response to wave loading are outlined, and recommendations for performing this type of work are given.

This work was supported by the Department of Energy through the Offshore Energy Technology Board.

Sponsored by: Delft University of Technology, The Netherlands
Massachusetts Institute of Technology, U.S.A.
The Norwegian Institute of Technology, Norway
University of London, England

Secretariat provided by: BHRA Fluid Engineering

Copyright: © BHRA Fluid Engineering
Cranfield, Bedford, England

INTRODUCTION

The new Christchurch Bay Tower was constructed for the National Maritime Institute for the purpose of obtaining information about the forces exerted on vertical cylinders by waves and currents. As part of the experiment, the Building Research Establishment was invited to monitor the behaviour of the tower, and this report deals with some aspects of the dynamic behaviour which have become apparent at this early stage.

The structure, which is depicted in Fig 1, has a mass of 830 tonnes, of which 90% is provided by the reinforced concrete base. It was fabricated at Calshot, in Hampshire, and in the spring of 1978 it was towed to Christchurch Bay and deployed in water of mean depth 8.4 m. A steel skirt around the periphery of the base was designed to penetrate the few centimetres of sand which covered the Barton clay at the selected site.

Tests of the dynamic properties of the tower were conducted both on land, at Calshot, and at sea, in Christchurch Bay. In each case, the response to forced excitation and to naturally induced excitation was monitored. The results have been assessed at the laboratories of the BRE.

Initially it was envisaged that a study of the tower's response at various sea states would form a significant part of the dynamic investigation. However, an aspect of data handling requirements has shown that, at present, insufficient wave excited data are available to assess these effects with any confidence. These data handling requirements are discussed here, and the implications for all data relating to dynamic performance of offshore structures are considered.

TEST PROCEDURE

(a) Forced vibration tests

This procedure is now well established, and has been reported in detail (Ref 1). If known forces are applied at known frequencies, and the response is monitored, then the major dynamic characteristic (resonant frequencies, damping ratios, modal stiffnesses) can be deduced.

In these tests a rotating mass vibration generator, mounted on a rotatable base, was attached to the deck of the tower (see Fig 2), and a servo-accelerometer was placed near to the generator to monitor motion in the direction of the generated force. At each resonance, the deflected form (or mode shape) was measured using a movable accelerometer to monitor motion at various heights (nine levels) up the main tower.

(b) Naturally induced excitation tests

The tower's response to naturally induced excitation (wind loading at Calshot and wave loading at Christchurch Bay) was monitored using three linear servo-accelerometers located at the first level below the platform. The analogue signal from each accelerometer was subjected to some signal conditioning (filtering and amplifying) and recorded on a Sangamo 3500 tape recorder. The signal conditioning was used to optimise the recorded signal to noise ratio. These recorded data were then returned to the BRE for analysis using spectral techniques.

DISCUSSION OF THE RESULTS OBTAINED FROM THE

STRUCTURE ON LAND (CALSHOT)

At Calshot, the tower was supported on a massive and very stiff wooden formwork. A caisson system was temporarily attached to the outside of the base (and surrounding the tower). The system was in place at the time of testing, as was a temporary access ladder, and these, inevitably, had some effect on the measured dynamic characteristics of the tower. In particular, the damping values were probably increased considerably, although tests performed on a model of a similar structure suggest that the resonant frequencies remained at essentially the same value.

The forced vibration tests were restricted to night-time during the period 7-9 November 1977, and this proved to be a windy period. Although this made testing more difficult, it did allow observations of the structure's response to wind excitation to be made. Three modes of vibration were investigated in detail. The two higher frequency modes (4.00 and 4.25 Hz) were fundamental translation modes, having classical cantilever type deflected shapes. The directions of these modes were ascertained by incrementally rotating an accelerometer (mounted on the deck of the tower) whilst the vibrator was producing a unidirectional sinusoidal force. The directions of the fundamental modes were found to be $105°$ and $15°$ (Fig 2) to an axis joining the centre of the wavestaff to the centre of the main tower.

The third mode investigated had a resonant frequency of 3.33 Hz and a direction of $160°$, which was in line with the principal axis of the wooden formwork.

Since the mode is not orthogonal to the other two modes, there appears to be a different system participating. Subsequent calculations produced a modal mass for the 3.33 Hz mode, which was six times greater than that calculated for the 4.00 and 4.25 Hz modes. Accordingly it has been assumed that, at this lower frequency, the tower and caisson were acting as one system, whilst at the higher frequencies only the tower was involved. This observation is not supported by direct measurements on the caisson, but it does fit the above data with the single exception that no orthogonal mode activity for the caisson and tower system was found. An investigation of the directional response of the tower and subsequent observation of wind-induced spectra taken from accelerometers located on the tower, did not show any orthogonal mode activity. With hindsight, it is a pity that the motion of the caisson was not monitored, but working at 5 am in heavy rain does tend to have a detrimental effect on experimental techniques.

The spectra (Power Spectral Densities) obtained from the recordings of the wind induced motion of the tower, confirmed the presence of the three modes. In each case the frequencies were slightly higher (3.60, 4.09 and 4.38 Hz) and the amplitudes of motion were smaller, than in the forced vibration tests. This corresponds with the amplitude dependent frequency characteristics which have been observed in tests of a model of a similar structure(Ref 1).

DISCUSSION OF THE TESTS CONDUCTED AT SEA (CHRISTCHURCH BAY)

The tower was subjected to forced vibration testing on the 21 and 22 September. The lowest frequency modes (3.92 and 4.20 Hz) were located and the modal directions were checked and found to be similar to those at Calshot. The vibration generator was then aligned with the two orthogonal directions in turn and the response monitored as the excitation frequency was increased.

Graphs of the excitation frequency against the response (frequency sweeps) are shown in Fig 3. Five modes were located, one of which was a torsional mode (approximately 4.8 Hz). This torsional mode is evident in both graphs in Fig 3 and it can be seen (from Table 1) that with the larger response the frequency is lower. The different responses result from the driving force being applied in different directions, with only the resolved part of the force, orthogonal to the direction joining the vibration generator to the torsional centre, causing a torque. For the calculation it was assumed that the torsional centre was at the centre of the main tower, but this is only a rough approximation.

The deflected shape for each mode was measured down the main tower and in each case what appeared to be a typical first mode cantilever deflection shape was obtained. It may at first seem surprising that the five modes appear to have similar mode shapes, but when it is realised that the tower also consists of a rigidly attached wavestaff and ladder, in fact more like a two bay portal, then this can easily be explained. This also gives an explanation for the small change of modal directions noted between the various modes, and is simply evidence of a more complex mode shape.

Proof that the wavestaff contributed significantly to the stiffness of the structure was obtained when it was raised. The frequency sweeps, shown in Fig 4, are taken from data recorded at this time, and show the tower to have different characteristics.

Table 1 lists the results obtained from the forced vibration tests at sea.

The response of the tower to wave loading was recorded for a 12-hour period over the night of 2/3 August 1978, and the recording was taken to the BRE for analysis. The data were found to form a non-stationary set, and were analysed merely with a view to noting the reason for the non-stationarity.

Fig 5 shows a power spectral density (PSD) for the 105° direction in the 0-10 Hz range. Fig 6 shows the response taken from two consecutive 2.84-hour periods, for the 15° position, in the region of the major reponses (3.75-6.25 Hz) and indicates that not only amplitudes of response but also frequencies of response were different in the two cases.

THE ANALYSIS OF RANDOM DATA OBTAINED FROM OFFSHORE STRUCTURES

A rigorous assessment of the dynamic properties of a structure can be obtained by forcing vibrations; however, it is usual to obtain some information by considering the response to wave and current action. This technique is used, at present, with large offshore rigs, as present day vibration generators could not provide sufficient force (at the fundamental frequencies) to make the forced response significantly larger than the ambient response. For this reason it is usual to use conventional spectral analysis techniques on the response of offshore structures to wave excitation for this assessment.

Before performing any statistical analysis on random data, it is a wise precaution to check whether the data being used form a stationary sample. This means that all statistical properties remain invariant with time, and spectral analysis techniques can then be used to give results in terms of mean levels, resonant frequencies, and damping values. Should the samples used not be stationary then it is possible to obtain misleading results.

The data recorded at Christchurch Bay formed a non-stationary sample, and it was apparent that it was not possible to obtain stationary data of sufficient length from a continuous record. (It is also evident that records obtained from structures in the sea have similar handling requirements and that spectra from continuous records must be treated with extreme caution.) In order to use standard spectral techniques, the data need to be preselected to form stationary samples, and one method of doing this is known as ensemble averaging.

Ensemble averaging involves the labelling and storing of data, and subsequent retrieval of like sections from the store. In the case of the Christchurch Bay tower the necessary labels are mean sea level, wave height, wave direction, current velocity, current direction, wind direction and wind speed. The spectra from short periods are stored on disk files with a series of labels attached to each section. Subsequently intervals for each label are selected and a search is made for spectra which have all labels within similar bands. These spectra are then added and averaged in the normal way. This technique forces the data to form stationary samples (with the one proviso that sections with large trends are excluded) and normal statistical inferences can be made. It is evident that this increases the data requirement by a very large amount.

SOIL-STRUCTURE INTERACTION, WATER-STRUCTURE INTERACTION
AND INTEGRITY MONITORING

A comparison of the towers characteristics at Calshot with those at Christchurch Bay shows only a small difference in resonant frequencies. The most probable explanation for this difference is that the resonant frequencies are amplitude dependent (amplitude softening), a larger force being used to excite the structure at Christchurch Bay than at Calshot. This implies that soil-structure interaction is of no consequence for the tower at its present site. This is confirmed by the fact that in each normal mode the motion at the base of the structure was negligible. (Linear motion in the two horizontal orthogonal directions, and angular motion in all three axis were investigated).

By performing forced vibration tests at both high and low tides (for similar sea states) it is theoretically possible to measure the 'added mass' of the water by measuring changes in resonant frequencies. A finite element calculation has been performed, and the effective mass of water was calculated (assuming the added mass to be equivalent to the mass of water displaced by the structure). For tide heights varying from 7.2 metres to 9.6 metres a variation in the lowest resonant frequency of 0.03 Hz was predicted, and also for water heights varying from 0 metres to 8.4 metres a 0.02 Hz change was predicted. The resonant frequencies were checked by tuning the frequency of the vibration generator for a maximum indicated response, each hour over a period of 12-hours, and although a small change was actually noted (up to 2.5%), it appeared to be totally independent of tide height.

Because the variation in applied natural forces to the tower necessitates using ensemble averaging before reliable results can be obtained from spectral analysis, it implies that such techniques as integrity monitoring (whereby damage may be detected by a small change in measured spectra), are not easily used with this type of structure, any change caused by damage being difficult to discern amongst the variations with wave height and direction.

FUTURE WORK CONCERNING THE MEASURED AND THE
PREDICTED RESPONSE OF THE TOWER

Perhaps the most important aspect of monitoring the behaviour of structures is that the measured behaviour can be compared with design guide prediction. At present the information from the tower is inadequate for the process, but it is intended to monitor the behaviour of the tower and record sufficient data to enable ensemble averaging to be used to provide accurate response spectra and sea spectra for various sea states.

Comparison between measured and predicted response of tall buildings to wind response (Ref 3) have shown that design guide predictions can sometimes be inadequate even when input parameters, like resonant frequencies and damping, are known.

It is quite likely that predictions of the response of offshore structures will also be inadequate, (probably suffering from similar misconceptions) but until data from such structures as the Christchurch Bay tower are readily available it is unlikely that any real advance will be made.

CONCLUSIONS

The dynamic behaviour of the new Christchurch Bay tower has been examined and it has been found that

1   The resonant frequencies are dependent upon the amplitude of vibration

2   Soil-structure interaction is not important in this system

3   Water-structure interaction has a neglible effect on the resonant frequencies of the system

4   Spectra obtained from sequential times can be significantly different, with the implication that

    (a) Integrity monitoring of offshore structures using spectral methods should be used with extreme caution.

    (b) Ensemble averaging techniques are necessary for statistically reliable data about response to be obtained.

ACKNOWLEDGEMENTS

The authors wish to acknowledge the valuable help and facilities provided by the National Maritime Institute during the course of this work, and to thank Russell Fry for his help during the forced vibration tests at Christchurch Bay.

The work reported in this paper forms part of the research programme of the Building Research Establishment and is published by permission of the Director. The construction and operation of the Christchurch Bay tower forms part of the programme of work of the Offshore Energy Technology Board which is funded by the Department of Energy.

REFERENCES

1   B R Ellis. A study of soil-structures - to be published

2   B R Ellis, A P Jeary, P R Sparks. Full-scale and model-scale studies of the dynamic behaviour of an offshore structure. Proceedings of International Conference on behaviour of slender structures. London, Sept 1977

3   A P Jeary, B R Ellis. The response of a 190 metre tall building and the ramifications for the prediction of behaviour caused by wind loading. Offered to the 5th International conference on Wind Engineering, Colorado 1979

TABLE 1   RESULTS OF THE FORCED VIBRATION TESTS ON THE COMPLETE TOWER AT CHRISTCHURCH BAY

Mode Direction	Resonant frequency (Forced vibration test) Hz	Damping % critical	Peak-peak Amplitude of motion at deck	Peak-peak Force at deck
105	3.92	1.29	3.62 ($\times 10^{-4}$ m)	642 N
105	4.36	-	1.77 ($\times 10^{-4}$ m)	794 N
15	4.20	1.32	1.36 ($\times 10^{-4}$ m)	737 N
15	5.12	1.47	1.95 ($\times 10^{-4}$ m)	1095 N
Torsion	4.77	1.31	2.87 ($\times 10^{-5}$ rads)*	5351 Nm* } Same
	4.81	-	1.11 ($\times 10^{-5}$ rads)*	3141 Nm* } mode

* Calculated assuming the centre of the main tower to be the torsion centre

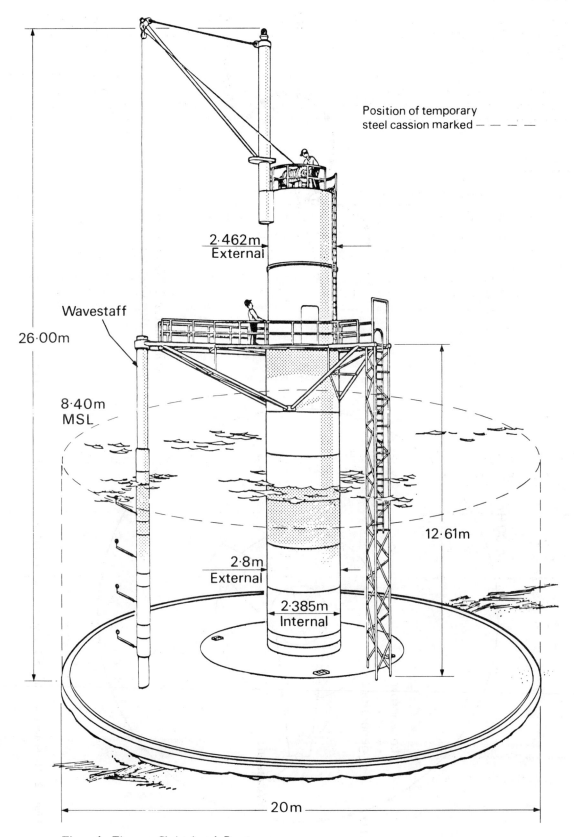

Figure 1 The new Christchurch Bay tower

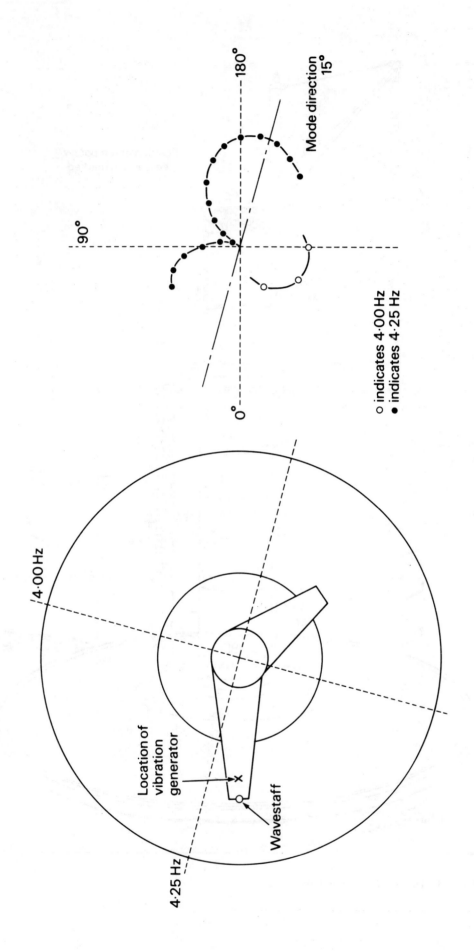

Figure 2 Mode direction shown on plan view of structure and results of mode direction test

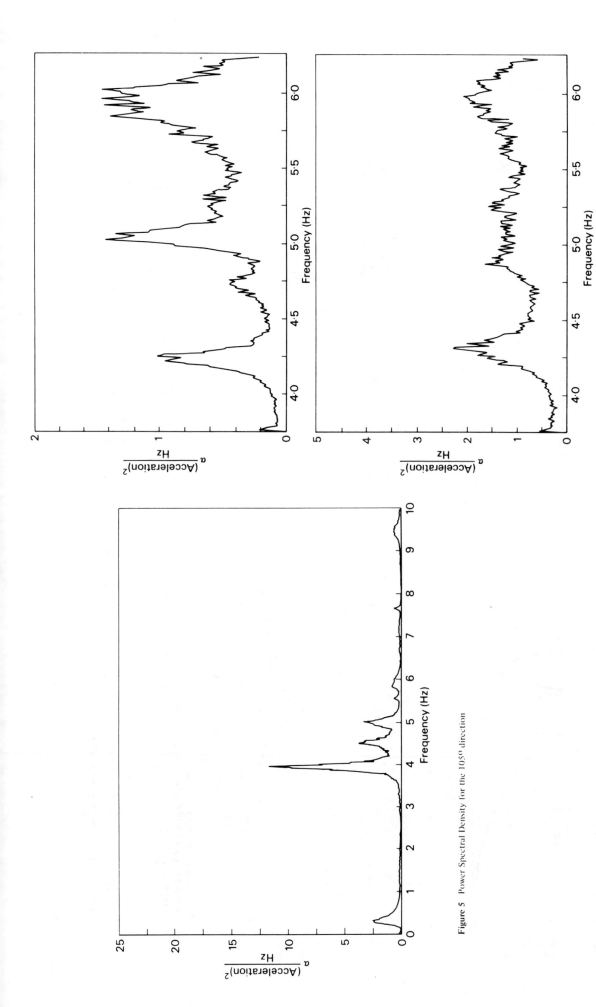

Figure 5 Power Spectral Density for the 105° direction

Figure 6 Power Spectral Densities for consecutive periods of data (15° direction)

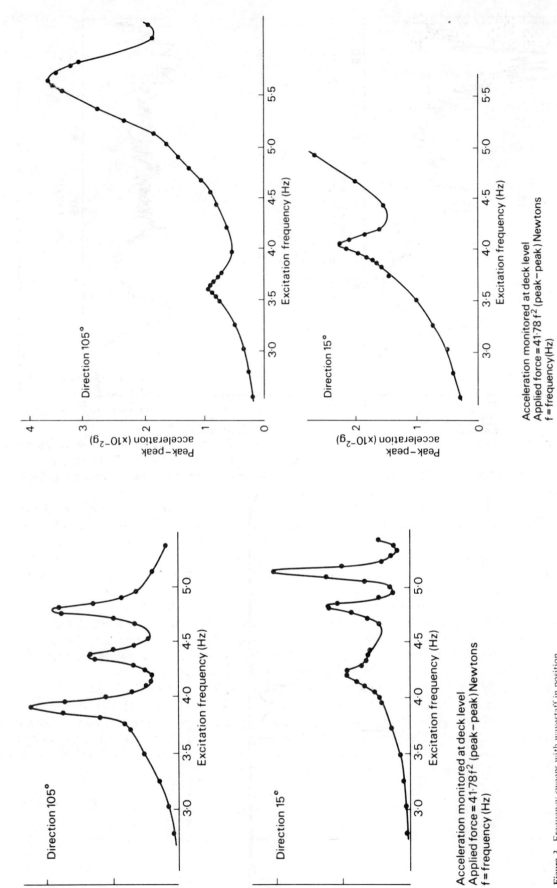

Figure 3 Frequency sweeps with wavestaff in position

Figure 4 Frequency sweeps with wavestaff removed

BOSS'79

PAPER 53

Second International Conference
on Behaviour of Off-Shore Structures

Held at: Imperial College, London, England
28 to 31 August 1979

USE OF GENERALIZED COORDINATES IN THE NON-LINEAR TIME-DOMAIN ANALYSIS OF STEEL JACKETS

A.K. Basu, BE, PhD, DIC, FICE, MIStructE, MRINA, MASCE, FIE (India)
and R.P. Singh, BSc(Engg), ME

Indian Institute of Technology, India

Summary

A computer method is presented for the time-domain analysis of steel jackets idealised as plane frames under the action of regular waves and currents. The flexibility of the soil-pile support system is represented in the model with the help of frequency independent impedance functions.

The fluid loading at the nodes of the jacket is calculated using the Morison equation assuming linear distribution along members of the hydrodynamic loading. The variation of the length of submergence of members near the sea surface with the passage of waves is taken into consideration and the nonlinearity due to the drag term is retained.

The mass matrix is generated by neglecting the rotatory inertia at all nodes except those at the base. The added mass contributions are calculated also on the basis of linear distribution of structural accelerations along the members, taking into account the variation of the length of submergence with time, where appropriate. The condensed stiffness matrix is formed corresponding to the translational degrees of freedom at each node along with the additional rotational degree of freedom at each of the base nodes.

The mass matrix, the stiffness matrix, the damping matrix and the fluid loading vector are then transformed to generalized coordinates. These generalized coordinates consist of the amplitudes of the first few normal modes of the fixed base structure together with three base degrees of freedom for each point of support. The reduced set of (nonlinear) equations of motion is numerically integrated by the Newmark beta method, with $\beta = \frac{1}{4}$.

A typical jacket (which is a simplified version of the BP Forties field "FD" jacket platform) has been analysed for wave heights of 3.05m (10 ft) and 15.24m (50 ft) with wave periods of 5 sec and 10 sec respectively, using different numbers of model co-ordinates. Solutions have also been obtained for the rigid-base condition for comparison. The effect of the number of mode shapes used in the analysis on the accuracy of the computed response for the two combinations of wave height and wave period is discussed.

INTRODUCTION

Penzien and Tseng (Refs.1,2) have developed a general method for dynamic analysis of steel jacket platforms, in which the foundation system is characterised by a set of frequency dependent impedance functions. The solution is obtained in the frequency domain. An alternative method that has been used (Refs.3,4) for the dynamic analysis under wave (and earthquake) loading involves the integration of the equations of motion in the time domain and has been adopted here. In this method the nonlinear drag term can be retained in its original form, since the effect of structural velocity on the drag force can be fully accounted for iteratively at every time step. The method can also incorporate the variation in the added mass due to fluctuation in water level through the modification of the loading term. Furthermore, this approach can, when necessary, handle both material and geometric nonlinearity in the structure-foundation system, unlike the frequency domain approach. However, in this method the foundation impedances have to be taken as independent of frequency.

The time-domain analysis presented here is applicable to jackets idealized as plane frames and subjected to regular waves and current. The vertical displacement at each node and the sway displacement at the level of each horizontal bracing member together with all the displacements at the base nodes are taken to constitute the dynamic degrees of freedom for the system. The computational effort is drastically reduced by expressing the frame displacements as the linear combination of the first few mode shapes of the jacket on rigid base together with the quasistatic displacements associated with the base degrees of freedom (Refs.1,2,5,6). Its use for the time-domain analysis of linear or nonlinear problems does not appear to have been reported so far.

EQUATIONS OF MOTION IN STRUCTURAL COORDINATES

Let the base degrees of freedom be denoted by $\{r_b\}$ and the translational degrees of freedom at the other nodes be $\{r_s\}$. The equations of motion of the structure-foundation system in structural coordinates in partitioned matrix form are then

$$\begin{bmatrix} M_{ss} & M_{sb} \\ M_{sb}^T & M_{bb} \end{bmatrix} \begin{Bmatrix} \ddot{r}_s \\ \ddot{r}_b \end{Bmatrix} + \begin{bmatrix} C_{ss} & C_{sb} \\ C_{sb}^T & C_{bb}+C_{bb}^f \end{bmatrix} \begin{Bmatrix} \dot{r}_s \\ \dot{r}_b \end{Bmatrix} + \begin{bmatrix} K_{ss} & K_{sb} \\ K_{sb}^T & K_{bb}+K_{bb}^f \end{bmatrix} \begin{Bmatrix} r_s \\ r_b \end{Bmatrix} = \begin{Bmatrix} P_s(t) \\ P_b(t) \end{Bmatrix} \quad (1)$$

where the square matrices from left to right are respectively the mass, the damping and the stiffness matrix and the corresponding vectors are respectively the nodal acceleration, velocity and displacement vector. The vector on the right hand side is that of the hydrodynamic loads at the nodes. The matrices and the load vector are formed as described in the following.

<u>Mass Matrix</u>: The structural mass matrix is formed by lumping at the nodes the member masses in air, including deck loads and water (upto still water level) contained in the members. The formation of the added mass matrix would be explained in the section, Hydrodynamic Loading. The contribution of the bracing members normal to the frame under consideration to the mass matrix is also taken into account.

<u>Stiffness Matrix</u>: The condensed stiffness matrix shown in Eqn.1 is obtained by inverting the flexibility matrix corresponding to the dynamic degrees of freedom, which in turn is generated from the overall stiffness matrix in the usual way. $[K_{ss}]$ is thus the condensed stiffness matrix of the fixed base structure and $[K_{bb}^f]$ is the stiffness matrix of the entire foundation system referred to $\{r_b\}$.

Damping Matrix: In the present work, which utilizes the modified mode superposition approach, the damping matrix need not be known explicitly, if the modal damping ratios are specified and if $\begin{bmatrix} C^f_{bb} \end{bmatrix}$, the damping matrix of the entire foundation system referred to $\{r_b\}$, is known.

Hydrodynamic Loading: The water particle velocities and accelerations due to waves are calculated at the undeflected positions of the nodes using the Airy wave theory with the theory assumed valid upto the actual water surface. The current velocity, if present, is vectorially added to the velocity due to waves.

The horizontal and vertical components of the hydrodynamic force per unit length at a point on a tubular member are found from the modified Morison equation as given by Chakravarty (Ref.7).

$$\begin{Bmatrix} F_x \\ F_y \end{Bmatrix} = 0.5\, C_d\, \rho\, A \left| \left\{ c_x(\dot{v} - \dot{d}_y) - c_y(\dot{u} - \dot{d}_x) \right\} \right| [T] \begin{Bmatrix} \dot{u} - \dot{d}_x \\ \dot{v} - \dot{d}_y \end{Bmatrix}$$

$$+ \rho V ([I] + (C_m - 1)[T]) \begin{Bmatrix} \ddot{u} \\ \ddot{v} \end{Bmatrix} - (C_m - 1)\rho V [T] \begin{Bmatrix} \ddot{d}_x \\ \ddot{d}_y \end{Bmatrix} \quad (2)$$

where $(\dot{u}, \dot{v})$ are velocities and $(\ddot{u}, \ddot{v})$ are accelerations of water particle, $(\dot{d}_x, \dot{d}_y)$ are velocities and $(\ddot{d}_x, \ddot{d}_y)$ are accelerations of the structure, all at the point and in the horizontal and vertical directions respectively; C_d is the drag and C_m the inertia coefficient; ρ is the mass density of water; A is the projected area and V the volume per unit length along the member; and $[T]$ is given by

$$[T] = \begin{bmatrix} c_y^2 & -c_x c_y \\ -c_x c_y & c_x^2 \end{bmatrix} \quad (3)$$

c_x and c_y being the direction cosines of the member. For horizontal and diagonal bracing members $A = D$ and $V = \pi D^2/4$, where D is the diameter of the member; but for the leg members and the vertical bracings A and V include the contributions of the members in the orthogonal plane.

For calculating the nodal loads the members are taken as pin-ended and F_x or F_y is assumed to vary linearly from one node to the other (for fully submerged members) or from the submerged node to the point of submergence (for partially submerged members). The hydrodynamic forces on the horizontal bracing members in the orthogonal plane are accounted for by assigning equivalent lumped volumes and areas to the appropriate nodes.

The last term on the r.h.s. of Eqn.2 gives rise to the added mass matrix which will have certain time-varying elements because of fluctuation in the sea level with the passage of the wave. In the present work this matrix corresponding to the still water level is taken to the left hand side of Eqn.1 to be added to the structural mass matrix. In the hydrodynamic load vector $\{P(t)\}$ a correction for the change in the added mass from that corresponding to the still water level is incorporated at every time step.

Foundation Impedance: The frequency-independent stiffness and damping properties of the soil-pile system at each pile cap have been computed here using Novak's formulation (Ref.9) corresponding to a nondimensional frequency of 0.3 assuming material damping to be zero in both soil and pile.

EQUATIONS OF MOTION IN GENERALIZED COORDINATES

Let $\{Z\}^T = \{Y\ r_b\}^T$ be the vector of generalized coordinates, where $\{Y\}$ is the normal coordinate vector of order (mx1). Note that m will generally be much smaller than the order of $\{r_s\}$. $\{r_s\}$ can now be written as

$$\{r_s\} = \{r_s^q\} + [\emptyset_s]\{Y\}, \qquad (4)$$

where $[\emptyset_s]$ is the matrix of the first m modes of the fixed base frame normalized w.r.t. the mass matrix $[M_{ss}]$, and $\{r_s^q\}$ is the vector of quasistatic displacements at nodes above the base associated with the imposed base displacements, and is given by

$$\{r_s^q\} = -[K_{ss}]^{-1}[K_{sb}]\{r_b\} = [\emptyset_b]\{r_b\} \qquad (5)$$

Then, the complete displacement vector can be written as

$$\begin{Bmatrix} r_s \\ r_b \end{Bmatrix} = \begin{bmatrix} \emptyset_s & \emptyset_b \\ 0 & I \end{bmatrix} \begin{Bmatrix} Y \\ r_b \end{Bmatrix} = [\emptyset]\{Z\} \qquad (6)$$

Eqn.1 is now premultiplied by $[\emptyset]^T$ and transformed to generalized coordinates using Eqn.6 to give

$$[\bar{M}]\{\ddot{Z}\} + [\bar{C}]\{\dot{Z}\} + [\bar{K}]\{Z\} = \{\bar{P}(t)\} \qquad (7)$$

where the first three matrices are respectively equal to the corresponding matrices in Eqn.1 premultiplied by $[\emptyset]^T$ and postmultiplied by $[\emptyset]$, and the loading vector $\{\bar{P}(t)\}$ is equal to $[\emptyset]^T\{P(t)\}$.

Because the damping actually present in the system cannot be precisely estimated $[\bar{C}]$ can be conveniently taken to consist of only two nonzero submatrices on the diagonal (Ref.1). The one corresponding to the normal coordinates consists of modal damping $2\eta_j\omega_j$, where ω_j and η_j are respectively the jth natural frequency and modal damping ratio of the fixed base structure, and the other submatrix is $[C_{bb}^f]$.

SOLUTION OF TRANSFORMED EQUATIONS OF MOTION

The equations of motion in generalized coordinates (Eqn.7) are solved by Newmark $\beta = 1/4$ method (Ref.9) for successive time steps to generate the time history of response. Given the response at the ith time step ($\{Z_i\}$, $\{\dot{Z}_i\}$, $\{\ddot{Z}_i\}$ etc.) the response at the next time step is found by iteration. For an assumed $\{Z_{i+1}\}$, the vectors $\{\ddot{Z}_{i+1}\}$ and $\{\dot{Z}_{i+1}\}$ are computed using Newmark's formulae and are then transformed into nodal accelerations and velocities. The hydrodynamic force vector is now found, first in structural and then in generalized coordinates. The solution of a set of linear equations then yields a new $\{Z_{i+1}\}$. The process is repeated if the individual displacements in the $\{Z_{i+1}\}$ assumed and computed $\{Z_{i+1}\}$ are not sufficiently close.

NUMERICAL SOLUTIONS AND DISCUSSION

The present method has been used to analyse the plane frame idealization of the B.P.Forties field "FD" platform shown in Fig.1 for 3.05m (10ft)/5 sec and 15.24m (50 ft/10 sec waves in the absence of currents, and using 1 to 5 lowest modes of the fixed base structure in succession. Solutions have been found for both rigid and flexible base conditions. The nondimensional frequency of 0.3 for the soil-pile system here corresponds to a period of about 0.19 secs.

The time integration was continued till the steady state condition was reached. The amplitude of the steady state deflection at the top of the deck obtained with the different number of fixed-base modes is shown in Fig.2, expressed as a multiple of the "exact" value, taken to correspond to the solution with 5 modes. The deflections with 4 modes are practically identical to (within 0.01% of) the exact values, as also are the base shears and base moments (not shown). The error in the top deflection for lesser number of modes is seen to be greater with the rigid base structures. For example, using only one mode this deflection is overestimated by about 6% in the rigid base and 1% in the flexible base, case under the 3.05m 5 sec waves, the corresponding figures for the 15.24m/10 sec waves being 10% and 7% respectively. The errors in the base shear and base moment for the single mode solution were in all cases within 0.5%.

It is interesting to note that whereas the increase in the deck deflection, base shear and base moment due to foundation flexibility for the 15.24m/10 sec wave was respectively about 45,25 and 15 percent, the corresponding figures for the 3.05m/5 sec waves were respectively 230,300 and 170 percent. This happened because the second harmonic of the wave loading in the latter case had a significant amplitude and a period much closer to the fundamental period of the flexible-base structure (2.44 sec) than of the fixed-base structure (2.07 sec). The effect of this resonance on the steady state deck deflection response is vividly demonstrated in Fig.3, where the results for both the fixed base and the flexible base conditions are shown.

CONCLUSIONS

For the chosen deep water platform with flexible base the results obtained with the first four normal modes of the rigid base structure (together with the base degrees of freedom) are practically exact, and those found using only one mode can be considered sufficiently accurate for design purposes. The same conclusions are valid when the base is taken as rigid.

REFERENCES

1. Penzien,J: "Structural dynamics of fixed offshore structures".Proc. Boss '76,International Conference on the Behaviour of Offshore Structures,Vol.1,Trondheim,Norway (August 2-5,1976).

2. Penzien,J. and Tseng,S: "Three-dimensional dynamic analysis of fixed offshore platforms". pp.221-243 in "Numerical Methods in Offshore Engineering" by Zienkiewicz,O.C. et al (ed.),John Wiley & Sons (1978).

3. Selna,L.G. and Cho, M.D: Nonlinear dynamic response of offshore structures".Proc. 3rd Offshore Technology Conference. Paper OTC 1402. Houston, Texas. (April 19-21,1971).

4. Burke,B.G. and Tighe,J.T: "A time series model for dynamic behaviour of offshore structures". Proc. 3rd Offshore Technology Conference. Paper OTC 1403, Houston, Texas. (April 19-21,1971).

5. Chopra,A.K. and Gutierrez,J.A: "Earthquake response analysis of multistory buildings including foundation interaction". International Journal of Earthquake Engineering and Structural Dynamics,3, pp.65-77.(1974).

6. Eatock Taylor,R: "A preliminary study of the structural dynamics of gravity platforms". Proc.7th Offshore Technology Conference. Paper OTC 2406. Houston, Texas. (May 5-8,1975).

7. Chakravarti,S.K., Tam,W.A. and Wolbert,A.L: "Total forces on a submerged randomly oriented tube due to waves". Proc. 8th Offshore Technology Conference. Paper OTC 2495. Houston, Texas. (May 3-6, 1976).

8. Novak,M: "Dynamic stiffness and damping of piles". Canadian Geotechnical Journal, 11, pp.574-598. (1974).

9. Bathe,K.J. and Wilson,E.L: "Numerical Methods in finite element analysis." pp.322-324. Prentice-Hall,Inc., Englewood Cliffs, N.J. (1976).

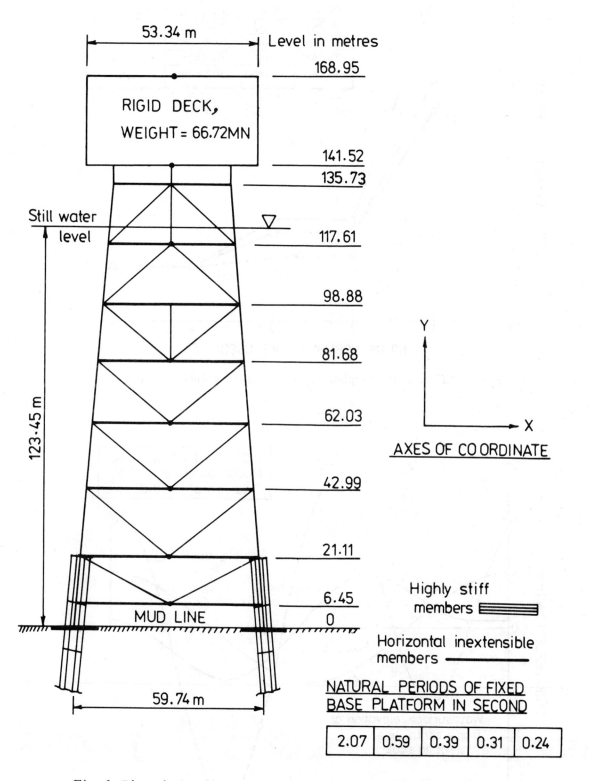

Fig. 1 Plane frame idealisation of B.P. Forties Field Platform 'FD'

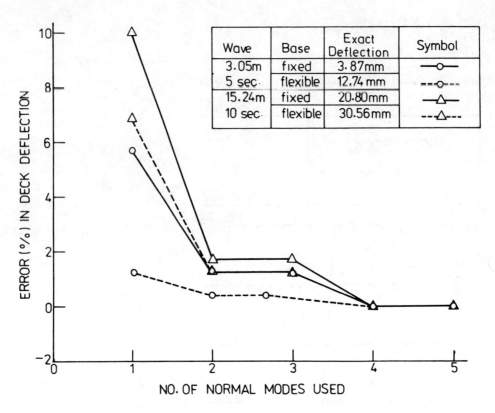

Fig. 2 Effect of the number of modes on the solution accuracy

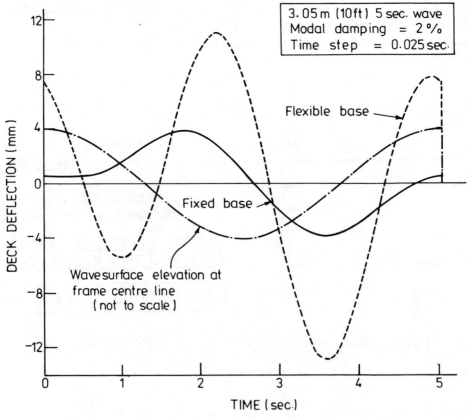

Fig. 3 Effect of superharmonic resonance on deck deflection

BOSS'79

PAPER 55

Second International Conference on Behaviour of Off-Shore Structures

Held at: Imperial College, London, England
28 to 31 August 1979

HYDRODYNAMIC COEFFICIENTS OF RECTANGULAR BARGES IN SHALLOW WATER

J.A. Keuning

Delft Hydraulics Laboratory, The Netherlands

and W. Beukelman.

Delft University of Technology, The Netherlands

Summary

Forced oscillation tests in vertical and horizontal direction have been carried out with a 2m model of a rectangular offshore pontoon and a cutter dredge pontoon on shallow water. Scale ratio 1 : 32.5. The waterdepths were 4.5, 1.75 and 1.2 times the draft of the model.

The frequency ranged from 0.2 Hz up to 2.0 Hz. Moreover the amplitude of oscillation was systematically varied to obtain a check on linearity.

All experimental results are presented in Figures to show the influence of water-depth, amplitude and frequency on added mass and damping.

Some of these experimental results have also been compared with corresponding calculated values.

The calculation methods used are based on strip theory and a diffraction model.

Sponsored by: Delft University of Technology, The Netherlands
Massachusetts Institute of Technology, U.S.A.
The Norwegian Institute of Technology, Norway
University of London, England

Secretariat provided by: BHRA Fluid Engineering

Copyright: ©   BHRA Fluid Engineering
Cranfield, Bedford, England

Nomenclature.

A_V	wave amplitude ratio
a,b,c,d,e,g	hydrodynamic coefficients
a	added mass, transformation coefficient, subscript for amplitude
B	beam
b	damping coefficient
C_V	dimensionless sectional added mass
$F_{1,2}$	force exerted by oscillator rod
F_{nh}	coefficient for influence of waterdepth
$\bar{F}_{Vr}$	real part of the vertical hydrodynamic force
$\bar{F}_{Vj}$	imaginair part of the vertical hydrodynamic force
g	acceleration due to gravity
h	waterdepth
I	mass moment of inertia of pontoon or cutter
k	wave number for deep water
k_o	wave number
L	length of pontoon or cutter
l	distance between oscillator legs
N	number of transformation coefficients
T	draft of pontoon
t	time
$\bar{V}$	vertical velocity
$\left.\begin{array}{l}x\\y\\z\end{array}\right\}$	right hand coordinate system
$\left.\begin{array}{l}x\quad\text{surge}\\y\quad\text{sway}\\z\quad\text{heave}\end{array}\right\}$ displacement	
ε	phase angle between force and motion
ζ_a	wave amplitude
θ	pitch angle
λ	wave length
ρ	density of water
ϕ	roll angle
ψ	yaw angle
ω	circular frequency of oscillation
∇	volume of the displacement of pontoon

1. Introduction.

For the calculation of ship motions in waves with the commonly used methods an
accurate knowledge of the hydrodynamic coefficients, i.e. added mass and damping,
as function of the frequency is essential.
For average ship forms in deep not restricted water these coefficients are fairly
well-known and reliable calculation methods do exist. This is not the case for some
situations which are nowadays more frequently met and in which the motions of
floating constructions are important. These are:
- rectangular pontoons or barges, anchored or moving at slow speed,
- ships at shallow water, i.e. moored ships and ships approaching harbours with
  small under keel clearance,
- cutter dredgers working in exposed areas.
All these situations cause special problems with respect to the forementioned calculation methods and large discrepancies between the different methods occur:
- pontoons give rise to problems due to both the rectangular shape and the large
  beam to draft ratio (B/T),
- **shallow water effects can until now only be calculated with a limited number**
  of methods especially for the case of small waterdepth to draft ratio (h/T); they
  are mathematically complex and lack extensive experimental verification,
- cutter dredgers pose a special problem due to the presence of a slot in the
  pontoon body, which makes it partly a double hull ship.

Experimental values for the hydrodynamic coefficients in all these situations are
scarce.

Therefore the Delft Hydraulics Laboratory and the Delft University of Technology
decided on a large experimental project to gain some insight into these problems and
to obtain a set of reliable experimental results.
For this reason forced oscillation tests and wave force measurement tests with
both a pontoon model and a cutter dredge pontoon model in three different waterdepths have been carried out.
In this paper the experimental values for added mass and damping are presented
and compared with the results of some currently used calculation methods.

1.1. Review.

For deep water the calculation methods to determine the hydrodynamic coefficients
are well-known (Ref. 1,2,3,4) and mainly based on the work of Ursell (Ref. 5) for
oscillating cylinders in a free surface.
The greater part of the experimental verification has been performed for normal
ship forms by forced oscillation model tests (Ref. 6,7) and only a small part was
related to cylinders with rectangular cross-sections. For this case extensive
experiments have been carried out by Vugts (Ref. 7,8) with respect to heaving,
swaying and rolling motions and also by Takaki (Ref. 15).
Generally the experimental results show a good agreement with the calculations
except for the rolling motion where viscous effects play a dominant role.
For the case of finite waterdepth several calculation procedures have also been
developed to determine the hydrodynamic coefficients. It is worthwhile to mention
in this respect the work of Porter (Ref. 1), Kim (Ref. 9), Keil (Ref. 10,11), Van
Oortmerssen (Ref. 12) and many others (Ref. 13.14.15.16.17.18.19,24,26).
A special method to determine the sway added-mass coefficients for rectangular
sections in shallow water has been presented by Flagg and Newman in (Ref. 20).
An important conclusion from the calculated results is that for a water depth/
draft ratio smaller than 4, the hydrodynamic coefficients gradually increase
with respect to the values for infinite water depth, while for a water depth/
draft ratio of 2 the changes are very substantial. Very few experiments on shallow
water have been carried out so far.
Forced oscillation tests with a ship model in shallow water have been conducted
by Tasai (Ref. 21), while motion measurements for sway, yaw and roll have been
performed by Fujii and Takahashi (Ref. 22).
The influence of shallow water on heave and pitch has been tested by Freakes and
Keay (Ref. 23) for a ship model, while it also has been discussed by Van Sluijs and

Tan in (Ref. 27).
Experiments with circular cylinders in shallow water have been carried out by Yu and Ursell (Ref. 24) to determine the damping by measuring the wave amplitude ratio.
Up to now no forced oscillation tests in shallow water are known for determining the hydrodynamic coefficients of cylinders with rectangular sections and a minimum beam/draft ratio of about 4. These values of the beam/draft ratio are often exceeded in the case of pontoons.

2. Experiments.

The experiments have been carried out in a basin of the Delft Hydraulics Laboratory. The dimensions of the basin are: length 40 m, breadth 35 m and maximum attainable waterdepth 0.70 m. On each side of the basin wave absorbing beaches made of gravel with a slope of 1:10 were construced which guaranteed minimal reflection of the generated waves. The bottom of the basin has been extensively smoothed and made horizontal within small tolerances.
In the middle of this basin a frame with cylindrical piles of small diameter supported the mechanical oscillators. The supporting frame could be adjusted in height to suit the different waterdepths. The resonance frequencies of the frame were all very large compared with the forced oscillation frequencies as used during the experiments. For the test two oscillators of the Planar Motion Mechanism type were used: one for the vertical motions i.e. heave, pitch and roll and one for the horizontal motions i.e. sway, yaw and surge. Two models have been used: one of a rectangular pontoon and one of a cutter-dredge-pontoon. Both models were constructed of aluminium and were identical except for the slot in the cutter dredge pontoon. The two models had sharp edges. For the main particulars reference is made to Figure 1.
Due to constructional problems it was not possible to fix the axis of rotation during the pitch and the roll modes through the centre of gravity of the models; consequently some surge and sway motion was present during these tests.
The models were connected to the oscillators by means of strain gauge dynamometers, measuring the forces in the direction of motion. These forces were reduced into an in phase component and a 90 degrees out-of-phase component with the motion, by means of an analogue Fourier Transformer, mechanically linked with the oscillators as described in (Ref. 6).
With these force components and the known particulars of the models the added mass and damping could be calculated using the formulae given in Appendix 1.
With the two models a series of tests has been performed using 12 oscillation frequencies in the range from 0,2 Hz upto 2.0 Hz with three amplitudes for a check on the linearity; and with three different waterdepths, i.e. 4.5, 1.75 and 1.2 times the draft of the model to investigate the influence of the waterdepth on added mass and damping.

3. Calculations.

The forced oscillation experiments with the model pontoons at a waterdepth of 0.5 m are considered to have no influence from the bottom. For this reason calculations assuming infinite waterdepth have been carried out.
After the determination of the sectional added mass and damping use was made of the well-known strip theory as presented in (Ref. 6) and by integration of the two dimensional cross sectional values over the length of the pontoon, results have been obtained for the pontoon.
The hydrodynamic coefficients with respect to deep water have been determined in two ways viz.:

1. the close-fit mapping method as described by de Jong in (Ref. 3) and based on Ursell's solution as presented in (Ref. 5) for the problem of an infinitely long cylinder oscillating in the free surface.
   With the aid of a conformal transformation of the pontoon-section to the unit circle the hydrodynamic forces could be computed and the results are shown in Fig. 6-7 for heave, pitch, sway and yaw in a dimensionless form as denoted in Appendix 1.
   The transformation formulae which were used, are given by:

$$z = \sum_{n=0}^{N} a_{2n-1} \zeta^{-(2n-1)} \qquad (1)$$

where: N = 2 for the three coefficient Lewis transformation
N = 5 for the case considered.

2. the <u>Frank close-fit method</u> (Ref.4) uses a distribution of pulsating sources along the contour of the two dimensional cross-section.
This method is based upon the determination of the velocity potential obtained for a distribution of source singularities over the submerged section. The sources which satisfy the linearized free-surface condition and the kinematic boundary condition on the surface of the section-cylinder have to be infinite in number to obtain a continuous source distribution. Frank came to an approximation by using a finite source distribution in such a way that a constant strength was considered for some straight segments replacing the original section. The hydrodynamic pressures are obtained from the potential by means of the linearized Bernoulli equation. Integration of these pressures over the submerged section cylinder gives the hydrodynamic forces.
The accuracy of this solution for normal ship forms depends on the chosen number of source segments. For the pontoon-section considered a number of 12 segments was used.
The hydrodynamic forces represented as added mass and damping have been calculated with a computer-program described and presented in (Ref. 4).
The results are shown in Fig. 6-7 with respect to the same motions and presented in the same way as denoted for the forementioned close-fit mapping method.

For shallow water <u>Keil</u> developed a method in (Ref. 10,11) to determine the sectional added mass and damping for both vertical and horizontal motions.
Porter first indicated in (Ref. 1) a way to find the velocity potential for a cylindrical body performing harmonic motions in the free surface of a fluid with finite depth. This method was developed for circular cylinders by Yu-Ursell in (Ref. 24) and for Lewis-form sections by Kim in (Ref. 9).
Keil extended this work and made also use of Lewis-transformation according to the indication N=2 in expression (1).
A velocity potential was synthesized from appropriate functions satisfying the boundary conditions on the free surface, the bottom and the body-surface while delivering radiated waves far from the body. Afterwards the pressure distribution was determined and integration of these pressures over the body-surface gave the hydrodynamic forces in a fluid of infinite depth.
As input for the computerprogram a set of wave-lengths is used from which the frequency is given by:

$$\omega = \sqrt{k_o g \, \text{thg}(k_o h)} \qquad (2)$$

in which $k_o = \frac{2\pi}{\lambda}$ = wave number
λ = wave length

The real part of the vertical hydrodynamic force may be written as the product of sectional added mass and acceleration and is written as

$$\overline{F}_{Vr} = a'_{zz} \overline{a} \qquad (3)$$

with: a'_{zz} = sectional added mass
$\overline{a}$ = the vertical acceleration

The sectional added mass for the vertical motions is expressed in a dimensionless way with:

$$C_V = \frac{a'_{zz}}{\rho \frac{\pi}{8} B^2} \qquad (4)$$

in which:

B = breadth of the section at the waterline.

The imaginary part of the vertical hydrodynamic force is equal to the product of the sectional damping and the velocity and may be written as:

$$\overline{F}_{vj} = b'_{zz} \overline{V} \qquad (5)$$

in which: b'_{zz} = the sectional damping
$\overline{V}$ = the vertical velocity

This force may also be expressed in terms of the squared amplitude ratio of the **radiated waves and the motion** by:

$$\overline{F}_{vj} = A_V^2 \rho \omega \overline{V} \frac{1}{k^2 F_{nh}} \qquad (6)$$

in which:

$$A_V = \frac{\overline{\zeta}_a}{z_a}$$

with: $\overline{\zeta}_a$ = amplitude of radiated wave
z_a = vertical motion amplitude
$k = \frac{\omega^2}{g}$ = wave-number for deep water

$$F_{nh} = \frac{\cosh^2(k_o h)}{k_o h + \sinh(k_o h) \cosh(k_o h)} \qquad (7)$$

From (5) and (6) follows the sectional damping coefficient:

$$b'_{zz} = A_V^2 \rho \omega \frac{1}{k^2 F_{nh}} \qquad (8)$$

Analogous expressions are derived by Keil in (Ref. 11) for the horizontal motions in shallow water. The results determined with the computer-program according to Keil are shown in Fig. 6-7 for heave, pitch, sway and yaw, related to the three water-depths considered. This presentation is also in a non-dimensional form.
For the cutter-dredge pontoon no calculations have been carried out up to now, but it seems reasonable to perform this in the future by considering a part of the pontoon as a catamaran.
To solve this problem for the deep water condition reference should be made to the work of De Jong in (Ref. 25), where he presents a method of determining the hydrodynamic coefficients of two parallel identical cylinders oscillating in the free surface.
In all calculations the rolling coefficients have been left out of consideration because comparison with the measurements is not possible due to the fact that the centre of rotation was not situated at the centre of gravity. Corrections should be necessary for the position of the centre of gravity and to eliminate the parasitic swaying motion which took place during the rolling experiments. However the accuracy of these corrections might hardly be sufficient to obtain reliable final results.
<u>Van Oortmerssen</u> (Ref. 12) formulated a method in which the three dimensional character of the flow around a ship on water of restricted depth can be taken into account. The method is based on linear potentional theory and supposes therefore an ideal fluid and small amplitudes for both the waves and the motions and uses a three dimensional source technique. With aid of the linearized free surface condition and the boundary conditions on the sea floor, the ship's surface and at infinity the velocity potential is found. The pressure on the ship's surface is than calculated using Bernoulli's theorem. Integration over the ship's surface yields the hydrodynamic forces. For the calculations an average number of ten sources over one wave length has been used. Results proved to be consistent for changes in the source distribution. Due to lack of time and the high costs involved only a limited number of calculations has been made.
The calculations were made for the pontoon on deep water and for the cutter-dredge pontoon on the waterdepth of 1.75 times T.

The results of these calculations are presented in the Figs. 6-7 in a dimensionless form, as described in Appendix 1.

4. Discussion of the results.

4.1. Experimental results.

The results of the tests are presented in a dimensionless form by using the formulae given in Appendix 1.
The added mass and damping for all motion components for both the pontoon and the cutter dredge pontoon are presented in Fig. 2-3 and on basis of the dimensionless frequency parameter:

$$\omega \sqrt{\frac{B}{2g}}$$

for one amplitude of oscillation and the three different waterdepths.
In Fig. 4 the dependency of added mass and damping on waterdepth is given for the most important motion components for three different frequencies. In Fig. 5 the dependency of the same hydrodynamic coefficients on the oscillation amplitude is shown for the three different waterdepths investigated and for the considered three frequencies of oscillation.
These last two figures are based on the results of the test with the pontoon only.
For the cutter-dredge pontoon the figures are the same in a qualitative way.

From these figures it can be seen that for the heave motion added mass increases considerably with decreasing waterdepth, giving the steepest increase for waterdepth to draft ratio's smaller than 2.
For large frequencies the added mass tends to become constant as shown in Fig. 2. The dependency of the added mass on the amplitude of oscillation is small for all three waterdepths investigated. The damping curve tends to zero for both very large and very small frequencies for the deep water situation only. With decreasing waterdepth the damping increases with the increase of oscillation frequency. This increase is most evident for the smallest waterdepth. The damping is strongly dependent on the amplitude of oscillation, as shown in Fig. 5.
Over the whole frequency range the added mass and damping for the cutter dredge pontoon is less than for the pontoon. This difference increases with the decrease of waterdepth.
The same qualitative tendencies can be seen for the hydrodynamic coefficients in the case of the pitch motion.

For the sway motion the influence of the waterdepth on added mass and damping increases with the decrease of the oscillation frequency. Especially for the low frequencies the influence of waterdepth on added mass is significant. For the damping this tendency is much less pronounced. These damping curves however, show only a small tendency to become zero for even the highest frequencies investigated, although small damping was to be expected for much lower frequencies.
Obviously viscous effects play an important role at the higher frequencies. At these frequencies a growing amount of eddy shedding from the sharp corners could be observed. Nevertheless both added mass and damping remain remarkably independent on the oscillation amplitude, see Fig. 5. In general the pontoon has less added mass than the cutter dredge pontoon contrary to the vertical motion.
For the damping these differences are fairly small. As far as surging is concerned, the influence of the waterdepth is generally small on both added mass and damping, although the latter tends to increase slightly for the smallest waterdepth. The independence on both added mass and damping on the oscillation amplitude
is fair for the lower frequencies but tends to decrease with increasing frequency. Here also the damping increases with the increase of oscillation frequency, and no tendency to decrease to zero damping for high frequencies can be observed. This tendency is equal for all three waterdepths investigated and appears to be even more pronounced than for swaying. In general the differences between the pontoon and the cutter dredge pontoon are small. For yawing the hydrodynamic coefficients show the same tendencies as for swaying. In analysing the results of the roll tests it should be emphasised that during these tests the axis of rotation did not go through the centre of gravity of the model. This implies, apart from corrections on

the moment of inertia of the oscillated models, that the model performed a simultaneous **roll and sway motion. A correction method applied to the results to eliminate** the effects of this sway motion on the forces measured gave no satisfactory results especially for the lowest waterdepth. Therefore the original results are presented in Figs. 2 and 3. From these figures it can be seen that the hydrodynamic coefficients for roll show little dependency on waterdepth. Here also the damping does not tend to zero for higher frequencies. Both added mass and damping show an evident dependency on the oscillation amplitude.

In general the pontoon has less added mass and damping compared with the cutter dredge pontoon.

4.2. Comparison of experiments and calculations.

For the deep-water situation it appears from Figs. 6-7 that there is a good agreement between the results of both calculation methods considered and the experiments. The differences between the results of the close-fit mapping method and the Frank close-fit method appeared to be rather small.

For the higher frequencies the prediction of the damping coefficient according to both mentioned calculation methods is somewhat too low which might be due to viscous effects which might increase with the frequency.

The results of the calculation method of Keil showed for deep water over the greater part of the frequency range good agreement with both other calculation methods and also with the experimental results. For the low frequencies, however, there appeared to be a rather strong deviation from the experiments and the other calculated results. This holds especially for the vertical damping coefficients.

For the second waterdepth it appears from Fig. 6-7 that the prediction according to Keil's method is rather satisfactory for the added mass except for the lower frequency range.

The agreement for damping, however, is rather poor especially for the low frequencies where there is an inexplicable tendency to infinity.

For the smallest waterdepth the deviations from the experiments are even greater particular for the damping coefficients of the vertical motions.

It is remarkable that with respect to the calculated results according to Keil's method the strong increase of damping occurs at higher frequencies when the waterdepth decreases.

It appears from Fig. 6 and 7 that the calculations according to the method of Van Oortmerssen give both for deep water and shallow water satisfactory results. For shallow water only the second waterdepth for the cutter dredge pontoon has been considered. It is an advantage that this method is able to take into account the "twin hull" part of the pontoon. The discrepancies are only significant at the highest frequencies, especially in the case of added mass for the surging motion. However, it should be kept in mind that it is to be expected that for these high frequencies viscous effects grow more important.

5. Conclusions and recommendations.

From the preceeding experiments and calculations the following conclusions and recommendations may be derived with respect to the hydro-dynamic coefficients of a pontoon in shallow water:

1. The influence of waterdepth on added mass and damping is most important for the vertical motions. Both added mass and damping increase with decreasing waterdepth.

2. The dependency of added mass and damping on the oscillation amplitude is more evident for the vertical motions than for the horizontal motions, while the dependency increases with decreasing waterdepth.

3. The damping appears to increase with frequency for the vertical motions at the lowest waterdepth ($h/T = 1.2$) and increases with frequency for surging at all waterdepths.

4. The cutter dredge pontoon has lower added mass and damping than the pontoon for the vertical motions. For the horizontal motions the opposite statement holds true.

5. The calculated values of added mass and damping for deep water agree very well with the experimental values for the highest waterdepth considered (h/T = 4.5).

6. The calculated results with respect to shallow water according to the method of Van Oortmerssen show a good agreement with the experimental values, while those according to Keil's method needs improvement especially for damping.

7. The experimental values for added mass and damping of the rolling motion needs further analysis for a good comparison with the calculated results. This comparison with respect to damping, however, will remain difficult on account of the dominant viscous influence for the rolling motion.

6. Acknowledgements.

Special thanks are due to the various members of the staff of the Delft Hydraulics Laboratory and Ship Hydromechanics Laboratory for their assistance in running the described experiments.

7. References.

a) Journal Articles.

1. Porter, W.R.: "Pressure distributions, added mass and damping coefficients for cylinders oscillating in a free surface." University of California, Inst. of Eng. Res., Berkeley. (July 1960).

2. Tasai, F.: "On the damping force and added mass of ships heaving and pitching." Rep. of Res. Inst. for Appl. Mech., Kyushu University, Vol. VII, no. 26. (1959).

3. Jong, B. de: "Computation of the hydrodynamic coefficients of oscillating cylinders."
Report 174 A of the Ship Hydromech. Lab., Delft University of Technology. (November 1969).

4. Frank, W. and Salvesen, N.: "The Frank Close-fit Ship Motions Computer Program." David Taylor Naval Ship Research and Development Center, U.S.A., Rep. 3289 (1970).

5. Ursell, F.: "On the heaving motion of a circular cylinder on the surface of a fluid." Quart. J. of Mech. and Appl. Math., Vol. 2, pp. 218-231 (1949).

6. Gerritsma, J. and Beukelman, W.: "The distribution of the hydrodynamic forces on a heaving and pitch shipmodel in still water." Int. Ship. Progress, Vol. 11, no. 123, pp. 506-522 (November 1964).

7. Vugts, J.H.: "The hydrodynamic forces and ship motions in oblique waves." Neth. Ship Res. Centre, T.N.O., report 150S. (December 1971).

8. Vugts, J.H.: "The hydrodynamic coefficients for swaying, heaving and rolling cylinders in a free surface." Neth. Ship Res. Centre T.N.O., report 112S, (May 1968).

9. Kim, C.H.: "Hydrodynamic forces and moments for heaving, swaying and rolling cylinders on water depth." Journal of Ship Research, Vol. 13, (1969).

10. Keil, H.: "Hydrodynamic mass and damping coefficient of a heaving cylinder in still water" (Hydrodynamische Masse und Dämpfungskonstante Tauchender Zylinder auf flachem Wasser). Schiffstechnik, Vol. 23, (1976).

11. Keil, H.: "The hydrodynamic forces at a periodic motion of a two-dimensional body in the still water surface" (Die hydrodynamische Kräfte bei der periodischen Bewegung zwei-dimensionaler Körper an der Oberfläche flacher Wasser). Institut für Schiffbau, Bericht Nr. 305,(February 1974).

12. Van Oortmerssen, G.: "The motions of a moored ship in waves." Thesis, Delft University of Technology (1976).

13. Ursell, F.: "On the virtual mass and damping coefficients for long waves in water of finite depth." Journal of Fluid Mech., vol. 76, part I (1976).

14. Ikebuchi, T.: "Wave induced forces and moments in shallow water." Journal of the Kansai of Naval Architects of Japan, no. 161,(1976).

15. Takaki, M.: "On the hydrodynamic forces and moments as acting on the two-dimensional bodies oscillating in shallow water." Res. Inst. of Appl. Mech., Kyushu University, Japan, vol. 25, no.78, (1977).

16. Kan, M.: "The added mass coefficient of a cylinder oscillating in shallow water in the limit $k \to 0$ and $k \to \infty$." Ship Res. Inst., Japan, no. 52,(1977).

17. Kwang June Bai: "The added mass of two-dimensional cylinders heaving in water of finite depth." Journal of Fluid Mech., vol. 81, part I (1977).

18. Hwang, J.H. and others: "Hydrodynamic forces for heaving cylinders on water of finite depth." Soc. of Naval Arch. in Korea, vol. 13, no. 3 (1976).

19. Tuck, E.O.: "Ship motions in shallow water." Journal of Ship Res. vol. 14, (1970).

20. Flagg, C.N. and Newman, J.N.: "Sway added mass coefficients for rectangular profiles in shallow water." Journal of Ship Res., vol. 15, no. 4 (December 1971).

21. Tasai, F. and others: "Ship motions in restricted waters, Part I - Tank tests". Res. Inst. of Appl. Mech., Kyushu University Japan, vol. XXVI, no. 81, (July 1978).

22. Fujii, H. and Takahashi, T.: "Measurement of the derivatives of sway, yaw and roll motions by forced oscillation technique." Journal of the Kansai of Naval Architects of Japan, vol. 130, pp. 169-183,(December 1971).

23. Freakes, W. and Keay, K.L.: "Effects of shallow water on ship motion parameters in pitch and heave." Massachusetts Institute of Technology, Department of Naval Architecture and Marine Engineering, Report no. 66-7 (1966).

24. Yu, Y.S. and Ursell, F.: "Surface waves generated by an oscillating circular cylinder on water of finite depth: theory and experiment." Journal of Fluid Mechanics, vol. 11. (1961).

25. Jong, B. de: "The hydrodynamic coefficients of two-parallel identical cylinders oscillating in the free surface." Report 268 of the Ship Hydromechanics Lab., Delft University of Technology,(June 1970).

b. Conference papers and proceedings.

26. Sayer, P. and Ursell, F.: "On the virtual mass, at long wave lengths of a half-immersed circular cylinder heaving in water of finite depth." 11th Symposium Naval Hydrodynamics, London, (March 1976).

27. Sluijs, M.F. van and Tan Seng Gie: "The effect of water depth on ship motions." Appendix 7 of Seakeeping Committee Report, 14th International Towing Tank Conference (1975).

8. Appendix I.

The equations of motion for forced oscillations in the six modes are given by:

surge
$$a(\rho\nabla + a_{xx})\ddot{x} + b_{xx}\dot{x} = (F_1 + F_2)\sin(\omega t + \varepsilon_x) \tag{A1}$$

sway
$$(\rho\nabla + a_{yy})\ddot{y} + b_{yy}\dot{y} + d_{y\psi}\ddot{\psi} + e_{y\psi}\psi = (F_1 + F_2)\sin(\omega t + \varepsilon_y) \tag{A2}$$

heave
$$(\rho\nabla + a_{zz})\ddot{z} + b_{zz}\dot{z} + c_{zz}z + d_{z\theta}\ddot{\theta} + e_{z\theta}\dot{\theta} + g_{z\theta}\theta = (F_1 + F_2)\sin(\omega t + \varepsilon_z) \tag{A3}$$

roll
$$(I_{xx} + a_{\phi\phi})\ddot{\phi} + b_{\phi\phi}\dot{\phi} + c_{\phi\phi}\phi = (F_1 - F_2)\frac{1}{2}\sin(\omega t + \varepsilon_\phi) \tag{A4}$$

pitch
$$(I_{yy} + a_{\theta\theta})\ddot{\theta} + b_{\theta\theta}\dot{\theta} + c_{\theta\theta}\theta + d_{\theta z}\ddot{z} + e_{\theta z}\dot{z} + g_{\theta z}z = (F_1 - F_2)\frac{1}{2}\sin(\omega t + \varepsilon_\theta) \tag{A5}$$

yaw
$$(I_{zz} + a_{\psi\psi})\ddot{\psi} + b_{\psi\psi}\dot{\psi} + d_{\psi y}\ddot{y} + e_{\psi y}\dot{y} = (F_1 - F_2)\frac{1}{2}\sin(\omega t + \varepsilon_\psi) \tag{A6}$$

Substituting the known motion of the model and its derivatives and reducing the measured forces into a component in phase and one 90 degrees out of phase with the motion it can be shown that:

$$a_{zz} = -\frac{1}{z_a\omega^2}\left\{(F_1 + F_2)\cos\varepsilon_z + c_{zz}z_a\right\} - \rho\nabla$$

$$b_{zz} = \frac{1}{z_a\omega}\left\{(F_1 + F_2)\sin\varepsilon_z\right\}$$

$$a_{yy} = -\frac{1}{y_a\omega^2}\left\{(F_1 + F_2)\cos\varepsilon_y\right\} - \rho\nabla$$

$$b_{yy} = \frac{1}{y_a\omega}\left\{(F_1 + F_2)\sin\varepsilon_y\right\}$$

$$a_{xx} = -\frac{1}{x_a\omega^2}\left\{(F_1 + F_2)\cos\varepsilon_x\right\} - \rho\nabla$$

$$b_{xx} = \frac{1}{x_a\omega}\left\{(F_1 + F_2)\sin\varepsilon_x\right\}$$

$$a_{\phi\phi} = -\frac{1}{\phi_a\omega^2}\left\{(F_1 - F_2)\frac{1}{2}\cos\varepsilon_\phi + c_{\phi\phi}\phi_a\right\} - y_{xx}$$

$$b_{\phi\phi} = \frac{1}{\phi_a\omega}\left\{(F_1 - F_2)\frac{1}{2}\sin\varepsilon_\phi\right\}$$

$$a_{\theta\theta} = -\frac{1}{\theta_a\omega^2}\left\{(F_1 - F_2)\frac{1}{2}\cos\varepsilon_\theta + c_{\theta\theta}\theta_a\right\} - I_{yy}$$

$$b_{\theta\theta} = \frac{1}{\theta_a\omega}\left\{(F_1 - F_2)\frac{1}{2}\sin\varepsilon_\theta\right\}$$

$$a_{\psi\psi} = -\frac{1}{\psi_a \omega^2} \left\{ (F_1 - F_2) \frac{1}{2} \cos\varepsilon_\psi \right\} - I_{zz}$$

$$b_{\psi\psi} = \frac{1}{\psi_a \omega} \left\{ (F_1 - F_2) \frac{1}{2} \sin\varepsilon_\psi \right\}$$

The cross coupling coefficients are omitted here because the experimental results need further analysis to be presented.

In the presentation of the results the added mass and damping coefficients have been made dimensionless using the following formulae:

For the <u>translations</u> heave, surge, sway:

    added mass :   divided by mass of the models $\frac{1}{\rho\nabla}$

    damping   :   divided by the mass of the models $\rho\nabla$ and multiplied with the square root of half the breadth divided by gravitational acceleration:

$$\frac{1}{\rho\nabla} * \sqrt{B/2g}$$

For the <u>rotations</u> pitch and yaw:

    added mass moment of inertia :   divided by the mass of the models times the length squared

$$\frac{1}{\rho\nabla L^2}$$

    damping   :   divided by the mass of the models times the length squared and multiplied with the square root of half the breadth divided by the gravitational acceleration:

$$\frac{1}{\rho\nabla L} * \sqrt{B/2g}$$

For the <u>rotation roll</u> : the same as for pitch and yaw except for L^2 in both terms which has been replaced by the breadth of the models squared i.e.:

$$\frac{1}{\rho\nabla B^2} \quad \text{and} \quad \frac{1}{\rho\nabla B^2} * \sqrt{B/2g}$$

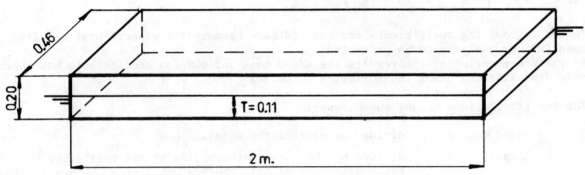
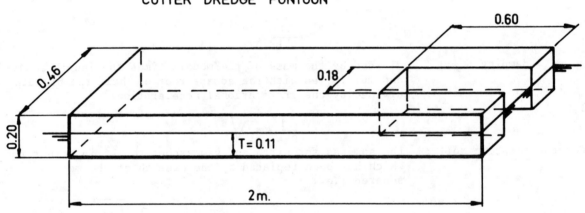

Fig. 1   Used barge models

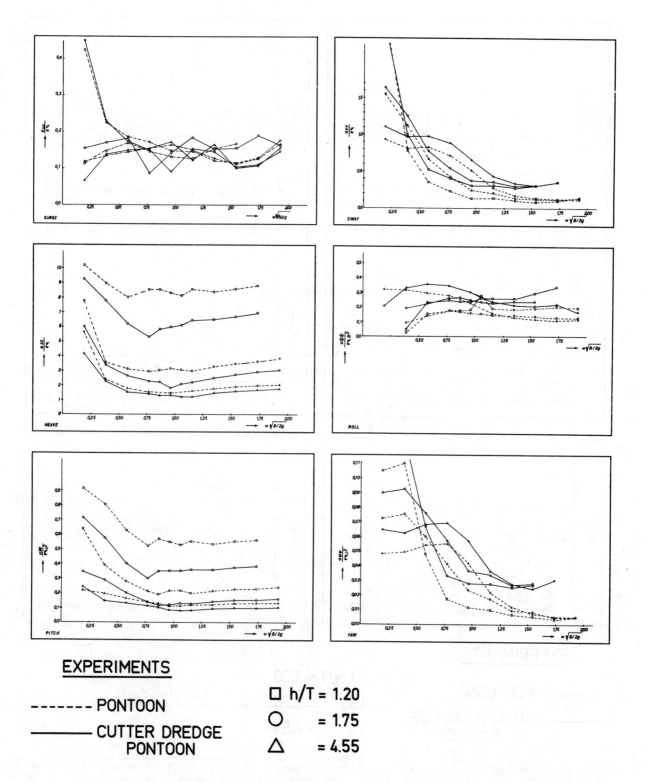

Fig. 2 Added mass related to frequency of oscillation for six modes of motion

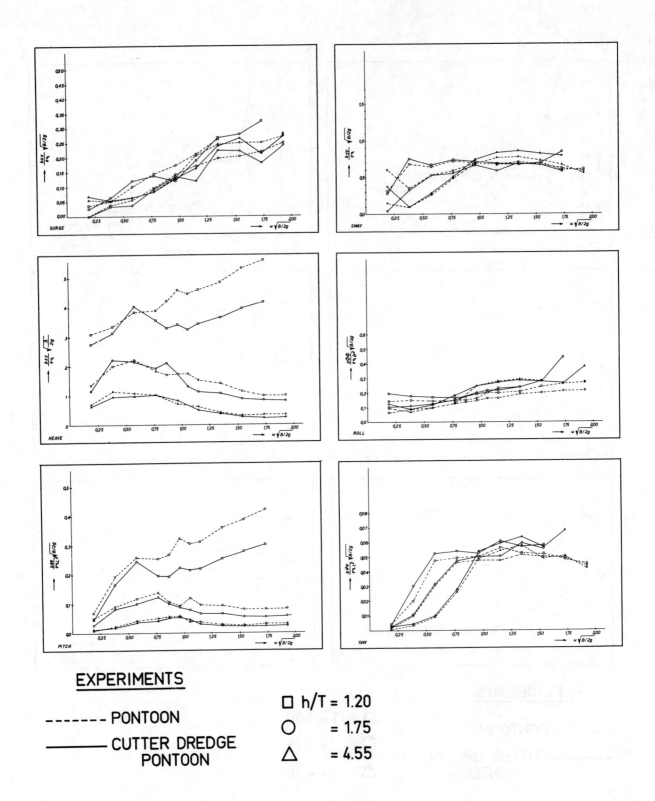

Fig. 3 Damping related to frequency of oscillation for six modes of motion

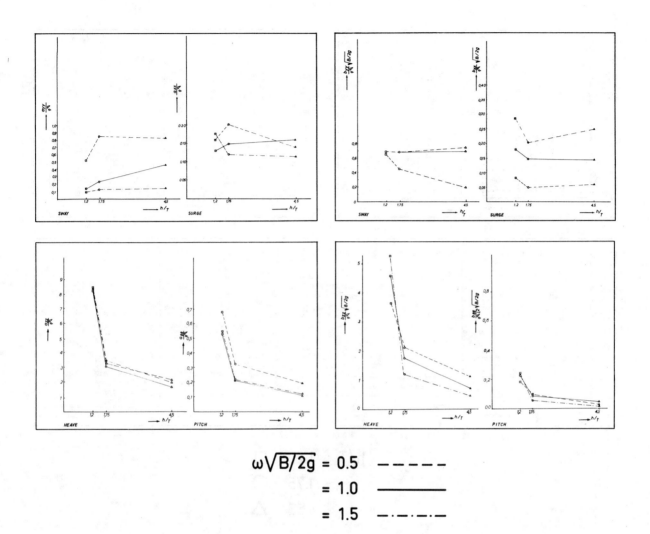

Fig. 4 Added mass and damping related to waterdepth draft-ratio for the pontoon
($\omega \sqrt{B/2g}$ = 0.5 , 1.0 , 1.5)

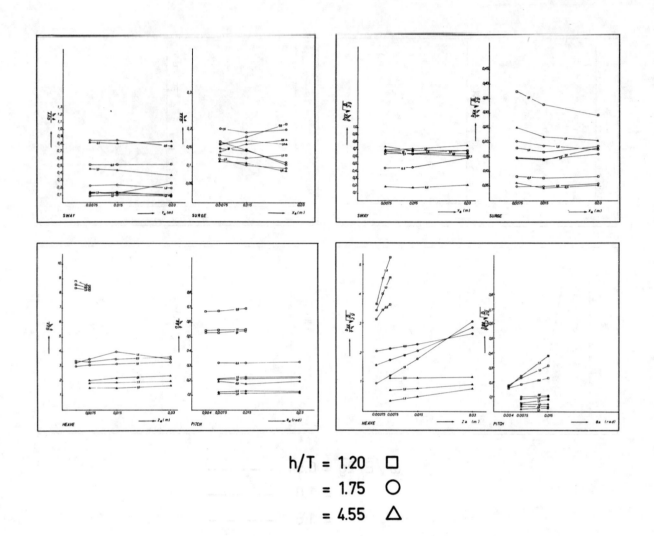

Fig. 5 Added mass and damping related to amplitude of oscillation for the pontoon

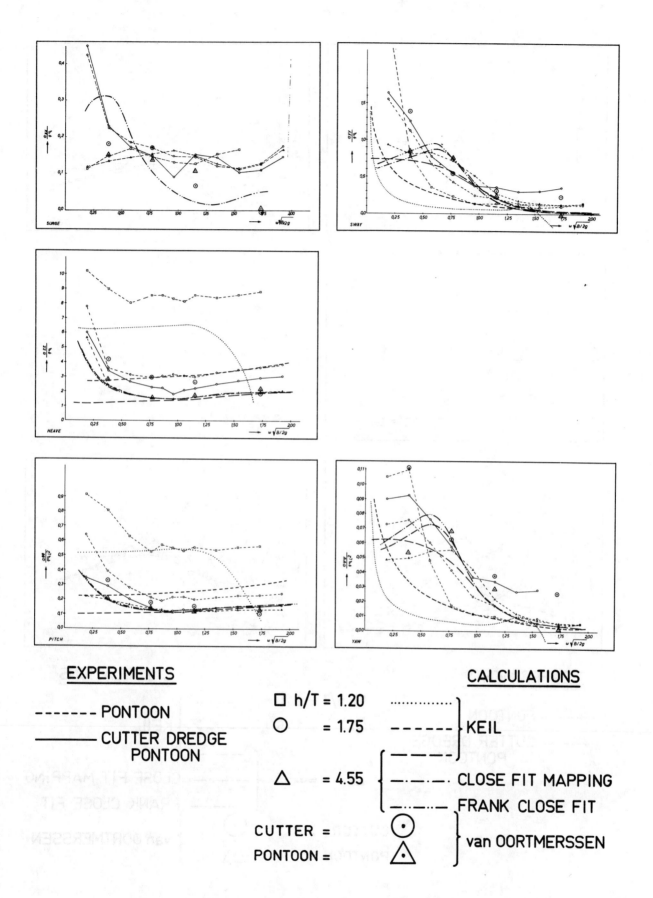

Fig. 6 Measured and calculated added mass related to frequency of oscillation

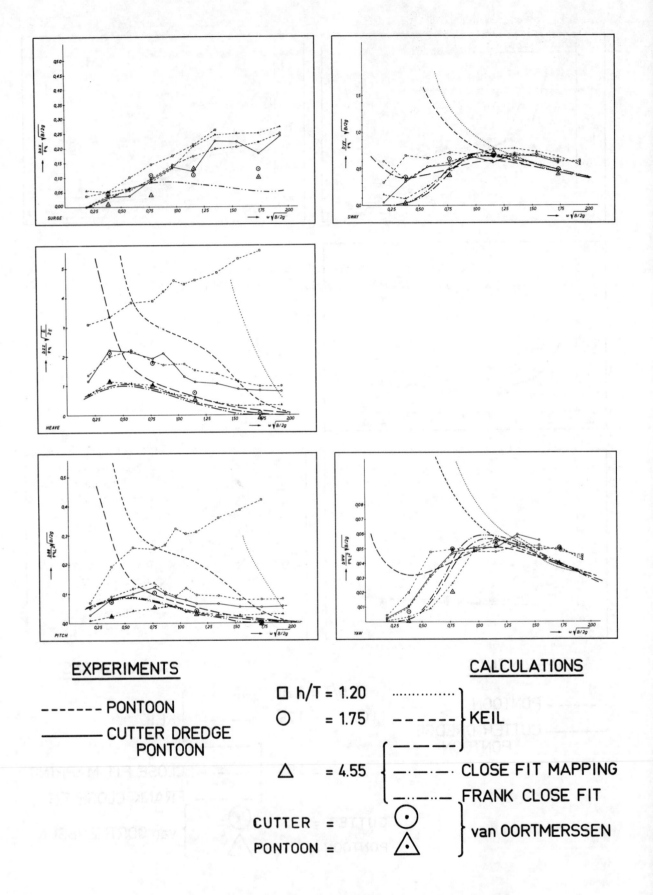

Fig. 7 Measured and calculated damping related to frequency of oscillation

BOSS'79

PAPER 56

Second International Conference
on Behaviour of Off-Shore Structures
Held at: Imperial College, London, England
28 to 31 August 1979

RECIPROCITY RELATIONS FOR OFFSHORE STRUCTURES VIBRATING AT ACOUSTIC FREQUENCIES

C.C. Mei

Massachusetts Institute of Technology, USA

Summary

In order to provide convenient checks for numerical programs designed for offshore structures, it is desirable to have certain identities which are implied by a consistently formulated theory. Following the existing works in water waves and elastic waves such identities are first deduced for an elastic structure vibrating in the sea. It is assumed that the structure rests on a rigid base and the sea bed is also rigid. The forcing derives from either underwater sound waves or excitation at the base by earthquakes. The frequency range considered is so high that gravity waves can be ignored, but acoustic waves are important. We then give an example that it is possible that such identities provide necessary, but not sufficient, checks of a numerical program. The example chosen is the hybrid-element method of Mei, Foda and Tong (1970). After casting the method in weak formulation it is shown that such identities are satisfied irrespective of discretization error. Thus the identities only check the over-all correctness, but not the accuracy of the numerical procedure. This theoretical feature deserves attention when other numerical methods are examined.

Sponsored by: Delft University of Technology, The Netherlands
Massachusetts Institute of Technology, U.S.A.
The Norwegian Institute of Technology, Norway
University of London, England

Secretariat provided by: BHRA Fluid Engineering

Copyright: ©   BHRA Fluid Engineering
Cranfield, Bedford, England

NOMENCLATURE
(Partial List Only)

a = amplitude

F_a^b = a-th component of reacting force on the base due to forced unit motion in the b-th generalized direction

F_a^D = a-th component exciting force due to diffraction acting on the base

h = water depth

$H_n^{(1)}$ = Hankel function of first kind

$H_n^{(1)\prime}$ = derivative of $H_n^{(1)}$ with respect to the argument

k = wave number

K_m = modified Bessel function of the second kind

K_{ab} = radiation stiffness matrix

k_n = cut-off wave number of the nth mode

n_j = unit normal vector positive if out of fluid or into structure

p = water pressure

Im = imaginary part of

Re = real part of

S_{BS} = surface bordering bottom (B) and structure (S)

S_{WS} = surface bordering water (W) and structure (S)

S_{AS} = surface bordering on air (A) and structure (S)

V_S = volume of structure

V_W = volume of water

W_a = generalized component of displacement of the rigid base

ε_{ij} = strain tensor in elastic structure

λ_{ab} = radiation damping matrix

τ_{ij} = stress tensor in elastic structure

ω = frequency

ϕ = velocity potential in water

$\psi^{(\ )}$ = test functions

$(\ )^a$ or $(\ )_a$ = a-th generalized component

$(\ )^S$ = of scattering problem

$(\ )^I$ = of incident wave

$(\ )^R$ = of radiation problem

$(\ )^D$ = of diffraction problem

$(\ )^*$ = complex conjugate

$(\ )^{(1)}$ = of problem 1

I. INTRODUCTION

Offshore structures are now constructed in such great depths that the dynamic interaction between the structure, the surrounding water and the supporting foundation has become an important topic of applied mechanics. Among those of physical origin, perhaps the most difficult obstacles to date are a meager understanding of soil behavior under dynamic loadings and the forces due to waves of large amplitude. Mathematical difficulties have limited the continuum mechanics approaches to largely linearized theory valid for infinitesimal motions.

Many offshore structures must be designed with due consideration to earthquake hazards. Now seismic waves are important in a rather high frequency range (0.01 < T < 1 sec, say) in comparison with surface water waves (5 < T < 20 sec) so that the deformation of the structure and of the ground is essential. Furthermore, due to the high frequencies involved, compressibility of water is not always negligible. In a sea of finite depth it is well known that compressibility implies the existence of guided acoustic waves in water which provides one mechanism of radiation damping. For a sea of depth h the guided waves are possible if the frequency is above certain cut-off frequencies: $\omega_n = k_n c$ where $k_n = (n + 1/2)\pi/h$, $n = 0,1,2,\ldots$ and c is the sound speed in water. Taking the depth to be 150 m and c = 1500 m/sec, the first cut-off frequency is $\omega_1 = \pi c/2h = 15.7/\text{sec}$ which corresponds to a period of $T_1 = 2\pi/\omega_1$ = 0.40 sec and falls within most of the design spectra of earthquakes. Only for sufficiently shallow sea or long period waves is compressibility of water not important.

Many simplifications have been made in the literature for the sea bed, which has been variably assumed to be perfectly rigid, or a discrete system with a finite degree of freedom, a continuum of an elastic layer or a viselastic layer. Sophisticated numerical schemes have been developed by many investigators. Because of the complexity of the mathematics, the computer programs are necessarily involved and it is highly desirable to have ways of checking the correctness of the computational scheme. This is, of course, a mathematical matter separate from but of equal importance as the validity of physical assumptions. As exact analytical solutions are rare in this field, one cannot easily find gauges for comparison. However, general identities which relate quantities of physical interest and which must be satisfied by a consistently formulated theory are helpful. Such theorems are, for example, symmetry of forces, energy theorems, etc., and has been well explored in wave problems of a single medium, e.g., water waves (see Wehausen (Ref. 1) and Newman (Ref. 2) for water, and Mei (Ref. 3) for elastic waves and references quoted therein).

It is the purpose of this paper to extend these identities to an elastic structure vibrating in ocean. Several identities are well known in more conventional topics; their validity in the present context nevertheless requires proof, while several others may be new to the structural mechanics literature.

To avoid undue mathematics we shall demonstrate the proofs for an elastic structure on a rigid foundation whose motion is assumed to be prescribed. This is, of course, a drastic simplification since in reality the sea bed is an active participant of dynamic interaction. We shall also derive the results for three-dimension only; the two-dimensional counterparts are straightforward to obtain. Extension to treat the sea bed as an elastic layer should be possible, but the proof would be much more involved as one must account for waves which propagate through the foundation and along the water-soil interface. The remaining task is left for the future.

2. FORMULATION

Consider an elastic structure S surrounded by a water layer of depth h which may be variable near S. The structure is assumed to be an isotropic and homogeneous elastic solid and is welded to a perfectly rigid foundation. In subscript notations for surfaces and volumes we shall use A for air, B for bottom of the sea, S for structure and W for water. Thus the volumes of water and structure are denoted by V_W and V_S and the interfaces are S_{AS}, S_{WS}, S_{BS}, S_{WB} and S_{WA}. (See Figure 1.)

Restricting to simple harmonic motions only we shall factor out the time dependence in the manner $\tilde{f}(\vec{x},t) = \text{Re } f(\vec{x})e^{-i\omega t}$. Within the elastic structure we have

$$\tau_{ij,j} + \rho_s \omega^2 u_i = 0 \quad \text{in } V_S \tag{2.1}$$

where

$$\tau_{ij} = \lambda \varepsilon_{kk} \delta_{ij} + \mu \varepsilon_{ij} \qquad \varepsilon_{ij} = \frac{1}{2}(u_{i,j} + u_{j,i}) \tag{2.2}$$

The notations are conventional: τ_{ij} for stress, u_i for displacement, ε_{ij} for strain and λ and μ are elastic constants. The boundary conditions are

$$u_i = \text{given} \quad \text{on } S_{BS} \tag{2.3}$$

$$\tau_{ij} n_j = 0 \quad \text{on } S_{AS} \tag{2.4}$$

The unit normal points <u>into</u> the elastic structure. In water we assume the motion to be compressible but irrotational so that the velocity potential ϕ satisfies

$$\nabla^2 \phi + k^2 \phi = 0 \quad \text{in } V_W \tag{2.5}$$

where $k = \omega/c$. The boundary conditions are

$$\phi = 0 \quad \text{(zero pressure)} \quad \text{on } S_{WA} \quad z = 0 \tag{2.6}$$

$$\frac{\partial \phi}{\partial n} = 0 \quad \text{(zero normal velocity)} \quad \text{on } S_{WB} \quad z = -h \tag{2.7}$$

The unit normal points <u>out of</u> water. At infinity, the radiation condition must be imposed. At the structure-water interface, continuity of normal velocity requires that

$$\frac{\partial \phi}{\partial n} = -i\omega u_i n_i \quad \text{on } S_{SW} \tag{2.8}$$

Decomposing the surface traction $\vec{T}(T_i = \tau_{ij} n_j)$ into normal and tangential components

$$\vec{T} = (\vec{T} \cdot \vec{n}) \cdot \vec{n} + \vec{n} \times (\vec{T} \times \vec{n}),$$

we require that normal stress be equal to the adjacent fluid pressure and that the tangential stress be zero so that

$$\vec{T} = \vec{T} \cdot \vec{n} = \tau_{ij} n_i n_j = p = -i\omega\rho\phi \quad \text{on } S_{SW} \tag{2.9}$$

One may now distinguish two kinds of problems: (i) diffraction problem where prescribed incident underwater waves ϕ^I strikes the structure; this is of interest in underwater explosions, and (ii) radiation problem where the foundation S_{BS} is forced to oscillate in a prescribed manner. The boundary condition (2.3) may be explicitly stated as follows:

(i) Diffraction problem: An incident wave impinging on an elastic structure on rigid and stationary footing

$$u_i = 0 \quad \text{on } S_{BS} \tag{2.10}$$

(ii) Radiation problem: The footing is forced to oscillate

$$u_i = U_i \quad \text{on } S_{BS} \tag{2.11}$$

The details of U_i are given below.

As is usual in linear problems, it is convenient to decompose first the general motion of footing into six generalized components. Let

$$\{W_a\} = \{\xi_1, \xi_2, \xi_3, \Omega_1, \Omega_2, \Omega_3\}, \quad a = 1, 2, \ldots 6 \tag{2.12}$$

be the generalized displacements of the rigid footing S_{BS} with $\{\xi_1, \xi_2, \xi_3\}$ representing components of translation along x,y,z axes, and $\{\Omega_1, \Omega_2, \Omega_3\}$ representing components of rotation about axes passing the point X_i^o. Mei (Ref. 3) has shown that

$$U_i = N_{ia} W_a \tag{2.13}$$

with

$$[N_{ia}] = \begin{bmatrix} 1 & 0 & 0 & 0 & z-Z_o & -(y-Y_o) \\ 0 & 1 & 0 & -(z-Z_o) & 0 & (x-X_o) \\ 0 & 0 & 1 & y-Y_o & -(x-X_o) & 0 \end{bmatrix} \tag{2.14}$$

It is now natural to decompose the radiation problem u_i^R into six generalized modes u_i^a so that

$$u_i^R = \sum_a (u_i^{a'} + 2u_i^{a''}) W_a \tag{2.15}$$

where $u_i^{a'} = \text{Re } u_i^a$ and $u_i^{a''} = \text{Im } u_i^a$. Because N_{ia} is real, cf. (2.14), it follows that u_i^a satisfies

$$u_i^{a'} = N_{ia} \qquad u_i^{a''} = 0 \qquad \text{on } S_{BS} \tag{2.16}$$

Otherwise u_i^a and its corresponding ϕ^a satisfy all other conditions for u_i^R and ϕ^R respectively.

The radiation conditions must be imposed at infinity. In case (i) let ϕ^S = scattered wave and ϕ^I = incident wave so that total $\phi = \phi^I + \phi^S$. In case (ii) let ϕ^R = radiated wave. We then require

Two dimensions: $\begin{matrix} \phi^S \\ \phi^R \end{matrix}$ are outgoing plane waves as $|kx| \gg 1$ \hfill (2.17.a)

Three dimensions: $\begin{matrix} \phi^S \\ \phi^R \end{matrix}$ are radially outgoing as $kr \gg 1$ \hfill (2.17.b)

Following Mei in treating rigid inclusions in an elastic medium, we may define, for the diffraction problem, the exciting force in the a-th generalized direction acting on the structure by the foundation at S_{BW} to be

$$F_a^D = -\int_{S_{BS}} N_{ia} \tau_{ij}^D n_j dS \tag{2.18}$$

with the indices $a = 1,2,3$ for forces and $= 4,5,6$ for moments. The total reacting force on the structure in the b-th generalized direction due to forced motion of the foundation can be expressed as

$$F_b^R = \sum_a W_a (K_{ba} - i\lambda_{ba}) \equiv \sum_a W_a F_{ba} \tag{2.19}$$

where K_{ba} and λ_{ba} are defined respectively as the stiffness and damping matrices:

$$\begin{matrix} K_{ba} \\ -\lambda_{ba} \end{matrix} = - \begin{pmatrix} Re \\ Im \end{pmatrix} \iint_{S_{BS}} N_{ib} \tau_{ij}^a n_j dS \qquad (2.20)$$

The complex sum $K_{ba} - i\lambda_{ba}$ is the b-component of the reacting force due to unit generalized forced motion of B in a-direction.

3. PRELIMINARY IDENTITIES

Consider two dynamic fields $u_i^{(1)}$ and $u_i^{(2)}$ for the same ω. By applying Green's theorem over V_S together with Eqs. (2.1), (2.2), and (2.4) we get

$$\iint_{S_{BS}+S_{SW}} [\tau_{ij}^{(1)} u_i^{(2)} - \tau_{ij}^{(2)} u_i^{(1)}] n_j dS = 0 \qquad (3.1)$$

Similarly we can also obtain an identity for $u_i^{(1)}$ and $u_i^{(2)*}$ which is the complex conjugate of $u_i^{(2)}$.

$$\iint_{S_{BS}+S_{SW}} [\tau_{ij}^{(1)} u_i^{(2)*} - \tau_{ij}^{(2)*} u_i^{(1)}] dS = 0 \qquad (3.2)$$

Apply Green's theorem for $\phi^{(1)}$ and $\phi^{(2)}$ over V_W and using (2.5), (2.6) and (2.7) we get

$$\iint_{S_{SW}+S_\infty} [\phi^{(1)} \frac{\partial \phi^{(2)}}{\partial n} - \phi^{(2)} \frac{\partial \phi^{(1)}}{\partial n}] dS = 0 \qquad (3.3)$$

and similarly

$$\iint_{S_{SW}+S_\infty} [\phi^{(1)} \frac{\partial \phi^{(2)*}}{\partial n} - \phi^{(2)*} \frac{\partial \phi^{(1)}}{\partial n}] dS = 0 \qquad (3.4)$$

On the structure-water interface S_{SW} we may decompose $\vec{T}^{(1)}$ and $\vec{u}^{(2)}$ into normal and tangential parts and use (2.9) to rewrite

$$\tau_{ij}^{(1)} n_j u_i^{(2)} = \vec{T}^{(1)} \cdot \vec{u}^{(2)} = (\vec{T}^{(1)} \cdot \vec{n})(\vec{n} \cdot \vec{u}^{(2)}) = \tau_{ij}^{(1)} n_i n_j u_k^{(2)} n_k = -i\omega\rho \phi^{(1)} \frac{\partial \phi^{(2)}}{\partial n}$$

We may now combine (3.1) and (3.3) to get

$$\iint_{S_{BS}} [\tau_{ij}^{(1)} u_i^{(2)} - \tau_{ij}^{(2)} u_i^{(1)}] n_j dS = -i\omega\rho \iint_{S_\infty} [\phi^{(1)} \frac{\partial \phi^{(2)}}{\partial n} - \phi^{(2)} \frac{\partial \phi^{(1)}}{\partial n}] dS \qquad (3.5)$$

Similar combination of (3.2) and (3.4) gives

$$\iint_{S_{BS}} [\tau_{ij}^{(1)} u_i^{(2)*} - \tau_{ij}^{(2)*} u_i^{(1)}] dS = -i\omega\rho \iint_{S_\infty} [\phi^{(1)} \frac{\partial \phi^{(2)*}}{\partial n} - \phi^{(2)*} \frac{\partial \phi^{(1)}}{\partial n}] dS \qquad (3.6)$$

Eqs. (3.5) and (3.6) are the basis of many interesting identities when the boundary conditions on S_{BS} and S_∞ are further specified.

We note first that if $\phi^{(1)}$ and $\phi^{(2)}$ are two outgoing waves at infinity then

$$\iint_{S_\infty} [\phi^{(1)} \frac{\partial \phi^{(2)}}{\partial n} - \phi^{(2)} \frac{\partial \phi^{(1)}}{\partial n}] dS = 0 \qquad (3.7)$$

This can be easily proven by using the following representation for an outgoing wave

$$\phi = \sum_{m=0}^{\infty} \{ \sum_{n=1}^{N} (a_{mn} \cos m\theta + b_{mn} \sin m\theta) H_m^{(1)}(\sigma_n r) \cos k_n z$$

$$+ \sum_{n=N+1}^{\infty} (a_{mn} \cos m\theta + b_{mn} \sin m\theta) K_m(\gamma_n r) \cos k_n z \qquad (3.8)$$

in three dimensions, or

$$\phi = \sum_{n=1}^{N} a_n e^{i\sigma_n x} \cos k_n z + \sum_{n=N+1}^{\infty} a_n e^{-\gamma_n x} \cos k_n z \qquad (3.9)$$

in two dimensions, where

$$k_n = (n + \tfrac{1}{2})\pi/h \ , \quad \sigma_n = \sqrt{k^2 - k_n^2} \ , \quad \gamma_n = \sqrt{k_n^2 - k^2} \qquad (3.10)$$

and N is the last integer for which $k^2 > k_N$ so that the N-term series represents the propagating modes where the remaining series the evanescent modes.

It can be shown that $\{\cos k_n z\}$ forms an orthoganal set in the range $z < h$; the set $\{\cos m\theta\}$ being orthogonal over $0 < \theta < 2\pi$. For three-dimensional problems we let S_∞ be vertical cylinder of *any* radius; the integral in (3.7) becomes

$$\int_0^{2\pi} d\theta \int_0^h dz \sum_{m=0}^{\infty} \{ \sum_{n=1}^{N} \sigma_n H_m^{(1)'} \cos^2 k_n z [\cos^2 m\theta (a_{mn}^{(1)} a_{mn}^{(2)} - a_{mn}^{(2)} a_{mn}^{(1)})$$

$$+ \sin^2 m\theta (b_{mn}^{(1)} b_{mn}^{(2)} - b_{mn}^{(2)} b_{mn}^{(1)})]$$

$$+ \sum_{n=N+1}^{\infty} \gamma_n K_m' \cos^2 k_n z [\text{same}] \qquad (3.11)$$

which is clearly zero. Use has been made of the fact $\partial/\partial n = \partial/\partial r$ on S_∞. The proof of (3.7) for two dimensions is similar.

4. TWO RADIATION PROBLEMS

Let $u^{(1)} = u_i^a$ and $u^{(2)} = u_i^b$ be two normal modes a and b of oscillation due to external forcing at the base, and consider (3.5). Because both are radiation problems the associated potentials $\phi^{(1)}$ and $\phi^{(2)}$ satisfy (3.7). Making use of the boundary values (2.16) we get from (3.5)

$$\iint_{S_{BS}} [\tau_{ij}^a N_{ib} - \tau_{ij}^b N_{ia}] n_j dS = 0 \qquad (4.1)$$

By definition (2.22) it is clear that the reacting force and hence, the stiffness and damping matrices, are both symmetric, i.e.,

$$F_{ab} = F_{ba} \ , \quad K_{ab} = K_{ba} \ , \quad \lambda_{ab} = \lambda_{ba} \qquad (4.2)$$

whatever the shape of the base or of the structure.

We next prove that the damping matrix is positive-definite.

The time-averaged rate at which energy is transferred from the base through the structure and finally to the fluid is

$$\left\langle \frac{dE}{dt} \right\rangle = \frac{1}{2} \text{Re}(-i\omega \iint_{S_{BS}} \tau_{ij} u_i^* n_j dS) = -\frac{i\omega}{4} \iint_{S_{BS}} (\tau_{ij} u_i^* - \tau_{ij}^* u_j) n_j dS \qquad (4.3)$$

Substituting

$$u_i = \sum_a W_a u_i^a \quad \text{and} \quad \tau_{ij} = \sum_a W_a \tau_{ij} \qquad (4.4)$$

we get

$$\left\langle \frac{dE}{dt} \right\rangle = -\frac{i\omega}{4} \sum_a \sum_b W_a W_b^* \iint_{S_{BS}} (\tau_{ij}^a u_i^{b*} - \tau_{ij}^{b*} u_i^a) n_j dS \qquad (4.5)$$

Upon using the boundary conditions (2.16) and the reality of u_i^b and u_i^a, we get

$$\left\langle \frac{dE}{dt} \right\rangle = -\frac{i\omega}{4} \sum_a \sum_b W_a W_b^* \iint_{S_{BS}} (\tau_{ij}^a N_{ib} - \tau_{ij}^b N_{ia}) n_j dS$$

$$= -\frac{i\omega}{4} \sum_a \sum_b W_a W_b^* (F_{ba} - F_{ab}^*)$$

$$= -\frac{\omega}{2} \sum_a \sum_b W_a W_b^* \lambda_{ab} \qquad (4.6)$$

Now the second integral in (4.3) is the same as the left hand side of (3.6). The right hand side of (3.6) is, by substituting (3.8) and using orthogonality

$$\int_0^{2\pi} d\theta \int_0^h dz \sum_{m=0}^{\infty} \{ \sum_{n=1}^{N} \sigma_n (H_m^{(1)} H_m^{(1)*'} - H_m^{(1)*} H_m^{(1)'}) |a_{mn}|^2 \cos^2 k_n z$$

$$+ \sum_{n=N+1}^{\infty} \gamma_n (K_m K_m' - K_m' K_m) |a_{mn}|^2 \cos^2 k_n z \} \cos^2 m\theta \qquad (4.7)$$

In the curley bracket the second series vanishes; in the first series one may note that $H_m^{(1)*} = H_m^{(2)}$ and use the Wronkian identity

$$H_m^{(1)} H_m^{(2)'} - H_m^{(2)} H_m^{(1)'} = -4i(\pi \sigma_n r)^{-1} \qquad (4.8)$$

Therefore Eq. (4.3) becomes

$$\left\langle \frac{dE}{dt} \right\rangle = -\frac{\omega}{\pi r} \sum_{m=0} \sum_{n=0} |a_{mn}|^2 \cos^2 m\theta \cos^2 k_n z \leq 0 \qquad (4.9)$$

Equality holds when $N = 0$, i.e., when the driving frequency is below the lowest cut-off frequency, hence there is no energy radiated away from the base to infinity. Eq. (4.6) then implies that in general

$$\lambda_{ab} > 0 \quad \text{if} \quad N \geq 1; \quad = 0 \quad \text{if} \quad N = 0 \qquad (4.10)$$

In particular, the diagonal terms of λ_{ab} are individually positive-definite for $N \geq 1$.

5. TWO DIFFRACTION PROBLEMS

Let $(u_i^{(1)}, \phi^{(1)})$ and $(u_i^{(2)}, \phi^{(2)})$ be two diffraction problems.

5.a Energy Theorem

Consider (3.6). Due to (2.10) the left hand side of (3.6) vanishes, thus

$$-i\omega\rho \iint_{S_\infty} [\phi^{(1)} \frac{\partial \phi^{(2)*}}{\partial n} - \phi^{(2)*} \frac{\partial \phi^{(1)}}{\partial n}] dS = 0 \qquad (5.1)$$

As a special case we let the two states be the same, i.e., there is only one state $u^{(1)} = u^{(2)} = u^D$, $\phi^{(1)} = \phi^{(2)} = \phi^D$. Eq. (3.1) becomes

$$i \iint_{S_\infty} (\phi^D \frac{\partial \phi^{D*}}{\partial n} - \phi^{D*} \frac{\partial \phi^D}{\partial n}) dS = \text{Im} \iint_{S_\infty} \phi^D \frac{\partial \phi^{D*}}{\partial n} dS = 0 \qquad (5.2)$$

Since the fluid pressure is $-i\omega\rho\phi$ and the fluid velocity is $\partial\phi/\partial n$ the left hand side is just the averaged energy flux across S_∞. Thus (5.2) implies conservation of energy. To be more explicit we write

$$\phi^D = \phi^I + \phi^S \qquad (5.3)$$

Let us assume that the incident wave is a plane acoustic wave travelling in the θ_I direction. In general, it can contain N propagating modes. Without loss of generality one can take just the n-th mode as being representative

$$\phi^I = a_n^I \cos k_n z \exp[i\sigma_n r \cos(\theta - \theta_I)] \qquad (5.4)$$

We shall now take S_∞ to be a vertical cylinder of great radius, only the asymptotic form of (3.8) is then needed for the scattered waves on S_∞

$$\phi^S \cong \sum_{q=1}^{N} a_q^S(\theta) \cos k_q z \sqrt{\frac{2}{\pi \sigma_q r}} e^{i\sigma_q r - i\pi/4} \qquad (5.5)$$

where

$$a_q(\theta) = \sum_{m=0}^{\infty} (a_{mq} \cos m\theta + b_{mq} \sin m\theta) e^{-im\pi/2} \qquad (5.6)$$

Substituting (5.3), (5.4) and (5.5) into (5.2), we get

$$\text{Im} \int_0^{2\pi} r d\theta \int_0^h dz [a_n^I e^{i\sigma_n r \cos(\theta-\theta_I)} \cos k_n z + \sum_{q=1} a_q^S(\theta) \sqrt{\frac{2}{\pi \sigma_q r}} e^{i\sigma_q r - i\pi/4} \cos k_q z]$$

$$[-i\sigma_n \cos(\theta - \theta_I) a_n^I e^{i\sigma_n r \cos(\theta-\theta_I)} \cos k_n z$$

$$- \sum_{p=1} i\sigma_p a_p^{S*}(\theta) \sqrt{\frac{2}{\pi \sigma_p r}} e^{i\sigma_p r - i\pi/4} \cos k_p z] \qquad (5.7)$$

Expanding the integrand and using orthogonality of $\cos k_n z$, we get

$$\text{Im} \iint = \int_0^{2\pi} d\theta \{-i\sigma_n r \cos(\theta - \theta_I) - \frac{2i}{\pi} \sum_{q=0} |a_q^S(\theta)|^2$$

$$+ r a_n^I (-) i\sigma_n \sqrt{\frac{2}{\pi \sigma_n r}} [a_n^{S*} e^{i\sigma_n r[-1+\cos(\theta-\theta_I)] + i\pi/4}$$

$$+ a_n^S \cos(\theta - \theta_I) e^{-i\sigma_n r[-1+\cos(\theta-\theta_I)] - i\pi/4}]\} \qquad (5.8)$$

Since $\text{Im } if = \text{Im } if^*$, the last two terms in the integrand may be combined

$$\text{Im} \int_0^{2\pi} d\theta\, a_n^I a_n^S(-i\sigma_n r) \sqrt{\frac{2}{\pi \sigma_n r}} e^{-i\pi/4} [1 + \cos(\theta - \theta_I)] e^{i\sigma_n r[1-\cos(\theta-\theta_I)]} \qquad (5.9)$$

which may be evaluated by the method of stationary phase to give

$$-\text{Im } 2ia^I a_n^S(\theta_I) = \text{Re } 2a^I a_n^S(\theta_I) \tag{5.10}$$

The final result from (5.8) is

$$\sum_{q=0}^{N} \int_0^{2\pi} d\theta |a_q^S(\theta)|^2 = -\pi \text{ Re } a_n^I a_n^S(\theta_I) \tag{5.11}$$

Since the integrand on the left is a measure of the total scattered energy, we conclude that if the incident plane wave consists of only the n-th mode, the scattered energy in all direction is directly related to the scattered wave amplitude, also of the n-th mode, in the forward direction ($\theta = \theta_I$).

5.b Symmetry

Consider (3.5). Again using (2.10) we get

$$\iint_{S_\infty} (\phi^{(1)} \frac{\partial \phi^{(2)}}{\partial n} - \phi^{(2)} \frac{\partial \phi^{(1)}}{\partial n}) dS = 0 \tag{5.12}$$

Upon substituting (5.4) and using (3.7) for $\phi^{(1)S}$ and $\phi^{(2)S}$ we get

$$\iint_{S_\infty} [\phi^{(1)I} \frac{\partial \phi^{(2)S}}{\partial n} + \phi^{(1)S} \frac{\partial \phi^{(2)I}}{\partial n}] dS = \iint_{S_\infty} [\phi^{(2)I} \frac{\partial \phi^{(1)S}}{\partial n} + \phi^{(2)S} \frac{\partial \phi^{(1)I}}{\partial n}] dS \tag{5.13}$$

We also suppose that the two incident waves $\phi^{(1)I}$ and $\phi^{(2)I}$ consist of the mode n only and are distinguished by their incident angles $\theta^I = \theta_1, \theta_2$ respectively. Let us denote the far-field amplitudes of the scattered waves by $a_n^{(1)S}$ and $a_n^{(2)S}$. The far-field potential of the scattered wave are expressible in the form of (5.5). By an analysis similar to that leading to (5.10) it can be shown that

$$a_n^{(1)S}(\theta_2 + \pi) = a_n^{(2)S}(\theta_1 + \pi) \tag{5.14}$$

Note that in this reciprocity relation only the n-th mode (the same mode as the incident wave) is involved although the scattered waves may contain other modes. If the incident wave consists of N propagating components, the scattered wave due to each component satisfies an equation similar to (5.14).

6. A RADIATION AND A DIFFRACTION PROBLEM: A RELATION OF HASKIND-HANAOKA TYPE

Consider Equation (3.5) with $u_i^{(1)} = u_i^a$, $\phi^{(1)} = \phi^a$ and $u_i^{(2)} = u_i^D$, $\phi^{(2)} = \phi^D$. Again the incident wave in the diffraction problem involves the n-th mode only. Using the boundary condition on S_{BS} and Eq. (3.7) for ϕ^D and ϕ^a we get

$$-\iint_{S_{BS}} \tau_{ij}^D N_{ia} n_j dS = -i\omega\rho \iint_{S_\infty} (\phi^a \frac{\partial \phi^I}{\partial n} - \phi^I \frac{\partial \phi^a}{\partial n}) dS \tag{6.1}$$

But the left hand side is, by definition (2.20), F_a^D, thus the a-th generalized component of the exciting force felt at the base of the structure can be found as soon as the incident wave is prescribed and the normalized radiation problem of the a-th generalized mode is solved. Furthermore, using the asymptotic expression (5.5) for ϕ^a the right hand side can be expressed in terms of $a_n^a(\theta_I + \pi)$ after an analysis similar to that leading from (5.13) to (5.14). The result is remarkably simple

$$F_a^D = -\omega\rho h \, a_n^a(\theta_I + \pi) \tag{6.2}$$

This formula implies that if the radiated wave amplitude $a_n^a(\theta)$ is known either by computation or by measurement, the exciting force on the fixed base due to the incident wave (also of n-th mode) is immediately found from the value of a_n^a in the backward direction. It is also possible to derive a relation of Newman-type (Ref. 2,3), but will be omitted for brevity.

7. A HYBRID ELEMENT METHOD AND THE RECIPROCITY RELATIONS

While an application of the deduced identities is to provide checks for computational methods, nevertheless, these checks may be necessary but not always sufficient, in that satisfaction of them assures one the correctness but not accuracy. This property has been shown theoretically by Aranha, Mei and Yue (Ref. 4) for the hybrid element method of Chen and Mei (Ref. 5) for water waves. Now the same hybrid element method has been extended to the present problem of elastic structures in water by Mei, Foda and Tong (Ref. 6). It is therefore possible to give a similar proof here. In (Ref. 6) the method was formulated in terms of a localized variational principle. Following (Ref. 4) we first recast it in the weak form.

7.a Weak Formulation of the Hybrid Element Method

Assume that a vertical cylinder C can be drawn in the fluid so that the uneven part of the sea bottom is entirely within C. Let us denote the fluid region inside C by V_W and that region outside C by $\bar{V}_W$. Within C, the equations (2.1-4) must be satisfied in the structure and (2.5-7) by ϕ in the fluid, and (2.8) and (2.9) at the interface S_{WS}. Outside the cylinder C the fluid potential is to be denoted by $\bar{\phi}$ which must satisfy Eqs. (2.5-7) and the radiation condition. In addition, across C the following conditions of continuity must be required

$$\phi = \bar{\phi} \qquad \frac{\partial \phi}{\partial n} = \frac{\partial \bar{\phi}}{\partial n} \qquad\qquad (7.1,a,b)$$

where n points outward from the interior of the region confined by C.

The first step of the hybrid element method is to represent the fluid potential $\bar{\phi}$ in $\bar{V}_W$ analytically in the form of Eq. (3.8). The coefficients a_{mn}, b_{mn} are of course unknown. Clearly they satisfy Eqs. (2.5-7) and (2.17). We shall say a function f is in $C^\infty(\bar{V}_W)$ if f can be expanded into such a series in the domain $\bar{V}_W$.

The second step is to seek a weak formulation of the problem for ϕ and u_i defined in V_W and V_S and by (7.1,a,b). Instead of these differential equations, we require that for certain test functions $\psi^{(o)}, \psi_i^{(1)} \ldots \psi_i^{(4)}, \psi^{(5)} \ldots \psi^{(8)}$ to be further defined, the following integral vanishes,

$$I = \iiint_{V_W} (\nabla^2 \phi + k^2 \phi)\psi^{(1)} + \iiint_{V_S} (\frac{\partial \tau_{ij}}{\partial x_j} + \rho\omega^2 u_i)\psi_i^{(1)} + \iint_{S_{AS}} \tau_{ij} n_j \psi_i^{(2)}$$

$$+ \iint_{S_{WS}} (\tau_{ij} n_j - i\omega\rho\phi n_j)\psi_i^{(4)} + \iint_{S_{WS}} (\frac{\partial \phi}{\partial n} - i\omega u_i n_i)\psi^{(5)} + \iint_{S_{WB}} \frac{\partial \phi}{\partial n} \psi^{(6)}$$

$$+ \iint_C (\frac{\partial \phi}{\partial n} - \frac{\partial \bar{\phi}}{\partial n})\psi^{(7)} + \iint_C (\phi - \bar{\phi})\psi^{(8)} \qquad\qquad (7.2)$$

For brevity we have omitted the obvious symbols dV and dS. By partial integration it can be easily shown that

$$I = \iiint_{V_W} (-\nabla\phi \cdot \nabla\psi^{(o)} + k^2\phi\psi^{(o)}) + \iiint_{V_S} (-\tau_{ij}\frac{\partial \psi_i^{(1)}}{\partial x_j} + \rho\omega^2 u_i \psi_i^{(1)})$$

$$+ \iint_{S_{SA}} \tau_{ij} n_j (\psi_i^{(1)} + \psi_i^{(2)}) + \iint_{S_{BS}} \tau_{ij} n_j \psi_i^{(1)}$$

$$+ \iint_{S_{SW}} [\tau_{ij} n_j (\psi_i^{(1)} + \psi_i^{(4)}) - i\omega\rho\phi n_i \psi_i^{(4)}]$$

$$+ \iint_{S_{SW}} [\frac{\partial \phi}{\partial n} (\psi^{(o)} + \psi^{(5)}) - i\omega u_i n_i \psi^{(5)}] + \iint_C \frac{\partial \phi}{\partial n} (\psi^{(o)} + \psi^{(7)})$$

$$- \iint_C \frac{\partial \bar{\phi}}{\partial n} \psi^{(7)} + \iint_C (\phi - \bar{\phi}) \psi^{(8)} \qquad (7.3)$$

For finite element approximation we shall represent the test functions and the unknowns in Sobolev spaces of the lowest possible differentiability. Specifically, we allow $u_i \in H^2(V_s)$* where with the <u>essential</u> constraint

$$u_i = U_i \qquad \text{on } S_{BS} \qquad (7.4)$$

and $\phi \in H^1(V_W)$. To avoid difficulties in defining derivatives along the boundaries, we require that:

$$\psi_i^{(1)} + \psi_i^{(2)} = 0 \quad \text{on } S_{SA}, \quad \psi^{(o)} + \psi^{(7)} = 0 \quad \text{on } C$$

$$\psi_i^{(1)} + \psi_i^{(4)} = 0 \quad \text{and} \quad \psi^{(o)} + \psi^{(5)} = 0 \quad \text{on } S_{SW} \qquad (7.5)$$

Note that $\psi_i^{(1)}$ itself is so far unconstrained on S_{BS}. Using these relations (7.3) becomes

$$I = \iiint_{V_W} (-\nabla\phi\cdot\nabla\psi^{(o)} + k^2\phi\psi^{(o)}) + \iiint_{V_S} (-\tau_{ij} \frac{\partial \psi_i^{(1)}}{\partial x_j} + \rho\omega^2 u_j \psi_j^{(1)}) + \iint_{S_{BS}} \tau_{ij} n_j \psi_i^{(1)}$$

$$+ \iint_{S_{WS}} i\omega\rho\, n_j \psi_j^{(1)} + \iint_{S_{WS}} i\omega\, u_j n_j \psi^{(o)} + \iint_C \frac{\partial \bar{\phi}}{\partial n} \psi^{(o)} + \iint_C (\phi - \bar{\phi})\psi^{(8)} \qquad (7.6)$$

$\psi^{(8)}$ is so far arbitrary; we shall choose it to be $\psi^{(8)} \in C^\infty(\bar{V}_W)$. Thus the weak formulation may be restated as follows:

Find $u_i \in H^2(V_S)$ with the essential condition (7.4), $\phi \in H^1(V_W)$ and $\bar{\phi} \in C^\infty(\bar{V}_W)$ so that

$$\forall\, \psi^{(o)} \in H^1(V_W): \quad \iiint_{V_W}(-\nabla\phi\cdot\nabla\psi^{(o)} + k^2\phi\psi^{(o)}) + \iint_{S_{SW}} i\omega u_j n_j \psi^{(o)} + \iint_C \frac{\partial \bar{\phi}}{\partial n}\psi^o = 0$$

$$\forall\, \psi_i^{(1)} \in H^2(V_S): \quad \iiint_{V_S}(-\tau_{ij}\frac{\partial \psi_i^{(1)}}{\partial x_j} + \rho\omega^2 u_j \psi_j^{(1)}) + \iint_{S_{BS}} \tau_{ij} n_j \psi_i^{(1)} + \iint_{S_{SW}} i\omega\rho n_j \psi_j^{(1)} = 0$$

$$\forall\, \psi^{(8)} \in C^\infty(\bar{V}_W): \quad \iint_C (\phi - \bar{\phi})\psi^{(8)} = 0 \qquad (7.7.a,b,c)$$

We next prove that this weak formulation and the variational formulation of Mei et al. are equivalent. Their stationary functional J is as follows

---

*These symbols are standard in literature on finite elements: $f \in H^n(V)$ means that f is defined in V and has square-integrable n-th derivatives.

$$J = \iiint_{V_S} (W - \frac{1}{2}\rho\omega^2 u_i u_i) + \frac{\rho}{2}\iiint_{V_W} [(\nabla\phi)^2 - k^2\phi^2]$$

$$- \iint_{S_{SW}} i\omega\phi u_j n_j - \rho\iint_C (\frac{1}{2}\bar{\phi} - \phi)\frac{\partial\bar{\phi}}{\partial n} \qquad (7.8)$$

where W is the strain energy. Taking the first variation and using the fact that

$$\frac{\partial W}{\partial \varepsilon_{ij}}\delta\varepsilon_{ij} = \tau_{ij}\frac{\partial \delta u_i}{\partial x_j} \qquad (7.9)$$

we find after some arrangements that

$$\delta J = \iiint_{V_S}(\tau_{ij}\frac{\partial \delta u_i}{\partial x_j} - \rho\omega^2 u_i \delta u_i) + \iiint_{V_W}\rho(\nabla\phi\cdot\nabla\delta\phi - k^2\phi\delta\phi) - \iint_{S_{SW}} i\omega\delta\phi u_j n_j$$

$$- \iint_{S_{SW}} i\omega\phi\delta u_j n_j + \rho\iint_C \delta\phi\frac{\partial\bar{\phi}}{\partial n} - \rho\iint_C (\bar{\phi} - \phi)\frac{\partial\delta\bar{\phi}}{\partial n} - \frac{\rho}{2}\iint_C (\delta\bar{\phi}\frac{\partial\bar{\phi}}{\partial n} - \bar{\phi}\frac{\partial\delta\bar{\phi}}{\partial n}) \qquad (7.10)$$

Now the last integral vanishes by applying Green's theorem to $\delta\bar{\phi}$ and $\bar{\phi}$ for the region $\bar{V}_W$. Clearly if we identify

$$u_i = -\psi_i^{(1)}, \qquad \delta\phi = -\psi^{(o)}, \qquad \frac{\partial \delta\bar{\phi}}{\partial n} = \psi^{(8)} \qquad (7.11)$$

Eq. (7.10) and Eq. (7.6) are then identical; the equivalence is proved.

7.b Reciprocity Relations and the Weak Solution

It remains to show that the weak solution to the problem defined by (7.7a-c) satisfies the identities in §§4-6 identically. It suffices to demonstrate the proof for one of them, say Eq. (4.2).

Consider first Eq. (7.7.a). Let us choose among admissible functions $(\phi,\bar{\phi},\psi^{(o)})$ = $(\phi^a,\bar{\phi}^a,\phi^b)$, then $(\phi,\bar{\phi},\psi^{(o)}) = (\phi^b,\bar{\phi}^b,\phi^a)$, and take the difference of the resulting equations, keeping in mind that ϕ^a and ϕ^b are weak solutions. We obtain

$$i\omega\iint_{S_{SW}}(u_j^a n_j \phi^b - u_j^b n_j \phi^a) = \iint_C (\phi^a \frac{\partial\bar{\phi}^b}{\partial n} - \phi^b \frac{\partial\bar{\phi}^a}{\partial n}) \qquad (7.12)$$

where u_j^a, u_j^b are also weak solutions. Into (7.7.c) we substitute $(\phi,\bar{\phi},\psi^{(8)}) = (\phi^a,\bar{\phi}^a,\bar{\phi}^b)$ and then $(\phi^b,\bar{\phi}^b,\bar{\phi}^a)$, in turn and take the differences of the resulting equations, yielding

$$\iint_C (\phi^a - \bar{\phi}^a)\frac{\partial\bar{\phi}^{(b)}}{\partial n} = 0, \qquad \iint_C (\phi^b - \bar{\phi}^b)\frac{\partial\bar{\phi}^a}{\partial n} = 0 \qquad (7.13.a,b)$$

Finally we substitute into (7.7.b) $u_i = u_i^a$, $\psi_i^{(1)} = u_i^b$ and $\phi = \phi^a$, and then $u_i = u_i^b$, $\psi_i^{(1)} = u_i^a$ and $\phi = \phi^b$. The difference of the resulting equation is

$$\iint_{S_{BS}} \tau_{ij}^a n_j u_i^b - \iint_{S_{BS}} \tau_{ij}^b n_j u_i^a = 0 \qquad (7.14)$$

after making use of the symmetry of τ_{ij}, and Eqs. (7.12) and (7.13).

In view of the essential boundary conditions satisfied by u_i^a and u_i^b and the definition (2.22) the symmetry of the reacting forces is proved, i.e., Eq. (4.2).

Other identities can be proved in much the same way by proper choice of the solutions and the test function.

Since u_j^a and u_j^b correspond to the weak solutions (for example, by finite elements) which are in practice approximate solutions, their identical satisfaction of (7.14) cannot be used to check the discretization error. This peculiar property may be a feature of the hybrid element method in question and has been verified numerically by Aranha et al. However, it serves to caution that care is needed in claiming accuracy of other numerical methods if only these identities are used as checks.

Acknowledgement: It is a pleasure to acknowledge the financial support of the Fluid Dynamics Program, Office of Naval Research (NR 062-228) and the Earthquake Engineering Program, National Science Foundation (ENV-7710236).

REFERENCES

(1) Wehausen, J.V.: "The motion of floating bodies". Annual Review of Fluid Mechanics, 3, pp. 237-268. (1971).

(2) Newman, J.N.: "The interaction of stationary vessels with regular waves". Proc. 11th Symp. on Naval Hydrodynamics, London. (1976).

(3) Mei, C.C.: "Extensions of some identities in elastodynamics with rigid inclusions". J. Acous. Soc. Amer., 64, 5, pp. 1514-1522. (1978).

(4) Aranha, J.A., Mei, C.C. and Yue, D.K.P.: "Some theoretical properties of a hybrid-element method for water waves." Submitted for publication. (1979).

(5) Chen, H.S. and Mei, C.C.: "Oscillations and wave forces in a man-made harbor in the open sea." Proc. 11th Symp. on Naval Hydrodynamics, London. pp. 573-594. (1976).

(6) Mei, C.C., Foda, M. and Tong, P.: "Exact and hybrid-element solution for the vibration of a thin elastic structure seated on the sea floor." To appear in Applied Ocean Research. (1979).

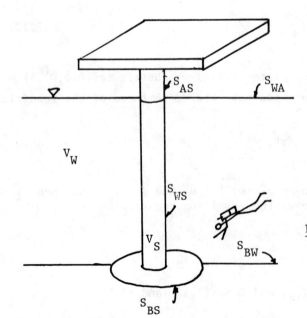

Figure 1. Definition sketch.

DYNAMIC RESPONSE OF TETHERED PRODUCTION PLATFORM IN A RANDOM SEA STATE

C. L. Kirk and E. U. Etok

Cranfield Institute of Technology, U.K.

Summary

A dynamic response spectral analysis is presented for a tethered production platform of the AKER TPP-41 type, subjected to a random sea state specified by the J.O.N.S.W.A.P. wave height spectrum.

The analysis considers uncoupled motions of the TPP in heave, surge, sway, roll and pitch for three wave directions and three depths of submersion of the platform. The results give r.m.s. amplitudes of motion and r.m.s. tensions in the mooring tethers from which the expected maximum values can be estimated. Slow drift oscillations of the TPP were also studied and found to be small compared with the direct wave induced motion and the drift due to wind and current.

NOTATION

a, b	=	spacing of corner columns
a', b'	=	effective lengths of hulls and cross braces, respectively
a_i	=	$\pi \rho C_m D_i^2 / 4$
a_x	=	$\pi \rho C_m d_h^2 e^{-kE}(\cos\alpha)/8$
a_z	=	$\pi \rho C_m D_h^2 e^{-kE}(\sin\alpha)/8$
a_1	=	$\pi \rho C_m D_h^2 e^{-kE}/4$
A	=	cross sectional area of one tether (3 cables)
A_x	=	$\pi \rho g C_m d_h^2 (\cot\alpha)/2$
A_z	=	$\pi \rho g C_m D_h^2 (\tan\alpha)/2$
A_c	=	effective base area of one corner column
b_i	=	$\pi \rho C_m g D_i^2 / 8$
C_m, C_{ma}	=	fluid inertia and added mass coefficients, respectively
C_{ac}, C_{ah}	=	fluid added mass coefficients for corner columns and hulls, respectively
$C_y, C_{\theta z}$	=	damping coefficient for heave and pitch, respectively
d_h, D_h	=	diameter of cross bracing and hulls, respectively
D_c	=	diameter of corner columns
D_e	=	equivalent diameter for non-circular hulls
D_i	=	diameter of ith column
E	=	depth of centre line of hulls below M.W.L.
E_c	=	Young's modulus of cables
F_i	=	fluid inertia force on ith column
F_c	=	total force on all columns
F_d	=	slowly varying drift forces due to waves
$F_{1,2_z}$	=	force on hulls 1 and 2 in z direction
F_z	=	total horizontal force on hulls in z direction
$F_{c_{x,z}}$	=	x and z components of forces on columns, respectively
$F_{x,z}$	=	x and z components of forces on main hulls and cross braces, respectively
$F_{xH}(k)$	=	$e^{-kE}\sin(k\beta b')\cos(k\gamma')$
$F_{zH}(k)$	=	$e^{-kE}\sin(k\beta a')\cos(k\gamma)$
F_{x,z_T}	=	total forces on columns, hulls and cross braces respectively.
F_{c_V}	=	total force due to dynamic pressure on base of corner columns
$F_{h_{1,2,3,4}}$	=	vertical forces on hulls and cross braces
F_y	=	$a_1 \dfrac{\sin(k\beta a')}{\beta}\cos(k\gamma) + a_4 \dfrac{\sin(k\beta' b')}{\beta'}\cos(k\gamma')$ = transfer function for for vertical wave force
F_g	=	centrifugal force due to angular motion θ of tethers
δF_b	=	variation of buoyancy force due to θ
g	=	acceleration due to gravity
$G_{x,z}(\omega)$	=	spectral density functions of displacements in x and z directions respectively.
$G_{c_y}(\omega)$	=	spectral density function of tether force due to heave

$G_{c_{p,r}}(\omega)$	=	spectral density functions of tether force due to pitch and roll
H, H_s	=	wave height and significant wave height respectively.
h_i	=	distance from the still water level to base of the corner columns
$H(\omega)$	=	frequency response function
I_p, I_{a_x}, I_{a_z}	=	moments of inertia of structure and fluid added mass about K.
$\overline{KG}, \overline{KB}$	=	distances from K to centres of gravity G and buoyancy B, respectively
k	=	wave number $(2\pi/L)$
K	=	T_o/L_c, effective lateral stiffness due to total cable pretension
K_c	=	AE_c/L_c, longitudinal stiffness of one tether
L, L_c	=	wave length and length of cable respectively
M_p	=	mass of TPP structure and equipment
$M_{a_{x,y,z}}$	=	fluid added mass in x,y and z directions respectively
M_{f_h}	=	mass of fluid displaced by hulls
M_y	=	$M_p + M_{a_y}$
$M_{c_{iz}}$	=	moment about K of horizontal forces on ith corner column
M_{p_z}	=	pitch moment about K of pressure forces on base of corner columns
M_{v_i}	=	pitch moments about K due to vertical inertia forces on hulls, i = 1,2,3,4
n, N	=	number of cables and waves respectively
p_i	=	dynamic pressure on base of ith corner column
P_i		(see eq.49)
$R_{x,y,z}$	=	transfer functions of wave forces in x,y, and z directions, respectively
R_p	=	transfer function of pitch moment
r	=	radius of vertical cylinder
$\bar{r}$	=	ratio (expected peak value/r.m.s. value)
$R^2(\omega)$	=	reflection coefficient for vertical cylinder
$S_H(\omega)$	=	spectral density function of wave height
S	=	distance measured along horizontal members
$S_{F_d}(\omega), S_{x_d}(\omega)$	=	spectral density functions of slow drift force and displacement, respectively.
t	=	time
T_o	=	initial pretension for all cables.
$\dot{u}, \dot{u}_n, \dot{v}$	=	horizontal, normal and vertical components of fluid acceleration, respectively.
x	=	coordinate distance measured in direction of wave propagation
x_o, y_o, z_o	=	amplitudes of surge, heave and sway, respectively
y	=	vertical distance below M.W.L.
α	=	direction of wave propagation relative to x axis (fig.1)
β, β'	=	$\cos\alpha$ and $\sin\alpha$, respectively
$\beta_x, \beta_y, \beta_p$	=	damping ratios in surge, heave and pitch respectively
γ, γ'	=	$(b\sin\alpha)$ and $(a\cos\alpha)$, respectively

ε	=	spectral width parameter
θ_x, θ_z	=	roll and pitch angles respectively
ρ	=	fluid density
σ	=	r.m.s. value
$\phi_x, \phi_y, \phi_z, \psi_y, \psi_z$	=	phase angles
ϕ	=	fluid velocity potential
ω	=	wave circular frequency
$\omega_x, \omega_z, \omega_y, \omega_p, \omega_r$	=	angular frequencies of surge, sway, heave, pitch and roll respectively
λ_b	=	$(W\overline{KG} - F_b\overline{KB})$
	=	angular stiffness due to weight and buoyancy.

INTRODUCTION

It is essential for the designer of a tension-leg platform to take into account the following dynamic factors in order to produce a reliable structure:

- overstressing and fatigue of the tether cables.
- wear of the tether attachment points.
- effect of platform motion on personnel and operations.
- excitation of marine risers due to platform motion.
- possibility of a slack tether leading to kinks and hence tether replacement as well as large snatch loads.
- influence of slow drift oscillations on riser stress.

The purpose of this paper is to present a frequency domain spectral analysis of the response of the AKER TPP-41 tethered production platform, excited by random waves specified by the J.O.N.S.W.A.P. wave height spectrum. Wave forces on the vertical columns and the hulls of the TPP are calculated by means of the inertial wave force component in Morison's formula, the drag force being considered negligible by comparison for the large diameter members considered. For the TPP-41 in 160 m water depth it is unnecessary to include drag forces due to the relative motion from the point of view of hydrodynamic damping, since the sway and surge periods of oscillation are of the order of 60 sec, with heave, pitch and roll being about 3 sec, thus direct resonance due to ocean wave excitation is extremely unlikely. For deeper waters however, the cables would be more flexible and resonance in heave, pitch and roll more likely. In that case linear hydrodynamic damping should be included. In the calculation of slow drift random oscillations however, which occur primarily at the natural period of surge, a linear hydrodynamic damping ratio of 0.03 was selected, being obtained from model tests.

The mathematical model of the TPP was chosen to have five degrees of freedom; surge, sway, heave, pitch and roll. In a time domain dynamic analysis presented by Beynet (et al), (Ref.1) in which the wave forces were modelled by means of Morison's formula, the maximum yaw motions of a TPP similar to the AKER TPP-41, were found to be extremely small and hence are not considered in the present paper. By deriving transfer functions for motion in the various degrees of freedom the power spectral density functions of motion and hence the dynamic tension in the tethers are established. Application of the results due to Ochi (Ref.2) enables the most probable extreme platform displacements and cable tension to be calculated using the r.m.s. values obtained from the spectral analysis.

The AKER TPP-41 was selected for study, being typical of various proposed designs having a rectangular deck, eight vertical columns, two or four hulls (or pontoons) and twelve tether cables (three per corner). The TPP is illustrated in Fig.1.

In the present study the following assumptions have been made:

1. Wave forces are calculated at the static equilibrium position using linear wave theory.
2. The effects of wave diffraction and sheltering in calculating wave forces are negligible.
3. Current and wave induced drag forces in Morison's equation are negligible in dynamic response calculations
4. Pre-tension in the mooring tethers remains constant.
5. Each of the five degrees of freedom can be considered uncoupled.
6. Integration of fluid inertial forces on the surface piercing columns is carried out up to the mean water level.
7. Inertia relief in pitch and roll due to horizontal motion can be superposed separately in calculating cable tension changes.

In justifying assumption No.4 it is recognised that the tether tensions will fluctuate about the total pre-tension T_o because of the vertical wave force on the TPP, variation in depth of submersion due to angular motion of the tethers about the sea bed and also because of roll and pitch motions. In the case of the TPP-41 the hull and vertical column dimensions have been selected so that the opposing vertical forces on the column bases and the hulls cancel at a wave period of T = 16 sec, which is almost coincident with the peak in the J.O.N.S.W.A.P. spectrum. In the spectral analysis of surge and sway motion it is therefore considered that tension variation due to vertical forces will be small in comparison with T_o. It is clear that a spectral analysis could not be carried out on a nonlinear or time varying system. Variations in buoyancy force due to angular tether motion were subsequently found to be small. Finally the horizontal resultant component of tension changes caused by roll/pitch motions will be zero since a tension increase on one side of the platform will be cancelled out by a corresponding decrease on the other side.

Assumption No.5 can be justified fully for heave, pitch and roll. For sway and surge however, the horizontal structural inertia forces of the platform have moments about K, the point of rotation (see Fig.1). Thus surge is coupled to pitch and sway to roll. The degree of coupling can be shown to be small for wave periods T < 25 sec and its effect has been ignored in this paper since the predominant wave periods in the J.O.N.S.W.A.P. spectrum are less than 25 sec. Uncoupled equations of motion also simplify computation and the interpretation of results. The reduction in pitch and roll moment due to horizontal motions is considered separately.

The following sections of the paper deal with the evaluation of wave forces and moments and the determination of transfer functions or response amplitude operators (R.A.O.'s) for the five degrees of freedom of the platform. As already stated the equations of angular and horizontal motion are assumed uncoupled.

SURGE AND SWAY

The inertial wave forces are calculated using Morison's equation with the horizontal fluid accelerations calculated at the centre lines of the columns and hulls. A broad crested periodic wave is assumed to propagate at an angle α to the X-X axis (see Fig.1) and the forces are evaluated taking account of the spatial coordinates of the members relative to the origin of the wave coordinate system. Wave forces are calculated on four horizontal members number 1, 2, 3, 4 and eight vertical members numbered 5, 6, ... 12.

Since all the members of the TPP are within 35 m of mean water level the horizontal fluid acceleration at depth y can be calculated from the deep water approximation

$$\dot{u} = \omega^2 \frac{H}{2} e^{-ky} \sin(kx - \omega t) \qquad \ldots (1)$$

where $k \simeq \omega^2/g$, H = wave height, $\omega = 2\pi/T$, T = wave period.

Forces on Columns

If 2a and 2b denote the spacings of the corner columns in the surge and sway directions respectively, the coordinates are given by $x_5 = b \sin\alpha - a \cos\alpha$, $x_6 = a \cos\alpha + b \sin\alpha$, $x_7 = -x_6$, $x_8 = -x_5$, $x_9 = b \sin\alpha$, $x_{10} = -x_9$, $x_{11} = -a \cos\alpha$, $x_{12} = a \cos\alpha$. $\qquad \ldots (2)$

The total force on the ith column is given by

$$F_i = a_i \int_0^{h_i} \dot{u}_i \, dy \qquad \ldots (3)$$

where $a_i = \frac{\pi}{4} \rho C_m D_i^2$, $C_m = 2.0$ (the inertial coefficient) and D_i = diameter of ith column.

Substituting from Eq. 1 in Eq. 3 and integrating yields

$$F_i = b_i H(1-e^{-kh_i}) \sin(kx_i - \omega t) \qquad \ldots (4)$$

where $b_i = \pi \rho C_m g D_i^2 / 8$. The total forces on all columns is then given by

$$F_c = H \sum_{i=5}^{12} b_i (1-e^{-kh_i}) \sin(kx_i - \omega t) \qquad \ldots (5)$$

The x and z components of F_c are then given by $F_{c_x} = F_c \cos\alpha$ and $F_{c_z} = F_c \sin\alpha$.

Forces on Hulls

To illustrate the procedure for deriving wave forces we will consider hull no.1 in Fig.2. Ignoring the intersections of the intermediate vertical columns, the effective hull lengths are approximated by $2a' = 2a - D_c$, $2b' = 2b - D_c$.

Z-Direction

For hull no.1 the normal component of fluid acceleration at point p, distance S from o', for the section o' - o'' is

$$\dot{u}_n = \omega^2 \frac{H}{2} e^{-kE} \sin\alpha \sin(kx - \omega t) \qquad \ldots (6)$$

where E = depth of hull centre line below the surface.

From geometrical considerations the coordinates S and x are related by $x = \beta S + \gamma$ where $\beta = \cos\alpha$ and $\gamma = b \sin\alpha$. Similarly for hull section o' - o''' the wave coordinate is given by $x = -\beta S + \gamma$. Substituting for x in Eq.6, using Morison's equation and integrating with respect to S over the length of the hull yields the total force in the z-direction as

$$F_{1_z} = 2a_z gH \sin(k\gamma - \omega t) \frac{\sin(k\beta a')}{\beta} \qquad \ldots (7)$$

where $a_z = \frac{\pi}{8} \rho C_m D_h^2 e^{-kE} \sin\alpha$ and D_h = diameter of hull.

For a rectangular hull of depth d and width w, the approximate diameter of a equivalent circle is given by

$$D_e = 2\sqrt{w \cdot d(1 + C_{m_a})/\pi C_m}$$

where C_{m_a} is the added mass coefficient. For a horizontal member at four diameters below M.W.L., Ref. 3 gives $C_{m_a} = 1.6$ which for $C_m = 2$, w = 9.5m, d = 12.6m gives $D_e = 14$m. In the limit when $\alpha = \pi/2$ (a beam sea), $\beta = 0$ and $\gamma = b$. Using L'Hospital's rule on eq.7 then gives

$$F_{1_z}(max) = 2a_z a' H \omega^2 \sin(kb - \omega t)$$

By symmetry the force on hull no.2 is

$$F_{2_z} = -2a_z gH \sin(k\gamma + \omega t) \frac{\sin(k\beta a')}{\beta} \qquad \ldots (8)$$

Combining Eqs. 7 and 8 gives the total force on the hulls in the z-direction as

$$F_z = -4a_z Hg \frac{\sin(k\beta a')}{\beta} \cos(k\gamma) \sin\omega t \qquad \ldots (9)$$

x-Direction

The forces in the x-direction due to hulls no.3 and no.4 are evaluated in a similar manner to the previous section. Referring to Fig.2 the S coordinate of a point on o'-o''' on hull no.4 is $x = \gamma' - \beta'S$, and for o' - o'' is $x = \gamma' + \beta'S$, where $\gamma' = a\cos\alpha$ and $\beta' = \sin\alpha$. Substituting for x in Eq.1, the normal component of fluid inertia force is obtained from Morison's equation by integration along the length of the hull. Repeating the procedure for hull no.3 and adding the results yields the total force in the x-direction as

$$F_x = -4a_x Hg \frac{\sin(k\beta'b')}{\beta'} \cos(k\gamma')\sin\omega t \qquad \ldots (10)$$

where $a_x = \frac{\pi}{8} \rho C_m d_h^2 e^{-kE} \cos\alpha$.

Equations 9 and 10 can be written in the form

$$F_z = -H A_z F_{z_H}(k) \sin\omega t \qquad \ldots (11)$$

$$F_x = -H A_x F_{x_H}(k) \sin\omega t \qquad \ldots (12)$$

where $A_z = \frac{\pi}{2}\rho g C_m D_h^2 \tan\alpha$, $A_x = \frac{\pi}{2}\rho g C_m d_h^2 \cot\alpha$, $F_{z_H}(k) = e^{-kE}\sin(k\beta a')\cos(k\gamma)$, $F_{x_H}(k) = e^{-kE}\sin(k\beta'b')\cos(k\gamma')$.

Total Force in x and z directions due to Columns and Hulls

From Eqs. 5, 11, 12 the total force in the x and z directions is found as follows,

$$F_{x_T} = H\left[\sum_{i=5}^{12} b_i(1-e^{-kh_i}) \sin(kx_i-\omega t)\cos\alpha - A_x F_{x_H}(k)\sin\omega t\right] \ldots (13)$$

$$F_{z_T} = H\left[\sum_{i=5}^{12} b_i(1-e^{-kh_i}) \sin(kx_i-\omega t)\sin\alpha - A_z F_{z_H}(k)\sin\omega t\right] \ldots (14)$$

Expanding the $\sin(kx_i-\omega t)$ term, Eqs.13 and 14 can then be written

$$F_{x_T} = H\left[\sum b_i(1-e^{-kh_i})\cos\alpha \cos\omega t \sin(kx_i) \right. \qquad \ldots (15)$$
$$\left. -\sin\omega t\left\{A_x F_{x_H}(k) + \sum b_i(1-e^{-kh_i})\cos\alpha \cos(kx_i)\right\}\right]$$

$$F_{z_T} = H\left[\sum b_i(1-e^{-kh_i})\sin\alpha \cos\omega t \sin(kx_i) \right. \qquad \ldots (16)$$
$$\left. -\sin\omega t \left\{A_z F_{z_H}(k) + \sum b_i(1-e^{-kh_i})\sin\alpha \cos(kx_i)\right\}\right]$$

Finally Eqs.15 and 16 can be expressed in the form

$$F_{x_T} = H \cdot R_x \cos(\omega t + \phi_x) \qquad \ldots (17)$$

$$F_{z_T} = H \cdot R_z \cos(\omega t + \phi_z) \qquad \ldots (18)$$

where $R_x = \left[A_x F_{x_H}(k) + \sum b_i(1-e^{-kh_i})\cos\alpha \cos(kx_i)\right]$

and $R_z = \left[A_z F_{z_H}(k) + \sum b_i(1-e^{-kh_i})\sin\alpha \cos(kx_i)\right]$

It is noted in the expressions for R_x and R_z that the $\sin(kx_i)$ terms sum to zero, hence the phase angles ϕ_x and ϕ_z are 90°

Transfer Functions of Wave Force per Unit Wave Height

From Eqs. 17 and 18 the transfer functions of wave force in the x and z directions are obtained as

$$\left|\frac{F_{xT}(k)}{H}\right| = R_x(k) \qquad \ldots (19)$$

$$\left|\frac{F_{zT}(k)}{H}\right| = R_z(k) \qquad \ldots (20)$$

Equations of Motion for Surge and Sway

For the direct wave excitation of the TPP, hydrodynamic damping forces as discussed in the introduction, can be neglected. The equations of motion in the x and z directions are then

$$(M_p + M_{a_x})\ddot{x} + Kx = F_{xT}(\omega, t) \qquad \ldots (21)$$

$$(M_p + M_{a_z})\ddot{z} + Kz = F_{zT}(\omega, t) \qquad \ldots (22)$$

where M_p = structural mass of platform and equipment, M_{a_x} and M_{a_z} = fluid added mass for columns and hulls in the x and z directions respectively, $K = T_o/L_c$ is the effective stiffness of the tethers in the horizontal plane, T_o = initial tension (platform buoyancy force - weight) in the tethers and L_c = length of tether (water depth - draft of TPP).

Fluid Added Mass

The fluid added mass for the columns and hulls is calculated from the following expressions,

$$M_{a_x} = \rho\frac{\pi}{4}\left[\left\{C_{a_c}\sum_{i=5}^{12} D_i^2 h_i\right\} + 4(C_{a_h} d_h^2 b')\right] \qquad \ldots (23)$$

$$M_{a_z} = \rho\frac{\pi}{4}\left[\left\{C_{a_c}\sum_{i=5}^{12} D_i^2 h_i\right\} + 4(C_{a_H} D_h^2 a')\right] \qquad \ldots (24)$$

where C_{a_c} = added mass coefficient for columns (usually $\simeq 1.0$)

$C_{a_{h,H}}$ = added mass coefficient for hulls corresponding to horizontal motion at depth E below surface. For square or rectangular section hulls the coefficients determined experimentally by Chung (Ref.3) were used.

Transfer Functions

From Eqs. 21 and 22 the steady rate response solutions for motion in the x and z directions are given by

$$H_x(\omega) = \frac{x_0}{F_{x_T}} = \frac{1}{M_x(\omega_x^2 - \omega^2)} \quad \ldots (25)$$

$$H_z(\omega) = \frac{z_0}{F_{z_T}} = \frac{1}{M_z(\omega_z^2 - \omega^2)} \quad \ldots (26)$$

The natural circular frequencies in the x and z directions are given by $\omega_x = \sqrt{K/M_x} = \sqrt{T_0/L_c M_x}$ and $\omega_z = \sqrt{K/M_z} = \sqrt{T_0/L_c M_z}$, where M_x, M_z are the total masses (structure and fluid) respectively. For a typical TPP, ω_x and ω_z in Eqs. 25 and 26 are usually considerably smaller than wave frequency ω and hence can be ignored.

Combining Eqs. 19, 20, 25 and 26 yields the transfer functions of displacement

$$\frac{x(\omega)}{H} = R_x(\omega)H_x(\omega) \quad \ldots (27)$$

$$\frac{z(\omega)}{H} = R_z(\omega)H_z(\omega) \quad \ldots (28)$$

Spectral Analysis

The p.s.d. functions in the x and z directions are obtained from

$$G_{x,z}(\omega) = |R_{x,z}(\omega)H_{x,z}(\omega)|^2 S_H(\omega) \quad \ldots (29)$$

where $S_H(\omega)$ denotes the wave height spectrum. The mean square responses are calculated from

$$\sigma_{x,z}^2 = \int_{\omega_1}^{\omega_2} G_{x,z}(\omega)\,d\omega \quad \ldots (30)$$

The integration limits for the J.O.N.S.W.A.P. spectrum were take as $\omega_1 = 0.2$, $\omega_2 = 1.0$ rad/sec.

HEAVE

The vertical wave forces consist of (a) the inertia force on the hulls due to vertical fluid acceleration (b) Dynamic pressure on the base of the four corner columns.

The vertical acceleration of the fluid at the hull centre lines, distance E below M.W.L. is

$$\dot{v} = -\omega^2 \frac{H}{2} e^{-kE} \cos(kx - \omega t) \quad \ldots (31)$$

Pressure Force

The dynamic pressure on the base of the ith corner column at depth h_i is

$$p_i = -\rho\frac{\partial\phi}{\partial t}\bigg|_{y=h_i} = \rho\frac{gH}{2} e^{-kh_i} \cos(kx_i - \omega t) \qquad \ldots (32)$$

Assuming that the column diameters are small compared with the wavelength, the pressure can be assumed to be constant over the area of the base, hence the total vertical force on the columns is

$$F_{cv} = \frac{\pi}{4} \sum_{i=5}^{8} p_i D_i^2 = \rho\frac{gH\pi}{8} e^{-kh_i} \sum_{i=5}^{8} D_i^2 \cos(kx_i - \omega t) \qquad \ldots (33)$$

Vertical Inertial Force on Hulls

From Fig.2 the vertical fluid acceleration at point p is

$$\dot{v} = -\omega^2 \frac{H}{2} e^{-kE} \cos[k(\beta S + \gamma) - \omega t] \qquad \ldots (34)$$

where $\gamma = b\sin\alpha$ and $\beta = \cos\alpha$

Integrating for hull no.1 gives the vertical force as

$$F_{h_1} = -a_1 Hg \frac{\sin(k\beta a')}{\beta} \cos(k\gamma - \omega t) \qquad \ldots (35)$$

Similarly for hull no.2

$$F_{h_2} = -a_1 Hg \frac{\sin(k\beta a')}{\beta} \cos(k\gamma + \omega t) \qquad \ldots (36)$$

where $a_1 = \frac{\pi}{4}\rho C_m D_h^2 e^{-kE}$

For hulls no. 3 and no.4 we obtain

$$F_{h_3} = -a_4 Hg \frac{\sin(k\beta' b')}{\beta'} \cos(k\gamma' + \omega t) \qquad \ldots (37)$$

$$F_{h_4} = -a_4 Hg \frac{\sin(k\beta' b')}{\beta'} \cos(k\gamma' - \omega t) \qquad \ldots (38)$$

where $a_4 = \frac{\pi}{4}\rho C_m d_h^2 e^{-kE}$, $\gamma' = \cos\alpha$, $\beta' = \sin\alpha$.

Summing equations 35, 36, 37 and 38 gives the total vertical force on the hulls as

$$F_{h_T} = -2H \cos\omega t \, F_y(k) \qquad \ldots (39)$$

where $F_y(k) = a_1 \frac{\sin(k\beta a')}{\beta} \cos(k\gamma) + a_4 \frac{\sin(k\beta' b')}{\beta'} \cos(k\gamma')$

Equation of Motion for Heave

The buoyancy stiffness of the column is negligible compared with the stiffness of the tethers and will be neglected. Thus the equation of motion in heave is written

$$(M_p + M_{a_y})\ddot{y} + M_{f_h} c_y \dot{y} + 4nK_c y = F_{h_t} + F_{cv} \qquad \ldots (40)$$

where c_y = damping coefficient due to the hulls
M_{a_y} = fluid added mass of hulls
$4n$ = total number of tethers (n per leg)
M_{f_h} = mass of fluid displaced by the hulls
K_c = stiffnesses of one tether = AE/L_c

Substituting Eqs. 33 and 39 into the right hand side of eq.40 gives

$$M_y \ddot{y} + M_{f_h} c_y \dot{y} + 4nK_c y = \left[\rho g \frac{H}{8} e^{-kh_i} \sum_{i=5}^{8} D_i^2 \cos(kx_i - \omega t)\right] - 2H \cdot F(k) \cos \omega t \quad \ldots (41)$$

where $M_y = M_p + M_{a_y}$

Expanding the $\cos(kx_i - \omega t)$ term and collecting terms in $\cos\omega t$ and $\sin\omega t$ Eq. 41 can be written

$$M_y \ddot{y} + M_{f_h} c_y \dot{y} + 4nK_c y = H \cdot R_y \cos(\omega t - \phi_y) = H \cdot R_y e^{-i\omega t} e^{i\phi_y} \quad \ldots (42)$$

where $R_y^2 = \left[\left\{\rho g \frac{\pi}{8} e^{-kh_i} \sum_i D_i^2 \cos(kx_i)\right\} - 2F(k)\right]^2 + \left[\rho g \frac{\pi}{8} e^{-kh_i} \sum_i D_i^2 \sin(kx_i)\right]^2$

and $\tan\phi_y = \dfrac{\sum_i D_i^2 \sin(kx_i)}{\left\{\sum_i D_i^2 \cos(kx_i)\right\} - \dfrac{16}{\rho g \pi} F(k) e^{kh_i}}$. Since the summation of $\sin(kx_i)$ is zero, $\phi_y = 0$.

Assuming a steady state solution of Eq.42 in the form $y = y_o e^{-i\omega t} e^{i\phi_y} e^{i\psi_y}$ yields

$$\frac{y_o}{H} e^{i\psi_y} = \frac{R_y}{(4nK_c - M_y \omega^2) - M_{f_h} c_y i\omega} = \frac{R_y}{4nK_c} \left[\frac{(1-\omega^2/\omega_y^2) + 2\beta_y i\omega/\omega_y}{(1-\omega^2/\omega_y^2)^2 + (2\beta_y \omega/\omega_y)^2}\right] \quad \ldots (43)$$

where $\beta_y = M_{f_h} c_y \omega_y / 8nK_c$ is the damping ratio for heave.

From Eq.43 the transfer function of tether force due to heave is

$$\frac{K_c y_o}{H} = \frac{R_y(\cos\psi_y - i\sin\psi_y)}{4n\left[(1-\omega^2/\omega_y^2)^2 + (2\beta_y \omega/\omega_y)^2\right]^{\frac{1}{2}}} = G_{cy}(\omega) \quad \ldots (44)$$

and the phase angle is given by

$$\tan\psi_y = \frac{2\beta_y \omega/\omega_y}{1-(\omega/\omega_y)^2}$$

Although included for completeness, damping in heave, pitch and roll was not included in computations.

The natural circular frequency in heave is

$$\omega_y = \sqrt{\frac{4nK_c}{M_y}} \quad \ldots (45)$$

Spectral Analysis

The mean square force per tether due to heave is given by

$$\sigma_{F_y}^2 = \int_{\omega_1}^{\omega_2} S_H(\omega) |G_{cy}(\omega)|^2 d\omega \quad \ldots (46)$$

PITCH

It is assumed that pitching motion takes place about a horizontal axis passing through K, as shown in Fig.3.

The following fluid accelerations produce moments about K:

(a) horizontal acceleration $\dot{u}$ on all vertical columns

(b) horizontal acceleration $\dot{u}$ on hulls (3) and (4) acting at height $D_h/2$ above K.

(c) vertical acceleration $\dot{v}$ on hulls 1, 2, 3, 4.

There is also the moment due to the dynamic pressure $p = -\rho \partial \phi / \partial t$ on the bases of the four corner columns. The relief in pitch moment due to horizontal motion is calculated separately and then superposed.

COLUMNS

The horizontal inertia force on the ith column is given by Eq.3 hence the moment of the column forces about K is given by

$$M_{c_{iz}} = a_i \cos\alpha \cdot \omega^2 \frac{H}{2} \sin(kx_i - \omega t) \int_0^{h_i} (h_i - y) e^{-ky} dy \qquad \ldots (47)$$

where
$$a_i = \frac{\pi}{4} \rho C_m D_i^2 \qquad \ldots (48)$$

Thus
$$M_{c_{iz}} = H \cdot P_i \sin(kx_i - \omega t) \qquad \ldots (49)$$

where
$$P_i = \frac{\pi}{8} D_i^2 \rho g C_m \left[h_i - \frac{1}{k} + \frac{e^{-kh_i}}{k} \right] \cos\alpha \qquad \ldots (50)$$

and the total moment for the columns is

$$M_{c_z} = H \sum_{i=5}^{12} P_i \sin(kx_i - \omega t) \qquad \ldots (51)$$

Hulls No.1 and No.2

Referring to Fig.2, to obtain the moment about K due to $\dot{v}$, consider for example an element dS at point p on hull no.1. Integrating the vertical fluid inertia force over the length of the hull gives the moment as

$$M_{v1} = -a_1 H \sin(k\gamma - \omega t) \left[\frac{\sin(k\beta a')}{(k\beta)^2} - \frac{a' \cos(k\beta a')}{k\beta} \right] \qquad \ldots (52)$$

where $a_1 = (\pi/4) \rho C_m D_h^2 \omega^2 e^{-kE}$, $\beta = \cos\alpha$ and $\gamma = b\sin\alpha$.

For hull no.2 the following moment is obtained

$$M_{v2} = a_1 H \sin(k\gamma + \omega t) \left[\frac{\sin(k\beta a')}{(k\beta)^2} - \frac{a' \cos(k\beta a')}{k\beta} \right] \qquad \ldots (53)$$

The pitch moment due to hulls no.1 and no.2 is therefore

$$M_{v1} + M_{v2} = 2a_1 H \left[\frac{\sin(k\beta a')}{(k\beta)^2} - a' \frac{\cos(k\beta a')}{k\beta} \right] \cos(k\gamma)\sin\omega t \quad \ldots (54)$$

This moment is a maximum when $\gamma = \alpha = 0$ and $\omega t = \pi/2$, with the wave in the antisymmetric position. As $\alpha \to \pi/2$ the moment approaches zero which can be proved by applying l'Hospital's rule to Eq.54.

Hulls no.3 and no.4

In this case the moment arm is constant and equal to 'a'. The vertical forces are given by Eqs.37 and 38 thus the pitch moment due to hulls no.3 and no.4 is

$$M_{v3} + M_{v4} = 4a_4 H g a \frac{\sin(k\beta' b')}{\beta'} \sin(k\gamma')\sin\omega t \quad \ldots (55)$$

Horizontal Inertia Force on Hulls

We now consider hulls no.3 and no.4 for which the horizontal forces are given by Eq.10. The moment arm about K is $D_h/2$ hence the moment of the horizontal forces is

$$M_{h_3} + M_{h_4} = -4a_x H g (D_h/2)\frac{\sin(k\beta' b')}{\beta'} \cos(k\gamma')\sin\omega t \quad \ldots (56)$$

Pressure Force

The moment about K due to the dynamic pressure on the base of the corner columns (5, 6, 7, 8) is

$$M_{p_z} = -\rho g \frac{H}{2} A_c e^{-kh_i} a \sum_{i=5}^{8} (-1)^i \cos(kx_i - \omega t) \quad \ldots (57)$$

where A_c = effective area of base of a column.

Total Pitch Moment

Adding moments given by Eqs.50, 54, 55, 56 and 57 the total pitch moment can be expressed as

$$M_Z = H \left[R(k)\sin\omega t + \sum_{i=5}^{12} P_i \sin(kx_i - \omega t) - \frac{\rho}{2} g A_c e^{-kh_i} a \sum_{i=5}^{8} (-1)^i \cos(kx_i - \omega t) \right] \quad \ldots (58)$$

where $R(k) = 2a_1 \left[\frac{\sin(k\beta a')}{(k\beta)^2} - a' \frac{\cos(k\beta a')}{k\beta} \cos(k\gamma) \right]$

$$+ 4g \frac{\sin(k\beta' b')}{\beta'} \left[a_4 a \sin(k\gamma') - (D_h/2) a_x \cos(k\gamma') \right]$$

Expanding the term $\cos(kx_i - \omega t)$ and collecting terms in $\cos\omega t$ and $\sin\omega t$ enables Eq.58 to be written as

$$M_Z = H . R_p \cos(\omega t - \phi_p) \quad \ldots (59)$$

where

$$R_p^2 = \left[R(k) - \sum P_i \cos(kx_i) - \frac{\rho}{2} gA_e a e^{-kh_i} \sum (-1)^i \sin(k\ x_i)\right]^2$$

and

$$\phi_p = 90°$$

Equations of Motion for Pitch

In addition to the elastic cable restoring moment about K, $M_b = -\lambda_b \theta_z$ due to buoyancy and weight, where $\lambda_b = (W.\overline{KG} - F_b.\overline{KB})$. The equation of motion, neglecting the moment $M_x \overline{KG}.\ddot{X}$ due to surge motion, is

$$(I_p + I_{a_z})\ddot{\theta}_z + M_{f_h} C_{\theta_z} \dot{\theta}_z + 4(nK_c a^2 - \lambda_b/4)\theta_z = H.R_p e^{-i\omega t} e^{i\phi_p} \quad \ldots (60)$$

The transfer function of pitch is obtained from the steady state solution to Eq.60 as

$$\frac{\theta_{z_0}}{H} = \frac{R_p(\cos\psi_z - i\sin\psi_z)}{4(nK_c a^2 - \lambda_b/4)\left[(1-\omega^2/\omega_p^2)^2 + (2\beta_p \omega/\omega_p)^2\right]^{\frac{1}{2}}} \quad \ldots (61)$$

where $\tan\psi_z = \dfrac{2\beta_p \omega/\omega_p}{1-(\omega/\omega_p)^2}$ and $\beta_p = \dfrac{C_{\theta_z} M_{f_h} \omega_p}{8(nK_c a^2 - \lambda_b/4)}$ is the damping ratio.

The natural circular frequency of pitch is given by

$$\omega_p = \sqrt{\frac{4(nK_c a^2 - \lambda_b/4)}{I_p + I_{a_z}}} \quad \ldots (62)$$

and the transfer function for tether force due to pitch is

$$G_{C_p}(\omega) = \frac{K_c \theta_{z_0} a}{H} \quad \ldots (63)$$

The mean square tether force due to pitch is found in the same manner as for heave, eq.46.

ROLL

The derivation of the equation of motion for roll about the x-axis is identical to that described for pitch, except that $\cos\alpha$ is replaced by $\sin\alpha$ and the hull dimensions are interchanged. The transfer function for tether force due to roll is denoted by $G_{C_r}(\omega)$. The natural circular frequency of roll is denoted by ω_r and is calculated from Eq.62, using I_{a_x} (moment of inertia of added mass for roll) and replacing 'a' by 'b'. For the AKER TPP-41, a = b

Spectral Analysis of Resultant Tether Force due to Heave, Pitch and Roll

The resultant mean square tether force variation due to pitch and roll is calculated from the expression

$$\sigma^2 F_{p,r} = \int_{\omega_1}^{\omega_2} S_H(\omega) \cdot G_{c_{r,p}}(\omega) \, d\omega \quad \ldots (64)$$

The total mean square force variation per tether is then obtained by adding Eq.46 and Eq.64 to give the r.m.s. force as

$$\sigma_F = \left[\sigma^2_{F_p} + \sigma^2_{F_r} + \sigma^2_{F_y} \right]^{\frac{1}{2}} \quad \ldots (65)$$

EXTREME VALUE OF TETHER FORCE VARIATION

It has been shown by Ochi (Ref.2) that for a random process with a spectral width parameter $\varepsilon < 0.9$, the most probable extreme value to occur during N observations is insensitive to ε. From Fig.3 of Ref. 2 the ratio of the extreme value to the r.m.s. during N observations, for $\varepsilon = 0$ and a probability of exceedance of 0.01 can be expressed approximately as

$$\bar{r} = \frac{F_{max}}{\sigma_F} = 3.46 + 0.444 \log_{10} N \quad \ldots (66)$$

For a storm of 12 hours duration, using a peak wave spectrum period of 17 sec, the average number of waves would be approximately N = 2540. Eq.66 then yields a ratio of 4.97 therefore a value of $\bar{r} = 5$ is used. The total tension per tether is the sum or difference of the initial tension (T_0/n) and the extreme value of the dynamic tension variation, $5\sigma_F$. The designer is concerned with (a) overstressing of a tether, and (b) loss of tension leading to snatch loads. For (a) to be avoided it is essential for $T_y > (T_0/n) + 5\sigma_F$, where T_y denotes the specified yield force per tether. To prevent (b) the initial tension must be high enough so that $(T_0/n) > 5\sigma_F$. It can be seen that the designer is therefore restrained between two conflicting requirements, since an increase in pre-tension to avoid (b) may lead to (a).

SLOW DRIFT OSCILLATIONS

Slow drift oscillations of surface piercing cylinders arise as a result of 2nd order terms in Bernoulli's equation. Havelock (Ref. 4) has shown that the slow drift force for regular waves of height H, on a cylinder of radius r is

$$F_d = \rho g \frac{r}{8} H^2 R^2(\omega) \quad \ldots (67)$$

where $R^2(\omega)$ is a reflection coefficient given by $R^2(\omega) = 1.3(r\omega^2/g)$ for $r\omega^2/g \leq 1$, or has a value of 1.3 for $r\omega^2/g > 1$.

The spectrum of slow drift force for a random sea state can be derived by use of Eq.67 and by reference to the analysis of noise through square law devices due to Rice (Ref. 5, pp.263/65). It has also been given by Pinkster (Ref. 6). The slow drift force is caused by cross-modulation effects between wave components in the wave spectrum, and its spectrum extends down to very low frequencies. Thus the possibility of 'resonant' random oscillations at the low frequency of surge motion exists. The spectrum of slow drift motion is given by

$$S_{F_d}(\omega_k) = \frac{(\rho g)^2 r^2}{2} \int_0^\infty S_H(\omega) S_H(\omega+\omega_k) R^4(\omega + \frac{\omega_k}{2}) d\omega \quad \ldots (68)$$

For slow drift oscillations it is essential to include hydrodynamic damping in the surge/sway transfer function. Thus the spectrum of slow drift motion is given by

$$S_{x_d}(\omega) = |H(\omega)|^2 S_{F_d}(\omega) \quad \ldots (69)$$

where
$$|H(\omega)|^2 = \frac{1}{M_x^2\left[(\omega^2-\omega_x^2)^2 + (2\beta_x\omega\omega_x)^2\right]} \qquad \ldots (70)$$

The mean drift displacement is calculated for the four corner columns as

$$\bar{x} = \frac{\bar{F}_d}{K} = \frac{\rho g r}{(T_0/L_c)} \int_0^\infty S_H(\omega) R^2(\omega) d\omega \qquad \ldots (71)$$

The r.m.s. slow drift force is obtained by integrating Eq.68 between limits $0 \leq \omega \leq \omega_c$, where ω_c denotes the frequency range of the wave spectrum. Since $H(\omega)$ is narrow banded it is found that the spectrum of slow drift is also narrow banded about the natural surge frequency ω_x.

DATA FOR EVALUATION OF PLATFORM MOTIONS

Platform data

The following data were used for the AKER TPP-41: Buoyancy = 44,527t for h_i = 35m Mass of deck equipment = 18,000t, mass of one corner column = 2300t, mass of one main hull + ballast = 2000t, total mass of TPP in air = 31,200t structural and fluid added mass in heave = 56,000t, structural and fluid added mass in surge = 64,100t, structural and fluid added mass in sway = 82,700t, structural and fluid added mass moments of inertia in roll and pitch = 1.49 x 10^8 and 9.68 x 10^7 tm^2 respectively, diameter of corner columns = 16m, spacing of corner columns = 70m, length of hulls (1,2) = 54m, depth x width of rectangular hulls (1,2) = 13 x 9.5m, diameter of hulls (or cross braces, 3,4) = 6m, diameter of small columns = 3.5m, height of platform centre of gravity above base = 41.7m. Natural periods for h_i = 35m; pitch and roll = 4s, heave = 3.7s, sway = 63s, surge = 55s.

Tether data

The platform is moored by 12 steel cables distributed between four legs: Number of wires per cable = 400, diameter of wire = 7 mm, area of wire per leg = 46,180 mm^2, Young's modulus = 2 x 10^{11} N/m^2, cable lengths = 125,130,135m corresponding to drafts of 35,30 and 25m respectively, stiffnesses per leg (3 cables) for 125 m length = 7.6 x 10^7 N/m (heave), 2.05 x 10^5 N/m (sway and surge), 9.3 x 10^{10} Nm/rad (pitch and roll), $T_0/4$ = pre-tension per leg = 2560t for 35 m draft.

Environmental data

The random sea state was described by the J.O.N.S.W.A.P. wave height spectrum for a significant wave heights 4,6,15 m. The surface current was taken as 1.34 m/sec, assumed constant over the depth of the platform and was used to calculate the current induced drift. The wind drift force on the platform was calculated using a sustained maximum wind speed of 56 m/sec. A water depth of 160 m was taken.

DISCUSSION OF RESULTS OF SPECTRAL ANALYSIS

Surge and Sway

The response amplitude operators for surge ($\alpha = 0°$) and sway ($\alpha = 90°$) shown in Fig.4 indicate that for T ≥ 10s an increase in the draft leads to higher amplitudes. This implies that the increase in column forces at maximum draft outweighs the decrease in hull forces. For h_i = 25 m, T = 17s, the sway response is about 20% greater than surge. It should be noted that zero amplitude, and hence zero resultant wave force, occurs when T = 9.5, 5.5, 4.2 sec corresponding to wavelengths of L = 140, 47, 28 m respectively. This is due to force cancellation on the columns and hulls.

For T > 17s the amplitude of response increases almost linearly with wave period and for the 'design' wave H = 30 m, T = 17s is equal to 12.6 m. This value will be compared with the maximum expected amplitude obtained from the spectral analysis.

Table No.1 gives the r.m.s. amplitudes of motion for $\alpha = 0°$, $45°$ and $90°$.

Draft (m)	$\alpha = 0°$ Surge (m)	$\alpha = 90°$ Sway (m)	$\alpha = 45°$ Surge (m)	$\alpha = 45°$ Sway (m)
25	2.32	2.67	1.65	1.92
30	2.45	2.68	1.74	1.92
35	2.54	2.68	1.81	1.92

Table 1

r.m.s. Amplitudes of Surge/Sway Motion

The maximum r.m.s. amplitude of 2.68 m occurs for a beam sea and is independent of draft. However for a quartering sea ($\alpha = 45°$) mean square superposition of surge and sway gives an r.m.s. amplitude of 2.64 m, which is almost as great as for the beam sea. The maximum expected amplitude, using Ochi's criterion is 13.4 m which is equal to the amplitude produced by a regular wave train of height H = 30 m and period T = 18s. The corresponding experimental amplitude given in Ref. 7 is 13.5m.

Surge and Sway Acceleration

The surge and sway response spectra were found to be narrow banded with a peak frequency equal to the peak in $S_H(\omega)$, i.e. $\omega = 0.37$ rad/s. The maximum expected peak acceleration of the platform can then be calculated as $\ddot{x} = \omega^2 a/g$. Taking 'a' as 13.4 m, this gives $\ddot{x} = 0.186$ g. Fig. 4 shows a local maximum response at T = 6.7s (L = 70 m). If we consider a regular wave of height H = 3.5 m (H/L = 1/20) the amplitude of response would be 0.6 m giving a corresponding platform acceleration of 0.05g which is about 1/3 of the expected peak acceleration. Thus it can be seen that in low sea states, although the platform displacement is small, the accelerations can be significant from the point of view of personnel.

Heave

The R.A.O. for heave is given in Fig.5. For T = 16s the response is zero indicating that the column and hull dimensions have been selected so that the hull inertial forces cancel the dynamic pressure forces on the base of the columns. A very large peak response occurs at T = 9.5s (L = 140 m) in which case the nodes of the wave coincide with the columns, giving zero pressure force while the inertia force on the hulls is a maximum. Thus a wave spectrum having a peak at T = 9.5s (ω_p = 0.66 rad/sec) with H_s = 4.6 m was used to determine the heave forces in the tethers to compare with the forces for the fully developed sea state (H_s = 15 m). For a draft of 35 m and $\alpha = 0°$ the r.m.s. stretch of the tethers due to heave was 0.0227 m for H_s = 15 m and 0.012 m for H_s = 4.6 m. Hence it can be seen that the tether stretch corresponding to the lower H_s is 55% of that due to the larger H_s because of the 'tuning' of the wave height spectrum to the peak in the heave R.A.O. It is thus clear that in carrying out a random fatigue analysis of the tethers, a range of wave height spectra should be used, each having a different H_s corresponding to various stages of development of a sea.

Roll and Pitch

The angles in figures 6 and 7 neglect the inertia relief due to horizontal motion which is calculated from

$$\theta_r = (M_p \overline{KG} + M_a \overline{KA}) \ddot{\xi}_{x,z} / (I_p + I_{a_{x,z}})(\omega_n^2 - \omega^2)$$

where $\overline{KA}$ = height of centre of added mass above K, ω_n = natural frequencies in roll or pitch and $\ddot{\xi}$ = surge or sway acceleration.

Combined Heave, Roll and Pitch

To obtain the R.A.O. for tether tension, the R.A.O.'s for heave, roll and pitch were combined, taking account of the phase differences, as shown in fig. 8, for $\alpha = 0°$, $45°$, $90°$ and $h_i = 35$ m. The p.s.d. functions for tension were calculated for two different sea states showing that the spectrum for $H_s = 15$ m is less narrow banded than for $H_s = 4.6$ m, indicating that the fully developed sea state will yield the greatest number of maxima/sec for tether tension.

The r.m.s. tensions per tether (3 cables) for α = 0, 45 and 90 degrees and $h_i = 35$ m are given for the two postulated sea states in table no.2 where the inertia relief moment due to horizontal motion has been included.

α (deg)	r.m.s. tension per tether (tonne) H_s (m)	
	4.6	15.0
0	218	183
45	188	328
90	304	253

Table 2 - r.m.s. Tether Tension

The previous linear dynamic analysis tacitly assumes that there are no discontinuities in pitch and roll stiffness due to cables becoming slack. If however the roll or pitch moment gives rise to slack tethers for part of a wave cycle, steady state response analysis is inapplicable and numerical integration must be applied to the resulting non-linear system. It is clear that since the buoyancy forces are much smaller than the cable elastic forces, the angular motion of the TPP will be influenced by alternate stiff and soft restoring moments when cables become slack.

While the linear constant cable stiffness analysis is valid for computing r.m.s. motions which do not lead to slack cables, it is not valid if the expected maximum roll motion gives rise to slack cables. If Ochi's factor of 5 x r.m.s. is taken as the maximum expected value of tension, a value of 1640 tonne is obtained which being 0.65 x initial tension implies taut tethers. This peak tension is comparable to the value of 1500 tonne given in Ref. 7. The r.m.s. tensions for both sea states are similar due to 'tuning' of the peak in the lower wave spectrum with peaks in the R.A.O. for tension.

For comparison the tether tension was obtained from figure 8 with $\alpha = 45°$ for a regular wave H = 30 m, T = 17 sec yielding 1324 tonne, in which case the tethers would remain taut. Comparing this value with the r.m.s. tension of 328 tonne indicates that a factor of about 4 x r.m.s. would be more realistic for design purposes. Assuming a minimum cable tensile strength of 1600×10^6 N/m^2 the maximum permitted force in a tether is 7387 tonne, hence Ochi's factor of 5 implies a reserve factor of 4.5, whereas the design wave approach gives a reserve factor of 5.58. For $\alpha = 45°$, Ref.7 gives a tension variation of 1500 tonne for H=30m, T=17 sec.

Centrifugal Force Effect

If the tethers are again assumed to be inextensible there will be a centrifugal force due to platform circular motion, arising from the angular motion of the tethers. For a periodic oscillation of the tethers of amplitude $\theta = \theta_o \sin\omega t$ the maximum centrifugal force is given by

$$F_g = M_p \dot{\theta}^2 (L_c + \bar{h}) = M_p \frac{(L_c + \bar{h})}{L_c^2} \omega^2 x^2 \quad \ldots (72)$$

where $L_c + \bar{h}$ is the height of the platform centre of gravity above the point of cable attachment at the sea bed. For a wave period of 17 sec ($\omega = 0.37$) and $x = 10$ m, $M_p = 31,200$ t, $L_c = 125$ m, $\bar{h} = 40$ m, Eq.72 gives $F_g = 450$ t or 37 t per cable. The maximum centrifugal force in the tethers which is small, occurs at the mid position of the horizontal motion and is therefore 90° out of phase with the buoyancy induced force, which is a maximum at the extremities of the horizontal motion.

Tether force due to variation of buoyancy

If the tethers are assumed to be inextensible then horizontal motion of the TPP will lead to draw-down and hence an increase in buoyancy force per tether given by

$$\delta F_b = \frac{\pi}{4} \rho g D_i^2 L_c (1-\cos\theta) \qquad \ldots (73)$$

where θ = angle or rotation of tethers corresponding to a platform horizontal displacement x. For small θ we can write $\cos\theta = 1-\theta^2/2$, $\theta = x/L_c$, whence Eq.73 becomes

$$\delta F_b = (\pi \rho g D_i^2 / 8 L_c) x^2 \qquad \ldots (74)$$

For a periodic wave the direct forced response of the TPP will give rise to tension variations in the tethers occurring at twice wave frequency, which should be considered in tether fatigue analysis.

The displacement x is composed of (1) directly excited motion due to wave inertia forces (2) slow drift motion + mean drift (3) static offset due to wind and current.

Numerical analysis shows that the maximum direct wave excited r.m.s. motion, $\sigma_{x_1} = 2.68$ m and the mean drift, $\bar{x} = 1.0$ m. The slow drift r.m.s. motion is $\sigma_{x_2} = 1.10$ m. Approximate calculations for wind and current drift were made for a surface current of 1.34 m/sec and a maximum sustained wind speed of 56 m/sec, giving an offset of 12 m. The r.m.s. resultant of $(\sigma_{x_1}^2 + \sigma_{x_2}^2)$ is 2.89 m.

The most severe case of offset will occur when the maximum expected value of the fluctuating motion is added to the sum of wind drift, current drift and mean slow drift, assuming that they all act simultaneously in the same direction. This leads to a resultant offset of approximately 28 m. The draw-down for this offset is 3.136 m and the increase in tension obtained from Eq.74 is 606 t per tether. The elastic extension of the tether under this load is 0.08 m which is negligible compared with $(x^2/2L_c)$, thus justifying the use of Eq.73. The expected maximum increase in leg tension is 25% of the 2560 t pre-tension. However in the spectral analysis of surge and sway, 25% increase in tension leads to an 11% increase in ω_x and ω_z in Eqs. 25 and 26, which will have little influence on the motion.

Effect of Inertia Relief on Tether Tension

The reduction of angular motion of the TPP due to the moment about K caused by horizontal motion is a major factor in the evaluation of tether tensions. For example if inertia relief is ignored tether tensions can be overestimated by as much as 100 percent As shown previously the inertia relief moment is proportional to $(M_p \overline{KG} + M_a \overline{KA})$, thus in principle it would be possible to arrange $\overline{KG}$ and $\overline{KA}$ so as to minimize the sum of the hydrodynamic and inertia force moments, and therefore the tension fluctuations. In practice this procedure might be difficult because of design constraints and because it would have to be carried out for a range of sea states.

CONCLUSIONS

1. A frequency domain analysis has been described for determining the sway and surge motions of a tethered production platform with parallel taut mooring cables and the loads in the tethers produced by heave forces, as well as roll and pitch moments on the TPP.

2. The greatest r.m.s. horizontal motion was found to occur for a beam sea being slightly greater than for a quartering sea.

3. Calculation of extreme displacements based on Ochi's factor of 5 r.m.s. predicts a peak sway motion of 13.4 m which is 8% of water depth. The corresponding expected peak lateral acceleration of the platform was 0.186 g.

4. Because of the coincidence of the wave spectrum peak at H_s = 4.5m with peaks in the R.A.O. for tension, the r.m.s. tensions were found to be comparable with values obtained for H_s = 15m. It is therefore recommended in design procedures to study tether tensions for a variety of sea states.

5. Application of Ochi's factor to tether tensions implies that the expected peak value of tension for H_s = 15 m is of the order of 60% of the pre-tension. This indicates that there is no possibility of the tethers becoming slack or the peak value exceeding the maximum permitted force.

6. The spectral density functions of tether tensions for H_s = 15 m and 4.6 m were narrow banded, the band width being greater for H_s = 15 m.

REFERENCES

1. Beynet, P.A., Berman, M.Y. and von Aschwege, J.T. "Motion, Fatigue and the Reliability Characteristics of a Vertically Moored Platform". Offshore Technology Conference, 10th Annual Proc. Houston, Texas, May 8-10, 1978, Paper OTC3304.

2. Ochi, M.K. "On Predictions of Extreme Values". Journal of Ship Research, March 1973, Vol.17, No.1, pp.23-37.

3. Chung, J.S. "Force on Submerged Cylinders Oscillating Near a Free Surface". Journal of Hydronautics, Vol.11, No.3, July 1977, 100, pp.100-106.

4. Havelock. T.H. "The Pressure of Water Waves Upon a Fixed Obstacle". Proc. Roy. Soc. Vol. 175A (July 1940), pp.409-421.

5. Rice, S.O. "Selected Papers on Noise and Stochastic Processes", Part IV, pp.263-265. Dover Publications (1954).

6. Pinkster, J.A. "Low Frequency Phenomena Associated with Vessels Moored at Sea", Society of Petroleum Engineers of AIME, paper presented at Spring meeting, Amsterdam, 1974, SPE paper, No.4837, pp.487-494.

7. Rowe, S.J., Fletcher, R.H. and Hedley, C. "The Model Testing of a Tethered Buoyant Platform and Its Riser System". Proceedings of a Seminar on Models and Their Use as Design Aids in Offshore Operations. Society for Underwater Technology, London, 4th May, 1978, paper no.5.

ACKNOWLEDGEMENT

This work was supported by S.R.C. Grant GR/A/04590.

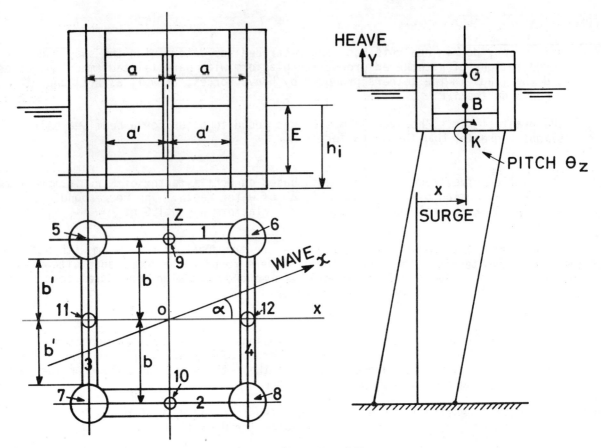

Fig. 1 Schematic of TPP.

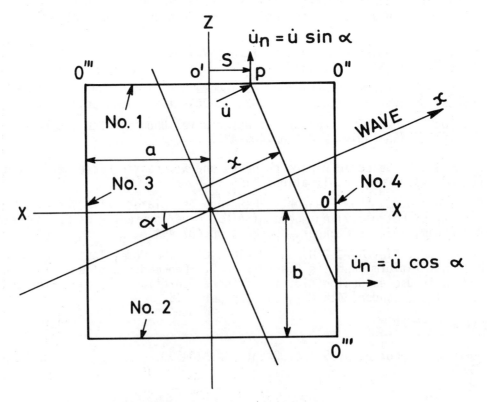

Fig. 2 Schematic of hull geometry.

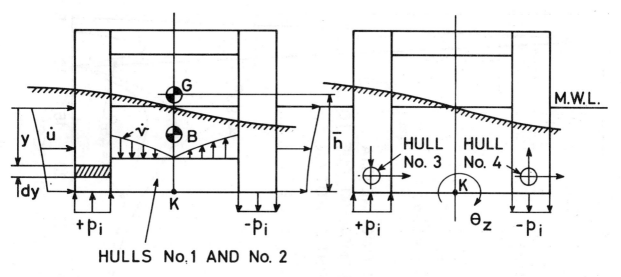

Fig. 3 Wave forces for pitch moment on TPP.

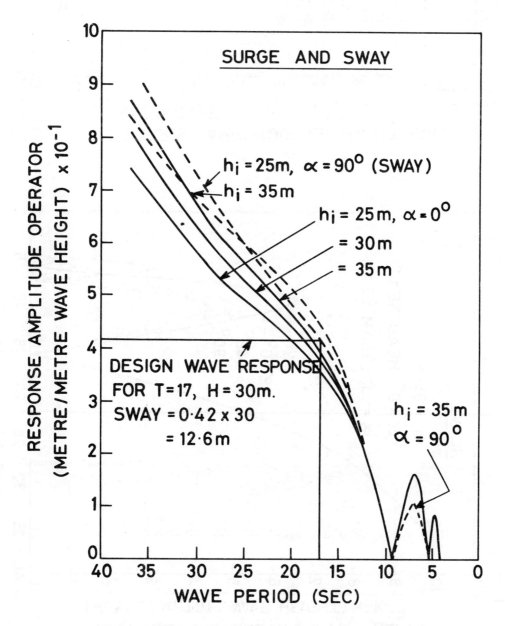

Fig. 4 R.A.O. for surge and sway.

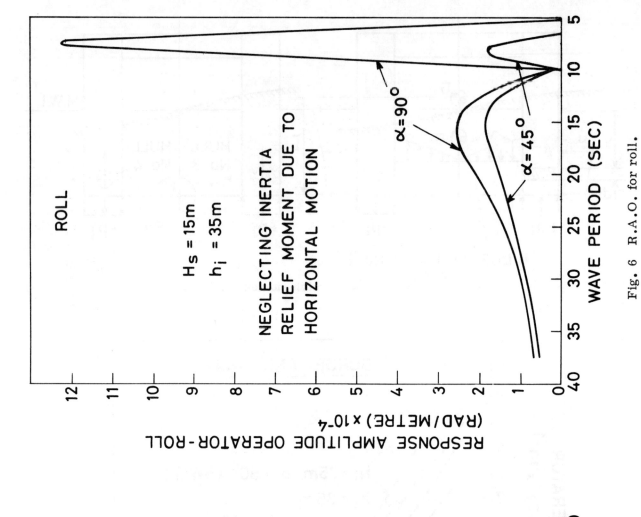

Fig. 5 R.A.O. for heave.

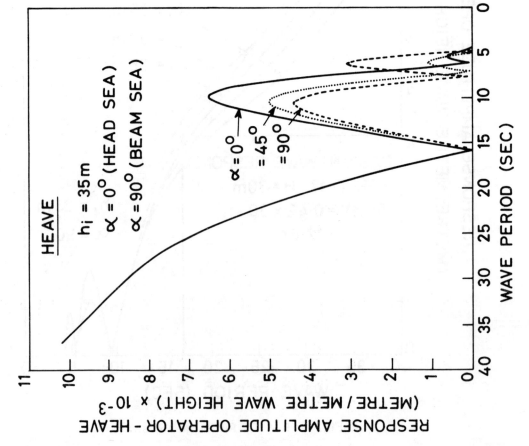

Fig. 6 R.A.O. for roll.

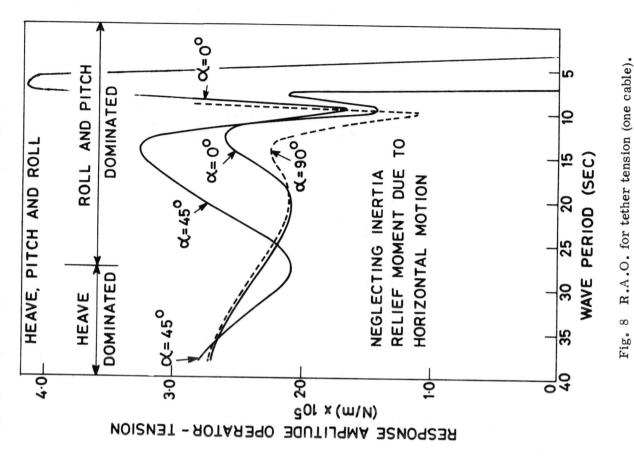

Fig. 8 R.A.O. for tether tension (one cable).

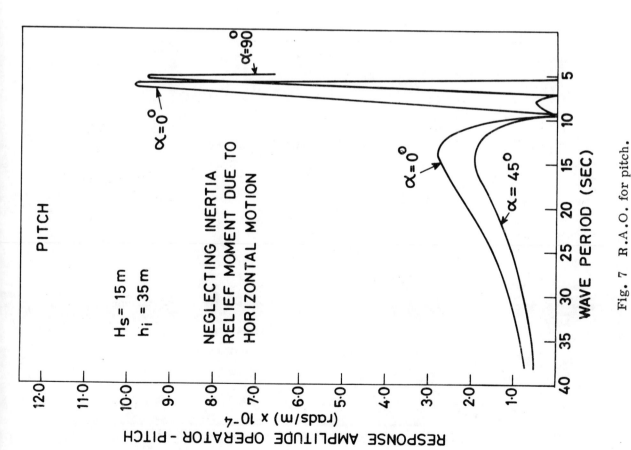

Fig. 7 R.A.O. for pitch.

163

BOSS'79

Second International Conference on Behaviour of Off-Shore Structures

Held at: Imperial College, London, England
28 to 31 August 1979

PAPER 58

HYDRODYNAMIC MODEL FOR A DYNAMICALLY POSITIONED VESSEL

Michael S. Triantafyllou, S.M., S.M., Ph.D.

Massachusetts Institute of Technology, U.S.A.

Summary

A formulation of the problem of drift motions is given by considering two different time scales. It is shown that no change is brought upon the existing theory of ship motions. The procedure is mathematically consistent while it provides a simple way to derive the equations of motion. The method has been applied in the case of a drilling vessel and some simulation results have been derived.

INTRODUCTION

A drilling vessel is subject to wave, wind and current forces. The wind and current forces are important at frequencies typically below 0.1 rad/sec, while the wave forces cover a large frequency range. Assuming small amplitude waves we can perform a perturbation analysis. It is shown in the next sections how the introduction of two time scales leads to a consistent theory of drift motions with no need to modify the existing theory of predicting the second order wave forces.

The present method can be used to derive the equations of drift motion from the non-linear equations of motion in a systematic way, and to obtain information on the frequency range of the motions that can be used for the design of the controller.

The method has been applied for the design of a specific vessel and the controller designed on such information has given very good simulation results.

FORMULATION OF THE PROBLEM

Assume the vessel subject to irregular unidirectional seas at angle θ_1 and current at θ_2 to the x axis of the vessel. The xz plane is the symmetry plane of the vessel, hence $Ixy = Iyz = 0$. Assume $Y_G = 0$, $Z_G = 0$, while $X_G \neq 0$ in general.

We know that pitch, roll and heave motions cause large restoring forces, so that they have a zero mean over a large period of time. The surge, sway, yaw motions cause no restoring forces so the uncontrolled ship is unstable in these motions, i.e. they can become arbitrarily large.

In this analysis, we will deal with the problem of a vessel which is restrained from steady drift (by mooring or dynamic positioning), but free to oscillate in the whole frequency range.

We will use three reference systems (see Fig. 1):

x_0, y_0, z_0 fixed on the undisturbed free surface

x, y, z fixed on the mean position of the ship

x_1, y_1, z_1 fixed on the ship

The problem will be formulated within potential theory.

FORCES ON THE VESSEL

The forces are due to wind, current and waves. The wind forces on a body are caused by the formation of a wake behind the body due to the separation of the boundary layer. The wavelength of the wind components should be therefore long enough to cover the whole body and form a substantial wake in order to produce a significant force. The vessel is a large body so the wavelength should be of the order of 30 m and above; this means that the wind frequencies are small (usually below 0.1 rad/sec)

The current is a quasi-steady phenomenon, so that the current forces change with periods of the order of a few hours.

The wave forces contain a wide range of frequencies due to the non-linear terms and boundary conditions involved which cause subharmonics and higher harmonics with respect to the wave frequencies.

The wave force on a body can be expressed as:

$$\bar{F} = \iint_{S_B} p \, \bar{n} \, ds \qquad (1)$$

where S_B is the wetted surface of the body, $\bar{n}$ is the local normal vector on the surface element ds; and p is the pressure.

Assuming the amplitude a of the incident wave small with respect to the wavelength λ, i.e. $a/\lambda = \varepsilon \ll \omega$ we obtain the incident wave potential ϕ_I satisfying the linearized free-surface boundary condition and which is of order ε [Wehausen and Laitone, Ref. (5)].

The disturbance caused by the body is expressed by:

(a) The diffraction potential due to the mere presence of the body and

(b) The radiation potentials caused by the six degree of freedom motions of the body.

By a perturbation analysis and considering that the body is moving around a steady mean position, we find that both diffraction and radiation potentials are of order ε

Returning to the expression for the force, we see that the pressure p, after subtracting the hydrostatic term is of order ε as shown by Bernoulli's equation:

$$p = -\rho \frac{\partial \phi}{\partial t} - \frac{\rho}{2}|V|^2 - \rho g(z-z_0) \qquad (2)$$

where V the speed of the fluid particle.

The wetted surface S_B is time dependent and its position can be uniquely determined by knowing the six motions of the body and the surface elevation. Both the motions and the elevation are of order ε, so the integral (1) can be expanded in a series of terms of orders $\varepsilon^0, \varepsilon^1, \varepsilon^2 \ldots$, so that the force $\bar{F}$ apart from a hydrostatic term of order 1 will consist of terms of order $\varepsilon^1, \varepsilon^2 \ldots$

DRIFT MOTIONS

If the support is not rigid (for example, using mooring, or thrusters), then the approach above must be modified to account for the drift motions and their contribution to the hydrodynamic forces.

It is well known that in the case of periodic motions, the unmodified perturbation method fails [Lin and Segel, Ref. (6)]. Poincaré developed a modified perturbation theory by expanding both the independent and the dependent variables in a series of the small parameter ε. A similar method is the multiple scale expansion: the dependent variable is assumed to be a function of distinct time scales $t, \varepsilon t, \varepsilon^2 t, \ldots$ with ε the small parameter.

A similar procedure will be used here: we will assume that the motions depend on two different time scales, t and εt.

The motions with restoring forces (heave, roll, pitch) will be of order ε, while the "softly" supported motions (surge, sway, yaw) will be of order ε^0:

$$\xi_j(t,\varepsilon) = \begin{cases} f_j(t,\varepsilon t,\varepsilon) = f_j^0(t,\varepsilon t) + \varepsilon f_j^1(t,\varepsilon t) + \ldots & j = 1,2,6 \\ f_j(t,\varepsilon t,\varepsilon) = \varepsilon f_j^1(t,\varepsilon t) + \ldots & j = 3,4,5 \end{cases} \qquad (3)$$

so that

$$\frac{\partial \xi_j}{\partial t} = \begin{cases} \dfrac{\partial f_j^0}{\partial t} + \varepsilon \left[\dfrac{\partial f_j^0}{\partial \tau} + \dfrac{\partial f_j^1}{\partial t} \right] + \varepsilon^2 \left[\dfrac{\partial f_j^1}{\partial \tau} + \dfrac{\partial f_j^2}{\partial t} \right] + \cdots \\ \qquad\qquad\qquad\qquad\qquad\qquad\qquad\qquad j = 1,2,6 \\ \varepsilon \dfrac{\partial f_j^1}{\partial t} + \varepsilon^2 \left[\dfrac{\partial f_j^1}{\partial \tau} + \dfrac{\partial f_j^2}{\partial t} \right] + \cdots \\ \qquad\qquad\qquad\qquad\qquad\qquad\qquad\qquad j = 3,4,5 \end{cases} \quad (4)$$

with $\tau = \varepsilon t$, and

$$\frac{\partial^2 \xi_j}{\partial t^2} = \begin{cases} \dfrac{\partial^2 f_j^0}{\partial t^2} + \varepsilon \left[2 \dfrac{\partial^2 f_j^0}{\partial t \partial \tau} + \dfrac{\partial^2 f_j^1}{\partial t^2} \right] + \\ \qquad + \varepsilon^2 \left[\dfrac{\partial^2 f_j^0}{\partial \tau^2} + 2 \dfrac{\partial^2 f_j^1}{\partial t \partial \tau} + \dfrac{\partial f_j^2}{\partial t^2} \right] + \cdots \\ \dfrac{\partial^2 f_j^1}{\partial t^2} + \varepsilon^2 \left[2 \dfrac{\partial^2 f_j^1}{\partial t \partial \tau} + \dfrac{\partial^2 f_j^2}{\partial t^2} \right] + \cdots \end{cases} \quad (5)$$

So by applying Newton's law with respect to a reference system fixed on the ship and by restricting our attention to surge sway and yaw, we have:

$$X = m \left\{ \ddot\xi_1 - \dot\xi_6^2 X_G - \dot\xi_2 \dot\xi_6 + \dot\xi_3 \dot\xi_5 - \dot\xi_5^2 X_G \right\}$$

$$Y = m \left\{ \ddot\xi_2 + \dot\xi_6^2 X_G + \dot\xi_4 \dot\xi_5 X_G - \dot\xi_3 \dot\xi_4 + \dot\xi_1 \dot\xi_6 \right\}$$

$$N = I_z \ddot\xi_6 + (I_y - I_x) \dot\xi_5 \dot\xi_4 + (\ddot\xi_4 - \dot\xi_4 \dot\xi_3) I_{zx}$$

$$+ m X_G (\ddot\xi_2 + \dot\xi_6 \dot\xi_1 - \dot\xi_4 \dot\xi_3) \quad (6)$$

where X is the total hydrodynamic surge force, Y the sway force, N the yaw moment. The hydrodynamic forces are of order ε, so that:

$$\frac{\partial^2 f_j}{\partial t^2} = 0 \qquad j = 1, 2, 6 \quad (7)$$

The first order wave forces as a result of linearity are of the same time scale as the first order wave potential, so that:

$$\frac{\partial^2 f_j^0}{\partial t \partial \tau} \text{ and } \frac{\partial^2 f_j^1}{\partial t^2} \text{ depend only on } t \quad (8)$$

The second condition implies that f_j^1 depends only on t, otherwise we obtain secular terms.

In addition, all quantities $\dot\xi_i \dot\xi_j$ should not contribute terms of order 1 so that:

$$\frac{\partial f_j^0}{\partial t} = 0 \qquad (9)$$

The final form of the expansion will be therefore in the form:

$$\xi_j = \begin{cases} f_j^0 (\varepsilon t) + \varepsilon f_j^1 (t) + \ldots & j = 1,2,6 \\ \varepsilon f_j^1 (t) + \ldots & j = 3,4,5 \end{cases}$$

The form of the surge, sway, yaw motions is shown in fig. (2) up to 1st order. On a slowly changing (the order of time scale is ε) large amplitude motion denoted by $x_a(\varepsilon t)$, a small amplitude fast changing motion denoted by $x_h(t)$ is superimposed.

HYDRODYNAMIC FORCES

Associated with the drift motions there are hydrodynamic coefficients (added mass and damping). We consider the total hydrodynamic boundary value problem and show the decomposition that results from the formulation above, between the problems related to motions x_a and x_h. The only two boundary conditions that change due to the drift motions are the one on the ship hull and the free surface condition. We assume no steady velocity (no current, no ship speed). In addition, we assume that yaw motion is small, i.e. we concentrate our attention on surge and sway. The assumption is not at all restrictive since yawing moments are small, so that yaw angles can be easily constrained below $10°$.

(i) **Boundary condition on the hull:**

$$\frac{\partial \phi}{\partial n} = U_n = \dot{\overline{r}}_0 \cdot \overline{n} \qquad \text{(see Fig. 1)} \qquad (10)$$

where if $\overline{r}_a(\varepsilon t)$ denotes the slow linear motions and $\varepsilon \overline{r}_h(t)$ the fast motions with $\varepsilon \overline{\alpha}(t)$ the angular fast motions then:

$$\dot{\overline{r}}_0 = \left\{ \varepsilon \frac{d\overline{r}_a}{d\tau} + \varepsilon \frac{d\overline{r}_h}{dt} + \varepsilon \frac{d\overline{\alpha}}{dt} \times \overline{s} \right\}$$

and $\overline{n}_0 = \overline{n} + \alpha \times \overline{n}$

so by retaining terms up to first order:

$$\frac{\partial \phi}{\partial n} = \varepsilon \dot{\overline{r}}_a \overline{n} + \varepsilon \dot{\overline{r}}_h \overline{n} + \varepsilon (\dot{\overline{\alpha}} \times \overline{s})\overline{n} \qquad (11)$$

The two last terms in the right hand side give the boundary condition when no drift is considered.

(ii) **Free surface condition**

The boundary condition relative to the x_0, y_0, z_0 system is $\phi_{tt} + g \phi_z = 0$ which relative to the x,y,z system becomes:

$$\left(\frac{\partial}{\partial t} + \frac{\partial x_a}{\partial t} \frac{\partial}{\partial x} + \frac{\partial y_a}{\partial t} \frac{\partial}{\partial y}\right) \phi + g \phi_z = 0 \qquad (12)$$

or $\qquad \left(\frac{\partial}{\partial t} + \varepsilon \frac{\partial x_a}{\partial \tau} \frac{\partial}{\partial x} + \varepsilon \frac{\partial y_a}{\partial \tau} \frac{\partial}{\partial y}\right) \phi + g \phi_z = 0$

so that for the fast motion potentials the boundary condition, to order ε, remains the same.

By considering the two boundary conditions we see that the fast motions problem can be decoupled from the slow motions. This is an important property because it means that we do not need to bring any changes upon the existing theory of small amplitude motions when drift motions are present.

We proceed now by considering the case when steady drift (or current) denoted by the vector $(U,V,0)$ is present. The two boundary conditions above become:

i) $\frac{\partial \phi}{\partial n} = (U,V,0)\,\bar{n} + \varepsilon \dot{\bar{r}}_a\,\bar{n} + \varepsilon[\dot{\bar{r}}_h\,\bar{n} + \bar{\alpha}(\bar{s} \times n) + (U,V,0)(\bar{\alpha} \times \bar{n})]$ (13)

ii) $(\frac{\partial}{\partial t} + U\frac{\partial}{\partial x} + V\frac{\partial}{\partial y})^2 \phi + g\phi_z = 0$ (14)

We see from the right hand side of the first condition that the fast motions are influenced by the $(U,V,0)$ vector in the same way as with the forward speed of the vessel [the treatment of the side speed V can be found in Triantafyllou ref. (4)].

The problem of the slow motions can be formulated now through a potential satisfying the conditions:

(i) $\frac{\partial \phi}{\partial n} = (U,V,0)\,\bar{n} + \varepsilon\,\dot{\bar{r}}_a$ on the hull (15)

(ii) $(\varepsilon\frac{\partial}{\partial \tau} + U\frac{\partial}{\partial x} + V\frac{\partial}{\partial y})^2 \phi_a + g\,\phi_z = 0$ at the free surface (16)

The conclusion is that the slow motions strongly depend on the steady speed (which can be steady drift or current) and more specifically they can be considered to be a first order pertubation of the steady translation boundary problem.

By using Bernoulli's equation we can show that the hydrodynamic forces in the case of the slow motions are of order ε^2:

$$\bar{F} = \iint_{S_B} p\,\bar{n}\,ds = -\rho \iint_{S_B} \frac{\partial \phi_a(\varepsilon t)}{\partial t}\,\bar{n}\,ds = -\rho \iint_{S_B} \varepsilon \frac{\partial \phi_a}{\partial \tau}\,\bar{n}\,ds = O(\varepsilon^2)$$ (17)

Next we prove that no interaction between the slow and the fast motions contributes to the second order force. The force is (considering only terms involving the potential):

$$\bar{F} = -\rho \iint_{S_B} (\frac{\partial \phi}{\partial t} + \frac{1}{2}|\nabla\phi|^2)\,\bar{n}\,ds$$

The time average of the force over a wave period will give the steady second order force. By substituting in the equation above the expression for ϕ:

$$\phi = \phi_h^1(t) + \phi_h^2(t) + \ldots + \phi_a^1(\varepsilon t) + \phi_a^2(\varepsilon t) + \ldots$$

we see that all interaction terms, i.e. $\nabla\phi_h^1 \cdot \nabla\phi_a^1$, contribute steady terms of order ε^3 because the time average of terms:

$A\,e^{i\varepsilon\omega t}\,e^{i\omega t}$ is of order ε. We conclude that the evaluation of the second order forces caused by the fast motions of the vessel remains the same. A more detailed discussion of the subject and of the proof is given in ref. (1).

APPLICATION

As we proved in the previous section, the hull condition of the slow motions is a perturbation of the boundary condition of the steady translation problem while the free surface condition is to first order the same.

A simple way therefore, to derive the hydrodynamic coefficients is to use published data on steady translation and consider the first order perturbation. The information will be particularly interesting as far as coupling among sway, surge, yaw is concerned. The added mass matrix cannot be obtained from such a perturbation, so it must be either obtained by considering the radiation problem as $\omega \to 0$ or from experimental results [see for example Motora Ref.(4a)].

The method has been applied to results given by English and Wise, Ref. (3) for a ship 94.49 m long:

$$C_X = 0.01 - 0.03 \cos(\psi) - 0.01 \cos(2\psi)$$

$$C_Y = [0.86 \sin(\psi) = 0.10 \sin(3\psi)]$$

$$C_N = 0 - 0.0563 \sin(\psi) + 0.083 \sin(2\psi)$$

$$C_X \frac{X}{\frac{1}{2}\rho V_c^2 LT} \quad C_Y = \frac{Y}{\frac{1}{2}\rho V^2 LT} \quad C_N = \frac{N}{\frac{1}{2}\rho V_c^2 L^2 T} \tag{18}$$

where V_c is the total speed, ψ the angle between $\overline{V}_c$ and the X axis, L is the length and T the draft of the vessel.

By considering the drift velocities to be a perturbation of the steady speed (U,V,0) i.e., if u is surge, v sway and $\dot\phi$ yaw velocity then:

$$\overline{V}_c = (U,V,0) + \varepsilon[(u,v,0) + (0,0,\dot\phi) \times \overline{s})] \tag{19}$$

With $\phi_0 = \tan^{-1}(V/U)$, we can derive a hydrodynamic model by introducing equations (18) and (19) in Newton's Law, equations (6). If $\phi_0 = 0$, we derive in non-dimensional quantities the hydrodynamic matrix:

$$\begin{Bmatrix} [1+\frac{a_X}{m}] s^2 - C_{X2} \cdot \cos(\phi_0) \cdot s & -C_{X2} \cdot \sin(\phi_0) \cdot s & -C_{X1} \\ -C_{Y2} \cdot \cos(\phi_0) \cdot s & [1+\frac{a_Y}{m}] s^2 - C_{Y2} \sin(\phi_0) \cdot s & -C_{Y1} + \frac{X_G}{L} s \\ -C_{N2} \cos(\phi_0) \cdot s & -C_{N2} \cdot \sin(\phi_0) \cdot s + \frac{X_G}{L} s^2 & \left[\frac{I + k_2^2 m}{mL^2}\right] s^2 - C_{N1} \end{Bmatrix} \tag{20}$$

where a_X the added mass in surge, a_Y in sway, K_2 the radius of gyration in yaw and s the complex frequency.

By letting $\psi = \tan^{-1}[(V_0 \sin(\phi) + v) / (V_0 \cos(\phi) + u)]$ we define by:

$$V_0 = \sqrt{V^2 + U^2}$$

$$C_{X_1} = \frac{\partial C_X}{\partial \phi} R_1 \qquad C_{Y_1} = \frac{\partial C_Y}{\partial \phi} R_1 \qquad C_{N_1} = \frac{\partial C_N}{\partial \phi} R_1 \text{ at } \phi = \phi_0, u = v = 0$$

$$C_{X_2} = C_X R_2 \qquad C_{Y_2} = C_Y R_2 \qquad C_{N_2} = C_N R_2 \qquad (21)$$

where $R_1 = \rho V_0^2 LT/mg \qquad R_2 = \rho V_0 LT/m\sqrt{g/L}$

Near $\phi_0 = 0$ the coupling terms become very important because the ship behaves as a lifting surface with small aspect ratio, so that even for small v the resulting force (lift) in y direction is large. We can write locally the hydrodynamic matrix:

$$\begin{bmatrix} (1 + \frac{a_X}{m})s^2 + 0.03 R_2 s & 0 & 0 \\ 0 & (1 + \frac{a_Y}{m})s^2 + 0.28 R_2 s & \frac{X_G}{L} s^2 + 0.28 R_1 \\ 0 & \frac{X_G}{L} s^2 - 0.0549 R_2 s & \frac{I+K_Z^2}{mL^2} s^2 - 0.0549 R_1 \end{bmatrix} \qquad (22)$$

IMPORTANCE OF THE MODEL

When designing a positioning system we are interested in three aspects of the controlled system:

    a) Stability

    (b) Error with respect to a desired position

    (c) Sensitivity of the error to variations of the parameters of the system.

The coupling terms of a multivariable system play an important role in all three aspects. At $\phi_0 = 0$ as we said before the ship presents significant coupling while due to its slenderness presents minimum steady resistance. If we want to keep the ship close to this angle, we have to design the system primarily to offset the coupling terms.

The present method was applied in Ref. (1) in conjunction with Rosenbrock's design theory of multivariable control systems in frequency domain [Ref.(7)] as extended by the author [Ref. (1)] to design the dynamic positioning system of a drilling vessel.

Typical simulation results are given in Figures (3), (4) and (5) which show small errors from the desired position. The design of the compensators for stability and error and sensitivity reduction was based on the results of the present formulation concerning coupling terms and frequency range. The simulated environmental conditions were: Irregular waves with significant wave height 4.9 m at $30°$ with respect to the x axis, current 3 knots at $10°$ and wind 65 km/hour at $25°$.

CONCLUSIONS

The subject of the drift motions has been formulated in a way that permits the use of the existing theory of small motions without modification.

Using the two time scale expansion, we can derive from equations (6) the drift equations of motion by keeping all terms to second order and then removing all terms of time scale 1. The derivation is straightforward and can be found in Ref. (1). The slow motions are influenced by the fast motions, so it is interesting to study the drift equations in their general form to determine which motions are the most significant.

When steady drift or current is present, then the fast motions are coupled with the steady speed, especially roll (Triantafyllou Ref. (4)]. The present analysis remains otherwise the same.

The slow motions problem is close to the steady translation problem and very useful information can be obtained from results on the second problem regarding the coupling terms. The present approach proved to be very useful to the author for the design of the controller while the simulated performance of the controlled vessel was very satisfactory.

ACKNOWLEDGEMENTS

The author is indebted to Professor C. Chryssostomidis of M.I.T. who gave many suggestions and advice on the topic. Thanks are also due to Professors F. Noblesse and R. W. Yeung, both of M.I.T. for their comments and suggestions.

The research was supported by a grant from N.S.F.

REFERENCES

(a) Journal Articles and Reports

(1) Triantafyllou, M.S., "Design of the Dynamic Positioning System of a Drilling Vessel", Sc.D. Thesis, Department of Ocean Engineering, M.I.T. (July 1978).

(2) Salvesen, N., "Second Order Steady State Forces and Moments on Surface Ships in Oblique Regular Waves", The Dynamics of Marine Vehicles and Structures in Waves, London (1974).

(3) English, J.A. and Wise, D.A., "Hydrodynamic Aspects of Dynamic Positioning", Trans. North East Coast Inst. of Engineers and Shipbuilders, Vol. 92 (1976).

(4) Triantafyllou, M.S., "Strip Theory of Ship Motions in the Presence of a Current", M.I.T. Report, Ocean Engineering Dept. (Nov. 1978).

(4a) Motora, S., "On the Measurement of Added Mass and Added Moment of Inertia of Ships in Steering Motion", 1st Symposium on Ship's Manoeureability, DTMB Report No. 1461, 1960.

(b) Books

(5) Wehausen, J.V. and Laitone, E. V., "Surface Waves", Handbuch der Physik, IX, Springer Verlag (1960).

(6) Lin, C.C. and Segel, L.A., "Mathematics Applied to Deterministic Problems in Natural Sciences", MacMillan Publishing Co. (1974).

(7) Rosenbrock, H.H., "Computer Aided Control System Design", Academic Press (1974).

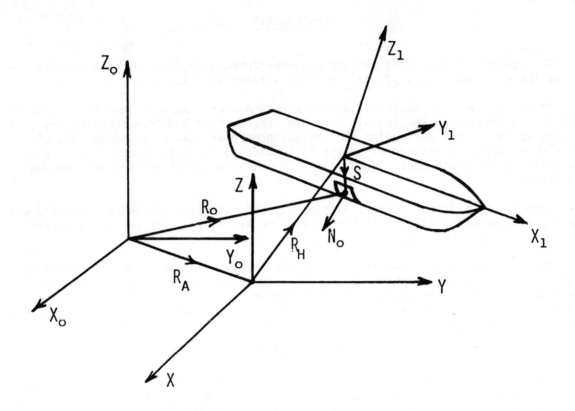

Fig. 1 Reference systems.

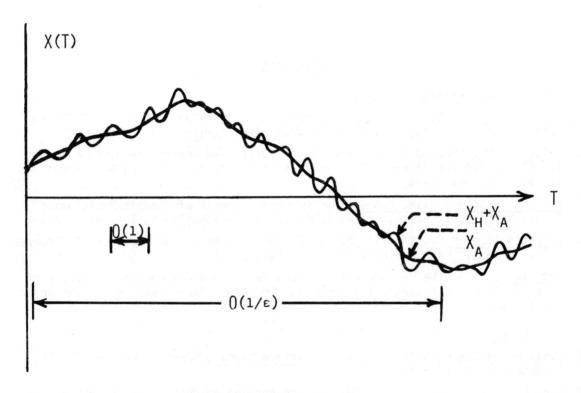

Fig. 2 Fast and slow motions.

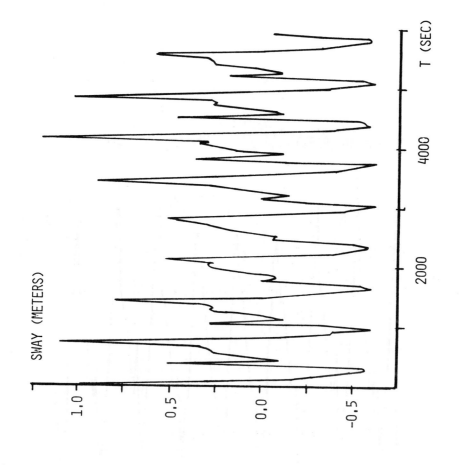

Fig. 3

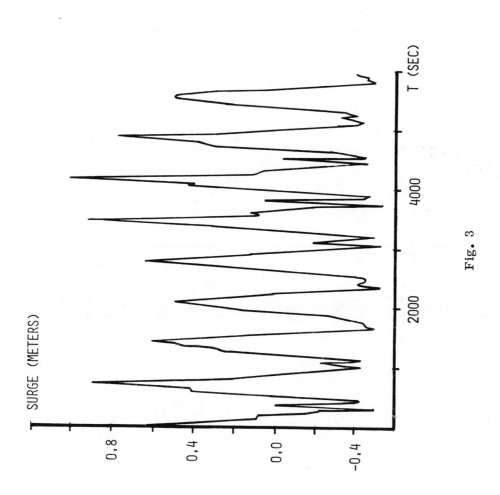

Fig. 4

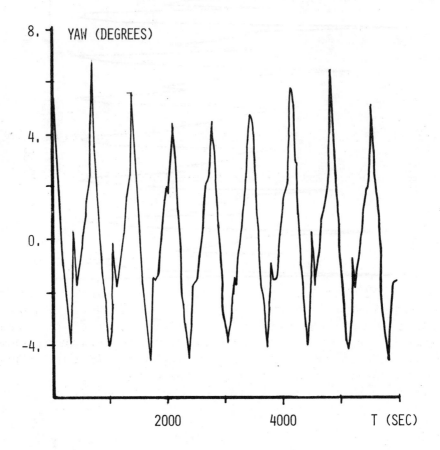

Fig. 5

BOSS'79

PAPER 59

Second International Conference
on Behaviour of Off-Shore Structures
Held at: Imperial College, London, England
28 to 31 August 1979

HYDRODYNAMIC ANALYSIS OF TANKERS AT SINGLE-POINT-MOORING SYSTEMS

O.M. Faltinsen, O. Kjærland, N. Liapis and H. Walderhaug

Norwegian Institute of Technology, Norway

Summary

The slow motions of a tanker moored to a buoy are studied. The current loads and the wave drift loads on a ship are discussed in detail. New experimental results are presented and show encouraging agreement with theoretical values. The current loads are presented for drift angles between 0° and 180°. The wave drift loads are presented for wave headings between 0° and 180° and for different wave lengths. An asymptotic theory for wave drift loads on a structure in regular waves of small wave lengths is presented.

A seven degrees of freedom linear stability equation system for the ship and the buoy is applied. A simpler equation system is used to make general conclusions about the stability of the system and the possibility of large motion slow drift oscillations of the ship and the buoy. Recommendations for design are presented. In particular the mooring arrangement between the ship and the buoy is discussed. The theoretical results are qualitatively compared with experimental results.

Sponsored by: Delft University of Technology, The Netherlands
Massachusetts Institute of Technology, U.S.A.
The Norwegian Institute of Technology, Norway
University of London, England

Secretariat provided by: BHRA Fluid Engineering

Copyright: © BHRA Fluid Engineering
Cranfield, Bedford, England

NOMENCLATURE

a	The negative of the x-coordinate of the fairlead
b	Characteristic length of structure. Used in Reynolds' number Rn and Strouhals' number Sn. For a cross-flow past a circular cylinder b is the diameter.
A_{jk}^{S}	Added mass values for the ship ($j,k = 1,2,\ldots,6$)
$C_{xs} = \dfrac{F_1^C}{\frac{1}{2}\rho V_c^2 \cdot L\,T}$	
$C_{ys} = \dfrac{F_2^C}{\frac{1}{2}\rho V_c^2 \cdot L\,T}$	
$C_{ns} = \dfrac{F_6^C}{\frac{1}{2}\rho V_c^2 \cdot L^2\,T}$	
f	Vortex shedding frequency
F_i^C	Current forces and moments on the ship. $i = 1$ refers to force component along the x-axis, $i = 2$ refers to force component along the y-axis and $i = 6$ refers to yaw moment about the z-axis
F_i^{SD}	Slow drift wave excitation forces and moments on the ship
F_i^{WD}	Wave drift forces and moments on the ship in regular waves
F_i^{W}	Wind forces and moments on the ship
F_o	Tension in the bow hawser
F_{crit}	See Equation 4
g	Acceleration of gravity
I_{66}^{S}	Mass inertia moment in yaw for the ship
k_{BH}	Restoring coefficient for the bow hawser
ℓ	Length of the bow hawser. Different meaning in Figure 25
ℓ_{crit}	See Equation 5
L	Length between perpendiculars of the ship
M^S	Mass of the ship
m_{yy}, $m_{\psi\psi}$	Defined after Equation 4
r	Yaw-velocity of the ship. Also used for radius of a circle
T	Used as average draught in C_{xs}, C_{ys} and C_{ns}
U	Free stream velocity
V_c	Current velocity
V_s	Ship velocity component along the y-direction
V_W	Wind velocity

NOMENCLATURE (cont.)

(x,y,z)   Coordinate system fixed to the ship. Origin in the centre of gravity. Positive z upwards. Positive x in the aft direction

β Angle between current direction and x-axis

λ Wave length of the incident waves

μ Angle between wave propagation direction and x-axis

ν Kinematic viscosity coefficient for water

ζ_a Wave amplitude of the incident waves

ρ Mass density of the water

ψ_c Angle between the current direction and the x-direction (see Figure 20) when the ship is in the average position

ψ_o Angle between the bow hawser and the negative x-direction of the ship (see Figure 20) when the ship is in the average position

$\dfrac{\partial M}{\partial \psi}$ The rate of change of the wind, current and average wave drift yaw moment on the ship with respect to the angle ψ at $\psi = \psi_c$ when the ship is rotating in positive yaw angle direction

$\dfrac{\partial F_y}{\partial \psi}$ Rate of change of the y-component of the wind, current and average wave drift force on the ship

1. INTRODUCTION

Offshore loading is an alternative to pipelines for transportation of oil and gas which becomes of more and more importance as the oilfield development moves into deeper and deeper water. Different types of single point mooring systems have been proposed (Koonce Ref. (1)). But we will in this paper consentrate on a floating buoy moored to the seabed by a spread mooring system.

Experience from the North Sea has pointed out that the mooring hawser between the buoy and the tanker is a critical part of the system. The mooring hawsers have required frequent replacement. This may be due to fatigue as well as high mooring loads. There are different reasons for the high mooring loads. One is the slowdrift large amplitude oscillations of the tanker which may be due to slowdrift excitation forces in waves, nonlinearities in the mooring system or unstabilities. Snap loads may also occur.

In this study we will consentrate on the slow motions of the tanker and the buoy, and assume that the wave induced motions of the tanker and the buoy with mean period in the vicinity of the mean period of the waves can be superimposed on the slow motions of the tanker and the buoy. We will in particular study the current loads and the wave drift loads on the tanker for different current angles and wave headings with respect to the ship, and extensive numerical and experimental results will be presented. The equations of motions of the tanker and the buoy are in particular used to study the stability of the system. If the system is unstable, large amplitude slowdrift oscillations may occur and cause high loads in the bow hawser. This problem has been studied in the case of steady wind and current by Wichers (Ref. 2). Wichers assumed that the motions of the buoy were completely determined by the motions of the tanker. We will consider a seven degrees of freedom model of the ship and the buoy in wind, current and waves. This procedure is compared with a simpler procedure.

We are also studying how different parameters affect the stability of the tanker. In particular, we are studying how the mooring arrangement may improve the stability. In this way we hope to contribute to increased operationability of an offshore loading system.

2. EXPERIMENTAL SETUP

In this chapter we will describe the experimental setup for the model tests we performed. In the model tests we measured current loads and wave drift loads on the ship and studied the stability of a ship moored to a buoy.

The model experiments were carried out in the No. 1 basin of the Ship Model Tank in Trondheim, which has the following dimensions: Length = 260 m, Breadth = 10,5 m, Depth = 5,75 / 10 m.

The wave generator used in our tests is of the plunger type and is capable of producing regular waves with a maximum height of 0.3 m, and irregular waves of 0.27 m significant wave height.

For ordinary towing tests with ship models the basin is equipped with a towing carriage of about 20 tons, but for tests with models of offshore constructions another lightweight carriage is linked to the main carriage, thus forming an open bay where the model setup may be overlooked from all sides.

For measuring forces strain gauges and mechanical balancing were used. Motions are usually measured by potensiometers in combination with pulleys or by the "Selspot" optoelectronic system.

The first stage of the experiments was performed with the ship model alone. A model of a 130 000 dwt tanker was chosen, the main dimensions and other particulars are given in Table 1. The main body is shown in Figure 1. Model scale was 1:50.

The ship model was given turbulence stimulation by applying vertical sand strips (one centimetre) to the fore and aft end, and two horizontal strips in the bilge area.

The ship model was made from wood and a stock propeller was fitted in order to establish astern thrust when the ship model was moored to the buoy. The following setup for measuring forces and motions was chosen (see Figure 2). A single pole having a 0.6 m diametre pulley on top was kept in position by 3 stays, the lower end of the pole attached to a gimbal in the model. On the pole was also mounted a potensiometer for measuring the heading of the model. The heave motion was measured by means of a thin wire running over a pulley carrying a potensiometer. The motions in the horizontal plane could also be measured by means of potensiometer carrying pulleys over which the mooring lines were running. The average yawing moment produced by waves or current loads was counteracted by weights attached to the wires from the pulley on top of the pole. Drift forces were balanced by weights so that the pole was kept as vertical as possible (see Figure 2). The forces and moments could then be measured either by counting weights, or by means of a strain gauge inserted in the wire (this was the actual method). The following items were measures: Heave, pitch, heading, horizontal motions, average yaw moment, average horizontal forces, carriage speed (when simulating current).

When measuring motions and forces caused by waves alone, the two carriages were standing halfway between the wave generator and the beach. In this case also the tankwalls were covered by wave absorbers of approximately the same length as the model in order to avoid disturbing reflections. When simulating current, the two carriages with the mooring suspensions were moving.

The second phase of the experiments was performed with the ship model moored to the buoy model. The model was made from 3 mm steel plate in an all welded construction. The dimensions and other data are given in Figure 3.

The buoy was moored to the two carriages by four horizontal mooring lines, the elasticity of which was simulated by prestrained, soft springs. The bow hawser was made up from one wire and two springs, the one of which was restricted to a certain maximum elongation in order to simulate the elasticity of the real hawser as shown on Figure 4. Separate spring combinations were made for each of the hawser lengths.

Only horizontal motions of the buoy and the ship were measured (by the Selspot system). The hawser tension was measured by means of a strain gauge (an aluminium ring fitted with four strain gauge filaments).

Astern thrust was provided by the aforementioned stock propeller. The propeller was not mounted on the tanker when current loads and wave drift loads were measured.

3. CURRENT LOADS

In this chapter we will discuss in detail current forces and moments on the ship. Experimental results will be presented. A procedure to scale the model test results is proposed.

We will use a coordinate system (x,y,z) fixed to the ship with origin in the centre of gravity of the ship. The z-axis is positive upwards. The x-axis is in the longitudinal direction of the ship and positive in the direction of the aft perpendicular.

The calculation of current forces and moments on the ship and the buoy is a difficult task and empirical formulas have to be used.

The longitudinal current force on the ship will be mainly due to viscous friction forces and was calculated by generalizing Remery and van Oortmerssen's (Ref. 3) procedure by introducing relative longitudinal velocity between the ship and the current.

The cross-flow principle was partly used to express the transverse viscous current force and current yaw moment on the ship. We can write

$$F_2^C = \frac{1}{2} \rho \int_L dx\ C_D(x) T(x) V_r |V_r| \qquad (1)$$

$$F_6^{VC} = \frac{1}{2} \rho \int_L dx\ C_D(x) T(x) V_r |V_r| \cdot x \qquad (2)$$

where the integration is over the length of the ship. F_2^C is the transverse current force and F_6^{VC} is the viscous current yaw moment on the ship. The total current yaw moment F_6^C is the sum of the Munk moment and the viscous current yaw moment F_6^{VC}. $C_D(x)$ is the drag coefficient for cross-flow past an infinitely long cylinder with a cross-sectional area equal to the sum of the ship cross-section at longitudinal co-ordinate x and its image above the free surface. $T(x)$ is the sectional draught and

$$V_r = V_c \sin\beta - V_s - xr,$$

where V_c is the current velocity, β is the instantaneous angle between the current direction and the x-direction of the ship, V_s is the velocity component of the centre of gravity of the ship along the y-axis, and r is the angular velocity component of the ship along the z-axis. Further ρ is the mass density of water.

The drag coefficient depends on the sectional form, the roughness characteristics of the section, a local Reynolds' number and a parametre similar to Keulegan-Carpenter number expressing the effect of the oscillatory wake on the structure.

The formulas (1) and (2) totally neglect three-dimensional end effects. The advantage of (1) and (2) is that they take proper care of the changing incident velocity due to yaw motion along the length of the ship.

In the steady flow case, the angle dependence β of the transverse current force and yaw moment is sometimes expressed in the form of Fourier-series in the angle β. In the case of pure sway motion one may be tempted to generalize this procedure, but it is not clear how to do it in the case of variable velocity along the ship.

According to the cross-flow principle (see Equations (1) and (2)) the angle dependence of F_2^C is $\sin^2\beta$ in the steady flow case. The angle dependence of the current moment is expressed as

$$F_6^C = R_N^{90} \sin^2\beta + \frac{1}{2}\rho V_c^2 (A_{22}^S - A_{11}^S) \sin 2\beta \qquad (3)$$

The last term is the Munk moment and the first term is the viscous current moment. The constant R_N^{90} has to be empirically determined. This procedure may not be too bad except when β is close to zero or 180°. In the latter cases it is possible that the transverse current force and yaw moment may be calculated by potential theory and lifting theory, by considering the ship as a low-aspect-ratio lifting surface (Jacobs (Ref. 4), Newman (Ref. 5)). One part of the yaw moment is due to the Munk moment which is a nonlifting effect. In the small angle of attack case the theoretical values will have a linear dependence in the angle β.

The current forces and moments on the buoy were calculated by the cross-flow principle.

Experimental and numerical results

The measured forces and moments on the model (See Chapter 2) in a homogeneous current are given in Figures 5, 6 and 7. The results are for both ballast and full-load conditions. The results are presented as a function of the drift angle β. Theoretical results are also plotted. For small angles of drift we estimated forces and moments following Jacobs (Ref. 4). For the rudder the slope of the lift coefficient curve is taken as $2\pi \cdot 0.9A/[1.8 + \sqrt{(A^2 + 4)}]$ with the effective aspect ratio A equal to twice the geometric aspect ratio.

The added mass in sway in the Munk moment was found from Prohaska's diagram by strip theory. The added mass value in surge was taken for an ellipsoid with corresponding length/breadth ratio. The added mass values were found to be in good agreement with zero-frequency added mass values obtained by three-dimensional sink-source technique (Faltinsen and Michelsen (Ref. 6)).

The calculated forces and moments by Jacobs' method are shown in Figures 5 and 7. According to the cross-flow principle, the transverse current force should vary like $C_D^{90} \cdot \sin^2\beta$. This curve is also plotted in Figure 5. Equation 3 is also plotted in Figure 7. The constant C_D^{90} and R_N^{90} are determined from experimental values at $\beta = 90°$.

From the figures we note that there is good agreement between lifting theory and experimental values for small angles of attack. Further the cross-flow principle combined with the Munk moment for the yaw moment seems to be reasonable when the angle of attack is not small.

In the experiments the model was towed sideways at speeds corresponding to Froude numbers (F_{nT}) based on mean draught from 0.12 - 0.31 in loaded conditions and from 0.19 - 0.75 in ballast. According to Norrbin (Ref. 7) the wave formation in loaded condition at $\beta = 90°$ and $F_{nT} = 0.32$ is so small that gravity in the free surface condition may be neglected. A check on this was run by cross-wise ($\beta = 90°$) towing of a plane board which had the form of the lateral plan of the model including the rudder. The board was towed at two draughts corresponding to loaded and ballasted model, and at the model speeds. In no case was there any measurable effect of wave resistance, the resistance coefficient being constant as expected for a flat plate at $\beta = 90°$ and zero Froude number.

The sideforce coefficient given in Figure 5 for loaded condition is approximately the same as found by English and Wise (Ref. 8) for a model without bilge keels. At ballast draught the sideforce coefficient is markedly reduced due to the more streamlined form of the hull sections at reduced draught.

To estimate the cross-wise ($\beta = 90°$) drag of the model by means of strip theory and two-dimensional drag data for cylinders is questionable due to lack of drag data for cylinders of ship section forms. For the same reason and also due to the uncertain effect of the free surface as a splitter plate at high Reynolds' number, it is difficult to expand the test results to full scale.

The splitter plate effect of the free surface is a function of Reynolds' number Rn and as shown in Figure 8 the Strouhals' number Sn expressing the vortex shedding frequency for a circular cylinder has been measured up to a Reynolds's number of 10^7, see Delany, N.K. and Sorensen, N.E. (Ref. 9). For the ship the Reynolds' number based on breadth and at a current speed of 2 knots is about $4 \cdot 10^7$. We might therefore expect that the ship sections in full scale are in the transcritical region. In this region we have a pronounced periodicity in the wake flow, and hence we assume that the splitter effect of the free surface is the same as in the subcritical region. The model tests were carried out in the supercritical region, and here we assume that the splitter effect of the free surface is negligible since the wake is free from periodic vortex shedding.

The model was fitted with sand strips as described in Chapter 2, and there is no doubt that the flow was indeed supercritical. We therefore suggest the following method of expansion of model test results for transverse current force to full scale

1. The relation between model drag and the drag of a circular cylinder is assumed the same in model- and full scale. The effect of cylinder length/diameter ratio is considered negligible in the supercritical region.

2. The relation between drag with- and without splitter plate is assumed the same in the subcritical and in the transcritical regions.

3. The effect of full scale hull roughness is considered negligible for a hull with a modern paint system.

The full scale two-dimensional drag coefficient of a circular smooth cylinder is assumed equal to 0.7.

4. The bilge keels increase the drag at $\beta = 90°$ with approximately 25% as taken from English and Wise (Ref. 8).

4. WAVE DRIFT LOADS

In this chapter we discuss the wave drift forces and moments on the ship. We will present experimental and numerical results for the tanker presented in Chapter 2.

According to Faltinsen and Løken (Ref. 10), it is possible to write the slowdrift excitation forces and moments on a ship in irregular waves in terms of the drift forces and moments in regular waves. To describe the wave drift forces and moments on the ship in regular waves we will use the method presented by Faltinsen and Michelsen (Ref. 6). This method is applicable for any structure, wave heading and water depth.

In Appendix A is derived an asymptotic formula for the wave drift forces and moments on any structure when the wave length of the incident regular sinusoidal waves are small compared to characteristic cross-dimensions of the body. These formulas will also be used.

Numerical and theoretical results

In the numerical calculations the computer program NV459 of Det norske Veritas was used. This computer program is based on the procedure of Faltinsen and Michelsen (Ref. 6). To describe the mean wetted surface of the ship a total of 208 elements were used in full load condition. In ballast condition 186 elements were used. The computations and the model tests were performed for λ/L = 0.35, 0.56, 0.68, 0.85, 1.02, where λ is the wave length of the incident regular waves and L is the length between perpendiculars of the ship. The computations were performed for a number of wave headings between $0°$ and $180°$. Model tests were also performed for λ/L = 0.175 and compared with the asymptotic theory presented in Appendix A.

The numerical and experimental results are plotted in Figures 9 through 19. The longitudinal and transverse drift force components F_1^{WD} and F_2^{WD} are nondimensionalized with respect to $\rho g \zeta_a^2 \cdot L$ where ζ_a is the wave amplitude of the incident waves. In the model tests proper care had to be taken so that the incident waves were measured without any effect from the ship. The yaw moment F_6^{WD} is nondimensionalized with respect to $\rho g \zeta_a^2 \cdot L^2$. The results are plotted as a function of μ which is the angle between the wave propagation direction and the x-axis of the ship.

We note from the results that drift forces and moments are quite sensitive to wave heading and wave length. For instance for λ/L = 0.35 and ballast condition, maximum transverse wave drift force is for $\mu = 90°$, while for λ/L = 0.56 and ballast condition maximum wave drift forces is maximum for $\mu = 45°$ and close to zero for $\mu = 90°$. We may also note that maximum wave drift moment does not necessarily occur for $\mu = 45°$. For instance when λ = 0.35 and ballast condition maximum wave drift moment occur for $\mu = 70°$.

We note that there is quite good agreement between experiments and theory. There is a certain spread in the experiments due to difficulties in measuring the drift forces and moments. In some cases we note a sudden increase in measured longitudinal force in the vicinity of $\mu = 0°$. This is not present in the theoretical results. The asymptotic theory presented in Appendix A has been compared with experimental results for λ/L = 0.175 in Figure 19. There is a certain agreement between theory and experiments especially for the transverse force, while there is larger disagreement for the longitudinal force. It should be noted that the asymptotic longitudinal drift force and yaw moment are quite sensitive to the form of the ship ends at the water line.

5. STABILITY INVESTIGATION

In this chapter we will apply a linear stability equation system for the ship and the buoy. The equation system has seven degrees of freedom. A simpler equation system will be used to make general conclusions about the stability of the system. Implications for design and operation are given. In particular the mooring arrangement between the ship and the buoy is discussed. The theoretical results are qualitatively compared with experimental results.

The slow motions of the tanker and the buoy can either be caused by instabilities, nonlinearities in the mooring line characteristics or slowdrift excitation loads in irregular waves. Slowdrift excitation loads due to unsteady wind is another possibility. To examine the instability problem we have to find the average position of the ship and the buoy and examine if this position is stable or unstable. This is done by neglecting external dynamic excitation effects, perform a small perturbation of the equilibrium position and examine if the ship and the buoy will return to their original position after a time period. If it is an unstable equilibrium position, slow motions of large amplitudes are likely to develop. Wichers (Ref. 2) has developed a procedure to investigate the stability of a ship moored to a single-point mooring system in current and wind. Wichers lets the motions of the buoy be completely determined by the motions of the ship. We generalized Wichers' procedure in the case the buoy is free to move in surge, sway, roll and pitch and the ship is free to move in surge, sway and yaw. Further we took into account waves in addition to wind and current. The experimental values given in Chapters 3 and 4 were used in the computations.

The average position of the ship and the buoy is defined in Figure 20, where ψ_C is the angle between the current and the x-direction of the ship, F_0 is the tension in the bow hawser and ψ_0 the angle between the bow hawser and the negative x-axis of the ship.

A computer program has been made to find the average position of the ship and the buoy. The computer program was found to be generally in agreement with the special cases studied by Wichers (Ref. 2).

A computer program to study the stability of the average position of the ship and buoy was also made. This computer program was compared with experimental values for stability and natural frequencies of a ship-buoy system. The experimental setup is described in Chapter 2. The computer program was found to be in qualitatively good agreement with the experiments. Details about the experimental results and theoretical procedure will be given in a future publication.

A simpler procedure to study the stability was also made and compared with the more complicated method. The simpler formulas were found to be quite reliable. The purpose of the latter procedure was to find simple formulas showing how main parameters like ship geometry, loading condition, bow hawser length, propeller thrust, equilibrium position, thrusters, bow hawser arrangement and system damping affected the stability of the system.

In the simple procedure we assumed that the buoy is fixed and the angle ψ_0 is zero. (see Figure 20). The reasons for choosing ψ_0 equal to zero is that ψ_0 often takes small values in reality. Further the larger ψ_0 is, the smaller the stability problem is.

When ψ_0 is equal to zero, and the buoy is fixed, we find that the surge motion of the ship is uncoupled from the sway and yaw motions. The natural frequency in surge will only depend on the restoring coefficient k_{BH} of the bow hawser characteristics. Further the ship is always stable in surge. Sway and yaw motions are independent of k_{BH}. That means that k_{BH} has no influence on the stability of the ship.

When ψ_0 is zero the linear viscous damping coefficients in sway and yaw are zero. In the sway-yaw modes of motions the frequencies of motions are very low and the potential damping due to wave generation is practically zero. If we set the coupled added mass in sway and yaw together with the damping in sway and yaw equal to zero, it is possible to show:

If the tension in the bow hawser $F_o > F_{crit}$, the ship is stable for any length ℓ of the bow hawser. If $F_o < F_{crit}$, the ship is stable for $\ell < \ell_{crit}$ and unstable for $\ell > \ell_{crit}$.

Here

$$F_{crit} = \frac{a\frac{\partial M}{\partial \psi} - \frac{m_{\psi\psi}}{m_{yy}}\frac{\partial F_y}{\partial \psi}}{\frac{m_{\psi\psi}}{m_{yy}} + a^2} \quad (4)$$

where $-a$ is the x-coordinate of the fairlead and

$$m_{yy} = M^S + A_{22}^S$$

$$m_{\psi\psi} = I_{66}^S + A_{66}^S$$

$$\frac{\partial M}{\partial \psi} = \frac{\partial}{\partial \psi}(F_6^C + \overline{F_6^{SD}} + F_6^W)\bigg|_{\psi=\psi_C}$$

$$\frac{\partial F_y}{\partial \psi} = \frac{\partial}{\partial \psi}(F_2^C + \overline{F_2^{SD}} + F_2^W)\bigg|_{\psi=\psi_C}$$

where the bar indicates mean values of the wave drift loads. Further M^S is the mass, A_{22}^S the added mass in sway, I_{66}^S the mass inertia moment in yaw, and A_{66}^S the added moment in yaw for the ship. F_i^C are current forces and moments on the ship. The index 2 stands for force component along the y-axis of the ship and the index 6 indicates yaw moment about the z-axis. F_i^{SD} are slowdrift wave excitation forces and moments on the ship and F_i^W are wind forces and moments. $\partial M/\partial \psi$ is the rate of change of the wind, current and wave drift yaw moment on the ship with respect to the angle ψ at $\psi = \psi_C$ when the ship is rotating in positive yaw angle direction. $\partial F_y/\partial \psi$ can be explained in a similar way. The normal case when $\psi_o = 0$ is that $\partial M/\partial \psi > 0$ and $\partial F_y/\partial \psi < 0$.

Further

$$\ell_{crit} = F_o \frac{(a^2 + \frac{m_{\psi\psi}}{m_{yy}})^2}{(\sqrt{-\frac{m_{\psi\psi}}{m_{yy}}(\frac{\partial M}{\partial \psi} + a\frac{\partial F_y}{\partial \psi})} + \sqrt{\frac{\Delta}{m_{yy}}})^2} \quad (5)$$

where

$$\Delta = -a[m_{\psi\psi}(F_o + \frac{\partial F_y}{\partial \psi}) + m_{yy}a(aF_o - \frac{\partial M}{\partial \psi})]$$

Experience from runs with the computer program says that the conclusions above are qualitatively valid for $\psi_o \neq 0$, too. Concepts like critical bow hawser length and critical bow hawser force have the same meaning as above, but their values are not given by (4) and (5).

ψ_o, $\partial M/\partial \psi$ and $\partial F_y/\partial \psi$ are the main parameters affecting F_{crit} and together with F_o the main parameters affecting ℓ_{crit}. Higher values of $|\psi_o|$, F_o and $\partial F_y/\partial \psi$ as well as lower values of $\partial M/\partial \psi$ make the ship more stable.

By using Equations (4) and (5) it is possible to show that current is the main source of instability and that the following parameters can affect the stability behaviour of the ship: ship geometry, loading condition, bow hawser length, propeller thrust astern, equilibrium position, thrusters, way of mooring and damping. This will be discussed in more detail below.

Influence of ship geometry on the stability

Equations (4) and (5) show how F_{crit} and ℓ_{crit} depend on ship geometry. They depend on $m_{\psi\psi}/m_{yy}$, $\partial M/\partial \psi$, $\partial F_y/\partial \psi$ and a.

According to (4) and (5) the fairlead should be located as fas away as possible from the centre of gravity of the ship. F_{crit} decreases and ℓ_{crit} increases when a increases. Other parameters that can significantly change by local changes of ship geometry are $\partial M/\partial \psi$ and $\partial F_y/\partial \psi$

A priori we expected that a number of fins - perhaps retractable - parallel to the ships symmetry plane, located astern under the ship and as far as possible from the centre of gravity of the ship (C.G.) should make the ship more stable. The critical length and critical force for the tanker, full loaded, with and without fins is shown in Figure 21. Indeed, F_{crit} is reduced, but ℓ_{crit} too is reduced a little. The reason is that each fin contributes with 13% reduction of the initial $\partial M/\partial \psi$, but at the same time with an 9% increase of $|\partial F_y/\partial \psi|$, which seems to dominate the variation of ℓ_{crit}. Notice that fins can be useful only in cases when we have current in our problem. In the case of waves only, fins are expected to have little effect on the wave drift loads on the ship and consequently on $\partial M/\partial \psi$ and $\partial F_y/\partial \psi$ in waves. It should be noted that the presence of the fins increase the damping forces, which has a positive effect on the stability of the ship.

Influence of loading condition on the stability

Relying on the fact that current is the main source of unstability and using (4) and (5) it is possible to show that a tanker in ballast condition is more stable than in fully loaded condition.

Concerning the trim, moving the load aft seems to decrease F_{crit}, but at the same time ℓ_{crit} too can decrease a little. That can be explained in a similar way as the effect of the fins on the stability.

These conclusions seem to agree with experience from running the computer program.

Influence of bow hawser length on the stability

The importance of the bow hawser length has been studied above. Very short ℓ_{crit} creates operational problems. One way of solving the stability problem would be to moor the ship by a short, flexible bow hawser to a light frame turning around a fixed horizontal axis on the buoy (Figure 22). In this mooring system the high frequency wave motions of the system do not create particular high forces, and in addition neither the ability of the frame to turn nor the values of k_{BH} have any influence on the stability.

By using (4) and (5) it is possible to show for a ship in regular waves that ℓ_{crit} is independent of the wave height ζ_a and F_{crit} increases with ζ_a^2. For a ship in wind ℓ_{crit} is independent of the wind velocity V_W and F_{crit} increases with V_W^2. For a ship in current ℓ_{crit} depends on Reynolds' number. That is why model tests may significantly overestimate ℓ_{crit}. F_{crit} in current increases with V_c^2. See Figure 23.

Influence of propeller thrust astern on the stability

If we increase the propeller thrust, the bow hawser tension F_0 will increase. This will make the system more stable. Propeller thrust will also affect the equilibrium position when $\psi_0 \neq 0$.

Higher F_0 is also wished in cases the environmental forces are not enough to keep the bow hawser stretched under the high or low frequency motions. Slack bow hawser should be avoided, so that snap loads do not occur.

On the other side, high level of bow hawser tension can be unnecessary or even undesirable. Besides propeller thrust costs energy.

In some cases the difference between F_{crit} and F_o is so high that unrealistic high values of propeller thrust are necessary to make the ship stable, for example in high current velocities. Propeller thrust applied without careful study of the system may not be any improvement at all. The direct increase of the line tension can be much higher than the benefit of a more stable system.

Influence of equilibrium position and thrusters on the stability

Ships used for offshore loading are often equipped with thrusters. These can be used in different ways depending on the environmental loads, to make the ship stable.

The thruster can be used to change the equilibrium position of the ship. That implies that the values of F_o, ψ_o, $\partial M/\partial \psi$, $\partial F_y/\partial \psi$ and damping forces change. The ship becomes stable for higher values of F_o, $|\psi_o|$, damping and $\partial F_y/\partial \psi$ and for lower values of $\partial M/\partial \psi$. None of these parameters are dominating, so that the effect of thrusters is usually a combination of the effects of all the above mentioned five parameters.

The effect of a thruster located at A.P. for the loaded condition of the full scale version of the tanker described in Table 1 in a 2 knots current is shown in Figure 24. We note that a 2 tons thrust gives a ℓ_{crit} that is more than 200 m. The line tension F_o is then less than 15 tons. For the same effect a 50 tons propeller thrust would have been required, and this would have resulted in a line tension higher than 50 tons.

The radical improvement with thruster in this case is because

a) ψ_C increases from 0 to about 12°. That makes the stabilizing effect of the damping stronger. For ψ_C = 0 we have almost no damping in yaw and sway. For ψ_C = 12° an appreciable viscous damping from the current appears. If this damping is ignored, the effect of the thruster is significantly reduced. (see the curve in Figure 24 marked "ℓ_{crit} without damping")

b) ψ_o increases from 0 to about 60°. That increases the stabilizing effect of the bow hawser. Since $\partial M/\partial \psi$ and $\partial F_y/\partial \psi$ change very little when ψ_C increases from ψ_C = 0° to ψ_C = 12°, the stabilizing effect of the bow hawser together with the increase of F_o can be considered the main reasons for the improvement shown by the curve marked by "ℓ_{crit} without damping" in Figure 24.

The same thrust produced by a thruster located under the fairlead has been shown to be much less effective, because ψ_C remains zero. This results in no changes of $\partial M/\partial \psi$, $\partial F_y/\partial \psi$ and damping. Further ψ_o is now only 17°, and the stabilizing effect of the bow hawser is now small.

A thruster close to A.P. used on the same ship in regular waves of wave length λ = 100 m seems to have a negative effect on the stability. That is because

a) There is only potential damping from waves and that has a very low value compared to the viscous damping from a current.

b) For λ = 100 m, $|\partial F_y/\partial \psi|$ shows an increase for increasing angle between ship's axis and wave propagation direction in the angle domain of interest.

Influence of bow hawser arrangement on the stability

Right mooring seems the most effective way to solve the stability problem and reduce the slowdrift motions of the ship-buoy system. It is equally effective in waves as in current or wind. Some proposed ways of mooring are shown in Figure 25. Calculations carried out for 25a) show that usually using a proper high value of restoring coefficient k_{BH} for the bow hawser, can make a ship-buoy system stable.

Model tests carried out for 25a, 25b in current, regular and irregular waves, verified the results of the calculations and showed that the slowdrift motions are significantly reduced in irregular sea.

The concepts of critical length and critical force that was introduced for one mooring line are also applicable in the case of cross-moorings. Calculations have shown that as a rule the line tensions from environmental forces alone are higher than the values of the critical force. This implies that usually, if relatively stiff wires are used, the ship-buoy system becomes dynamically stable, for any distance of the ship from the buoy.

Calculations have been carried out for the studied tanker fully loaded and cross-moored with k_{BH} = 25 tons/m. The tanker was shown to be stable for any distance from the buoy in current of V_c = 1 or 2 knots and in waves of λ = 100 m and wave amplitudes from 1 m up to at least 10 m. In current of V_c = 3 knots ℓ_{crit} was 560 m and in V_c = 4 knots ℓ_{crit} was 370 m. In the calculations D = 29 m and d = 14.5 m (see Figure 26) was used.

In Figures 26 - 27 the critical length for cross-mooring is shown as a function of k_{BH}, D and d.

Model tests for the studied tanker in fully loaded conditions, in a 3, 4, and 5 knots current, in regular and irregular seas have been carried out. The distances ℓ = 50, 100, 200 and 250 meters were tried. A value of k_{BH} = 25 tons/m was used. It was found that the ship was stable both in current and regular waves. In irregular waves some slow frequency motions were observed (ψ_o up to $10°$).

The model shown in Figure 25b has been tried too, with ℓ_1 = 30 m and ℓ = 100 m. The model seemed to be stable in a 3 knots current.

The model was tried moored by the traditional - one bow hawser - way too, in a 3 knots current. It was obviously unstable. Theoretically we found that ℓ_{crit} was 10 m for one bow hawser.

The calculation and model tests described here have been carried out for ψ_o = 0. One can generalize and suggest similar solutions in case we have more than one environmental load of different directions ($\psi_o \neq 0$) (see Figure 28).

Influence of damping on the stability

While damping has usually little effect on the natural frequencies of the ship, it can have a certain effect on the stability of the ship. In Figure 24 it was presented a case where viscous damping had a significant effect on the critical length of the bow hawser. In steady environmental loads the natural frequencies of the ship in yaw and sway are very low and the potential damping negligible. It was found to have no effect on the stability.

Influence of the buoy on the stability

Calculations using the seven degrees of freedom model of the ship-buoy system on the cases presented in Figures 23 and 24 have shown that the stability of the ship is the same whether the ship is moored to a buoy or to a fixed point.

CONCLUSIONS

The slow motions of a tanker moored to a buoy are studied. The current loads and the wave drift loads on a ship are discussed in detail. New experimental results are presented. It is found in the studied case that

1. Transverse current forces and yaw moments on the ship for small angles of attack are reasonable obtained by lifting theory combined with the Munk moment for the yaw moment. Transverse current forces and yaw moments on the ship for drift angles in the vicinity of $90°$ are reasonably obtained by the cross-flow principle combined with the Munk moment for the yaw moment.

2. Wave drift loads on a ship are very sensitive to wave heading and wave lengths, and no simple formulas can be derived showing the dependence on wave length, wave heading, structural form as well as wave induced motions of the ship. The method of Faltinsen and Michelsen is shown to give reliable results for drift forces and moments.

A seven degrees of freedom linear stability equation system for the ship and the buoy is applied. A simpler equation system is used to make general conclusions about the stability of the system and the possibility of large motion slowdrift oscillations of the ship and the buoy. Implications for design and operation are presented. It is particularly recommended to use:

1. Cross-mooring between the ship and the buoy.

2. Thrusters to change the equilibrium position of the tanker and thereby increase the stability of the ship-buoy system.

ACKNOWLEDGEMENT

The model tests were performed at the Ship Model Tank in Trondheim with the clever assistance of their staff. We will in particular acknowledge the work of Mr. Furunes and Mr. Trondsen.

REFERENCES

1. Koonce, K.T.:"Technology needs for deepwater operations", Proceedings of BOSS'76, Trondheim, 1976

2. Wichers, J.E.W.:"On the Slow Motions of Tankers Moored to Single Point Mooring Systems", OTC 2548, 1976

3. Remery, G.F.M. and G.van Oortmerssen.:"The Mean Wave, Wind and Current Forces on Offshore Structures and Their Role in Design of Mooring Systems", OTC 1941, 1973

4. Jacobs, W.R.:"Estimation of Stability Derivatives and Indices of Various Ship Forms and Comparison with Experimental Results", Davidson Laboratory, R-1035, Sept. 1964

5. Newman, J.N.:"Marine Hydrodynamics", The MIT Press, Cambridge, Massachusetts, 1977

6. Faltinsen, O.M. and Michelsen, F.C.:"Motions of Large Structures in Waves at Zero Froude Number", Proceedings of International Symposium of The Dynamics of Marine Vehicles and Structures in Waves, University College London, 1974

7. Norrbin, N.:"Forces in Oblique Towing of a Model of a Cargo Liner and a Divided Double-Body Geosim", The Swedish State Shipbuilding Experimental Tank, Publ. No.57 1965

8. English, J.W. and Wise, D.A.:"Hydrodynamic Aspects of Dynamic Positioning", Trans. North East Coast Inst. of Engineers and Shipbuilders. Vol. 92, No. 3, 1976, pp. 53-72

9. Delany, N.K. and Sorensen, N.E.:"Low-Speed Drag of Cylinders of Various Shapes", NACA, Techn. Note 3038, 1953

10. Faltinsen, O.M. and Løken, A.E.:"Slow Drift Oscillations of a Ship in Irregular Waves", Applied Ocean Research, 1, 1979

11. Maruo, H.:"The drift of a body floating in waves", Journal of Ship Research, 1969

APPENDIX A

ASYMPTOTIC THEORY FOR WAVE DRIFT LOADS

Consider a structure in incident regular waves. We assume the structure has vertical sides at the waterplane, and that the wave length is small compared to some characteristic length of the structure. In the case of a ship, the draught of the ship will be a suitable characteristic length. Due to the small-wave length assumption, the wave excitation forces will be small. This implies that the wave induced motions of the structure can be neglected. Due to the small-wave length assumption and the rapid exponential decay of the waves down in the fluid, it is only a small part of the structure close to the waterplane that will affect the flow-field. This implies that we may replace the structure by a stationary, vertical infinitely long cylinder with cross-section equal to the water plane area of the structure.

We will neglect all viscous effects and base the procedure on potential theory. In the analysis we neglect diffraction effects from sharp corners of the structure, and assume that the change in the waterplane area is small over a wave length.

We will write the incident wave potential as

$$\phi_I = \frac{g\zeta_a}{\omega} e^{kz} \cos(kx\cos\mu + ky\sin\mu - \omega t) \tag{A1}$$

where g is the acceleration of gravity, ζ_a the amplitude of the incident waves, ω the circular frequency of the waves, k the wave number, μ the wave propagation direction with respect to the x-axis (See Figure 29) and t is the time variable.

We will introduce local coordinate systems (n', s') along the waterplane area curve (see Figure 29). Here n' is orthogonal to the waterplane area curve and s' tangential to the waterplane area. We may then write the incident wave potential as

$$\phi_I = \frac{g\zeta_a}{\omega} e^{kz} \cos(kn'\sin(\theta+\mu) + ks'\cos(\theta+\mu) + kx_0\cos\mu + ky_0\sin\mu - \omega t) \tag{A2}$$

where (x_0, y_0) are the x-,y-coordinates of the origin of the local coordinate system defined in the figure. θ is also defined in the figure.

When the waves hit the structure they will be totally reflected. This implies that in certain regions around the structure there will be shadow regions, i.e. no flow at all (see Figure 29).

In the non-shadow areas the total velocity potential can be written as a sum of the incident wave and the reflected wave, i.e.

$$\phi = 2\frac{g\zeta_a}{\omega} e^{kz} \cos(ks'\cos(\theta+\mu) + kx_0\cos\mu + ky_0\sin\mu - \omega t) \cdot \cos(kn'\sin(\theta+\mu)) \tag{A3}$$

The force on the body can be obtained from the Bernoulli's equation

$$p = -\rho\frac{\partial\phi}{\partial t} - \rho g z - \frac{\rho}{2} V^2 \tag{A4}$$

where ρ is the mass density of the water and V is the fluid velocity. We may write the mean drift force components and yaw moment correct to second order in wave amplitude as

$$\bar{F}_i = \int_L \left\{ \frac{\rho}{g} \left(\frac{\partial\phi}{\partial t}\right)^2_{z=0} - \frac{\rho}{2g}\left(\frac{\partial\phi}{\partial t}\right)^2_{z=0} - \frac{\rho}{4k} V^2_{z=0} \right\} n_i \, d\ell \tag{A5}$$

where i is 1, 2 or 6 and $\bar{F}_i$ is the drift force component in the x-direction, $\bar{F}_2$ is the drift force component in the y-direction and $\bar{F}_6$ is the drift yaw moment with respect to the z-axis, and

$$n_1 = \sin\theta$$

$$n_2 = \cos\theta$$

$$n_3 = x_o\cos\theta - y_o\sin\theta$$

The integration in (A5) is along the non-shadow part L of the waterplane curve. The bar denotes time average. The index z = 0 indicates that the variable should be evaluated at z = 0. The first term in the brackets is due to the first term in Bernoulli's equation, the second term is due to the second term in Bernoulli's equation, while the last term is due to the last term in Bernoulli's equation. By substituting (A3) we will find that

$$\bar{F}_i = \frac{\rho g \zeta_a^2}{2} \int_L \sin^2(\theta+\mu) n_i \, d\ell \tag{A6}$$

This formula agrees with Maruo's formula (Maruo, Ref. 11) for the longitudinal force on a ship in head sea. Maruo derived his formula by using the equations for conservation of momentum and energy in the fluid.

Equation (A6) should be quite easy to apply in practise. We have worked out three special cases below.

Example 1
Infinitely long horizontal cylinder

$$\bar{F}_1 = 0, \quad \bar{F}_2 = \frac{\rho g}{2} \zeta_a^2 \sin^2\mu \cdot L, \quad \bar{F}_6 = 0$$

Example 2
Structure with circular waterplane area of radius r.

$$\bar{F}_1 = \frac{2}{3}\rho g \zeta_a^2 r \cos\mu, \quad \bar{F}_2 = \frac{2}{3}\rho g \zeta_a^2 r \sin\mu, \quad \bar{F}_6 = 0$$

Example 3
Structure with waterplane area consisting of two circular ends of radius r and a parallel part of length 2b (see Figure 30).

We find that

$$\bar{F}_1 = \frac{2}{3}\rho g \zeta_a^2 r \cos\mu, \quad \bar{F}_2 = \rho g \zeta_a^2 \{\frac{2}{3}r \sin\mu \pm b \sin^2\mu\}$$

$$\bar{F}_6 = -\frac{\rho g \zeta_a^2}{3} b r \sin 2\mu$$

where the plus-sign is valid when $0 \leq \mu \leq \pi$ and the minus-sign is valid when $\pi \leq \mu \leq 2\pi$. The formulas above give an indication of the angle-dependence of drift forces and moments on a ship in regular waves of small wave length.

Table 1.   MAIN PARTICULARS OF 130 000 DWT TANKER

PARAMETRE	UNIT	CONDITION	
		BALLAST	LOADED
Length between perpendiculars	m	285.60	285.60
Beam	m	46.71	46.71
Depth	m	20.35	20.35
Draught fore	m	4.84	13.82
Draught aft	m	7.04	13.82
Draught mean	m	5.94	13.82
Displacement	tons	61754.00	154980.00
Centre of gravity, longitud. from ⊗	m	+ 2.10	+ 6.46
Centre of gravity, transv. from baseline	m	9.73	11.03
Metacentric height	m	21.50	8.97
Pitch/yaw radius of gyration	m	71.40	71.40
Roll radius of gyration	m	16.35	16.35
Natural pitch period	s	9.80	9.80
Natural roll period	s	9.40	12.80

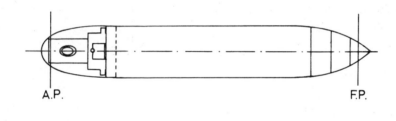

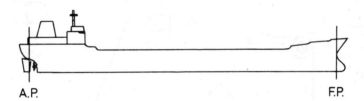

BODY PLAN

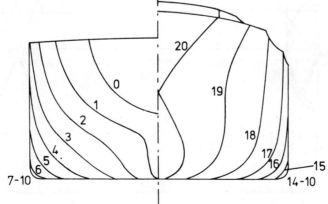

Figure 1   Body plan of tanker model

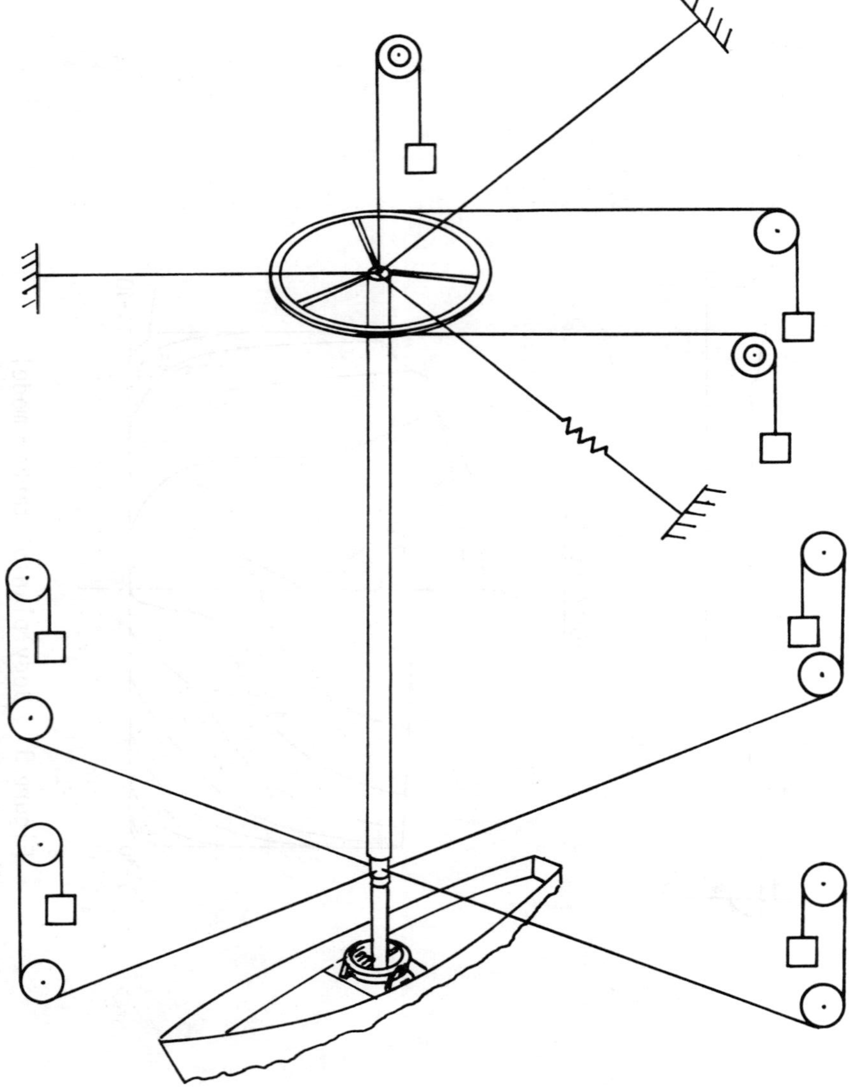

Figure 2 Experimental setup for testing of tanker model alone

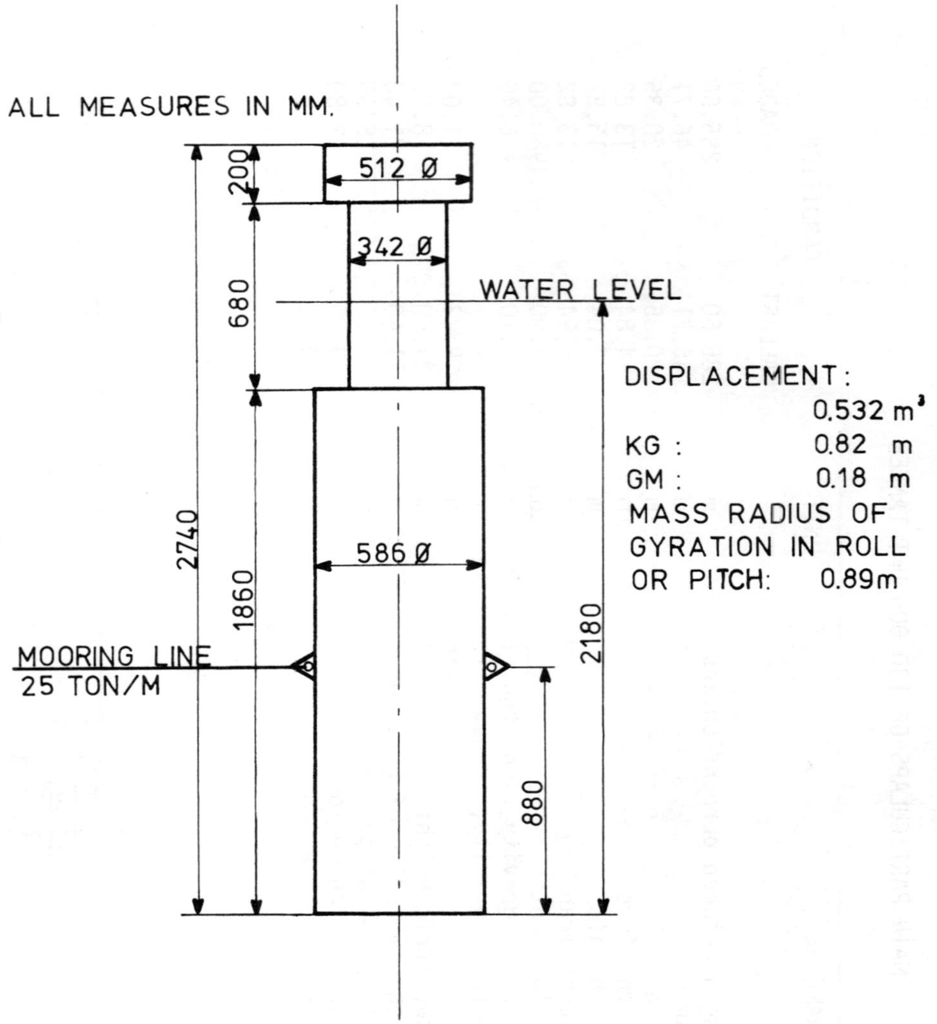

ALL MEASURES IN MM.
512 Ø
342 Ø
WATER LEVEL
200
680
2740
1860
586 Ø
MOORING LINE
25 TON/M
880
2180
DISPLACEMENT:
0.532 m³
KG : 0.82 m
GM : 0.18 m
MASS RADIUS OF
GYRATION IN ROLL
OR PITCH: 0.89m

Figure 3 Buoy model

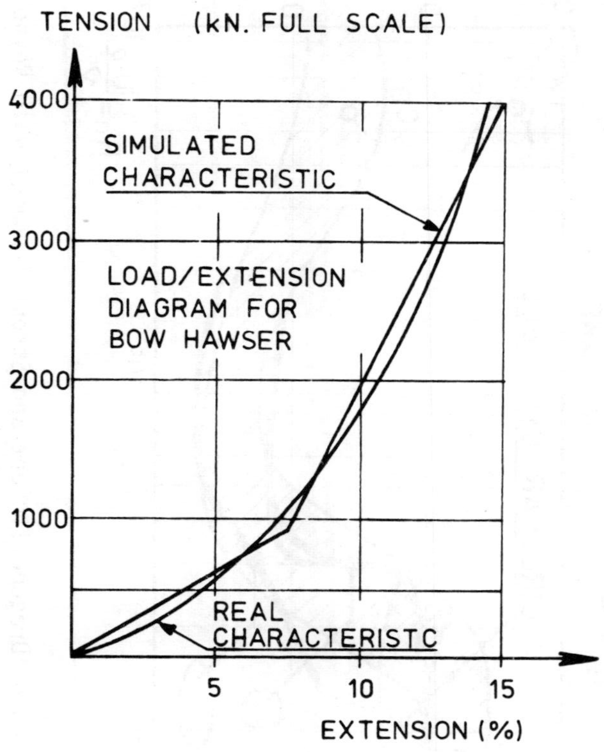

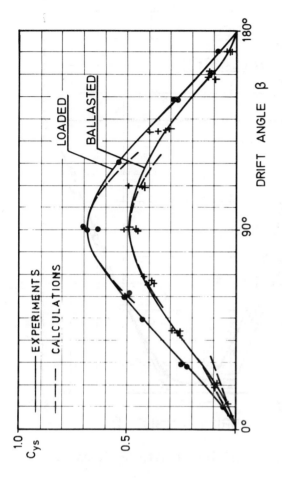

Figure 5 Variation of Transverse Current Force Coefficient with Drift Angle

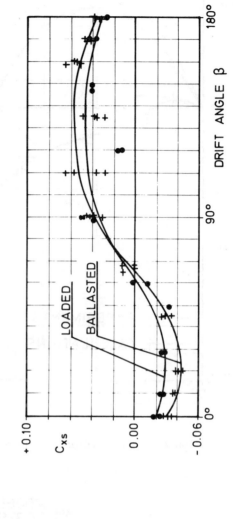

Figure 6 Variation of Longitudinal Current Force Coefficient with Drift Angle

Figure 4 Bow hawser characteristics

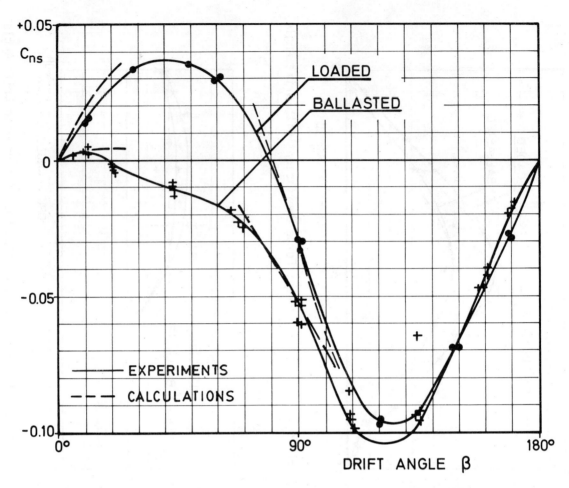

Figure 7   Variation of Current Moment Coefficient with Drift Angle

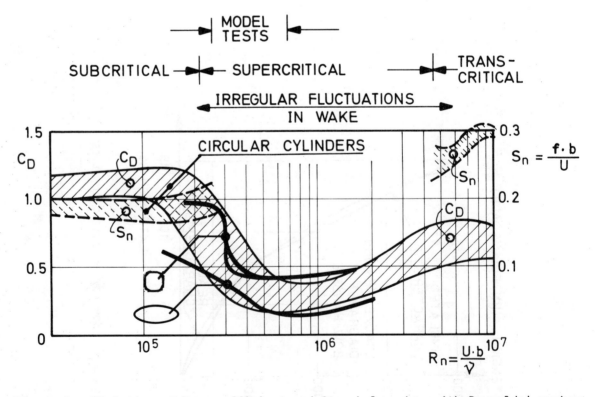

Figure 8   Variation of Dragcoefficient and Strouhal number with Reynolds' number

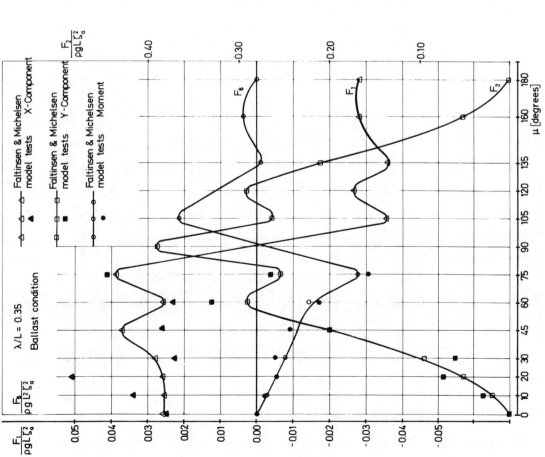

Figure 9 Wave drift loads as a function of wave heading

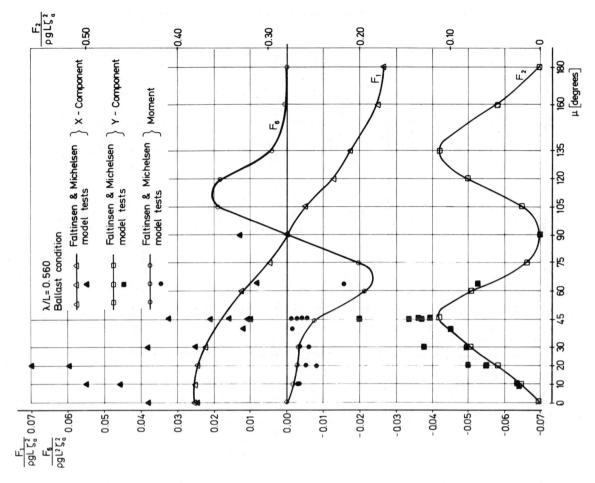

Figure 10 Wave drift loads as a function of wave heading

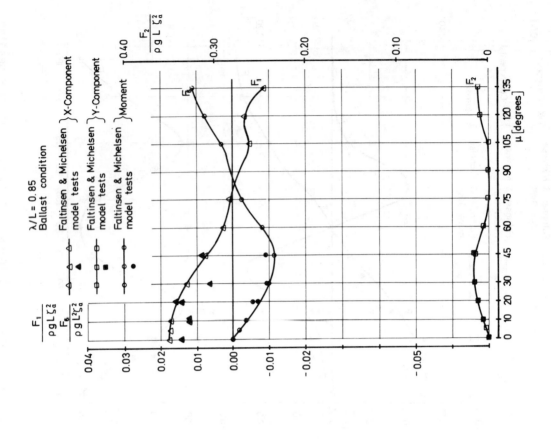

Figure 12 Wave drift loads as a function of wave heading

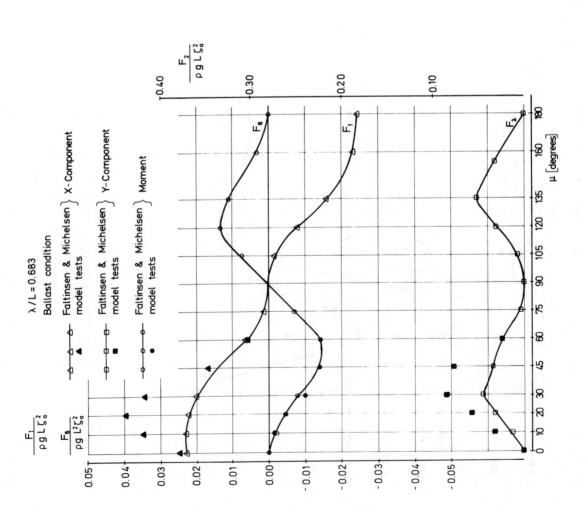

Figure 11 Wave drift loads as a function of wave heading

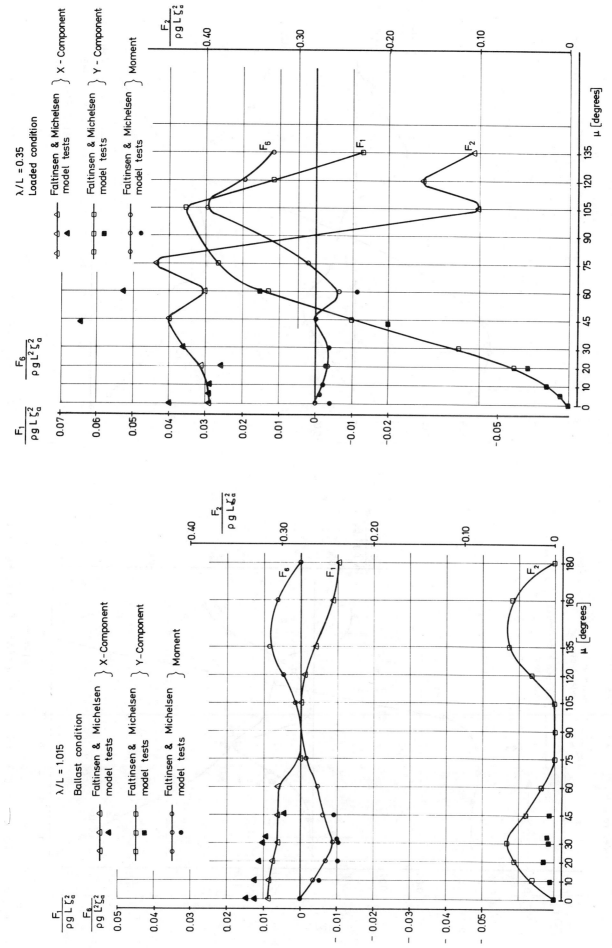

Figure 14 Wave drift loads as a function of wave heading

Figure 13 Wave drift loads as a function of wave heading

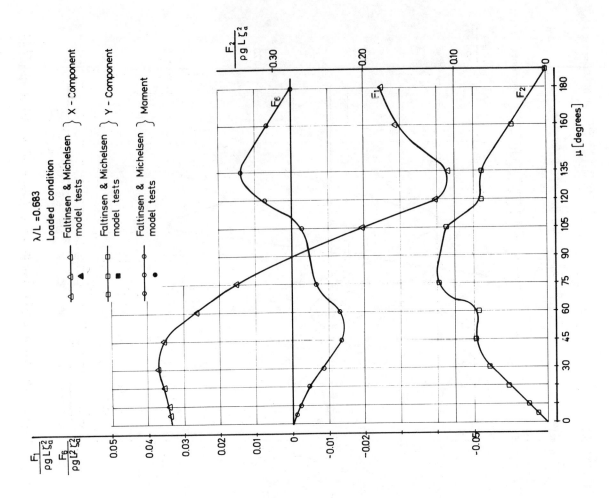

Figure 16 Wave drift loads as a function of wave heading

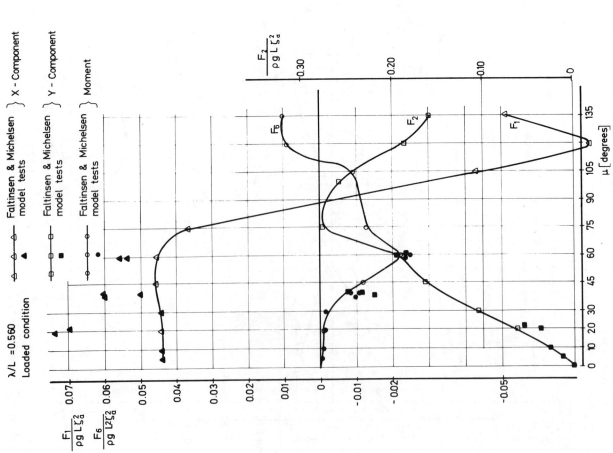

Figure 15 Wave drift loads as a function of wave heading

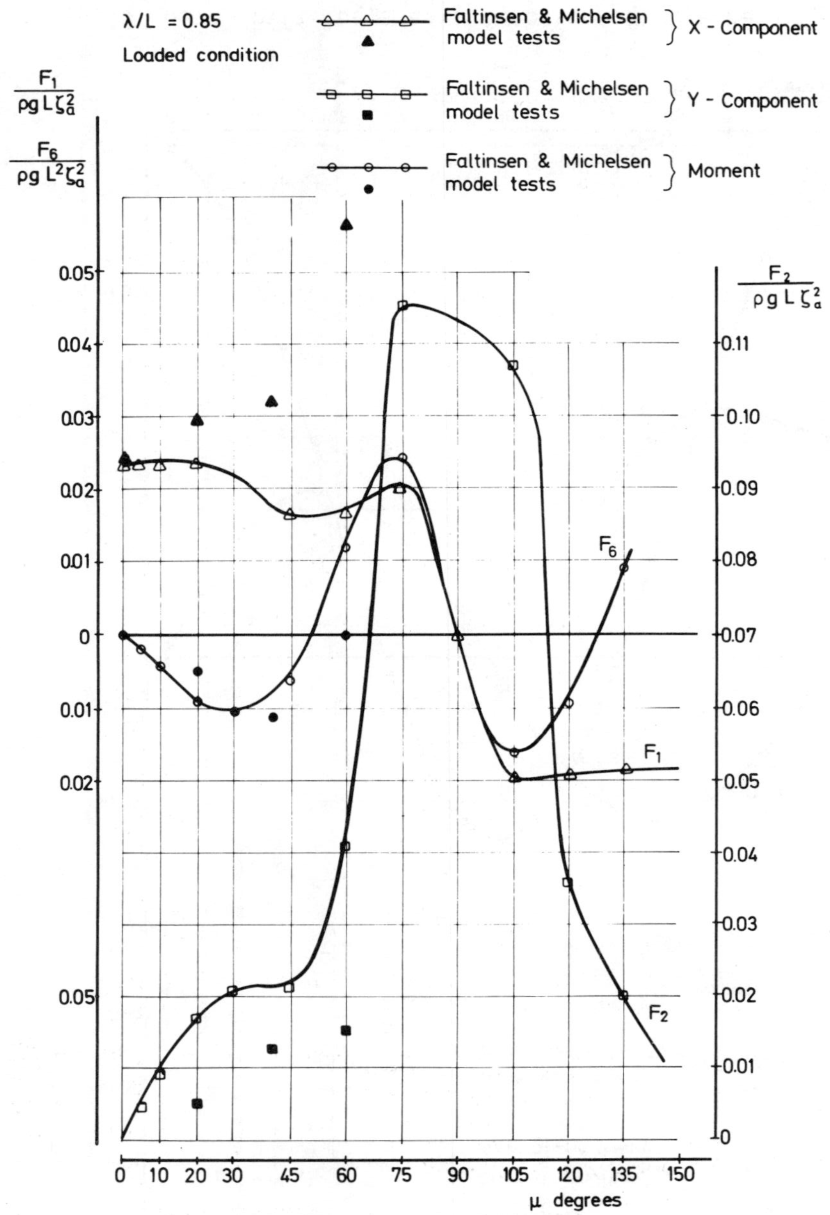

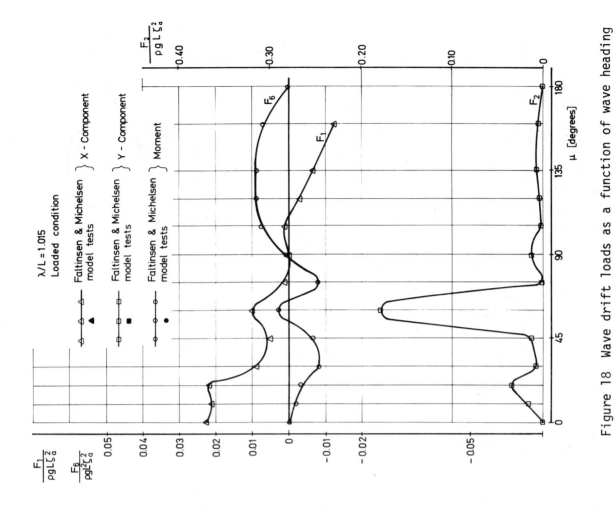

Figure 17 Wave drift loads as a function of wave heading

Figure 18 Wave drift loads as a function of wave heading

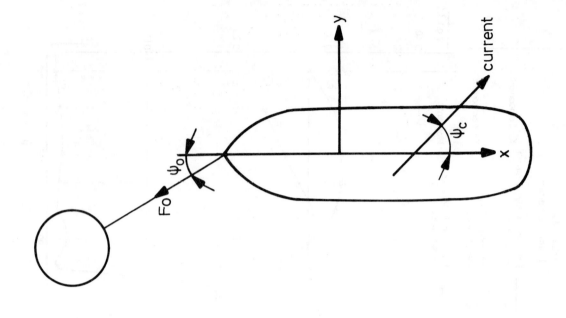

Figure 20 Definition of parameters in ship-buoy system.

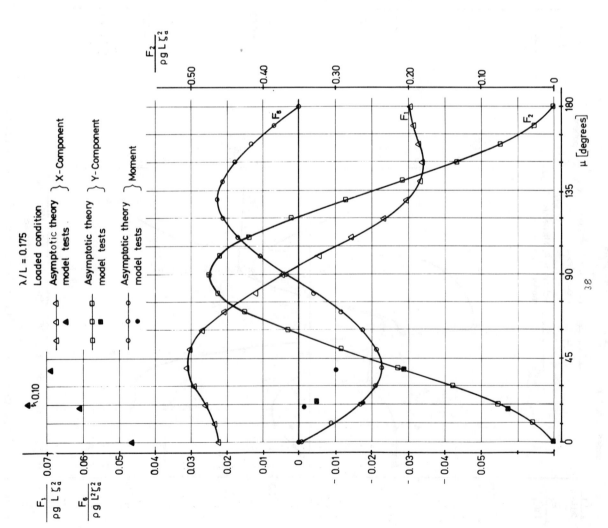

Figure 19 Wave drift loads as a function of wave heading

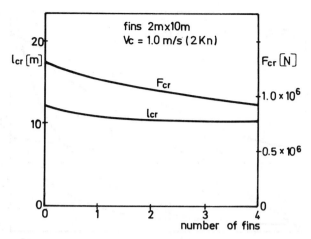

Figure 21 Critical length ℓ_{crit} and critical bow hawser tension F_{crit} for a 130 kDWT fully loaded tanker as a function of the number of fins

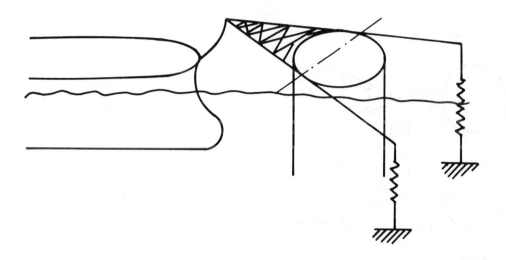

Figure 22 Mooring arrangement

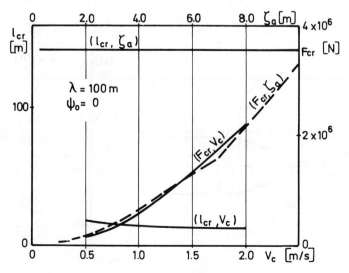

Figure 23 Critical length ℓ_{crit} for a 130 kDWT fully loaded tanker as a function of current velocity and wave amplitude

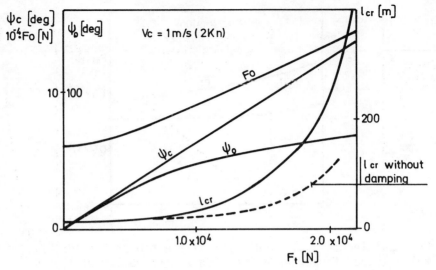

Figure 24 Critical length ℓ_{crit} for a 130 kDWT fully loaded tanker as a function of side thrust by a thruster located close to A.P.

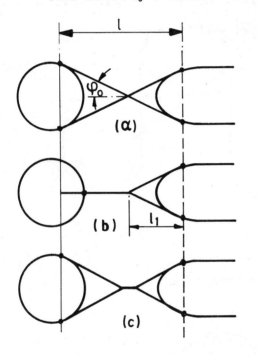

Figure 25 Mooring arrangement

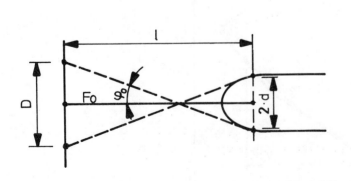

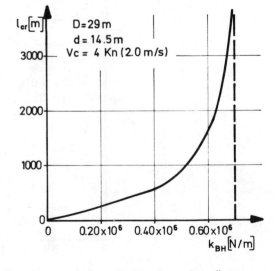

Figure 26 Critical distance ℓ as a function of bow hawser elasticity k_{BH} for "cross-mooring" of a 130 kDWT fully loaded tanker

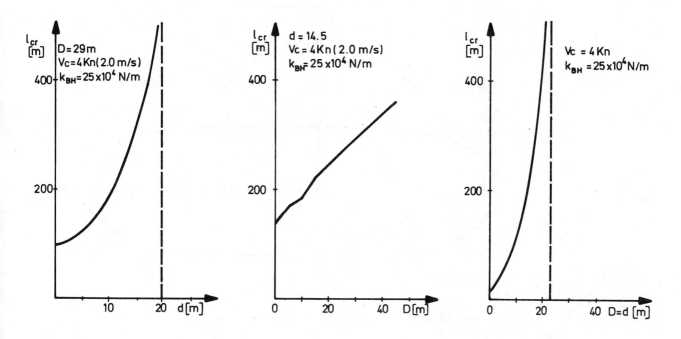

Figure 27 Critical distance ℓ as a function of D and d of a 130 kDWT fully loaded tanker

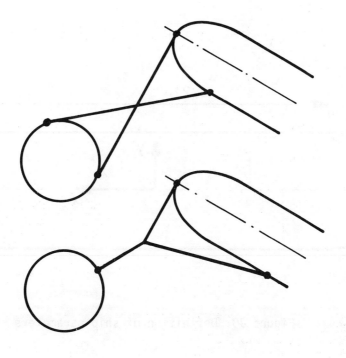

Figure 28 Mooring arrangement

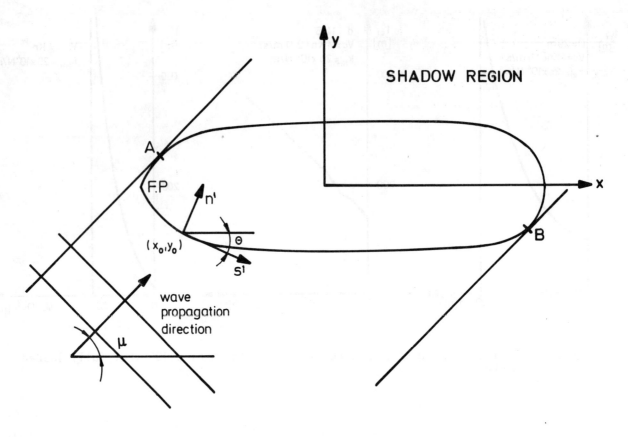

Figure 29 Definition of ship and wave parameters

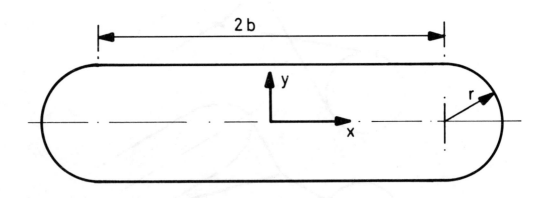

Figure 30 Definition of ship parameters

BOSS'79

Second International Conference on Behaviour of Off-Shore Structures

Held at: Imperial College, London, England
28 to 31 August 1979

PAPER 60

PRODUCTION RISER ANALYSIS

M.P. Harper, PhD.

Koninklijke/Shell Exploratie en Produktie Laboratorium, Rijswijk, Netherlands.

Summary

The dynamic analysis of a multibore production riser in a northern North Sea environment is discussed.
Results presented concern the bending moment response due to random wave loading and the sensitivity of these results to variations in hydrodynamic coefficients.

1. INTRODUCTION

A possible alternative for offshore production is a system comprising a floating production unit linked to a subsea manifold by a multibore riser (Fig. 1). This riser conducts live crude to the floater and returns processed crude to a pipeline or tanker mooring. Accurate prediction of production riser dynamics is a key item in the evaluation of these concepts.

It is now recognised that such analysis should be based on a realistic 'random sea state' model, where riser response is treated as a random variable characterised by statistical distributions. This approach has been adopted in studying the behaviour of the multibore riser of Fig. 1.

This note specifically concerns the standard deviation of bending moment response ($\sqrt{m_{oM}}$). Results are presented to illustrate the dependence of this variable on the significant wave height ($H_{1/3}$) and mean period (T_m) of the input sea spectra. In addition, the sensitivity of these results to variations in the empirical hydrodynamic drag and inertia coefficients (C_D, C_M) is investigated.

2. BACKGROUND

The structural representation of a production riser is straightforward. Assuming that the deflections of the central export riser and the peripheral flowlines are identical, the riser may be modelled as a single beam. For this analysis a two-dimensional, small displacement, finite element formulation was employed.

The interaction of the riser with the random sea environment is less clearly defined. The 'relative velocity' form of the Morison equation was assumed here for the external forcing function, viz. (for a two-dimensional model):

$$\frac{d\tilde{f}}{dz} = \frac{1}{2} \rho d\, C_D\, (\tilde{V}_w + V_c - \dot{\tilde{x}})\, |\, \tilde{V}_w + V_c - \dot{\tilde{x}}\, | + \frac{1}{4} \rho \pi d^2\, (C_M \tilde{a}_w - (C_M - 1)\, \ddot{\tilde{x}})$$

where $\frac{d\tilde{f}}{dz}$ is the normal force per unit length on a riser of representative diameter d, ρ is water density, V_w (a_w) is the normal component of water-particle velocity (acceleration) due to wave action, V_c is current velocity, and $\tilde{x}$ is the horizontal displacement of the riser. The dot notation indicates differentiation with respect to time and $\sim$ overbar denotes a random variable.

The validity of the above hydrodynamic forcing function is questionable. The drag/damping term is formed by the linear superposition of three velocity fields, one of which (V_w) is orbital in nature. It may be argued that the different turbulence fields associated with these velocity components will cause their interaction to be non-linear. Further work is required to determine the errors involved.

Difficulty also arises in the values assumed for C_D and C_M. Model tests on <u>fixed</u> multibore arrays reflect a strong dependence of C_D and C_M on both Reynolds' Number and Keulegan-Carpenter Number (Ref. 1, 2). There are no unique values for these flow parameters in a random sea and thus only wide ranges for C_D and C_M may be defined.

In view of these uncertainties, appreciation of the sensitivity of riser response to variation of the hydrodynamic coefficients is vital.

3. SOLUTION PROCEDURE

Due to the non-linearity of the 'relative velocity' Morison hypothesis, the riser problem has been solved by simulation in the time domain.

A Pierson-Moskowitz form, characterised by $H_{1/3}$ and T_m, is assumed for every unidirectional surface elevation spectrum. Linear wave theory is invoked. Correlated time series for water-particle velocities and accelerations and riser-top motions are realised by discretising the input spectrum into a finite number of randomly phased components. A pre-processor program, based on an inverse Fast Fourier Transform method, is used to generate and store these time series. A linear coordinate mapping is used to refer wave properties to the instantaneous surface elevation.

These input data are then accessed by the main riser program and a dynamic solution obtained directly via the Houbolt integration scheme. Mean values, standard deviations, and envelopes of riser response are obtained directly from this program. Complete output time series can be stored in a data file, and post-processor programs used for further comprehensive probabilistic/spectral analysis.

Experience has shown that simulation of 30 minutes of real time (in $\frac{1}{2}$ sec. timesteps) is sufficient to obtain reliable response statistics. Excluding downstream statistical analysis, total CPU time for one seastate is approximately 3 minutes on a UNIVAC 1100/82.

4. RESULTS

The ranges of $H_{1/3}$ and T_m used in this analysis reflect the environment of the northern North Sea. For results presented here current velocity was taken as zero and base values of $C_D = 1$ and $C_M = 2^*$ were assumed.

4.1. Variation of $\sqrt{m_{oM}}$ with $H_{1/3}$ and T_m

Distributions of $\sqrt{m_{oM}}$ along the length of the riser are shown in Fig. 2. Maxima are seen in the upper riser due to direct wave action, and in the lower riser due to dynamic response. It should be noted that modelling of wave properties up to the instantaneous surface elevation results in non-zero values for mean bending moment in the wave zone.

The variation of $\sqrt{m_{oM}}$ was investigated for two fixed points on the riser in the neighbourhood of the maxima, viz. points A (+146 m) and B (+64 m), Fig. 2. For the upper point A, results indicate non-linear dependence on $H_{1/3}$, Fig. 3. Lines of constant wave steepness** are also plotted, showing that at higher $H_{1/3}$ values the bending moment response at constant steepness varies little.

A similar plot of $\sqrt{m_{oM}}$ for the lower point B, Fig. 4, reveals less dependence on T_m and a more linear variation with $H_{1/3}$.

In both upper and lower riser $\sqrt{m_{oM}}$ increases with decreasing T_m. Also, bending moments at both points are of similar magnitudes for $H_{1/3} \leq 6$ m; for $H_{1/3} > 6$ m the response in the upper riser becomes the more severe.

4.2. Variation of $\sqrt{m_{oM}}$ with C_D and C_M

The sensitivity of $\sqrt{m_{oM}}$ to variations in C_D, C_M is illustrated by Fig. 5 (point A) and Fig. 6 (point B). Bending moment at the upper point A varies strongly and directly with C_D, the magnitude of the effect increasing with $H_{1/3}$. Conversely, for the lower riser the response is only weakly influenced by C_D, and then mostly in an inverse manner.

A higher C_M value results in increased $\sqrt{m_{oM}}$ values at both upper and lower riser levels. The magnitude of the increase is similar at both points, and for the upper point A does not vary significantly with $H_{1/3}$.

5. CONCLUDING REMARKS

In assessing the feasibility of a production riser system, the specific 'engineering' objectives of analysis are:

- a storm 'disconnect' policy and procedure;
- specifications for riser components;
- an estimate of long-term structural integrity and damage tolerance.

To meet these objectives, the statistics of many response variables must be considered, e.g. stress/bending moment, bottom angle, riser-floater relative displacement etc., and analysis should include both connected and free-hanging riser cases.

The work reported here forms part of a wider investigation aimed at establishing analytical methods for calculating the statistics of riser response and, perhaps more importantly, developing a rationale for interpretation of these statistics.

ACKNOWLEDGEMENTS

The author gratefully acknowledges work by Dr R.P. Nordgren and co-workers at Shell Development Company, Bellaire Research Centre, Houston, in developing the original riser mathematical model.

---

* C_D is based on the pitch circle diameter of the multibore array.
C_M is based on the sum of individual cylinder areas.
** Wave steepness is the ratio of $H_{1/3}$ to the wavelength of a regular wave with period T_m.

REFERENCES

1. Sarpkaya, T.: "Wave Loading in the Drag/Inertia Regime with Particular Reference to Groups of Cylinders", Proc. Symposium on Mechanics of Wave-Induced Forces on Cylinders, International Association for Hydraulic Research, University of Bristol, U.K., Sept. 1978.

2. Joint industry model test programme on hydrodynamic loading of multibore production riser sections. Results confidential to participants.

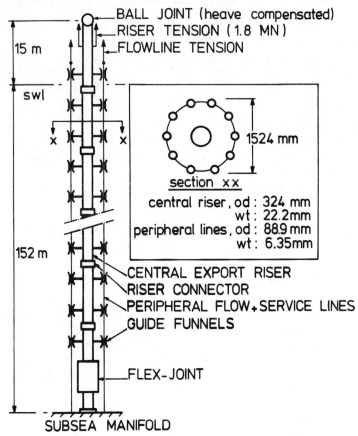

Fig. 1 Multibore production riser schematic

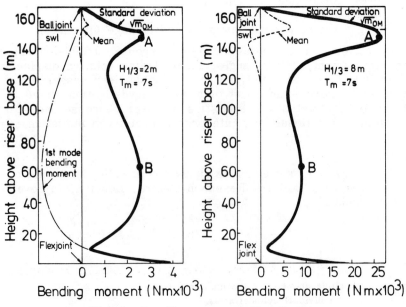

Fig. 2 Bending moment distributions

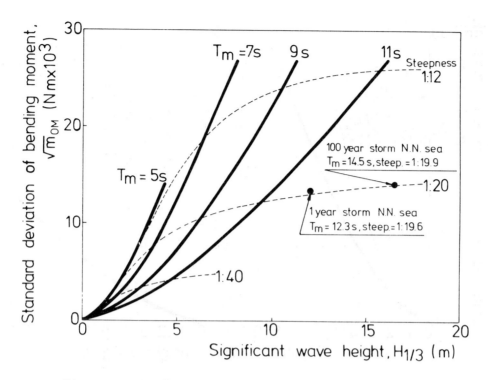

Fig. 3 Bending moment response in upper riser (point A)

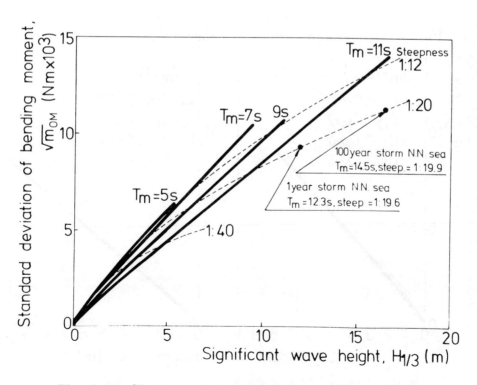

Fig. 4 Bending moment response in lower riser (point B)

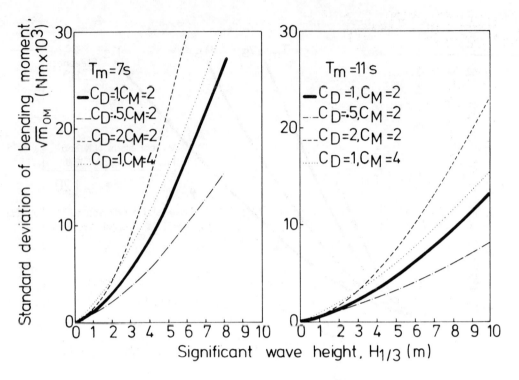

Fig. 5 Effect of C_D, C_M on bending moment in upper riser (point A)

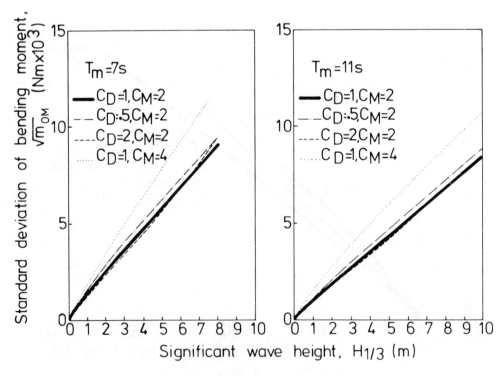

Fig. 6 Effect of C_D, C_M on bending moment in lower riser (point B)

BOSS'79

PAPER 61

Second International Conference on Behaviour of Off-Shore Structures

Held at: Imperial College, London, England
28 to 31 August 1979

ON THE HYDRODYNAMIC ANALYSIS OF ARBITRARILY SHAPED AND MULTIBODIED MARINE STRUCTURES – APPLICATION TO WAVE-ENERGY DEVICES AND SHIP/DOCK INTERACTION

M. Katory and A.A. Lacey

Hong Kong Polytechnic, Hong Kong   Institute of Mathematics, Oxford, U.K.

Summary

The motion of a multibodied marine structure in regular waves is described by one system of coupled simultaneous equations. The number of the equations in the system is determined by the number of the component bodies of the structure, and by the way in which the bodies are joined to each other.

The method can be used to deal with many types of offshore structures including tow-out arrangements, wave-energy devices and ship/dock interaction.

If the bodies are joined by articulation, the hinge forces can also be determined by this method. Two-dimensional sources of the Green's function type are adopted to determine the sectional hydrodynamic properties of the structure's multi-bodies, which are separated by water. Some or all of the component bodies may be asymmetric.

The usual strip-theory technique is then used to determine the coefficients of the equations of the structure's motion.

An application is made for two wave-energy devices and the ship/dock interaction.

Sponsored by: Delft University of Technology, The Netherlands
Massachusetts Institute of Technology, U.S.A.
The Norwegian Institute of Technology, Norway
University of London, England

Secretariat provided by: BHRA Fluid Engineering

Copyright: © BHRA Fluid Engineering
Cranfield, Bedford, England

1. Statement of the problem and introduction.

   The motion of a multibodied marine structure in regular waves can be described by one system of coupled simultaneous equations. The number of equations in the system is determined by the number of the component bodies of the structure, and by the way in which the bodies are joined to each other.

   The method can be used to deal with many types of offshore structures, including tow-out arrangements, wave-energy devices and ship/dock interaction.

   If the bodies are joined by articulation, the hinge forces can also be determined by this method. Two-dimensional pulsating sources of the Green's function type are adopted to determine the sectional hydrodynamic properties of the structure's multi-bodies, which are separated by water. Some or all of the component bodies may be asymmetric.

   The usual strip-theory technique is then used to determine the coefficients of the equations of the structure's motion.

2. The two-dimensional Green's function.

   The two-dimensional Green's function describes the velocity potential $G(z,\zeta;t)$ at a point z due to a pulsating source of unit strength at point ζ in the fluid domain (see Fig. 1), and is written in the following form (Ref. 1):

   $$G(z,\zeta;t) = \frac{1}{2\pi}\left[\log(z-\zeta) - \log(z-\bar{\zeta}) - 2PV\int_0^\infty \frac{e^{-ik(z-\bar{\zeta})}}{k-\nu}dk\right]\cos\omega t - e^{-i\nu(z-\bar{\zeta})}\sin\omega t \qquad (1)$$

   where ω = frequency of oscillation
   $\nu = \omega^2/g$
   g = acceleration due to gravity
   $i = \sqrt{-1}$
   t = time
   z, ζ and $\bar{\zeta}$ are as shown in Fig. 1.

   The symbol PV means that only the principal value of the integral is to be taken.

   This function, which is suitable for infinite depths of water, satisfies the conditions of irrotationality of the flow field, the boundary condition at the free surface and the radiation condition of the outgoing waves. A condition which has yet to be satisfied is that of the boundary condition on the surfaces of the bodies.

   2.1 Determination of the sectional hydrodynamic properties.

   If each of the two-dimensional bodies in Fig. 1 is forced to oscillate in each of the modes (sway, heave and roll) at a time when the other bodies are assumed fixed at their neutral position in the previously undisturbed calm fluid surface, the resulting velocity potential created at the point z can be expressed by

   $$\phi_S^i(z) = \int_S Q^i(\zeta) \cdot G(z,\zeta)\,dS \qquad (2)$$

   where $Q^i(\zeta)$ = source strengths to be determined
   superscript i refers to the mode of motion of the oscillating body
   S refers to the submerged surfaces of the bodies.

   The source strengths $Q^i(\zeta)$ are obtained by satisfying the boundary condition on the bodies' surfaces. The condition is that the fluid velocity normal to the surface of the body equals the velocity of the body surface, viz:

   $$n \cdot \nabla \phi_S^i = -\delta_{ij} \cdot n \cdot v_j \qquad (3)$$

where **n** is the outgoing unit vector normal
v_j is the velocity vector associated with the motion of the jth co-ordinate
δ_{ij} is the Dirac delta function
∇ is the differential operator.

Applying this condition at a number of points around the submerged contours of the bodies, Eqn (3) results in a system of simultaneous integral equations, the solution of which will yield the values of the source strengths $Q^i(s)$, and hence the velocity potential $\phi_s^i(s)$ at each of the points.

The associated pressure p_i acting on the surface can then be determined by employing a linearised Bernoulli's equation, viz:

$$p_i = -\rho \frac{\partial}{\partial t}[\phi_s^i(s)] \qquad (4)$$

where ρ = the relative mass density of the fluid.

Upon separating p_i into its real and imaginary parts and integrating over the surface of each body, the added mass and damping coefficients, including the coupling terms, can be obtained.

If the process is repeated for other modes of motion and for the other bodies, the rest of the sectional coefficients required for the equations of motion can be similarly obtained, i.e. we have

$$\left. \begin{array}{l} a_{ijmn} = -\frac{1}{\omega^2} \int_S Re(p_i)\, ds \\ b_{ijmn} = -\frac{1}{\omega} \int_S Im(p_i)\, ds \end{array} \right\} \qquad (5)$$

where a_{ijmn} = the added mass effect of the mth body in ith mode on nth body in jth mode
b_{ijmn} = the wave damping effect of the mth body in ith mode on nth body in jth mode.

2.2 Determination of sectional wave forces and moments.

The forces and moments exerted on the bodies due to a regular wave of known potential ϕ_w can also be obtained by employing Green's function. In this case the diffracted velocity potential ϕ_D due to fixing all the bodies at their mean position among the waves is determined by satisfying the boundary condition on the surfaces of the bodies, viz:

$$n \cdot \nabla \phi_D + n \cdot \nabla \phi_w = 0 \qquad (6)$$

The forces and moments are then obtained by employing Bernoulli's equation for the pressures, in a manner similar to that described in section 2.1.

3. Evaluation of the principal value integral.

To evaluate the principal value integral, Frank has suggested the use of the sine and cosine integral series forms (Ref. 2). However, for large arguments of the exponent, which occur in the calculations for many types of offshore structure, the series converges very slowly and a large number of terms would be required to attain acceptable solution accuracy. This would lead to an uneconomically long stretch of computing time being used.

A better way to evaluate the integral is to split it into two parts, finite and infinite. The finite part can easily be converted into a standard series with a very small number of terms. The infinite part can be converted into another series, the type of which depends upon the magnitudes of both the real and imaginary parts of the exponent, so that the number of terms in the series can be made small.

For example, if the magnitude of the real and imaginary parts of the exponent is large, the following method of expansion can be used.

The integral is rewritten in the form shown below:

$$PV = \int_0^\infty \frac{e^{-ik(z-\bar{z})}}{k-\nu} dk = \int_0^\infty \frac{e^{\alpha t}}{t-1} dt \qquad (7)$$

where $\alpha = -i\nu(z-\bar{z})$

Again,
$$\int_0^\infty \frac{e^{\alpha t}}{t-1} dt = e^\alpha \left[\int_{-1}^1 \frac{e^{\alpha t}-1}{t} dt + \int_1^\infty \frac{e^{\alpha t}}{t} dt \right] \qquad (8)$$

For the finite integral we have
$$\int_{-1}^1 \frac{e^{\alpha t}-1}{t} dt = \sum_{n=1}^\infty \frac{2\alpha^{2n+1}}{(2n+1)!(2n+1)} \qquad (9)$$

For the infinite integral we have
$$\int_1^\infty \frac{e^{\alpha t}}{t} dt = \left[\sum_{n=0}^5 a_n \alpha^n \log\alpha - \sum_{n=0}^6 b_n \alpha^n \right] / \sum_{n=0}^6 c_n \alpha^n \qquad (10)$$

(The values of the coefficients a_n, b_n and c_n are given in Table 1.)

4. Comparison with the old method of calculation.

Added mass and damping coefficients in the heave mode obtained by this method are compared (in Fig. 2) with results obtained by Nordenstrom et al. (Ref. 3) for a catamaran section. The agreement is very good.

5. Application to the motion of multi-bodied marine structures in waves.

As an application of the method to multi-bodied marine structures, we consider two wave-energy devices and ship/dock interaction.

5.1 The wave-contouring raft.

The motion of the Cockerell wave-contouring raft (Ref. 4) can be described by the following system of equations:

$$\left\{ M_{nj} + \sum_{m=1}^N \sum_{i=1}^3 A_{ijmn} \right\} \ddot{x}_{mj} + \sum_{m=1}^N \sum_{i=1}^3 B_{ijmn} \dot{x}_{mj} + \sum_{i=1}^3 C_{ijn} x_{nj} + H_{nj} = F_{nj} \qquad (11)$$

where A_{ijmn} and B_{ijmn} are respectively the total added mass (or mass moment of inertia) and wave-damping influences due to the mth raft in ith mode on the nth raft in the jth mode

C_{ijn} = the total restoring force or moment influence due to the ith mode on the jth mode for raft n

F_{nj} = the wave force or moment in the jth mode for raft n

H_{nj} = the hinge force in the jth direction for raft n

M = the mass or mass moment of inertia in the jth mode for raft n

N = the number of rafts in the system

x_{nj}, $\dot{x}_{nj}$ and $\ddot{x}_{nj}$ are respectively displacement, velocity and acceleration in the jth mode for raft n.

The compatibility condition at the hinges connecting the rafts can be expressed by the following system of equations:

$$\left. \begin{array}{l} x_{1(n)} - \bar{y}_{(n)}(1 - \cos x_{3(n)}) = x_{1(n+1)} + \bar{y}_{(n+1)}(1 - \cos x_{3(n+1)}) \\ x_{2(n)} - \bar{y}_{(n)} \sin x_{3(n)} = x_{2(n+1)} + \bar{y}_{(n+1)} \sin x_{3(n+1)} \end{array} \right\} \qquad (12)$$

where $n = 1, 2, \ldots\ldots N-1$

$x_{i(n)}$ = the motion displacement in the ith direction for the raft n (i = 1, 2 and 3 for sway, heave and roll respectively)

$\bar{y}_{(n)}$ = the distance between the centre of gravity of the raft n to its r.h.s. hinge.

Equations (11) and (12) can then be solved simultaneously to yield the motion responses and the hinge forces for the raft arrangement.

For the arrangement shown in Fig. 3, the motion responses and hinge force amplitudes calculated by this method are shown in Fig. 4, 5 and 6.

5.2 The oscillating water-column wave-energy device.

The heaving motion of the oscillating water-column device (Ref. 4) can be described by the following two simultaneous equations; the first equation describes the motion of the device, which comprises two bodies attached rigidly to each other, and the second describes the motion of the internal free surface (see Fig. 7):

$$(M + A_{1122} + A_{2222} + A_{1222} + A_{2122})\ddot{x}_d + (B_{1122} + B_{2222} + B_{1222} + B_{2122})\dot{x}_d$$
$$+ (A_{3122} + A_{3222})\ddot{x}_s + (B_{3122} + B_{3222})\dot{x}_s + C_d x_d = F_{12} + F_{22}$$
$$(A_{1322} + A_{2322})\ddot{x}_d + (B_{1322} + B_{2322})\dot{x}_d + A_{3322}\ddot{x}_s$$
$$+ B_{3322}\dot{x}_s + C_s x_s = F_{32} \qquad (13)$$

where A_{ijmn} and B_{ijmn} are as defined in section 5.1
C_d = heave restoring force coefficient for the device
C_s = heave restoring force coefficient for the internal free surface
F_{12}, F_{22} and F_{32} are the vertical wave forces acting on body 1, body 2 (of the device) and the internal free surface respectively.
M = mass of the device
x_d, $\dot{x}_d$ and $\ddot{x}_d$ are respectively the displacement, velocity and acceleration of the device
x_s, $\dot{x}_s$ and $\ddot{x}_s$ are respectively the displacement, velocity and acceleration of the internal free surface.

The motion of the internal free surface relative to the device is expressed as

$$x = x_d - x_s \qquad (14)$$

For the device with the geometry shown in Fig. 7, the motion amplitude of the internal free surface obtained by the new method is shown in Fig. 8.

5.3 Ship/dock interaction.

Interaction between ship and dock arises from their being near to each other in the presence of sea-waves. When they are floating in parallel their motion in regular waves can be determined by solving a system of six simultaneous equations for sway, heave and roll for both vessels.

Ohkusu (Ref. 5) has used an approximate method, which is applicable only to two bodies of rectangular section, to deal with this problem. Fig. 9 shows a comparison of results obtained by the new method presented here, with Ohkusu's analytical and experimental results, for one arrangement in beam waves.

6. Conclusion.

In conclusion it can be stated that the two-dimensional Green's function method can be used in conjunction with the present analytical method for the determination of the motion in waves of many types of multi-bodied marine structures. The numerical difficulties associated with the evaluation of the principal value integral can be overcome by the method illustrated in this paper.

References:

1. Wehausen, J.V. and Laitone, E.V. : "Surface waves, Handbuch der Physik". Encyclopedia of Physics, ed. S. Fluegge, Vol. 9, Fluid Dynamics 3, Springer-Verlag, Berlin, 1960.

2. Frank, W. : "Oscillation of cylinders in or below the free surface of deep fluids". Naval Ship Research and Development Centre, Report 2375, Washington, 1967.

3. <u>Nordenstrom, N. et al.</u> : "Prediction of wave-induced motions and loads for Catamarans". Det Norske Veritas, Publication, No. 77, Oslo, September 1971.

4. <u>Kenward, M.</u> : "Waves a Million". New Scientist, London, May 1976.

5. <u>Ohkusu, M.</u> : "Ship motions in vicinity of a structure". Proceedings of the International Conference on the Behaviour of Off-Shore Structures, Norwegian Institute of Technology, Trondheim, 1976.

Table 1

n	a_n	b_n	c_n
0	-1	0.5772111	0.76273617
1	0.23721365	0.8632023	0.28388363
2	-0.0206543	0.024131325	0.066786033
3	0.000763297	0.017319882	0.012982719
4	0.0000097687	0.0011190676	0.0008700861
5		0.0002990067	0.00029892
6			

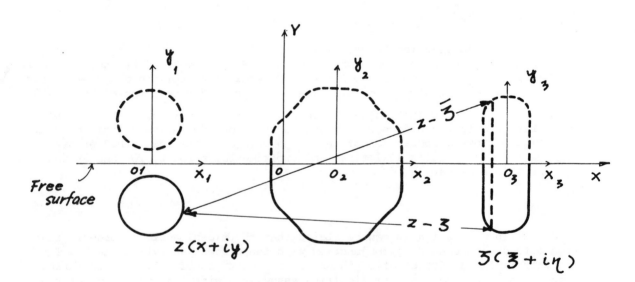

X, Y = Global Co-ordinates

$X_1, Y_2 \ldots Y_3$ = Local Co-ordinates

ξ, η, x and y are referred to global Co-ordinates

Fig 1 *Three bodies floating in fluid in the presence of a free surface*

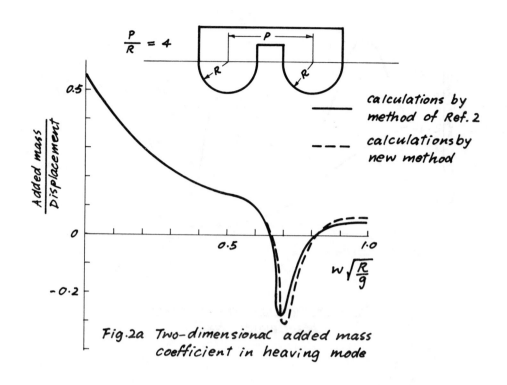

Fig.2a Two-dimensional added mass coefficient in heaving mode

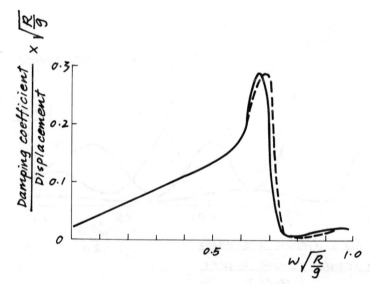

Fig 2b Two-dimensional damping coefficient in heaving mode

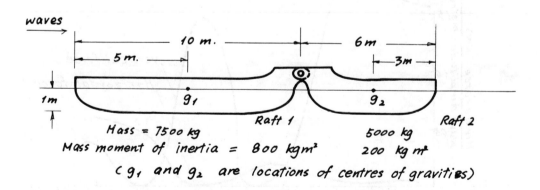

Fig 3 Wave contouring rafts

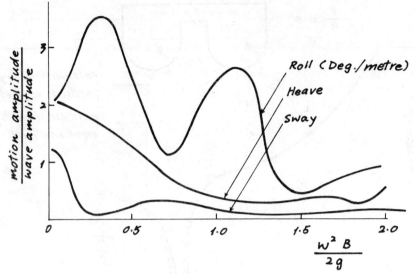

Fig. 4 Motion responses in regular beam waves - Raft 1

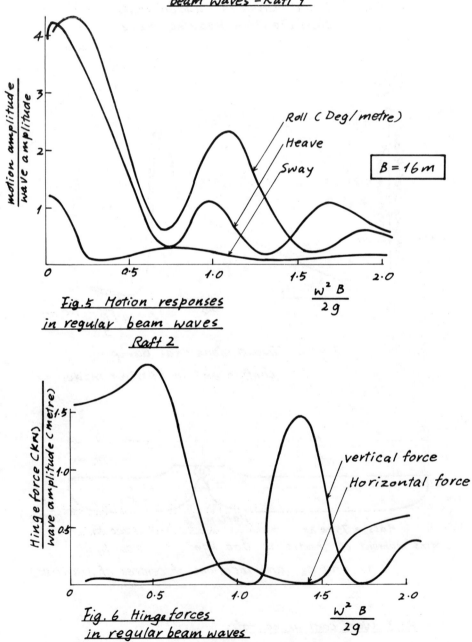

Fig. 5 Motion responses in regular beam waves Raft 2

Fig. 6 Hinge forces in regular beam waves

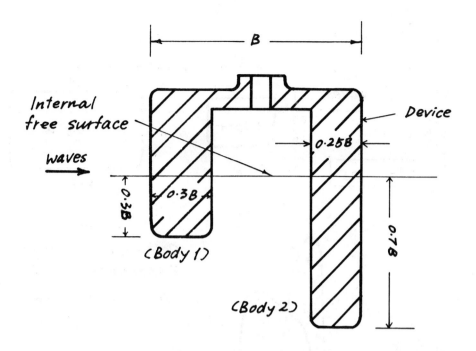

Fig 7 Oscillating water column

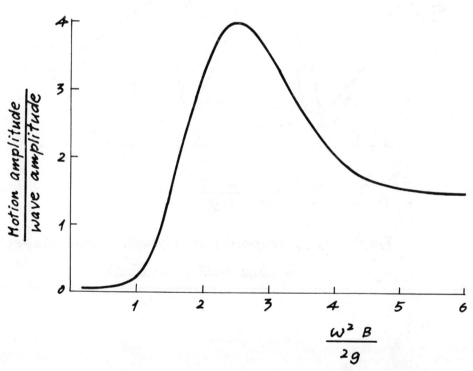

Fig 8 Motion of internal free surface relative to the device

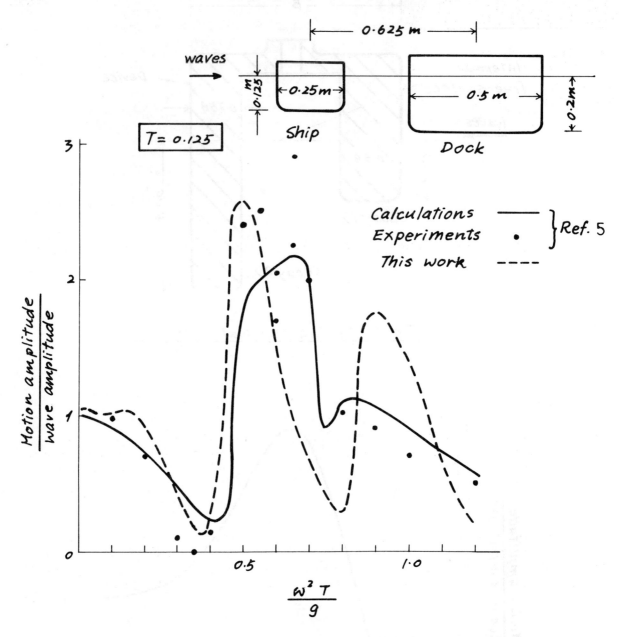

Fig 9 Sway response in regular beam waves for a ship next to a dock

BOSS'79

PAPER 62

Second International Conference
on Behaviour of Off-Shore Structures

Held at: Imperial College, London, England
28 to 31 August 1979

FLOATING BREAKWATERS AND TOPOLOGICALLY GENERATED CIRCULATION

Brian MacMahon, B.E., Ing., E.I.H. (Grenoble)

Imperial College, U.K.

Summary

The phenomenon of vortex formation when a body withdraws from a fluid, as exemplified in the action of a floating breakwater, is taken as a model for the development of a rotational flow about a body in a real fluid. The intermittent irrotational action of the wave crests on the breakwater elements suggests, in the general case, a similar periodic encountering by the body of separate fluid domains, as it makes displacements in time. This involves replacing the conventional three dimensional space framework embedded in a uniform stream of time by a four dimensional structure in which the time axis has a space like character. The irreversible continuous velocity fields that are ordinarily observed in fluids are, on this view, three dimensional sections of reversible four dimensional displacement patterns. Stokes' classic result for the resistance experienced by a sphere in slow translation through a viscous fluid is rapidly obtained on these assumptions. A major departure is the admission of the possibility of relative displacements in time, which appear to project into three dimensions as rotations.

Sponsored by: Delft University of Technology, The Netherlands
Massachusetts Institute of Technology, U.S.A.
The Norwegian Institute of Technology, Norway
University of London, England

Secretariat provided by: BHRA Fluid Engineering

Copyright: ©   BHRA Fluid Engineering
Cranfield, Bedford, England

Creation of a circulation in discontinuous motion

A common example of the sudden conversion of irrotational to rotational motion is the collapse of a moving cavitation bubble. The bubble forms in a low pressure region, is moved relative to the fluid by the pressure field, thereby acquiring a certain energy and momentum because of its hydrodynamic mass. The collapse or annihilation consequent on arrival in a region of higher pressure leaves behind a doublet pattern equivalent to a pair of vortices of opposite hand.

The breaking of a wave is another instance of a suddenly generated circulation in a largely irrotational motion. In the plunging type, so spectacularly effective in producing dissipation, the agent of topological change is the entrapped core of air.

The motion in the water due to a non heaving floating breakwater, composed for example of a series of two dimensional circular cylinders arranged with their axes perpendicular to the direction of wave propagation (Fig. 1), is in part a sequence of irrotational flows impulsively generated by successive crests as they submerge the breakwater sections. The subsequent arrival of the troughs produces a change in the topology of the flow domain boundary as the elements are uncovered. This allows the kinetic energy of the flow associated with each element to pass over into that of a vortex pair to be ultimately dissipated by viscosity. The increments or quanta of K.E. extracted at the wave frequency are readily calculable.

Calculation of the breakwater dissipation

In a wave on deep water of semi-amplitude a and frequency σ the constant magnitude of the directionally changing water velocity at the surface is $a\sigma$, giving the K.E. associated with the streaming motion due to the element as $\frac{M_H}{2}(a\sigma)^2$ where M_H is the hydrodynamic mass per unit width. This assumes negligible wake development - valid under certain combinations of period, amplitude and element dimension. Reflections of the first order wave and harmonic generation (Ref. 1) complicate the picture in practice. The rate of dissipation for N elements is the total kinetic energy per unit width divided by the period i.e. $\frac{NM_H}{4\pi}a^2\sigma^3$. The power transmission in the incident wave is $\frac{1}{2}\rho g a^2 \frac{g}{2\sigma}$ per unit crest length. To produce a given reduction in wave amplitude or energy the number of elements required is therefore proportional to $1/\sigma^4$ or to the fourth power of the wave period.

This is not an encouraging result. Other shapes, for example rectangular, would have a considerably greater hydrodynamic mass (Ref. 2) but would need to be more exactly matched to the wave conditions. The real value of the concept appears to the author not to lie in this particular application (there are other disadvantages e.g. mooring forces proportional to (amplitude)^2) but in the suggestive power of the assumed action.

Topological change in continuous motion

In view of the simplicity of the above mechanism of vortex generation one is tempted to ask whether other forms of fluid dissipation, such as wake formation, when a body is translated through a continuous fluid, might be explained in a similar way. This, one recalls, was the problem left by d'Alembert "for geometers to explain".

If the pattern in Fig. 2 represented a fluid at rest the resistance to the solid translating through the crests could be calculated from the rate of generation of impulse if the velocity were high enough not to allow time for boundary layer development. A nearer approach to motion through a continuous medium would be a displacement through a many fluid system separated by fluid free layers as in Fig. 3. The essential point that needs to be explained in the continuous case is how the irrotational doublet streamlines are "released" (as by the layers above) from a body so as to form, for example, a vortex wake.

Were we to assume that the time axis had a similar character to the space (x,y,z) axes, the motion of any body would consist in a passage through an infinite number of spaces - just as displacement along the z axes entails passing through an infinity of xy planes. As the body would be in each space for an instant only and collapse (like the bubble) to reappear in the following space, the resistance to motion could be calculated from the sum of the impulses required to generate the irrotational patterns in these separate spaces*. The picture would be as in Fig. 4.

A four dimensional hydrodynamic model for real fluids

The resistance corresponding to the impulsive generation of an infinite number of irrotational flows in a finite time would itself be infinite. However the discrete quantum like nature of eddies in real wakes suggests that the number of "spaces" or space-time domains encountered in unit time is finite. In the case, for example, of creeping flow past a sphere, the initial variables are the kinematic viscosity and the radius. The association of these two specifies a frequency proportional to ν/a^2, the assumption of very small Reynolds Numbers ensuring that the developed frequency, U/a, is negligible by comparison $\left(\frac{U/a}{\nu/a^2} = \frac{Ua}{\nu} \ll 1\right)$. This "rest spin" is on our hypothesis, the frequency of encounter by the sphere of the separate space-time domains when the motion is very slow. In the case of a sphere the geometry has no preferred orientation in 3 dimensional space. The appropriate value for the kinematic viscosity is therefore 9ν - corresponding to the nine velocity gradients ($\partial u/\partial x, \partial u/\partial y, \partial u/\partial z, \partial v/\partial x, \partial v/\partial y, \partial v/\partial z, \partial w/\partial x, \partial w/\partial y, \partial w/\partial z$) "which define the state of relative motion at any point of the fluid" (Ref. 3). The introduction of a frequency $9\nu/a^2$ by the sphere is not altogether surprising as a kinematic quality, ν, has already been assigned to the fluid. The units of ν are those of velocity multiplied by length so that a virtual circulation is specified in the rest state.

It appears to be generally true that a three dimensional view introduces spin. In the present interpretation the rotation comes from a three dimensional projection of a four dimensional **displacement** - rather as a two dimensional projection of a three dimensional motion gives rise to "divergence". The sphere in the example, although ostensibly at rest, is actually moving on in time. This is the basis of the paradoxical **notion** of "rest spin".

From the frequency of encounter it is simple to calculate the viscous resistance. In one second the impulse $\frac{2}{3}\pi\rho a^3 U$ must be generated $9\nu/a^2$ times giving rise to a force $\frac{2}{3}\pi\rho a^3 U \times \frac{9\nu}{a^2} = 6\pi\mu aU$ in agreement with the result obtained by Stokes.

The possibility opens up of bringing hydrodynamics to bear on some of the intractable problems of fluid motion - the price being a considerable increase in abstraction.

Acknowledgements

My thanks for helpful discussions are in particular due to Professor C.M. White, formerly of the Civil Engineering Department, Imperial College. I am also grateful to Professor T. Brooke Benjamin of the University of Oxford, Professor P. Holmes of Liverpool University, Miss S. Ormerod and Dr J. Sioris of the hydrodynamics research group at the Civil Engineering Department, Imperial College and Mr R.W.P. May of this group and the Hydraulics Research Station, Wallingford.

References

1. MacMahon, B. "The tow-out of a large platform". 10th International Liège Colloquium on Ocean Hydrodynamics. Edited by J.C.J. Nihoul. Elsevier 1979.

2. Wendel, K. "Hydrodynamic masses and hydrodynamic moments of inertia". ("Hydrodynamische Massen und Hydrodynamische Massenträgheitsmomente" Jahrb.d.STG Vol. 44, 1950). David Taylor Model Baisin Translation No. 260.

3. Stokes, G.G. "On the effect of the internal friction of fluids on the motion of pendulums". P. 14. Mathematical and Physical Papers, Vol. 3, Cambridge University Press, 1901.

*A problem simplified by this approach in a smaller number of dimensions is that of determining the lift on a 2 dimensional wing having a circulation Γ. The bound line vortex is visualised translating through U metres in unit time. A surface $U \times 1$ m^2 of air is affected by unit span and the impulsive pressure $\rho\Gamma$ of the moving circulation communicates a total impulse - $\rho\Gamma U$ per second to the fluid. The reaction to this is the lift force.

ADDENDUM

In the case for example of creeping flow about a sphere the important variables are the kinematic viscosity and the radius. The association of the two specifies a frequency proportional to ν/a^2, the assumption of very small Reynolds Number ensuring that the imposed frequency U/a is negligible by comparison ($\frac{U/a}{\nu/a^2} = \frac{Ua}{\nu} \ll 1$). This "rest spin" is, on our hypothesis, the frequency of encounter of the separate space time domains when the motion is very slow. In the case of a sphere at rest (or virtually so) the geometry has no preferred orientation in 3 dimensional space. The appropriate value for the kinematic viscosity is then 9μ corresponding to the nine velocity gradients ($\frac{\partial u}{\partial x}, \frac{\partial u}{\partial y}, \frac{\partial u}{\partial z}, \frac{\partial v}{\partial x}, \frac{\partial v}{\partial y}, \frac{\partial v}{\partial z}, \frac{\partial w}{\partial x}, \frac{\partial w}{\partial y}, \frac{\partial w}{\partial z}$). The definition of a frequency $9\nu/a^2$ by the sphere is not altogether surprising as a kinematic quality has already been assigned to the fluid. The units of ν are those of velocity multiplied by length so that a certain virtual circulation is specified in the rest state.

From the frequency of encounter it is simple to calculate the viscous resistance. In one second the impulse $\frac{2}{3}\pi\rho a^3 U$ must be generated $9a^2/\nu$ times giving a force $\frac{2}{3}\pi\rho a^3 U \times \frac{9\nu}{a^2} = 6\pi\mu a U$ in agreement with the result obtained by Stokes.

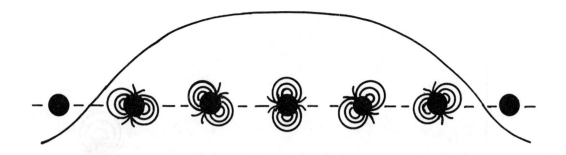

Fig. 1 Irrotational disturbance flow of breakwater elements

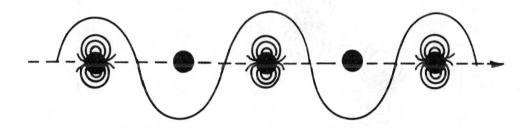

Fig. 2 Successive positions of a body translating through a frictionless liquid with a free surface. (ignoring gravity)

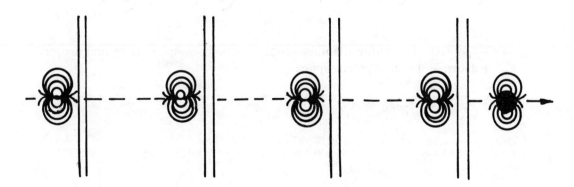

Fig. 3 "Wake" left behind by a body moving through a frictionless fluid partitioned by imaginary fluid free layers.

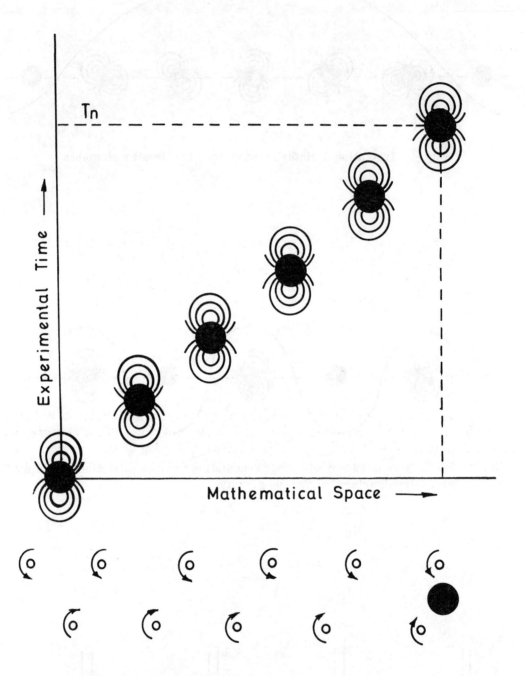

Fig. 4 Hypothetical quantum displacements of a body in space time and projected pattern at T_n in physical space.

BOSS'79

PAPER 63

Second International Conference
on Behaviour of Off-Shore Structures
Held at: Imperial College, London, England
28 to 31 August 1979

WAVE INDUCED MOTIONS OF MARINE DECK CARGO BARGES

W.P. Stewart, BSc, MSc and W.A. Ewers, MA, DPhil.

Atkins Research and Development, U.K.

Summary

The motion responses to environmental forces of a full scale marine deck cargo barge (100m x 30m) have been measured during a tow across the North Sea. Wind and wave measurements were also made enabling the response spectra and transfer functions presented in this paper to be derived. These results are compared with theoretically predicted values and the importance of roll damping is discussed.

Sponsored by: Delft University of Technology, The Netherlands
Massachusetts Institute of Technology, U.S.A.
The Norwegian Institute of Technology, Norway
University of London, England

Secretariat provided by: BHRA Fluid Engineering

Copyright: © BHRA Fluid Engineering
Cranfield, Bedford, England

Introduction

Until recently there has been a general lack of knowledge concerning the motion response of marine barges (with beam/draught ratios over 4) to waves where the natural roll periods of the vessels are near those of the waves. Computer programs are known to predict unrealistic motions near resonance primarily because of their inability to simulate the viscous damping associated with barge roll motions. Furthermore there is a considerable divergence of opinion as to the correct values of equivalent linear damping terms that may be used in order to limit theoretical responses at resonance.

Physical scale model testing is the solution to which most designers resort. However the scale effects concerning the hydrodynamic damping of roll motions are not quantifiable; indeed the damping mechanism itself is not clearly understood.

The Jasmine's Turtle Experiment

In order to overcome the problems associated with small scale model testing, a full scale experiment was carried out on a marine barge in the North Sea. This barge, the Jasmine's Turtle was scheduled to carry a cargo of 3 modules for the Shell Brent C platform from Newcastle to Norway. The geometric and mass properties of the loaded and ballasted barge are given in Fig. 1.

Atkins Research and Development installed a number of instruments on the barge before tow-out, and these are listed in the next section. In view of the poor reliability of unattended instrumentation, the key components in the system were duplicated where possible. All signals were telemetered to the tug (Starmi) using the EMI telemetry system 36, where they were recorded on two tape recorders. Fig. 2 shows the instrumentation on the barge.

When the tow reached the deep waters of the Norwegian trench Jasmine's Turtle was turned into the weather and held steady whilst a waverider buoy was deployed in order to measure wave heights. The environmental conditions during this period are summarised in Fig. 3. It must be stressed that these conditions are those that were visually observed. The figures for wave height and period shown are best estimates by experienced seamen of significant wave height, period and direction. They compare reasonably with the spectrum derived from measurements made at the same time with the waverider buoy.

Data

The data transmitted from the barge and monitored and recorded on the tug were as shown in Table 1. Two orthogonal pairs of accelerometers were installed to measure surge and sway accelerations. One pair was fixed well above the barge centre of gravity, high up on the cargo, and one pair well below, down in the bilges.

Analysis

The axes systems and equations used in the analysis are given in Fig. 4. The derived data are sufficient to fully describe the motions of the barge if yawing motion can be ignored, and the accelerometer time histories can be integrated to yield the corresponding velocity time histories. This integration was performed satisfactorily in the frequency domain. Substituting the velocities in equation (2) yields the roll and pitch angular velocities, and the height of the pitch and roll centres. (These being defined as those points which have zero velocity components in the surge and sway directions respectively). The roll and pitch angular velocities derived as above with the measured accelerations are substituted into equation (3). This yields the roll and pitch angular accelerations, as well as the surge and sway components of acceleration of the C.G. To check on this analysis, the roll angular velocity time histories were numerically integrated, and compared with the gyro outputs. For most of the data, good agreement was obtained.

Results

Fig. 5 shows time histories of the roll angular velocity, the sway acceleration of the C.G. and the height of the roll centre above the C.G.

The most noticeable features of these time histories are the large excursions (to $\pm\infty$) of the roll centre. These excursions occur at zero crossing points in the roll velocity and sway acceleration time histories, and consequently have no large forces associated with them. Numerically they arise from the division of large sway velocities by near zero values for the roll angular velocity. This is to be expected when the barge follows the water surface profile of long period waves. Experience has shown that this type of motion is a characteristic of large marine barges.

Fig. 6 shows the wave spectrum derived from the wave rider buoy signal and Figs. 7 and 8 show the roll and pitch response spectra derived from the gyro outputs at the same time (tow stationary). These spectra were used to compute the transfer functions shown in Figs. 9 and 10.

Discussion

Fig. 11 shows the wave spectrum, the roll spectrum and the roll transfer function plotted together against period. The wave spectrum is clean and has most energy around 8 seconds period (as estimated by the seamen) with **some evidence of the twelve** second swell. However, a fairly generous estimation of the energy under this area of curve indicates a significant swell height of only 0.6m (c.f. 2m estimate by seamen).

It is interesting to find that the roll response spectrum has twin peaks, each being well defined. The peak 6.2 seconds corresponds to the natural roll period of the barge. The second peak at 6.9 seconds is caused by response at the wave frequency. The corresponding peak in the roll transfer function is unexpected and may indicate that there was more energy in the waves at 6.9 seconds than indicated by the wave spectrum which has been smoothed. Alternatively there may have been less energy in the waves at 6.6 seconds (an unexpected trough in the roll transfer function) than indicated, which would have the same result.

Having obtained some results for a full scale barge in real offshore conditions it is important to compare these results with existing theory. Thus in Fig. 12 the Jasmine's Turtle (JT) transfer function is compared with transfer functions predicted for a similar barge by two computer programs. Table 2 gives the properties of both JT and the computer model. The first program, TRITON, is a linear diffraction-radiation program applicable to three-dimensional bodies of arbitrary shape. The method of solving the Laplace equation is by 8 noded brick type fluid finite elements. No external damping is added to represent viscous effects. The other program, TRITIR, is a non linear simulation program. Hydrodynamic characteristics are input from the linear analysis in TRITON but are then used under less restrictive conditions. In particular non-linear effects near the free surface are considered. A surface integral of drag force proportioned to the square of the relative velocity between surface elements of the body and the fluid is computed at each time step. Hence TRITIR, to some extent, simulates the viscous roll damping forces, whilst TRITON simulates only the radiation damping arising from the generation of surface waves. TRITIR has been found to predict scale model test results on barges extremely well. TRITON, with less damping, overestimates roll response at resonance.

It is encouraging to note that the JT result is in the same region as that predicted by the theoretical models. However, the first impression given by Fig. 12 is that the JT transfer function must be for a more lightly damped system that any of the theoretical curves, although the importance of wave direction must not be underestimated. The TRITON curves are for three angles of wave attack; $30°$, $60°$ and $90°$ (beam on) and for unit amplitude waves. The TRITIR curve is for beam waves only of 3m height (close to the significant height measured).

Evidently the barge response is much greater around resonance than predicted by theory for a wave attack angles of $30°-60°$. Between 7.5 and 10.5 seconds response is less than predicted by theory and from 10.5 seconds upwards measured and predicted responses are close.

Conclusion

Despite the problems of directionality with the wave spectra it is concluded that the full scale barge roll response is less damped in the random sea state measured than is predicted by a theoretical computer program calibrated against model test results.

Table 1

Instrument	Signal	Unit
Colnbrook Gyro	Roll Pitch Heave	degree degree metre
B.S.R.A. Gyro	Roll Pitch Heave	degree degree metre
Strain-Gauged Shackles	Port Tension Starboard Tension	tonnes force tonnes force
Accelerometers	Accelerometer 1 (Lower Sway) Accelerometer 2 (Lower Surge) Accelerometer 3 (Upper Surge) Accelerometer 4 (Upper Sway)	m/s² m/s² m/s² m/s²
Electronic Circuit	Test Voltage (2.5v)	volt
Battery Pack	Battery Supply Voltage	volt
Waverider Buoy (not on the barge)	Wave Height	metre

Table 2

Property/Barge	Jasmine's Turtle	Computer Model
Mass	6141 tonnes	6581 tonnes
Draught	2.62m (mean)	2.74m
Gyradius	8.92m	8.48m
KG	6.39m	4.88m
BG	5.08m	3.50m
GM	17.81m	19.36m
I_{xx}	4.89E5 tonne m²	4.73E5 tonne m²
T_{roll}	6.3 sec	6.0 sec

Mass Properties:-
r = 8.92m   ) including cargo
I_{xx} = 4.89E5 tonne m²) and ballast
)
Mass = 6141 tonnes

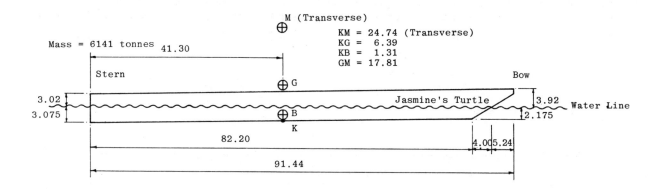

Barge is of uniform width, 27.43m, with constant rectangular cross section. Bow is raked at 30°. All hull plating is flat except at corners which have a radius of approx. 400mm.

Fig. 1 Side elevation of barge (cargo not shown) with principal dimensions given for as-towed configuration.

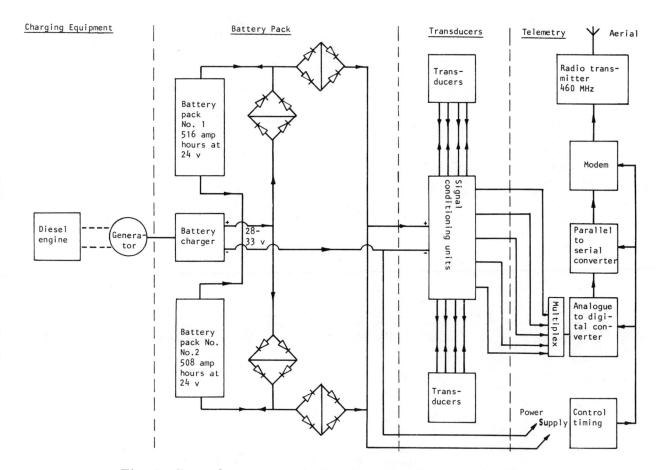

Fig. 2 General arrangement of equipment on Jasmine's Turtle.

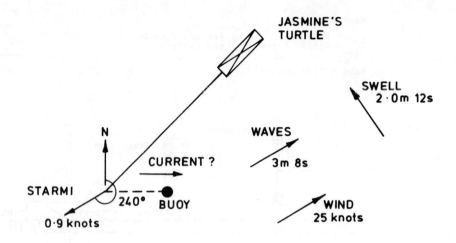

Fig. 3 The tug/barge configuration during the deployment of the waverider bouy on Wednesday 8th March, 1978.

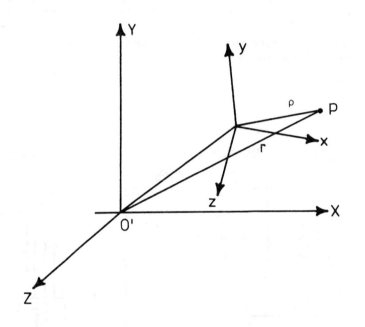

OXYZ = is the global axis system

O'xyz = is the local (barge) axis system with angular velocity $\underset{\sim}{\omega}$ relative to OXYZ

If P is fixed relative to the barge axis system (O'xyz)

$$\underset{\sim}{r} = \underset{\sim}{R} + \underset{\sim}{\rho} \qquad (1)$$

$$\underset{\sim}{\dot{r}} = \underset{\sim}{\dot{R}} + \underset{\sim}{\omega} \times \underset{\sim}{\rho} \qquad (2)$$

$$\underset{\sim}{\ddot{r}} = \underset{\sim}{\ddot{R}} + \underset{\sim}{\omega} \times (\underset{\sim}{\omega} \times \underset{\sim}{\rho}) + \underset{\sim}{\dot{\omega}} \times \underset{\sim}{\rho} \qquad (3)$$

Fig. 4 Axis systems for vessel motion analysis.

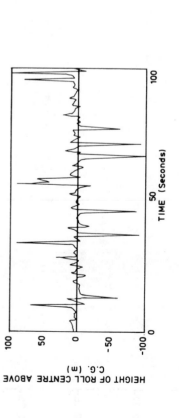

Fig. 6 Wave spectrum obtained using wavender buoy on Wednesday 8th March, 1978.

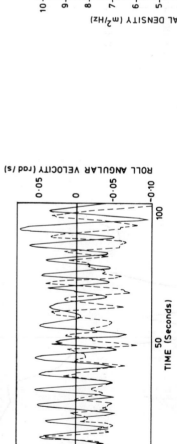

Fig. 5 Time histories of roll angular velocity, sway acceleration of C.G. and height of roll centre above C.G. (record 80, tape 3).

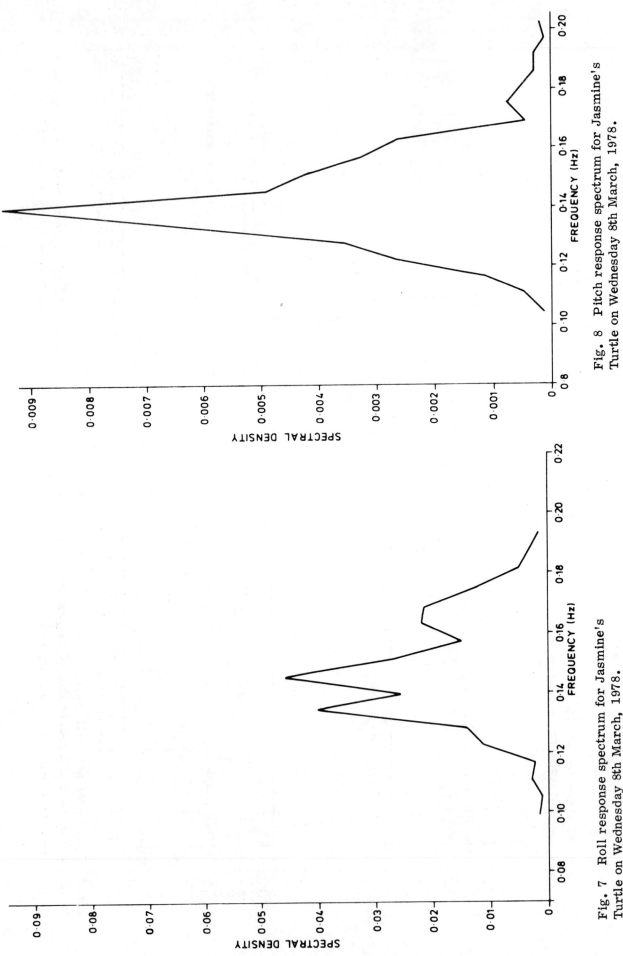

Fig. 7 Roll response spectrum for Jasmine's Turtle on Wednesday 8th March, 1978.

Fig. 8 Pitch response spectrum for Jasmine's Turtle on Wednesday 8th March, 1978.

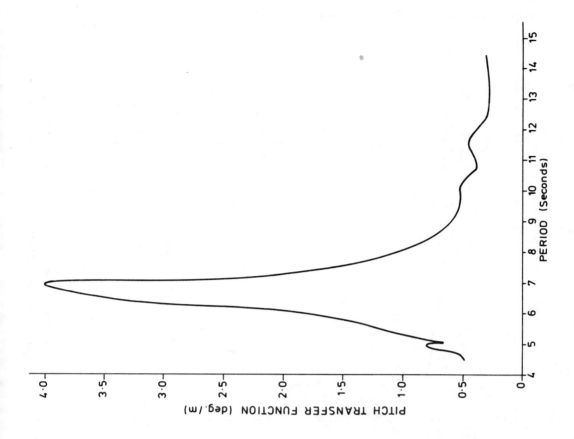

Fig. 10 Pitch transfer function for Jasmine's Turtle on Wednesday 8th March, 1978.

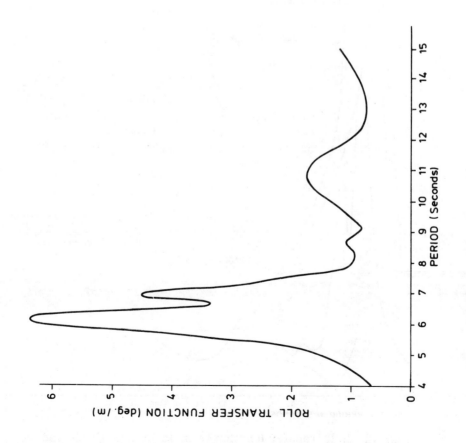

Fig. 9 Roll transfer function for Jasmine's Turtle on Wednesday 8th March, 1978.

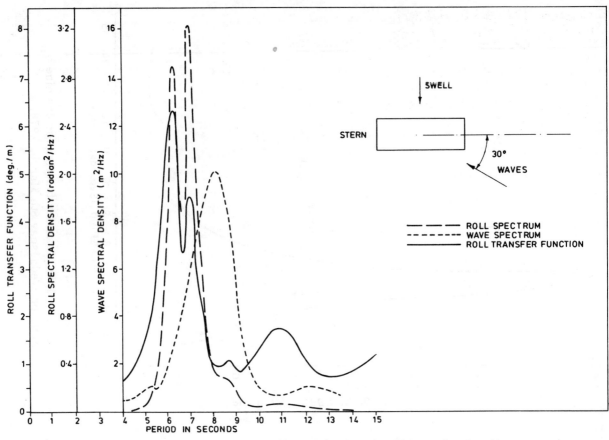

Fig. 11 Wave spectrum, roll spectrum and roll transfer function.

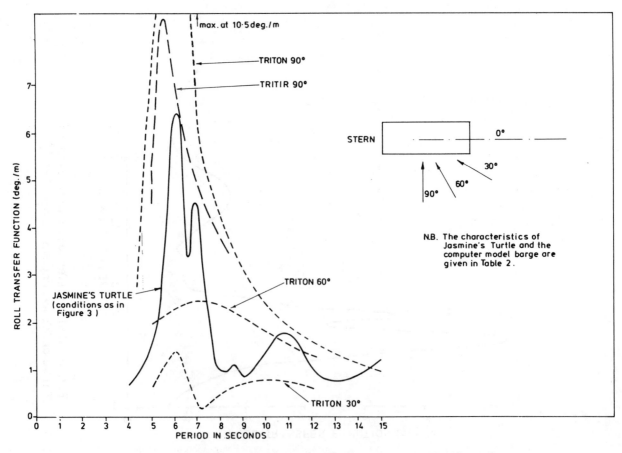

Fig. 12 Roll transfer function from Jasmine's Turtle and transfer functions predicted by programs TRITON & TRITIR.

BOSS'79

PAPER 64

Second International Conference
on Behaviour of Off-Shore Structures
Held at: Imperial College, London, England
28 to 31 August 1979

THREE GRAVITY PLATFORM FOUNDATIONS

I. Foss and J Warming

Det norske Veritas, Norway

Summary

The paper presents three case studies for concrete gravity structures of the Sea Tank design in the North Sea. For all structures the installation process was closely monitored, and instrumentation is installed for surveillance of the structures and their foundation during operation. The paper describes each structure and the soil conditions at the sites. The practical experiences from the installation phase, including the performance of the instrumentation system, are briefly described.

The penetration resistance of concrete skirts into the sea bed during installation is covered in some detail. Actual penetration resistance is compared to the predicted resistance, and a calculation procedure which gives results in reasonable agreement with the observation is presented.

From the operation of the platform only preliminary data are as yet available. Data regarding settlement, dynamic motions and scour are presented.

It is concluded that the foundation behaviour of the platforms has been satisfactory in most aspects. The installation operations, including underbase grouting, were successfully completed. The skirt penetration resistance was smaller than the conservatively calculated values. On TP-1 scour to a depth of 2m developed at two corners. Remedial measures have prevented further scour and re-established the sea bed. With a qualification regarding data available, there is no indication that long term behaviour of the structure is not in reasonably good agreement with the expectations.

Sponsored by: Delft University of Technology, The Netherlands
Massachusetts Institute of Technology, U.S.A.
The Norwegian Institute of Technology, Norway
University of London, England

Secretariat provided by: BHRA Fluid Engineering

Copyright: ©   BHRA Fluid Engineering
Cranfield, Bedford, England

1. Introduction

At present 13 concrete gravity structures have been installed on location in the North Sea and one (Statfjord B) is in the stage of construction. In principle only four concepts have been used: ANDOC (Dunlin A), CONDEEP (Beryl A, Brent B, Brent D, Frigg TCP-2, Statfjord A and Statfjord B), DORIS (Ekofisk 1, Frigg CDP-1, Frigg-Scotland Booster and Ninian Central) and SEA TANK (Frigg TP-1, Brent C and Cormorant A). The Ekofisk tank was the first platform to be installed in 1973, while eleven more structures were installed during the years 1975-1978. These structures may all be considered as prototypes as very limited experience with such giant gravity structures was available prior to and during the construction phases.

The purpose of this paper is to present data to add to the experience. Case studies for the three SEA TANK structures:
    Frigg TP-1 (TP-1)
    Brent C   (BC)
    Cormorant A (CA)
will be described, including presentation of data from the installation phase and data from the operation of the platforms to the extent they are available.

Det norske Veritas has for all three structures acted as Certifying Authority on behalf of the Department of Energy, UK. In this function Veritas has been engaged in the discussions regarding the need for instrumentation and data analysis for the installation phase operations as well as for the long term performance monitoring.

With respect to the companies involved in the design, fabrication, etc. of the structures please refer to the acknowledgements, Section 11.

The fabrication of the three structures started in 1974-1975 in the dry basins at the construction site Ardyne Point, Scotland. TP-1 was the first to be completed and was successfully installed in the Frigg field early summer 1976. BC and CA were both completed with respect to the substructures in 1977, then towed to Norway (Stavanger and Stord, respectively) for deck installation, module hook-up, etc. and both structures were finally installed early summer 1978. General information on the three structures is given in Table 1.

The installation phase was closely monitored, and the data gathered from the extensive instrumentation provide a good basis for back-calculations, e.g. the penetration resistance of the wedge shaped concrete skirts will be covered in some detail in Section 6.

With respect to the performance during platform operation the data so far available are less comprehensive due to the short period of data acquisition (BC and CA) and to some extent due to measuring difficulties. However, the trends observed will briefly be reviewed in the following sections, including our comments regarding the instrumentation system and the comparisons with expected behaviour.

2. Platform Design

The three gravity structures, designed by the French company SEA TANK CO. (STC) and built by Sir Robert Mc.Alpine & Sons Ltd., are of very similar design, especially regarding the foundation. They are, however, somewhat different in magnitude and overall lay-out, and there are significant differences with respect to the soil conditions.

Some main design data are presented in Table 2, and the geometry of the concrete substructure, the base and the skirts is illustrated in Figs. 1, 2, 3 and 4.

Regarding the foundation one should note the following main features:
- the base slab is flat and square
- the skirts are made of concrete, wedge shaped and wide compared to other designs.

The skirts were designed to penetrate into the foundation soil to a depth sufficient to ensure the stability of the structure under the extreme loading conditions, with allowance for a reasonable depth of scour. The skirt system also forms the confinements necessary for the underbase grouting. The concrete skirts have a significant strength to resist lateral loads. In particular, the concrete skirts have a larger capacity than steel skirts to accommodate the loads applied during the "touch-down" phase of the installation operation. Thus, the foundation system did not need dowels to bring the structure to a halt before touch-down of the skirts.

The base slabs are massive concrete rafts with thicknesses varying from 3m (TP-1 and BC) to 4m (CA). The slabs were designed to resist such high local contact pressures that might develop in the pregrouting phase due to base contact with an uneven seafloor. However, the pressures were not predictable with high accuracy, hence precaution and monitoring during the installation operations were deemed necessary.

The internal walls in the caisson, dividing the caisson into a large number of independant ballast compartments, were all plane wall constructions. Water level differences in adjacent cells were restricted by the capacity of the walls, for TP-1 8m difference could safely be accommodated, for BC and CA only about 5m. This condition necessarily limited the available ballasting eccentricity and, hence, the possibilities to cope with uneven skirt penetration resistance.

The foundation design included underbase grouting, the objectives of the grouting being:
- to prevent excessive long-term settlement and tilting of the structures, due to incomplete skirt penetration
- to ensure direct load transfer from base slab to seabed, with an even distribution of the soil reaction
- to improve platform stability.

The grout was composed of Portland cement, sodium silicate and seawater with w/c ratio exceeding 3, a slurry density of 12 KN/m^3 and a 28 days' cube strength of 1.5-2.5 N/mm^2.

3. Site and Soil Conditions

Site and soil conditions at the three sites were explored by extensive investigations both in the field and in the geotechnical laboratories involved.

The results of the soil investigations are summarized in Fig. 5: Average strength profiles used e.g. for stability calculations, and in Fig. 6: Average cone resistances in the upper 3-5m, used e.g. for skirt penetration calculations.

At all three sites the determination of undrained shear strength has been based on extensive laboratory testing of the soil samples with emphasis on the results of triaxial tests (UU and CU) and simple shear tests, both static and cyclic loading. However, the results show considerable scatter and, hence, also the great number of more simple measurements (pocket penetrometer, fall cone, torvane, etc.) have been used in the assessment of the shear strength profiles. In Fig. 5 only the results of the triaxial tests have been shown. In the upper soil layers the shear strength profiles used for design are mainly based upon the results of the CPTs interpreted with cone factors in the range $N_k \sim 15-20$.

The seafloor topography at all three sites is very even with seafloor undulations not exceeding 0.1-0.2m but with gently sloping seafloors as follows:
- TP-1 : 1.0% down towards North-West
- BC   : 0.4% down towards North-East
- CA   : 0.7% down towards South.

4. Instrumentation

All three platforms have been instrumented to various extent to control and monitor the installation operations and to monitor the operation performance of the structure and the foundation.

The detailed requirements to the instrumentation were to some extent only defined and agreed upon after the start of platform construction. Thus, instrumentation was added or modified during the construction phase, which might explain some of the operational difficulties.

Installation Instrumentation

The instrumentation for monitoring the installation operations is summarized in Table 3 and will briefly be commented on in the following.

The positioning was controlled by electronic range measuring systems (BC and CA) and by infrared optical measuring (TP-1), all surface systems. Thus, the platform position during final approach to the target area was recorded relative to nearby installations with well known coordinates. The systems worked satisfactorily. For BC and CA a system based on subsea acoustic transponders was provided as back-up. This system was never used and is in general considered less suitable for the actual purpose, as e.g. the acoustic noise introduced by the machinery in the structures and the tugs tends to black out the transponders.

The three platforms were all equipped with electronic inclinometers, of which the TP-1 system had a rather poor sensibility. Thus, on TP-1 (and as back-up on CA) the inclination was also monitored by a pair of U-tubes, i.e. waterfilled plastic tubes installed in the deck structure with manual recordings of the water levels in the two free ends. The system proved to be accurate, but rather slow to read.

Echo sounders with high resolution were used primarily to monitor keel clearance before touch-down and further provided measurements with respect to depth of skirt penetration in the initial penetration phase. The echo sounders were, however, not initially installed for platform installation purposes, and some of the echo sounders failed to operate after the long exposure to the environment during platform construction, towing, etc. Thus new echo sounders were installed on BC shortly before tow-out from Stord.

For TP-1 the skirt penetration was accurately monitored by optical levelling instruments from the nearby QP platform. On BC and CA closed hydraulic systems with pressure transducers on the seafloor were the primary source of information from shortly after touch-down to final penetration. The various systems for vertical control were calibrated (zero adjusted) by means of submarine inspections with observations of actual depth of skirt penetration. A measuring rod operated by the submarine manipulator arm was used to measure the distance from the base slab to seafloor at several positions around the base. For all three structures the decision to terminate the skirt penetration phase was based on this type of direct observation.

The differential water pressure in the skirt compartments relative to outside sea water needs monitoring during several stages of the installation process including the grouting operation. On BC and CA identical systems were installed comprising pressure transducers connected to stand pipes from each of the 15 respectively 16 skirt compartments and with reference transducers on each of the 4 shafts to correct for outside sea water level.

Identical systems were used on all three structures to monitor the skirt compartment water pressure during the grouting operation. These systems normally comprised four piezometer pipes connected to each of the skirt compartments. The water level in the pipes, reflecting the water pressure, was then measured by sounding lines with electrical level switches.

On TP-1 all data were logged and analysed manually. The process to convert raw data to engineering properties was time consuming and possessed sources of human error, especially when one considers the strain on the personnel involved in the installation operation. On BC and CA the data handling was considerably improved, as all raw data were fed into a programmable desk computer with facilities to print out the engineering data of immediate concern.

Instrumentation for Performance Monitoring

As a consequence of the need for long term surveillance of the structures including the foundation, all three installations are being monitored with respect to:
- settlements
- pore water pressure in the foundation soils
- dynamic motions.

The installed measuring systems and instruments are indicated in Table 4.

Settlements are being measured in three different ways. For TP-1 optical measurements are carried out from the nearby piled QP jacket.

CA is equipped with a system based on a steel casing drilled down to 55m below mudline. 10 radioactive bullets were shot from the casing into the surrounding soil at various levels. Settlements, including selective settlements of the different soil layers, are recorded by measuring the relative position of the bullets with a sensitive probe being lowered into the casing. The same casing is used to run a "digi-tilt" torpedo (inclinometer) for measuring possible lateral displacements of the foundation.

On BC a similar system could not be established as no sleeves were provided for drilling. Thus BC is equipped with two closed hydraulic systems with pressure transducers laid out on the sea bottom. The transducers are placed at a distance of approx. 70m from the platform base. The pressure being measured is equivalent to the level difference between the transducer and a hydraulic reservoir located on the cantilevered base slab.

Excess pore water pressure in the soil due to the weight of the platform and build-up of the pressure during storms are monitored by means of pore pressure gauges installed at various positions and depths. TP-1 had originally 4 gauges placed on 2m long "spears" attached to the raft. However, all 4 transducers have now ceased to function, only one of the gauges functioned during the first year after platform installation. Both BC and CA have the same type of "shallow" pore pressure gauges on spears, 10 on each structure. Two of the gauges on CA are dead at the time of writing, while the eight others and all of the gauges on BC give sensible signals. In addition to the shallow gauges, 5 deep pore pressure gauges have been installed on CA at various depths down to 20m. The installation comprises a casing housing the gauges with hydraulically inflated packers to isolate the single gauges. Contact with the pore water outside the casing is established through perforations in the casing, made at correct level by means of bullets shot from the inside.

Dynamic motions of the three platforms are measured by means of accelerometers; the numbers, types and positions are given in Table 4.

Originally, TP-1 was only equipped with 2 linear accelerometers at caisson roof level. In order to achieve a more comprehensive measure of the dynamic response, another package containing 2 linear and 2 angular accelerometers are due to be installed by the end of 1978.

To correlate the dynamic motions and possible pore water pressure built up during storms with the environmental conditions, wave data (height and period) are required for the actual recording periods. On TP-1 and CA the sea state is recorded by wave radars placed at extremities of the deck. Wave data in the Brent area are acquired from a Waverider buoy located in the vicinity of the Brent B platform, approx. 4 km south of BC.

On BC and CA the instrument signals are logged and to some degree analysed by automatically operating, computer based data acquisition systems. On TP-1 the instrument signals are sampled at specified intervals, and the raw data recorded on magnetic tape. None of the data loggers was yet installed by the end of 1978, but all are due to be installed in the beginning of 1979. In the intermediate period after platform installation, data have been logged by temporary and more or less manually operated systems.

5. Installation

TP-1

The installation of TP-1 took place on the 5th of June, 1976, with final touch-down at 3.30 p.m. This marked the end of a three week towing operation from Loch Fyne on the West coast of Scotland, routed north of the Shetlands.

Very narrow tolerances were given for the final positioning as TP-1 was to be placed approximately 100m north of the existing QP (quarters platform), the bridge to connect the two deck structures already fabricated.

The final operations to bring the structure to a halt at the correct location were not easily controlled, and some skidding probably added to the problems. After first touch-down with the southern corner the platform was moved horizontally and rotated. Thus, to some degree the skirts "bulldozed" in the top soil layers.

Skirt penetration was terminated on the 8th of June, approx. 72 hours after touch-down when the southern corner of the base slab made contact and the capacity to apply eccentric ballasting was exceeded. The base contact was registered by the extensive instrumentation with strain gauges in the raft, but the soil pressure did not develop significant stresses in the base. During skirt penetration the platform tilted, as expected, towards north-east, and the tilting of about 0.6% could only partially be counteracted by applying maximum allowable eccentric ballasting, see Fig. 8.

The grouting operation was commenced with the northernmost compartment. Problems were encountered during the entire grouting operation owing to the fact that some grout passed both to outside seafloor and to adjacent compartments. The "piping" is believed to be due to insufficient skirt penetration combined with erosion channels formed under the skirts in the intitial penetration phases. In total, 2550m^3 of grout slurry were injected, the theoretical volume being only 1900m^3. By the end of the grouting operation all leaks were sealed off, and the underbase grouting was successfully completed.

Cormorant A

CA had touch-down on the 7th of May, 1978, at 5.30 a.m. after a week towing operation from Stord, Norway. The final positioning encountered some of the same problems as for TP-1. The ballasting down to touch-down level was probably carried out a little too fast resulting in some skidding and unexpected horizontal movements of the platform. Soil disturbance was, however, negligible. The skirt penetration phase went very smoothly and was terminated on the 8th of May, at 2 a.m., i.e. approx. 20 hours after touch-down. By that time, the northern corner of the base slab was in contact with the sea floor, the minimum average skirt penetration of 2.5m was achieved, and the platform was level. As for TP-1, the base contact only resulted in insignificant rise in the stresses in the base slab.

With respect to the grouting operation, the situation was very similar to that of TP-1, i.e. some grout passed from the compartments being grouted to adjacent compartments and to the sea floor. A step-wise grouting procedure with 3-4 injection stages per compartment and with intermediate settling of the grout eventually resulted in successful completion of the operation. Total volume injected was $7500m^3$, the theoretical volume being only $4900m^3$.

Brent C

BC was installed on the 18th of June, 1978, with touch-down at 9.30 p.m. The towing operation beginning on the 11th of June, at Stord, Norway, was delayed a few days due to a short period with rough weather. The final positioning and ballasting to touch-down encountered no problems. The skirt penetration phase was executed smoothly and was successfully terminated approximately 57 hours after touch-down on the 21st of June. By that time, the south-west corner of the base made contact with the sea floor, average skirt penetration was 2.6m and the platform was level.

The grouting procedure was from the beginning adapted to meet the problems with possible leaks. Although some leakage did occur, the procedure worked well, and the amount of excess grout was kept low. Total volume of injected grout was $4500m^3$ compared with a calculated volume of $3900m^3$.

Summarizing Comments

The final positioning requires special attention for the actual type of foundation structure. Problems with skidding must be envisaged and should be considered in the installation procedure. However, the installation of BC proved that full control can be maintained provided the crew is well drilled and the vertical approach to touch-down is carried out very gently.

Piping problems for the grouting operation should as far as possible be avoided. To achieve this it is recommendable to carry out the touch-down and initial skirt penetration at a very low rate of ballasting and to aim at a minimum of 2m skirt penetration, especially where sea bed soils consist of sand.

Extensive and reliable instrumentation, including appropriate data acquisition and routine data analysis, is essential for a proper control of the installation operations. It is of great importance for the success that the planning of the various instrumentation systems is at an advanced stage of design when the platform construction is commenced.

6. Penetration of Skirts

The skirt penetration has been subject of comprehensive analyses in the design and planning phases. The penetration resistance was not easily assessed due to lack of experimental data and little experience with the actual type of penetrating member.

For BC and CA a minimum average skirt penetration was needed to ensure the required level of safety for the foundation stability.

For TP-1 stability considerations alone were not governing for the skirt penetration, although scouring might be a hazard, which deep skirt penetration would reduce. A certain penetration was, however, needed in order to provide confinements for the underbase grouting. With these demands and further, with a strong desire to keep platform tilt less than 0.5% (against a 1% sloping sea floor), it was important to assess the penetration resistance.

The range of calculated penetration resistances vs. penetration depths is indicated in Fig. 7.

Three methods of analysis were used for the wedge shaped skirts:

1. Tip resistance calculated by bearing capacity and skin resistance by earth pressure theory.
2. Tip resistance derived directly from the cone resistance (q_c), skin resistance calculated by earth pressure theory.
3. Total penetration resistance based on the cone resistance (q_c).

For TP-1 all three methods were applied. However, Method 1 - yielding lowest resistance - was thought to give the most reliable result. The range of results shown in Fig. 7 is for Method 1 with values of $\phi: 42° - 45°$, however with rather high values of N_γ, N_q and K_p.

For BC and CA Method 3 was used (however roughly checked against Method 1). The designer calculated the penetration resistance per unit length as the largest value of:

1) $q_{c,tip} \cdot b_{tip} + 2 \int_0^z f_s(d)\,dz$

2) $q_{c,av} \cdot b_{av} + 2 \int_0^z f_s(z)\,dz$

where q_c is cone resistance
f_s is sleeve friction.

No corrections were made for differences in shape, rate of penetration, surface roughness and relative depth of penetration between the cone and the skirt. The range of penetration resistance shown in Fig. 7 covers various interpretations of CPT data. The approach was believed to be conservative. However, the available penetration force was sufficient to achieve almost full skirt penetration, and hence more refined analyses were not really justified, especially considering the inaccuracies involved.

Knowing the actual penetration resistances, the above methods have been checked in an attempt to obtain compatibility between theory and actual performance.

It has been found that Method 1, i.e. tip resistance calculated by conventional bearing capacity theory and skin resistance calculated by earth pressure theory, yields by far the best correlation between theory and observed performance. This statement is valid for both dense sand and stiff clay. In such a back-calculation the following method was applied:

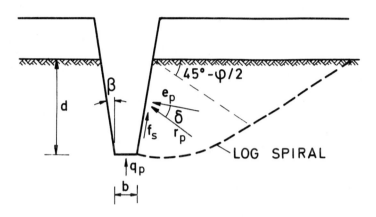

In sand: $q_p = \frac{1}{2} \cdot \gamma' \cdot b \cdot N_\gamma + \gamma' \cdot d \cdot N_q$

$\quad$ (N_γ, N_q according to Caquot-Kerisel, see e.g. /2/ Table F 1.1)

$e_p = K_p \cdot \gamma' \cdot z$ $\quad$ (K_p according to e.g. /2/ Table F 4.2)

$f_s = e_p \tan \delta$ $\quad$ ($\tan \delta \sim 2/3 \tan \phi$)

$$R = q_p \cdot b \cdot L + 2L \int_0^d e_p (\sin\beta + \tan\delta \cos\beta)\, dz$$

ϕ was taken as best estimate of angle of shearing resistance determined for drained and plain strain conditions.

In clay: $q_p = N_c \cdot c_u$ $\quad$ ($N_c = 5.7$)

$e_p = 2.5 \cdot c_u$

$f_s = \alpha \cdot c_u$ $\quad$ ($\alpha \sim 0.8$)

$$R = N_c \cdot c_u \cdot b \cdot L + 2L \int_0^d (2.5\, c_u \sin\beta + \alpha c_u \cos\beta)\, dz$$

c_u was taken as the "most probable" fit to the shear strength profile, determined from average values of q_c with a cone factor $N_k = 15$.

For BC and CA this approach gave good agreement between observed and calculated penetration resistances, as shown in Fig. 7.

The agreement is apparently less for TP-1. $\phi = 45°$ was used in the calculation, which is in accordance with $\phi_{dr} = 42°$ determined from triaxial CID tests, when correction for plain strain is taken into account.

However, there are several complicating factors for TP-1: As indicated in Fig. 7, the initial part of penetration (~0.6m) took place with negligible resistance. This might be explained by the disturbance of the top soil caused by the horizontal movements of the platform during touch-down and by some "piping" in the initial skirt penetration phase. The sloping sea floor and the ballast eccentricity (ref. Fig. 8) adds to the difficulties in the analysis.

In conclusion, the back-calculation indicates that reasonable results for penetration resistance are obtained in stiff, clayey soils. It is therefore proposed to use this method for the analysis of future foundations of similar type. Due to limited experience the proposed method of analysis should be applied with more caution for sandy soils.

7. Settlements

As mentioned above, settlements for TP-1 are being measured optically. Three different reference bases have been used in the course of observations. The measurements further show considerable scatter, thus no firm settlement curve can be drawn. However, the following results are indicated:

Rate of settlements:
- first month   : ~40mm
- after three months   : very small

The total settlement 2 years after the installation in June, 1976 was approximately 10cm, and no significant differential settlement has been observed. These data comply reasonably well with the predictions of total settlements during platform lifetime of 20-25cm, and the results also compare well with measurements from other gravity structures on similar soil conditions.

Data from Brent C are at the time of writing not available.

From the time of installation of CA settlements have been recorded daily with the "short term settlement" device deployed during the installation phase, i.e. a closed hydraulic system similar to the BC long term monitoring system. In September 1978, the permanent system (radioactive bullets) was made operational. Readings have since then been taken twice a month.

The two different systems for measuring settlements of CA are apparently not in agreement, and results are therefore not presented. However, measurements will be continued and analysed, and more consistent data are likely to be available at the time of the conference.

8. Dynamic Motions

At the time of writing, only data from TP-1 are available, and these data only refer to recordings on a strip-chart recorder during the first winter season (1976/77).

In sea-states up to $H_{1/3}$~8m platform resonance frequencies are measured to be 2.5s (0.4 hz) and 1.8s (0.56 hz) in W-E and N-S directions, respectively. This is in good agreement with the predicted dynamic response, and the difference in the resonance frequencies in the two perpendicular axes reflects, as expected, the difference in stiffness due to the two columns placed in line on the N-S axis.

Lateral displacements of the caisson are calculated to be less than ±2mm for $H_{1/3}$ less than 8m.

Preliminary analysis indicates that the soil shear modulus for the small strains recorded at the above sea state is very high, i.e.

several times higher than the shear modulus predicted for the design loading.

From a surveillance point of view, the principal purpose of measuring dynamic displacements is to establish possible long term effects in the foundation behaviour, notably variations in the stiffness of the foundation. The parameters of principal concern are the displacements of the caisson correlated to sea state at storm condition. However, the frequency content of the motions should also be recorded. Any long term effects on foundation stiffness will change the resonance frequencies of the system. Preliminary indications for caisson and tower type structures are, however, that the sensitivity of this parameter is considerably less than the sensitivity of the displacements, which should therefore be regarded as the principal parameter.

9. Scour

The square shaped base of the Sea Tank platform design is not favourable with respect to scour. For any shape of structure the sea bed currents, including the cyclic wave-induced currents, will be influenced by the structure. However, the increase in water particle velocity close to the sharp corners of the base is considerable.

From a study /6/ of the scour potential at several North Sea platforms the following figures may be quoted:

	D_{50} (mm)	τ_1 (N/m^2)	τ_{10} (N/m^2)	τ_{crit} (N/m^2)
TP-1	0.17	0.65	1.05	0.15 - 0.45
BC	0.60	0.45	0.55	0.25 - 0.75
CA	0.14	0.35	0.45	0.25 - 0.75

where D_{50} : The 50% grain size fraction for the sea floor soils

τ_1, τ_{10} : Calculated, actual shear stresses for combined current and wave action for the 1-year storm and 10-year storm, respectively

τ_{crit} : Critical shear stress, calculated range

As it can be seen from the above table, it was not unexpected that at TP-1 scour developed at and around the E- and W-corners of the base. Only a few months after platform installation, i.e. before the winter season, the first significant scour could be seen. The scour developed during the winter and extended by the end of the winter to a depth of approximately 2m locally. This means that the scour almost reached the tip of the skirts before remedial action was carried out. During the spring of 1977, the scoured grooves were filled in with gravel bags and with coarse gravel placed by divers. This was originally considered to be a temporary solution awaiting a more comprehensive scour protection with surface dumped gravel to be placed. However, later surveys have shown that the gravel bags around the corners have stabilized the situation, and for the time being this protection is deemed satisfactory when combined with further close monitoring.

For BC and CA the situation is more favourable as indicated in the above table. Further, the sea bed consists of clay (except for shallow deposits of sand at mudline), and the skirt penetration is approximately 2.5m. For the time being, no scour protection is considered necessary. However, a monitoring program was initiated soon after platform installation. In order to quantify the surveys of the sea bed next to the base slab, a number of marked stakes (12 at each of the structures) have been placed around the periphery of the base. One of the stakes at BC is shown on the photography, Fig. 9.

At the time of writing, no tendency to scour is observed at Brent C or Cormorant A.

10. Conclusions

In general, the foundation behaviour of the three Sea Tank platforms has been satisfactory in most aspects.

The complicated installation operations have been successfully completed in compliance with the specified procedures. With minor exceptions the instrumentation systems for the installation monitoring performed satisfactorily, which effectively secured the necessary control during the installation phases.

The skirt penetration resistance was in general less than predicted, the predictions were, however, all considered to be conservative. The experience achieved with the penetration of wedge shaped skirts has by back-calculations resulted in the confirmation of a generally applicable calculation method that appears reliable, at least for soil conditions of similar types as those encountered at the three sites. For sandy soil conditions more experience is desirable to draw final conclusions.

The underbase grouting was successful, although some problems were encountered through the escape of grout.

There is no indication that long term behaviour of the structures is not in reasonably good agreement with the expectations. A reservation in this respect is that the data available at present are rather scarce. Besides the fact that BC and CA have only been in operation for about half a year at the time of writing, the reason for the scarceness of data is that the instrumentation has taken more time to implement than anticipated. One must, however, appreciate that the installation of the instrumentation and the start-up of routing data logging is a delicate and time consuming task, that more often than not is in collision with other platform activities (of higher priority!)

On TP-1 scour to a depth of 2m developed at two corners. Remedial measures such as placing of sand bags and gravel have prevented further scour and re-established the original sea bed level.

The recording of data during platform installation and during platform operation is a must to assess the integrity of the structure including the foundation. With proper interpretation the data are, furthermore, essential to future planning and design of the actual type of structure.

The views presented in this paper, and the conclusions drawn above, solely represent the views of the authors and not necessarily Det norske Veritas, or the designers, or operators referred to.

11. Acknowledgements

The authors want to thank the many companies involved in the design, construction and operation of the three gravity platforms for their kind permission to present this paper and for their valuable advice.

This acknowledgement is dedicated to:

Elf Aquitaine Norge A/S, as operator of Frigg, TP-1.
Shell Expro UK, as operator of Brent C and Cormorant A, on behalf of Shell/Esso.
Sea Tank Co., as the main designer and contractor for all three platforms, including most of the instrumentation for the installation monitoring.
Sir Robert McAlpine & Sons Ltd., as main contractor on the concrete structures.
Mecasol, who acted as main geotechnical consultant to STC on all three platforms and also carried out a portion of the laboratory testing, in particular the cyclic triaxial testing.
Norwegian Geotechnical Institute (NGI), who acted as geotechnical consultant to Elf and to Shell/Esso. NGI also participated in the site investigations and carried out a considerable portion of the laboratory testing, including the cyclic simple shear tests. NGI further supplied part of the instrumentation as subcontractor to STC, i.e. the enclosed hydraulic systems for measuring skirt penetration (BC and CA) and for measuring settlements (BC).
Fugro Cesco, who carried out site investigations on all three sites, including laboratory testing.
McClelland, who made some of the borings at BC and CA, including laboratory testing.
Solmarine (Groupe Soletanche), who acted as design consultant and subcontractor to STC on the three structures in all aspects related to the underbase grouting.
Syminex, who acted as design consultant to Elf and to STC with repect to the long-term instrumentation on all three platforms. Syminex further supplied part of this instrumentation as subcontractor to STC, e.g. the pore pressure gauges, the settlement device on CA and the accelerometers.

12. References

1. **Det norske Veritas**: "Rules for the design, construction and inspection of offshore structures" (1977).

2. **Det norske Veritas**: "Rules for the design, construction and inspection of offshore structures. Appendix F, foundations" (1977).

3. **Derrington, J.A.**: Construction of McAlpine/Sea Tank gravity platforms at Ardyne Point, Argyll. Institution of Civil Engineers, London 1976.

4. **Wangensten, T.**: Full scale measurements on gravity platforms. Scour-prediction method and observations (11 platforms). DnV internal report No. 78-514 (1978).

INSTALLATION	TP-1	BRENT C	CORMORANT A
FIELD (ALL IN UK SECTOR)	FRIGG	BRENT	CORMORANT
FUNCTION	GAS TREATMENT	DRILLING/PRODUCTION/OIL STORAGE	DRILLING/PRODUCTION/STORAGE
OWNER	ELF AQUITAINE/TOTAL	SHELL/ESSO	SHELL/ESSO
OPERATOR	ELF AQUITAINE NORGE	SHELL EXPRO UK	SHELL EXPRO UK
PLATFORM DESIGN	SEA TANK CO.	SEA TANK CO.	SEA TANK CO.
MAIN CONTRACTOR (SUBSTRUCTURE)	MC.ALPINE & SONS LTD/ SEA TANK CO.	MC.ALPINE & SONS LTD/ SEA TANK CO.	MC.ALPINE & SONS LTD/ SEA TANK CO.
FABRICATION STARTED	1974	1974	1975
DATE OF INSTALLATION	5TH JUNE 1976	18TH JUNE 1978	7TH MAY 1978

TABLE 1
GENERAL INFORMATION

INSTALLATION	TP-1	BRENT C	CORMORANT A
COORDINATES	E 02°03'51.4" N 59°52'47.3"	E 01°43'19.0" N 61°05'46.5"	E 01°04'23.1" N 61°06'09.0"
WATER DEPTH (LAT)	103	141	150
AREA OF BASE (M^2)	5,200	10,500	10,000
AREA OF CAISSON (M^2)	4,900	8,100	9,800
HEIGHT OF CAISSON (M)	42	57	56
SKIRTS: DEPTH BELOW BASE (M)	2.0	3.0 - 2.0	3.0 - 2.0
TOTAL LENGTH (M)	535	1040 (780 + 260)[1]	980 (750 + 230)[1]
TIP AREA TOTAL (M^2)	160	310	195
THICKNESS (AT BASE/TIP) M	0.90/0.30	1.20 - 0.90/0.30	1.35 - 1.00/0.20
SUBMERGED WEIGHT (OPERATING) (MN)	1150	1800	2300
DESIGN WAVE CONDITIONS			
MAX. HEIGHT (M)	29.0	30.5	29.8
PERIOD (SEC)	15	15	17
DESIGN ENV. LOADS (WAVES + WIND)			
HORIZONTAL (MN)	340	470	540
MOMENT (MN·M)	7,500	17,000	26,000

TABLE 2
PLATFORM DESIGN DATA

[1] (a + b): a, LENGTH OF 3M SKIRTS. b, LENGTH OF 2M SKIRTS

INSTALLATION	TP-1		BRENT C		CORMORANT A	
	Installed	Used	Installed	Used	Installed	Used
Positioning						
Electronic range system	2	2	2	2	1	1
Acoustic transp.	0	0	4	0	4	0
Laser system	0	0	1	1	1	1
Optical	1	0	0	0	0	0
Inclination						
Electronic inclinometers	2	2	4	4	4	4
U-tubes	2	2	0	0	2	2
Draught						
Pressure transducer	2	2	4	4	4	4
Paint marks on columns	2	2	2	2	4	4
Bottom clearance and skirt pen.						
Echo Sounders	3	1	3	2	4	2
Closed hydraulic system	0	0	2	2	1	1
Optical	1	1	0	0	0	0
Ground switches	4	4	4	2	0	0
Submarine inspection		1		1		1
Ballast water level						
Pressure transducers	2 x 9	2 x 9	2 x 12	2 x 12	2 x 17	2 x 17
Skirt compartment water pressure						
Pressure transducers	0	0	15 + 4	15 + 4	16 + 4	16 + 4
Electrical sounding line (grouting only)		4		4		4
Base contact						
Strain gauges	76	67	0	0	10	10
Data acquisition						
Data logging	Manually		Manually		Manually	
Data processing	Manually		Desk computer		Desk computer	

TABLE 3
INSTALLATION INSTRUMENTATION

INSTALLATION	TP-1		BRENT C		CORMORANT A	
Settlement						
No. of devices	1		2		2	
Type	Optical		closed hydraulic system		1) Radioactive bullets 2) Closed hydraulic device	
Lateral displacement						
No. of devices	0		0		1	
Type					Digitilt (inclinometer)	
Pore pressure gauges						
No. of	0(4)		10		8(10) + 5	
Type	Shallow Vibrating wire gauges		Shallow Vibrating wire gauges		Shallow Deep Vibrating wire gauges	
Dynamic motions						
No. of accelerometers	2 x 4	2	2	4	2	4
Positions	Deck	Caisson	Top of shaft	Bottom of caisson	Top of shaft	Bottom of caisson
Type	$\ddot{x},\ddot{y},\ddot{\theta}_x,\ddot{\theta}_y$	$\ddot{x},\ddot{y}$	$\ddot{x},\ddot{y}$	$\ddot{x},\ddot{y},\ddot{\theta}_x,\ddot{\theta}_y$	$\ddot{x},\ddot{y}$	$\ddot{x},\ddot{y},\ddot{\theta}_x,\ddot{\theta}_y$
Stresses in base slab	0		0		10 vw strain gauges	
Sea state	Wave radar		Wave-rider buoy[1]		Wave radar	
Data acquisition	Computer based logging Analogue data recording on magnetic tape		Computer based logging processing, digitizing. Data recording on magnetic tape		Computer based logging processing, digitizing. Data recording on magnetic tape	

( ) Numbers in brackets refer to initially installed number of gauges
1) Located near Brent B platform, 4km south of actual platform position

TABLE 4 - INSTRUMENTATION FOR PERFORMANCE MONITORING

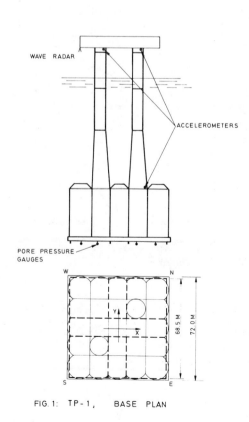

FIG. 1: TP-1, BASE PLAN

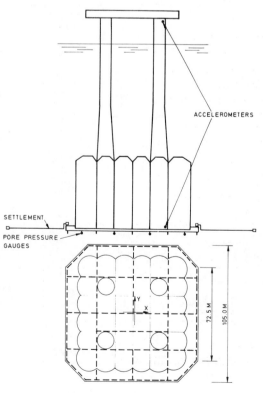

FIG. 2: BRENT C, BASE PLAN

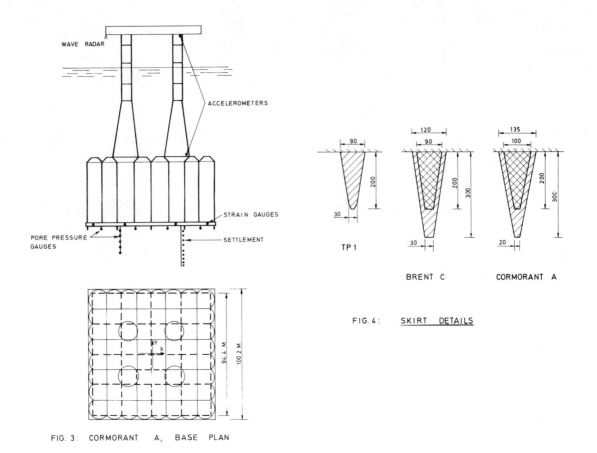

FIG. 4: SKIRT DETAILS

FIG. 3: CORMORANT A, BASE PLAN

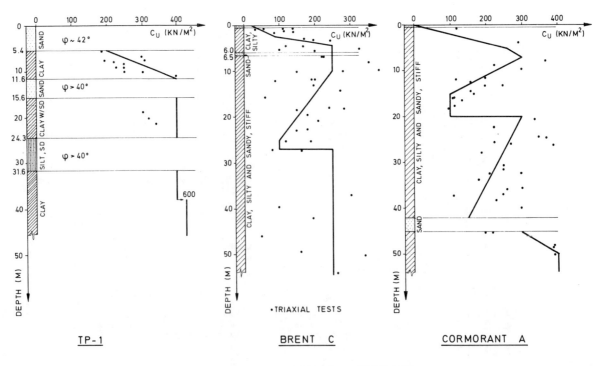

TP-1

BRENT C

CORMORANT A

FIG. 5: SOIL CONDITIONS, STRENGTH PROFILES

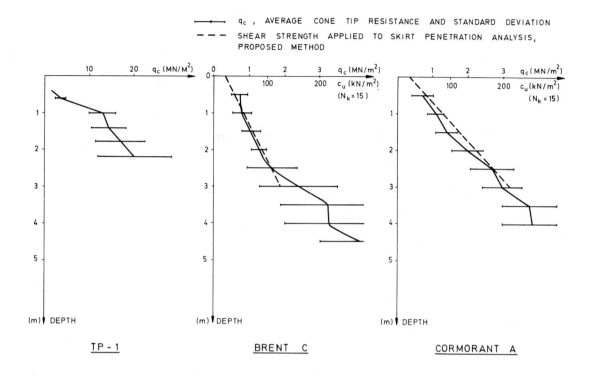

FIG. 6: SOIL CONDITIONS IN THE UPPER M'S AS RECORDED BY CPT'S

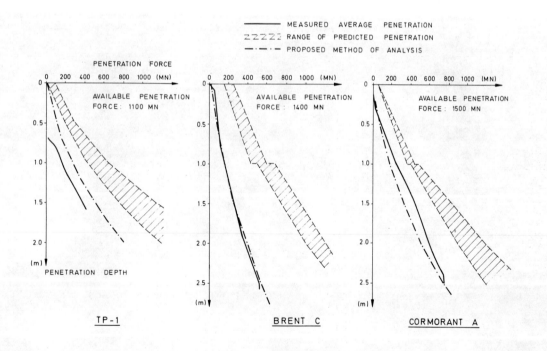

FIG. 7: PENETRATION RESISTANCES VS. PENETRATION DEPTH

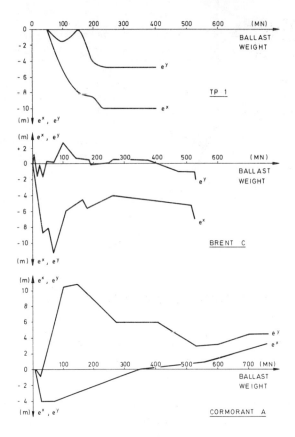

FIG. 8: BALLAST ECCENTRICITY VS. BALLAST WEIGHT

FIG. 9: SCOUR MONITORING STAKE

BOSS'79

PAPER 66

Second International Conference
on Behaviour of Off-Shore Structures

Held at: Imperial College, London, England
28 to 31 August 1979

LARGE MODEL TESTS FOR THE OOSTERSCHELDE STORM SURGE BARRIER

L. de Quelerij and J. K. Nieuwenhuis

Rijkswaterstaat, The Netherlands

M. A. Koenders

Delft Soil Mechanics Laboratory, The Netherlands

Summary

For the foundation design of the storm surge barrier in the Oosterschelde estuary, the Netherlands, large model tests (scale 1 : 10) have been performed.

The aim of these tests was to verify the prediction of the deformations of the supporting structures (pylers) of the barrier and to obtain information on the pore water pressure amplitudes in the subsoil underneath the pylers during cyclic loading.

The information received from the tests, combined with the results of an extensive soil investigation led to greatly improved predictions of the expected deformations. It also necessitated the redesign of the original foundation bed - because of the unexpected large hydraulic gradients.

Sponsored by: Delft University of Technology, The Netherlands
Massachusetts Institute of Technology, U.S.A.
The Norwegian Institute of Technology, Norway
University of London, England

Secretariat provided by: BHRA Fluid Engineering

Copyright: © BHRA Fluid Engineering
Cranfield, Bedford, England

NOMENCLATURE

a	- air content
c	- equivalent radiation damping constant
D_r	- relative density
e	- volume strain
f	- frequency
F	- force, time-dependent force
g	- acceleration of the earth gravity
G	- shear modulus
ΔH	- headloss load
ΔM	- wave load amplitude
$H_{ij,\phi}$	- transfer functions
i	- $\sqrt{-1}$
I	- mass moment of inertia
k	- permeability
K	- compression modulus of soil
K_w	- compression modulus of water
K_{air}	- compression modulus of air
l	- hysteretic constant
L	- length
m	- construction mass
n	- porosity
p	- pore water pressure
s	- equivalent spring constant
t	- time
V	- weight
x	- horizontal translation
z	- settlement
β	- hysteric damping ratio
γ	- shear strain
δ	- deformation
δ_{ij}	- Kronecker delta
Δ	- Laplacian ($\frac{\delta^2}{\delta x_i^2}$); increment
$\varepsilon, \varepsilon_{ij}$	- strain
$\lambda\alpha$	- scale factor for variable α
ρ	- mass density of water
σ	- total stress
σ'	- effective stress
ϕ	- rotation
ω	- circular frequency

Indexes:

^	- Fourier transform
a	- air included
m	- model
p	- prototype
xx	- horizontal mode of movement
xϕ	- coupling mode of movement
$\phi\phi$	- rocking mode of movement

INTRODUCTION

Since the end of 1974 extensive studies and investigations have been performed for
the foundation design of the Storm Surge barrier in the Oosterschelde estuary, the
Netherlands. Several alternatives of the barrier have been considered in the first
study-period and ultimately led to a structure, based on the concept that as much
construction work as possible should be independent of waves and tides, hence construction parts should be prefabricated on land. As a consequence of this concept
it was decided that all concrete work would be done in a large dry dock and that
all prefabricated parts would be transported to the location and assembled there.
This included the major concrete parts and steel parts of the barrier.
It is the same concept as used for the gravity structures in the North Sea.
However, a great deal of work has to be done at the location of the barrier still,
among others the preparation of the foundation bed and the construction of the so-
called sill, which is composed of small stones up to very large blocks weighing
10 tons.

a. Location and function of the barrier

The storm surge barrier is located in the mouth of the Oosterschelde estuary in the
South-Western part of the Netherlands (see fig. 1).
It will form the first and major defence against the threat of flood of the area
surrounding the Oosterschelde basin.
In that 9000 metres wide mouth are situated two artificial sand islets lying alongside
three channels, through which sea-water flows in and out the Oosterschelde basin.
The total tidal discharge is about 1100 million cubic metres.
Crossing these three channels over a total length of 3000 metres and at waterdepths
of 35 metres for the deepest channel the barrier will be constructed (see fig. 2).

The function of the barrier is to maintain the situation of tidal movements of salt
water, although the total discharge will be reduced, in order to preserve the natural
state and ecology of the estuary. Only in cases of severe storm conditions will the
barrier be closed temporarily by means of gates, cutting off the high water and high
waves coming from the North Sea (see fig. 3).
The forces which the closed barrier has to withstand during severe storms are
composed of wave-forces and the force resulting from the difference in water level
between the North Sea-side and Oosterschelde basin side, the headloss.
In accordance with other sea-defence structures in the Netherlands, like the sea-
dikes, the structure should withstand a storm, which may occur with an excess
frequency of 2.5×10^{-4} per year (so a 4000-year storm).
The probability of simultaneous occurrence of high wave forces and great headloss
force gave the maximum resulting force, which may be expected on the barrier during
its lifetime of 200 years.
Of these forces the wave force is the most interesting component.
Consideration has been given not only to the one-time maximum wave but also to the
less higher and more frequent waves.
From the statistical studies an expectation is obtained of the magnitude and
frequency of cyclic loads on the structure.
The major part of the load applies to the steel gates, which are suspended between
the so-called pylers, concrete piers, standing 45 metres apart centre to centre.
From the gates the loads are being transferred to the pylers and further to the
subsoil.
Transfer of loads of the pyler to the subsoil takes place through the base of the
pyler and through the coarse material of the sill which surrounds the pyler above
foundation level. Thereby the greater part of the loads will go through the base
to the ground (see fig. 4).
The soil at the sea bottom at the location of the barrier consists of Holocene sand
which is loose and contains a small percentage of silt and locally some clay.
Underneath the loose sand, the thickness of which varies from 0 to about 10 metres,
firmly packed Pleistocene sand is found which is pre-loaded and which also contains
silt.
Locally the silt percentage can be so high - 20 to 26 percent - that it can be unacceptable. Therefore replacement of that silty sand by clean sand becomes necessary,
where such sand is located close to the base of the pyler structure.

b. Geotechnical topics

It is evident that for any structure under cyclic loading loose sand will give unsurmountable problems. During the early study period a caisson-type barrier has been considered, which can tolerate relatively large deformations. But even such a structure will undergo too large a deformation, when nothing is done to the loose sand, as was proven by two tests on large caisson-models. These tests, one on loose sand and one on densified sand, have been described by De Leeuw (ref. 1).

For the final design consisting of prefabricated pylers the loose sand has to be densified anyway. But even in the situation that the granular subsoil is dense and firm the deformation of the pyler can still be critical and therefore has to be predicted very accurately.

Too much tilting of the pylers can cause malfunctioning of the gates and the upper structure as gates are suspended between the pylers.

As an interface between pyler and the sand of the sea-bottom a granular layer is required in order to prevent erosion of the sand at the edges of the pyler base because of the tidal water movement. When the pyler has been placed on this granular layer, the foundation bed, grout is injected in order to fill the voids and to create good friction between pyler base and foundation bed.

Next to and between the pylers the foundation bed is covered by an 8 - 11 metres thick filter construction, the sill, making a gradual transition to the top layer of the sill. This top layer consists of large blocks, which are necessary to withstand the very high flow conditions.

It is obvious that at the cyclic load transfer from the base to the subsoil the porewater pressures change. However, due to the difference in permeability and stiffness of the coarse foundation bed material and the sand, differences in pore pressures between these layers also occur. These differences cause very high, up to several hundreds of percents, cyclic gradients. The filter density of the foundation bed has to be proof against these gradients.

Briefly it can be stated that in view of the requirements of the structure there are two main geotechnical questions:

1) to predict accurately the stability and the deformations of the pylers under static and cyclic loading; and also to check the possibility of amplification of the deformations by wave energy at resonance frequency;

2) the pore-water pressure variation beneath the pylers during cyclic loading with respect to the filter stability of the foundation bed.

c. Research methods

There are several methods at our disposal to answer these questions, through calculations and through model-testing or a combination of both. Both have their limitation. Even the best solution theoretically, namely a 1 : 1 scale model test, would not have been adequate for this type of construction, which consists of many pylers at different locations. Most important therefore was to collect sufficient and correct soil data at each location, by preference through in-situ testing. That has been done in a very thorough way by means of static cone penetrometer tests, penetrating up to 50 metres in the sand, special continuous borings of 20 metre length, in-situ porosity measurements, in-situ pressuremeter tests and in-situ permeability tests. By using calculation models there always remains the crucial question, which are the correct soil parameters; no matter whether a simple calculation method is applied or a sophisticated 3-dimensional finite element method.

The use of finite element methods requires a more detailed description of the soil behaviour. For this purpose a lot of triaxial tests, both static and cyclic, have been performed on sand and sill materials.

Apart from the difference in soil parameters, derived from different types of tests, one has to take into account a statistical deviation of the real prototype parameters due to inhomogeneity of the soil.

In order to try to understand what exactly occurred during the test and to derive soil parameters from the observations made, the following method has been used. One starts from a theory which is simplified to such a degree that closed form solutions can be used to perform a least-squares fit. One might raise the question how good this approach actually is. On one hand errors are introduced because the theory is simplified, on the other hand any natural soil system seldom is a homogeneous object. The nice thing about least-squares routines is that they compute averages and standard deviations of parameters and with averages the engineer usually characterizes his soil, while the standard deviations may be viewed at as a substitute for the classical safety-factor. Indeed least-squares techniques are well-rooted in statistics and it is not very difficult to incorporate these results in a probabilistic approach.

A question that remains to be answered is of course to what degree solutions of an equation may be simplified so that the error is small enough for the theory to keep its predictive force. It is always possible to expand an expression in a Taylor series and for most stable processes one often can deduce a set of parameters such that a series converges. Furthermore finite element techniques provide solutions which can be compared with the approximated solutions. Both ways of estimating the error requires at least some insight in the numerical values of soil paremeters. Here laboratory tests come in handy. From all this it is obvious that in order to pass a verdict on the functioning of a constructor's design a whole chain of methods is to be used - one cannot do without the other.

2. ANALYSIS OF DEFORMATION

a. Introduction

Due to the complexity of time-fluctuating load conditions and the 3-dimensional geometry of the pier and corresponding stress distributions in the subsoil, neither analytical nor numerical calculation models provided a proper prediction of the behaviour of the prototype. Based on some simplifications, such as plane strain conditions, 100% drainage, hyperbolic shear stress-deformation relation and logarithmic compression behaviour, finite element calculations with the program CONSOL (ref. 2) have been carried out. The soil parameters have been deducted from tri-axial tests on the Oosterschelde sand and foundation-bed materials at representative stress paths.

Since the pier design was very sensitive to the maximum amount of deformation during the design-storm loadings effort was put in model tests in order to limit the range of predicted deformations.

Two types of tests can be distinguished, small and large ones. On the one side a large number of centrifuge tests have been carried out at a geometrical scale 1 : 120 at stress level 1 : 1 (120 g field) and fully drained. These tests were performed by Prof. P.W. Rowe, University of Manchester.

Because of their restricted modelling properties, especially scaling the foundation bed layer of about 1 m thickness in prototype, the tests were mainly used for design guidance with respect to design parameters such as base areas, depths of embedment and weight of the pier.

The relatively short preparation and execution time made it possible to run a sufficient number of centrifuge tests. The results (ref. 3) fit with parametric studies of Consol calculation. Figure 5 shows some typical centrifuge test results, related to the horizontal deformation δ_h at sill level.

On the other side more knowledge was needed on the absolute value of the deformations, together with the force distribution of the static and cyclic loadings to the subsoil. Also information on the pore pressure distribution, especially with respect to the cyclic gradients, **was** desired.

For these reasons relatively large scale model tests were performed at the location Kats. For practical reasons (facility and time) it was decided to use a geometrical scale 1 : 10. The results would be used for a direct transfer to prototype scale and to check the input of the f.e.m. program CONSOL.

When using physical models to predict prototype behaviour two main problems have to be solved. The first one is the correct modelling of the prototype conditions, from which the soil mechanic properties of the subsoil and sill and the loading scheme are most important. The requirement is that all model elements of the soil have the same shape of cyclic and static stress-strain behaviour and pore water pressures as in prototype situation. The second problem is, especially for 1g-models, the scaling of the model results to prototype stress level.

b. Model test 1 : 10

It was shown through in-situ cone resistance and porosity measurements that the piers with the largest loads, namely the middle piers of the deepest channel, the Roompot, were situated on Pleistocene sand with high cone resistance (> 30 MN/m²), starting at 10 m minus foundation level.

The 10 m layer between base and the Pleistocene sand consists of loose Holocene sand, which will be densified. These conditions together with the fact that both the foundation bed and the sill is artificially constructed, were the requirements for modelling.

The Kats model with base (ca. 2.5 × 5.0 m² and weight 288 kN (ref. 4)) was situated in a concrete tank, with dimensions 25 × 10 × 4 m³. Since the predicted deformations of adjacent piers are nearly equal there exists a plane of symmetry, halfway between two piers, where no shear stresses can be transmitted.

Based on this assumption only one pier model had to be built and two symmetrical planes, consisting of the concrete side walls of the tank covered by plastic, between which a lubricating oil layer has been put.

The Pleistocene sand was simulated by identical Oosterschelde sand and was densified by means of a vibrating plate in order to achieve the prototype high relative density ($D_r \approx$ 80 to 90%).

Similar to the prototype situations the Holocene sand layer was compacted by a vibrating poker up to a $D_r \approx$ 70%.

Since the air content of the soil is of importance for the pore pressure - total stress ratio (the real importance was found out <u>after</u> the test performance) all the sand has been poured under water.

As for scaling the sill materials priority has been given to the requirement that the deformation behaviour for the various layers in model and prototype had to be similar. As a consequence the grain-size distribution, grain shape and density must remain the same as much as possible. However, the average grain size d_{50} should not exceed 1/8 of the layer-thickness. Also of importance for the test was the influence of the seepage flow on the grain skeleton due to static headloss. According to the scale rules, see **below,** the seepage pressures in the model must be on geometric scale everywhere. For the foundation bed and lowest sill-layer with semi turbulent flow, the same material as in prototype was used, so the permeability was the same.

For the upper sill layers with prototype dimensions of 60 - 300 kg respectively 1000 - 3000 kg stones on a geometrical scale 1 : 10 were used.

The corresponding permeability of these model layers at turbulent flow were about 3 times lower compared to prototypes. However, finite-element quasi-static flow calculations showed that the difference in potential features between model and prototype were **negligible.**

Scale factors

In the Kats test much attention was paid to finding the correct scale factors in order to determine the frequency of external cyclic loads in the model and the transformation factors for the measured deformations. Based on the equilibrium equation and Biot's law:

$$\frac{\partial \sigma_{ij}}{\partial x_i} = \frac{\partial p}{\partial x_j} \tag{1}$$

$$\frac{k}{\rho g} \Delta p = \frac{\partial e}{\partial t} + \frac{n}{K_w} \frac{\partial p}{\partial t} \tag{2}$$

and for the same specific weight, porosity and permeability in the model and prototype the following similarity <u>theoretically</u> had to be fulfilled ($\lambda_\alpha = \alpha m / \alpha p$).

$$\begin{aligned}
\text{length} \quad & \lambda L = 1/10 \\
\text{effective stress} \quad & \lambda \sigma' = \lambda L = 1/10 \\
\text{pore pressure} \quad & \lambda p = \lambda L = 1/10 \\
\text{force, weight} \quad & \lambda F = \lambda_L^3 = 1/1000 \\
\text{strain} \quad & \lambda_\varepsilon \\
\text{displacement} \quad & \lambda s = \lambda \varepsilon \times \lambda L = 1/10\, \lambda_\varepsilon \\
\text{time} \quad & \lambda t = \lambda \times \lambda L = 1/10\, \lambda_\varepsilon \\
\text{air content} \quad & \lambda a = \frac{\lambda_\varepsilon}{\lambda L} = 10\, \lambda_\varepsilon
\end{aligned}$$

The air content scale factor follows from the relation:

$$K_{w_a} = \frac{K_{w_{a=0}}}{(1-a) + a K_{w_{a=0}}} \approx \frac{K_{air}}{a} \tag{3}$$

The consequence of the air content for the test results is discussed in another section of this paper.

It is obvious that the most important scale factor, which directly influences the load frequency and the displacement translation, is the strain ratio $\varepsilon m/\varepsilon p$. If every strain increment follows the elasticity theory then:

$$\sigma'_{ij} = 2\hat{G}(\varepsilon_{ij} - e \cdot \delta_{ij}) + 3K \cdot e \cdot \delta_{ij} \qquad (4)$$

This means that if $\lambda_G = \lambda_K$ for every change of strain there exists an absolute $\lambda\varepsilon$. Based on drained static triaxial tests it appears that K and G are proportional to $\sqrt{\sigma}$, so $\lambda\varepsilon = \sqrt{\lambda\sigma} = 1/\sqrt{10}$ and $\lambda_t = 1/10\sqrt{10} \approx 1/30$.

Based on this rule the cyclic load in the model was applied at a frequency 30 times higher than in prototype ($f_p \approx 0,1$ Hz; $f_m = 3$ Hz).

Additional research on scale factors based on cyclic, drained/undrained triaxial tests and centrifuge tests at different stress levels, indicated the need to distinguish different deformation scale factors for different deformation components i.e. static, plastic cyclic and elastic cyclic deformations.

The ideal situation would be if the model test results can just be multiplied by a specific scale factor to obtain prototype data. Practice proved to be more complicated. In particular the non-uniformity of the strain scale factor and the difference in air-content requires the use of calculation models and theoretical insight acquired from the model tests, in order to predict prototype behaviour.

Loading programme and instrumentation

The pier model is loaded by a combination of static headloss and sine-shaped wave loads by means of a hydraulic jack, in such a way that the load level (both mean and amplitude) is increased gradually. To simulate the seepage flow the headloss force is also partly exerted by a constant hydraulic headloss (hm/hp = 1/10) in the model. In order to determine the influence of load frequency on the drainage the storm loadings have been applied both with low ($\lambda_t = 1$; $f = 0,1$ Hz) and scaled ($\lambda_t = \frac{1}{10\sqrt{10}}$; $f = 3$ Hz) frequency.

Also a specific dynamic parcel ($f = 0,1 - 12$ Hz) was included in order to derive the dynamic transfer functions (see next section).

Registration equipment was placed on the model to determine the horizontal load - controlled by a function-generator - and the transfer of this load to the subsoil through the base and through the front and side walls of the construction. In order to register separately the base and side load the model was divided into two parts, connected by pressure cells. In the symmetric vertical plane of the pier the rotations and translations were measured. A total of 40 pore pressuremeters have been placed in the soil. In particular the scaling of the load frequency imposed heavy demands on the registration equipment for which the so-called pulse code modulation system was applied.

Test results

A typical registration of the deformation development during the storm parcels is shown in figure 8. Several deformation components can be distinguished:
1. direct increase of the deformation due to the static headloss load which remains practically constant during the static load;
2. when applying the cyclic wave loads the mean deformation increases; the vertical component is called the shakedown. This effect is large for the first cycle and decreases with next cycles. When another higher wave amplitude has been applied this picture is repeated;
3. the third component is the elastic cyclic deformation, the so-called swing. This remains practically constant during the number of cycles. It is shown in figure 9 that a reasonable linear relation exists between load amplitude and swing amplitude.

The partition of these three deformation components as a function of the load level is plotted in figure 10. The results pertain to the rotation. The translation components are nearly the same.

Figure 11 shows the results of the translation at base level, the rotation and mean vertical displacements during the top of the load of each parcel of the loading schemes A_1, A_2, C_1 and C_2. Notice that the design load is about 151 kN; $H_{static} = 89$ kN; $\Delta H_{cyclic} = 62$ kN and the vertical weight of the model $V = 288$ kN.

From the results it appears that the load frequency (0,1 to 3 Hz) has no significant influence on the deformation behaviour until the design load. When introducing higher

load levels this influence clearly becomes significant and causes a higher "failure" at higher frequencies, where a deformation of 15 mm translation is considered to be "failure".
The effect of preshearing already becomes significant for the translation after the design load was applied (compare curve A2 - A1). However, the effect of preshearing resulting in a more rigid soil behaviour, becomes most obvious for the rotation (see curve C2) when the model was loaded up to "failure".
The force transfer from the model to the subsoil during the time-scaled frequency (curve C1) is shown in figure 12. A distinction is made between front, footing and gate-force transfer. It was found that 70 to 80% of the total load is transferred through the footplate. Comparison of the predicted (Consol) and actual deformations (part A2 - C1) as plotted in figure 13, shows an over-estimation of both the translation and the rotation by a factor of 2. The predicted deformation mechanism and the force transfer agree well with the actual performance.
Based on the test results the Consol input had to be changed. Especially the G, K-moduli and the stress distribution were adjusted.
The improved Consol calculations agree better with the test results.
In order to extrapolate the test results to prototype conditions - taking into account the effect of the tank bottom - the measured rotations and translations should be multiplied by a factor of 5,0 and 45 respectively. If also some extra uncertainties are considered as a small difference in soil conditions between model and prototype and other schematic influences the multiplication factors will not exceed 7,6 and 80 respectively, see figure 14.

c. Dynamic behaviour

In the last decade half-space theories have become increasingly popular in describing the dynamical properties of soil-structure interaction. However, these theories yield a solution in the case of geometrically regularly shaped bodies such as a circular cylinder or an infinite strip. Especially for embedded footings the application of visco-elastic-half-space-theory becomes a difficult matter.
Beredugo and Novak (ref. 5) solved the problem for an embedded circular cylinder with radius r_o. When applying their results to footings with a rectangular base they derive an equivalent radius from the equality of footing bases.
In this paper a method is proposed in which the problem of choosing an equivalent radius is circumferenced. This can be done if results from a model test are available. From a linearized theory transfer functions are derived. Using a least squares routine the unknown parameters such as damping and equivalent spring constants can be found. Then these parameters can be scaled to prototype-conditions and a prediction is obtained.

Theory and scale rules

The reaction of a construction, which is loaded with a horizontal time dependent force $F(t)$ is described by a horizontal displacement $x(t)$ and an angular displacement $\phi(t)$. No vertical displacement is considered. In the first place it turned out that in the model test there hardly is any. In the second place when applying a horizontal load it is difficult to deal with it in a linear theory. The Fourier transform of $F(t)$ is denoted by $\hat{F}(\omega)$, the Fourier transform of the displacements by $\hat{x}(\omega)$ and $\hat{\phi}(\omega)$.
In first order linear approach $\hat{F}(\omega)$, $\hat{x}(\omega)$ and $\hat{\phi}(\omega)$ satisfy the relation:

$$\begin{pmatrix} \Omega & I_z \\ I_z & V \end{pmatrix} \begin{pmatrix} \hat{x} \\ \hat{\phi} \end{pmatrix} = \begin{pmatrix} \hat{F} \\ -b\hat{F} \end{pmatrix} \tag{5}$$

in which
b = the distance between the centre of gravity and the point of application of F
Ω = $-m\omega^2 + s_{xx} + i\, l_{xx} - i\omega\, c_{xx}$
I_z = $-s_{x\phi} - i\, l_{x\phi} + i\omega\, c_{x\phi}$
V = $-I\omega^2 + s_{\phi\phi} + i\, l_{\phi\phi} - i\omega\, c_{\phi\phi}$

s, c and l describe distinct physical phenomena. These are easily imagined when one focusses on an energy-picture. The loading force F(t) provides energy which is only partly being transferred in movement. This portion is measured by s. Part of the energy is lost in waves that run to infinity. This is described by the radiation damping c. Some energy is lost because of friction and the hysteretic constant l indicates how much.

The ratio of the energy that is associated with the horizontal and the rocking mode of movement clearly is a function of the frequency. If this were not so, no mixed index parameters would be needed. In fact it is quite simple to show that for very low frequencies only four constants s , l , s_{xx} and l_{xx} describe the construction's behaviour fairly well.

When choosing the construction-mass some extra mass should be included to simulate an amount of vibrating waves next to the pier. The subject is treated by Westergaardt (ref. 6).

Eq. (5) can be inverted. One finds transfer functions:

$$H_\phi = \frac{\hat{\phi}}{F} = \frac{-b\Omega - I_z}{\Omega V - I_z^2} \quad \text{and} \quad H_x = \frac{\hat{x}}{F} = \frac{V + bI_z}{\Omega V - I_z^2}$$

The parameters s, c and l are - strictly speaking - frequency dependent. However, it was shown by Beredugo and Novak that a very good approximation is obtained when constant parameters are used.

Suppose now that from a model-test m, I_o, s, c and l are known, how then can one make a prediction for the transfer functions for the prototype. This will be carried out here.

If the ratio between a characteristic length in the prototype and the corresponding characteristic length in the model is λ_L it follows:

$$m_p = \lambda_L^3 m_m$$
$$I_p = \lambda_L^5 I_m$$

Apart from the terms $-m\omega^2$ and $-I\omega^2$, Ω, I_z and V can be written as:

$$M_j(S_j + ia_o c_j^* + 2S_j \beta)$$

j = 1,2,3 apply to the horizontal, coupling and rocking mode of movement respectively. s_j and c_j^* are constants independent of the dimensions of the construction (they may depend on the geometry though).

a_o is the dimensionless frequency: $a_o = \dfrac{\omega r_o}{\sqrt{\dfrac{G}{\rho}}}$

G stands for the G-modulus of the soil and ρ for the density. M_j has the form $G \cdot r_o^j *$ (a function which depends only on relative dimensions). β represents the hysteretic damping ratio.

In order to derive scale-rules one considers that G^2 usually is taken proportional to the effective normal stress, hence:

$$G_p = \sqrt{\lambda_L} \cdot G_m$$

From this it follows easily that

$$s_{j,p} = \lambda_L^j \cdot \sqrt{\lambda_L} \cdot s_{j,m} \quad \text{and}$$
$$a_{o,p} = \lambda_L^{3/4} \cdot a_{o,m}$$

Scaling β is less simple. According to Hardin and Drnevich (ref. 7).

$$\beta = \beta_{max} \frac{\gamma/\gamma_r}{1 + \gamma/\gamma_r}$$

in which γ represents the shear-strain due to shear stress τ and γ_r some reference-strain:

$$\gamma_r = \frac{\tau_{max}}{G_{max}}$$

For G_{max} one takes the shear-modulus of statically-loaded soil: G_{max}^2 is again proportional to the effective normal stress.

$$\frac{\gamma_{r,p}}{\gamma_{r,m}} = \sqrt{\lambda_L}$$

In order to find $\dfrac{\gamma_p}{\gamma_m}$ the dynamical shear-strain is replaced by the statical shear-strain. This approach is good as long as the ratio between the amplification-factor for the model and the prototype is not far from 1. Thus:

$$\frac{\gamma_p}{\gamma_m} = \frac{\hat{\phi}_p(o)}{\hat{\phi}_m(o)}$$

and from (5) - neglecting the hysteretic damping - one arrives at:

$$\frac{\gamma_p}{\gamma_m} = \frac{\sqrt{\lambda_L}}{\lambda_L^3} \cdot \frac{\hat{F}_{x,p}(o)}{\hat{F}_{x,m}(o)}$$

If one performs the model-tests such that the expected loads in the prototype-case are greater by a factor λ_L^3 than the loads in the model test-situation (such scaling is in accordance with scale-rules derived from Biot's law) then:

$$(\frac{\gamma}{\gamma_r})_p \, (\frac{\gamma_r}{\gamma})_m = 1 \quad \text{and} \quad \beta_p = \beta_m$$

Application

In the test model a sinusoidal load was applied. Forces and displacements were measured. From the data the two transfer-functions H_ϕ and H_x were found (see fig. 15). With a least-squares routine the parameters s, l and c were found. After scaling - using the scale-rules above - the transfer functions of the prototype were predicted (see fig. 16). It is seen that the lowest resonant-frequency is decreased by almost a factor of 5 from 10 Hz to 2 Hz. Since the spectral intensity of the North Sea-waves is expected to be a narrow band around 0,1 Hertz the conclusion must be that the dynamical behaviour of the prototype construction will not be important for the deformation.

3. PORE PRESSURES

a) The pore pressures in the subsoil due to cyclic loading are important for two reasons. The first is to find out the amount of drainage and the generation of excess pore pressures in order to determine the influence on the deformation behaviour. The second is to understand and predict the cyclic gradients at the interface of the different foundation bed layers. These gradients jump discontinuously at the interface when a difference in soil properties (permeability and rigidity) of those layers exists.
Based on previous research (1 : 10 model tests at Kats, NGI simple shear tests and MIT, LGM analysis of triaxial tests) it was expected that in the densified Holocene no noticeable generation of excess pore pressures was to be expected.
The deformation of the prototype-structures could be predicted assuming nearly fully drained conditions.
The 1 : 10 pyler tests, as described above, confirmed both points of view.
Figure 8 shows different pore pressure components, the magnitude of which changes during a storm parcel, analogous to the deformation development.
It shows:
1) after increasing the static headloss the pore pressure changes gradually. At constant headloss the pore pressures dissipate. Since the ratio $\Delta p/\Delta \sigma$ is small (about 10%) the influence on the deformation is negligible;
2) during the cyclic loading the consolidation continues. Despite the relatively large consolidation time practically no generation of excess pore pressures occur up to design load level. At the beginning of a load parcel the average pore pressures change in the same way as the static load, probably because of dilatancy and extension. At higher load levels some interesting developments of generation of excess pore pressure were measured. A complete understanding of these developments is not available as yet;
3) a cyclic variation of the pore pressures due to the cyclic loading. A reasonable linear correlation between the wave load amplitude and the pore pressure amplitude was found independent from the static load level (see fig. 17).

b) Cyclic pore water pressures and cyclic gradients.

The behaviour of pore-pressures at low frequencies is described by Biot's law (see eq. 2).

For our problem, i.c. the determination of cyclic gradients in the foundation bed of
the model and prototype, we have chosen to solve it in both one and two dimensions.
The one-dimensional solution is used to investigate deviations of the average in a
multi-layered system, the two-dimensional one served us to determine the soil parameters from the model-test. Whatever the dimension, something has to be assumed for
the volume-strain. Usually this is done by simultaneously solving the equations of
equilibrium together with some choice for the stress-strain relationship. In our case
it is sufficient to note that finite-element computations show that the amplitude of
the total stress does not change appreciably with depth when the latter is not greater
than about a quarter the length of the construction. It follows that $\partial\varepsilon/\partial t$ is a function of x and t only. Only cyclic solutions are of interest so:

$$p = \hat{p}\,e^{iwt} \;;\; \varepsilon = S(x)e^{iwt}$$

For one-dimensional cases $S(x)$ is a simple delta-function, for two-dimensional cases
we have chosen for $S(x)$, a function as shown in figure 18 linear between the edge of
the construction and decreasing according to an e-power to zero outside it.
Next boundary conditions must be stated. They are:
1) continuity of pressure and flow everywhere;
2) no flow at infinity;
3) an impervious top underneath the construction and zero pore pressure outside it.

Condition no. 3 is hard to handle. As it turned out an imaginary impervious seal at
the top of the foundation bed extending outside the structure is not a bad approach
since the top layer's permeability is much greater than the bottom layers.
Using these conditions and approximations a solution in integral-form is obtained,
which serves us for the least-squares routine. A plot of equi-amplitude lines is
presented in figure 18.

Discussion of results

At low frequencies it is possible to find a set of soil parameters which satisfy the
derived equations. We find permeabilities of the two layers, the air content of the
water and the stiffnesses of the soil. The stiffness is related to the G-modulus.
From the S-function as computed by the least-squares method and an estimate for the
stresses and Poisson's ratio the shear modulus G is found. The latter is also found
from the shear wave velocity (Richart, Hall and Woods, ref. 8) and from the s_{pp} and s_{xx}
as computed in the previous section. Laboratory tests are available as well. For sand
all these are compared:
1. from the pore-pressures   : G =   27 MN/m^2
2. from the shear-wave velocity: G =   34 MN/m^2
3. from s_{pp} and s_{xx} : G = 20-27 MN/m^2
4. from laboratory tests   : G = 20-30 MN/m^2

As for the permeabilities they agree quite well with the parameters found in laboratory tests. The amount of air in the water of course is a factor not easily imitated
in the laboratory. However, it is an important quantity in the sense that it is responsible for the validity of the scale rules. Under prototype-circumstances hardly
any air is to be expected. In the top layer of the model (sea-gravel) an air content
of about 5% was found. In the sand a much smaller air content was backfigured.
The effect of gas in water is that it keeps the pore pressures at a small value, provided the gas comes in bubbles.
In figure 18 we drew an asymmetrical function. This is not a bad drawing but reality.
It was found in the model that on one side of the construction the amplitudes of pore
pressure were much larger than on the other side. This can be understood if one considers that there is a volume strain associated with both the horizontal mode of movement and the rocking mode. On the side of the construction where the load is applied
these two volume-strains are in phase; on the other side they are out of phase. The
phase-shifts are neatly reproduced by the theory.
The theory predicts the amplitude of the pore-water pressure to be linear with the
amplitude of the load. This is pictured in figure 17.
However, the theory does not predict all the effects observed in the test.
The most important effect not yet completely understood is the dependency of the
pore pressure amplitude on the frequency. From the theory one would expect the amplitude to go up at increased frequency in the bottom layer. In the test they go down.
After scaling the parameters the pore pressures and their gradients can be computed

for the prototype situation. The computed gradients across the sand-seagravel interface were considered too large for a stable filter.
Hence the foundation bed had to be modified for the prototype.

REFERENCES

1. de Leeuw, E.H.: "Results of Large Scale Liquefaction Tests".
   Proc. BOSS Conference '76, Vol. II, pp. 404.

2. Kenter, C.J. and Vermeer, P.A.: "Computation by Finite Elements".
   Proc. Foundation Aspects of Coastal Structures, oct. '78, Vol. 1, section III.2.

3. Rowe, P.W. and Craig, W.H.: "Predictions of caisson and pier performance by dynamically loaded centrifugal models".
   Proc. Foundation Aspects of Coastal Structures, oct. '78, Vol. 2, section IV.3.

4. Quelerij, L. de and Broeze, J.J.: "Model tests on piers scale 1 : 10".
   Proc. Foundation Aspects of Coastal Structures, oct. '78, Vol. 2, section IV.2.

5. Beredugo, Y.O. and Novak, M.: "Coupled Horizontal and Rocking Vibration of Embedded Footings".
   Can. Geotechnical Journal. $\underline{9}$, 477. (1972)

6. Westergaardt, H.M.: "Water pressures in Dams during Earthquakes".
   Trans. Am. Soc. of Civ. Eng. $\underline{S9}$, 8, III, 418 (1933).

7. Hardin, B.O. and Drnevich, V.P.: "Shear Modulus and Damping in Soils Design Equations and Curves".
   J. of the Mech. and Found. Div. Am. Soc. of Civ. Eng., SM.7, 667, July 1972.

8. Richart, F.E., Hall, J.R. and Woods, R.D.: "Vibrations of soils and foundations".
   Prentince-Hall, 1970.

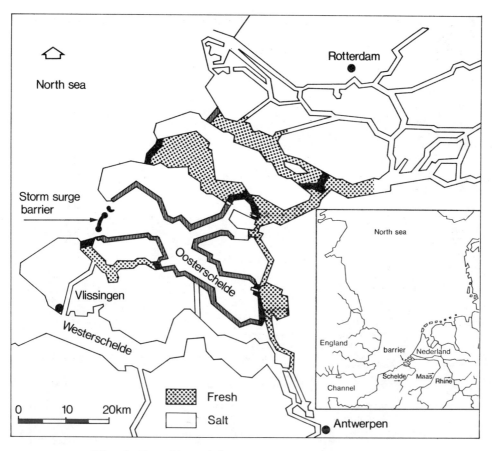

Fig. 1 Location of the Oosterschelde estuary

Fig. 2 Location of the storm-surge barrier

Fig. 3 The storm-surge barrier

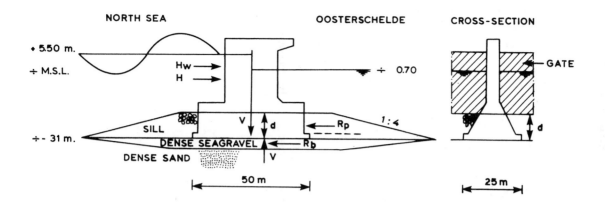

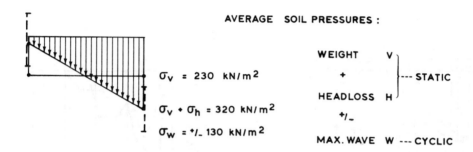

Fig. 4 Foundation of the pyler construction

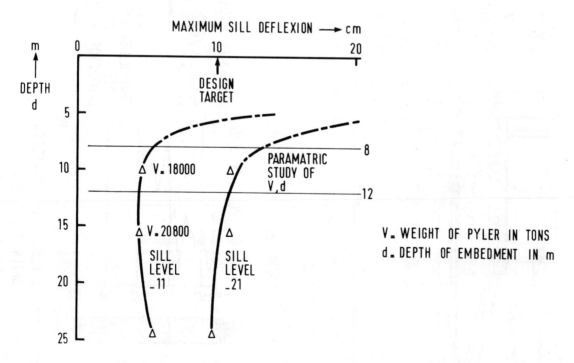

Fig. 5 Variation of sill deflexion with depth of embedment from centrifuge tests

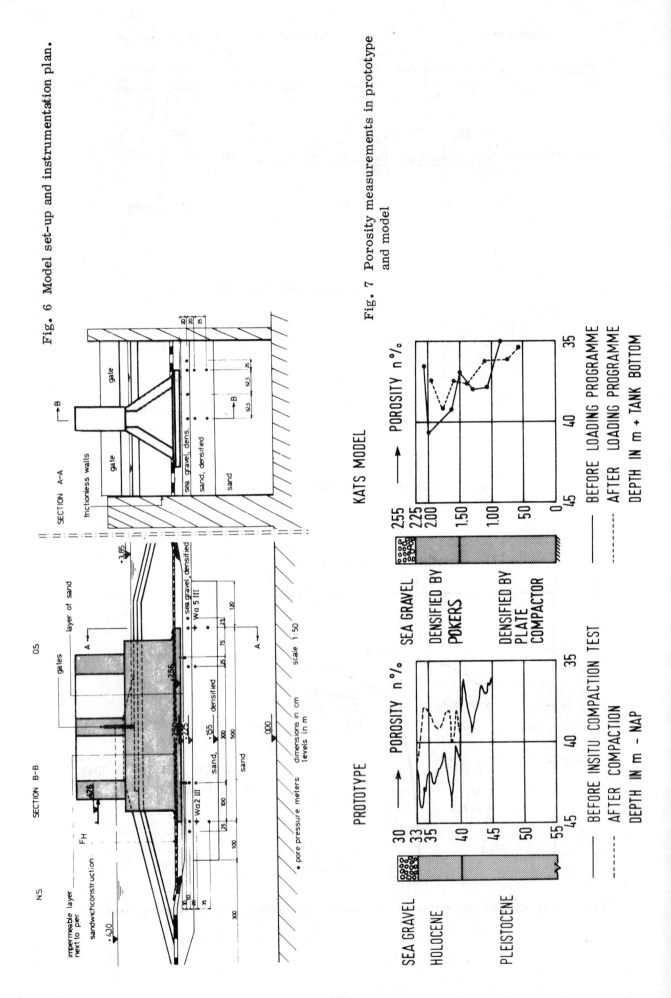

Fig. 6 Model set-up and instrumentation plan.

Fig. 7 Porosity measurements in prototype and model

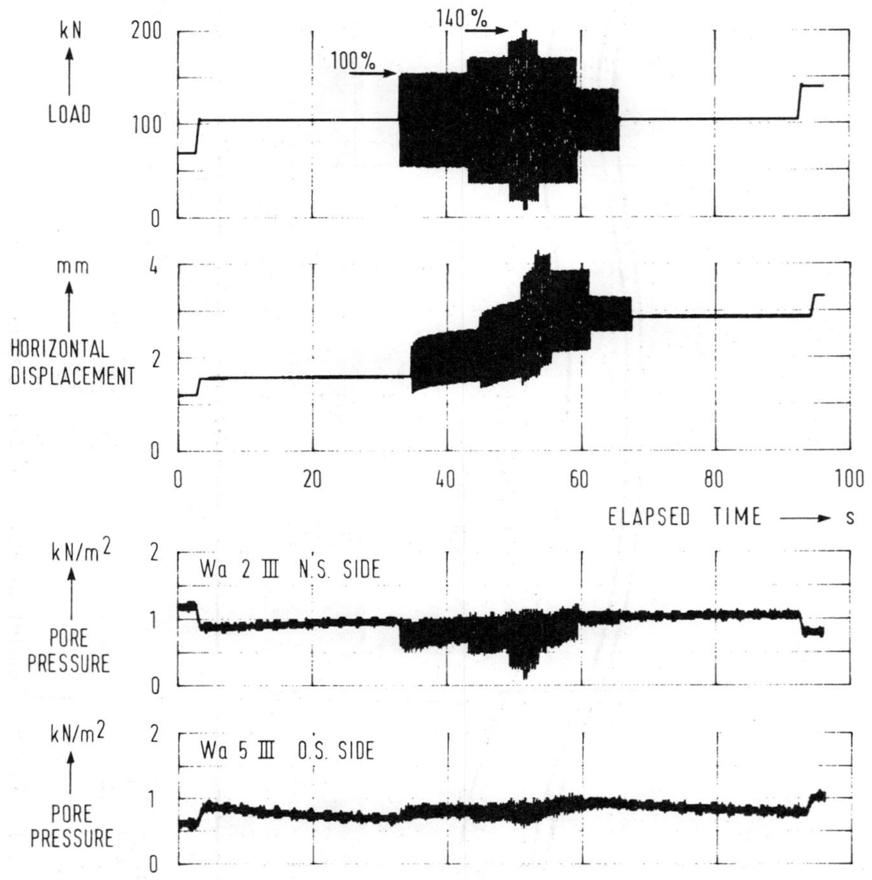

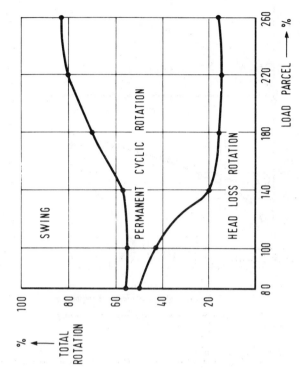

Fig. 9 Relation between rotation-swing and wave load amplitudes

Fig. 10 Contribution of rotation components at different load levels

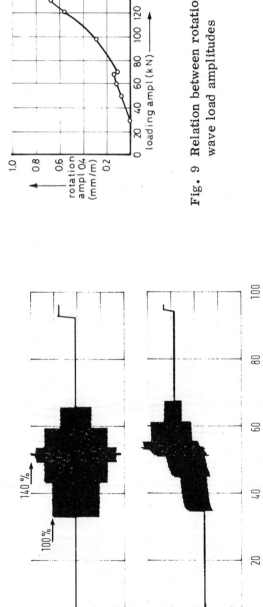

Fig. 8 Simultaneous records of load, deformation and pore pressure during a storm parcel

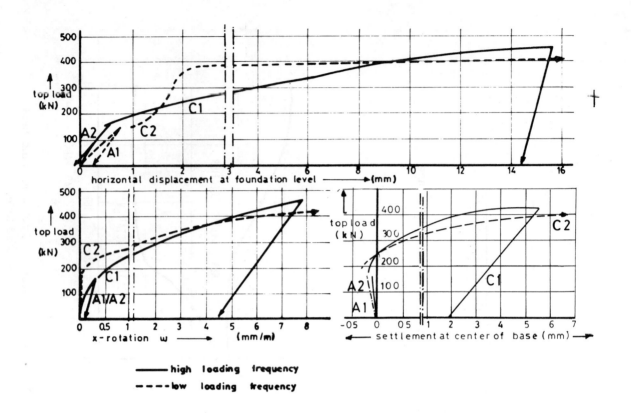

Fig. 11 Influence of frequency and preshearing on the deformations

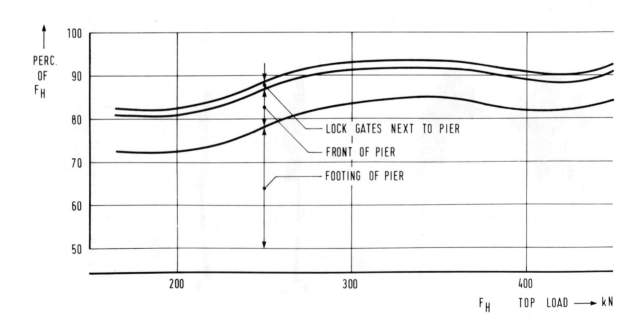

Fig. 12 Force transfer at different load levels

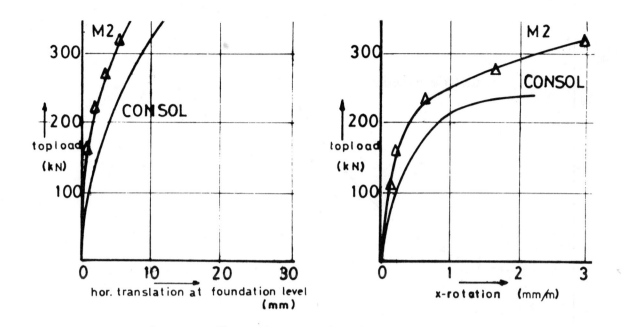

Fig. 13 Comparison of predicted and actual deformations

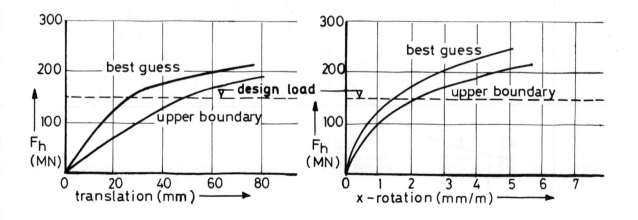

Fig. 14 Prototype predictions

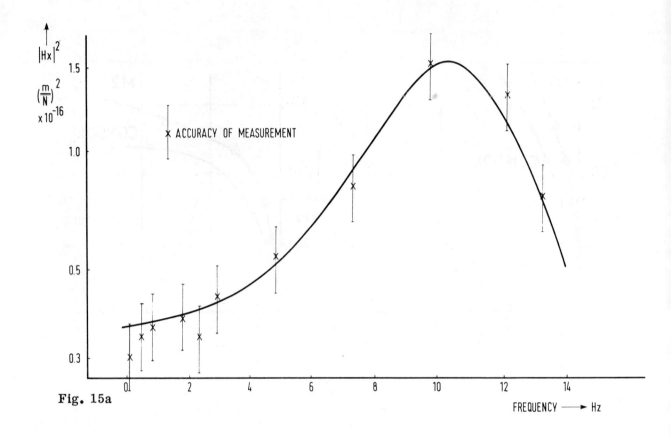

Fig. 15a

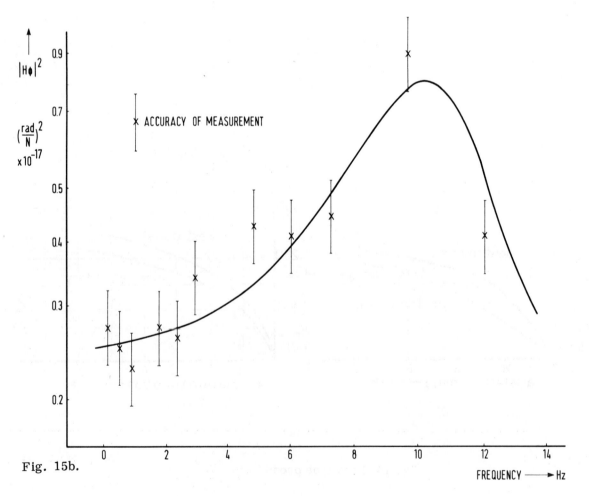

Fig. 15b.

Fig. 15 $|Hx|^2$ and $|H\phi|^2$ as a function of the frequency

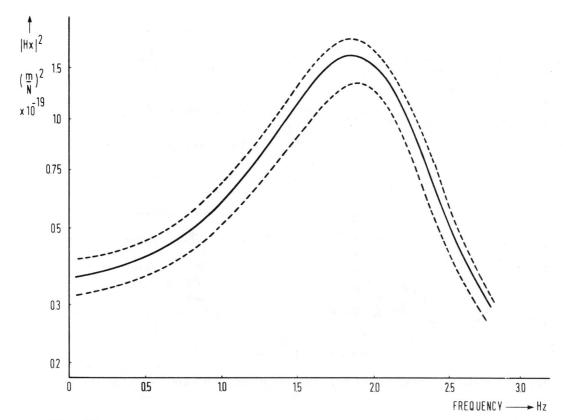

Fig. 16a

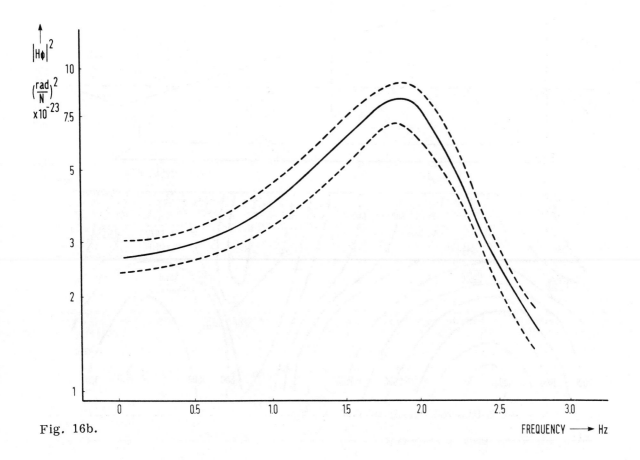

Fig. 16b.

Fig. 16 Predicted range of $|Hx|^2$ and $|H\phi|^2$ for prototype

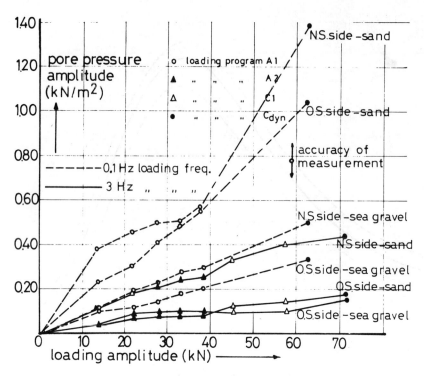

Fig. 17 Relation between pore pressure amplitudes and wave load amplitudes

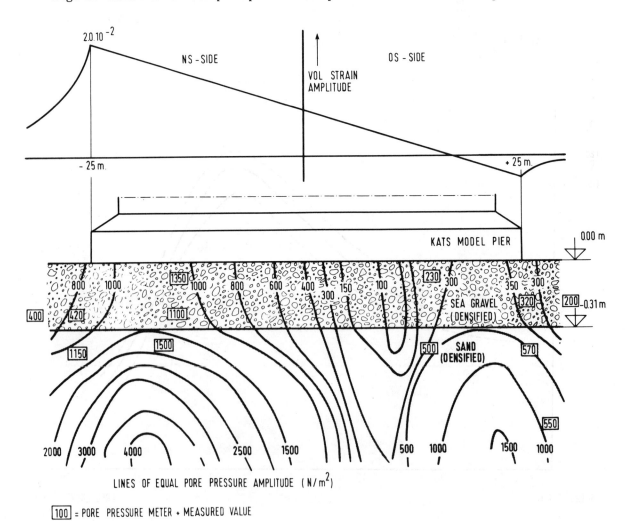

Fig. 18 Lines of measured and calculated pore pressure amplitudes

PAPER 67

BOSS'79

Second International Conference on Behaviour of Off-Shore Structures

Held at: Imperial College, London, England
28 to 31 August 1979

DEVELOPMENTS IN PILING FOR OFFSHORE STRUCTURES

W.J. Rigden, MSc, MICE.
BP Trading Ltd., U.K.

J.J. Pettit, MICE.
BSP International Foundations Ltd., U.K.

H.D. St. John, BSc(Eng), PhD.
Building Research Establishment, U.K.

and

T.J. Poskitt, DSc, FICE, FIStructE.
Queen Mary College, U.K.

Summary

This paper reports the results of driving and loading tests on two steel pipe piles driven into glacial till at the BRS Test Site at Cowden. The tests had two primary functions:-
(i) To field test the prototype of a new hydraulic hammer.
(ii) To assess theoretical predictions of the driving resistance of similar open ended and closed ended pipe piles and to compare their axial load carrying capacity.

The two piles were 0.45 m o.d. and were driven 9 m into a stiff till. One pile was open ended without a shoe and the other closed ended with a flat plate. Both were comprehensively instrumented with strain gauges, accelerometers, and pore water and total pressure cells.

The two piles were tested to failure under axial compression at a constant rate of penetration about one month after driving.

Wave equation studies of the pile and hammer are presented and comparisons made between measured and predicted axial capacity and stresses during driving.

Sponsored by: Delft University of Technology, The Netherlands
Massachusetts Institute of Technology, U.S.A.
The Norwegian Institute of Technology, Norway
University of London, England

Secretariat provided by: BHRA Fluid Engineering

Copyright: © BHRA Fluid Engineering
Cranfield, Bedford, England

1.0   INTRODUCTION

The past few years have seen the introduction of hydraulically operated drop hammers for pile driving. The mechanism of such hammers provides automatic control over the impact energy and the stroke rate. This makes them attractive for offshore applications. Over water and under water developments are envisaged. The first objective of the work described here was to field test a new hydraulically activated hammer by driving two test piles.

Theoretical calculations suggest that driving closed ended piles through clay would be easier than driving the same piles open ended, and that the former would have a higher capacity. However this has not been confirmed by full scale field tests. Therefore the two test piles were taken open and closed ended respectively and after driving were load tested to failure.

The final objective of the present work was to develop robust and reliable instrumentation for use offshore in the evaluation of pile installation and in-service behaviour.

2.0   THE HAMMER

The hammer used for driving the test piles was a BSP H.A. type having a ram mass of 3.5 tonnes. This type of hammer presents a driving system which is a new approach to basic pile driving principles. It provides the piling contractor with the simplest and most reliable of piling tools - the drop hammer together with modern technology control.

The drop hammer or driving mass is lifted by a loop that is actuated over a sheave at the top of the leaders, Fig. 1. The rope then passes down to the lower portion of the mast and round a sheave at the bottom of the actuator. The end of the rope is then fixed to the hammer guide carriage. As the hydraulic ram in the actuator forces the sheave downwards the hammer lifts and when the ram returns the hammer drops under free fall.

The system offers many advantages that are self evident such as simple drop mass; isolation of mechanism from impact area; easy visual control of drop; low front end weight and soft driving ability. In addition there are advantages that are embodied in the actuator technology. These include: automatic remote control and the ability to vary the relationship between drop mass, engine horsepower and blow rate. This means that the H.A. user has a piling system that can readily adapt to suit changes of pile contracts.

The energy is related to the drop mass and a range of hammer masses are available. The stroke is infinitely variable between limits of 200 to 1370mm. Generally the hammer works at 40 blows per minute when it produces its maximum energy, but this can be increased to 60 blows per minute when approximately half the maximum energy is produced. The hammer is controlled by a remote electro hydraulic system allowing the stroke to be adjusted. It can be controlled for single blow operation or for a fully automatic sequence.

The system is suitable for soft driving as the loop of rope allows the hammer to follow the pile without restriction. Piles can be driven with rakes up to $30°$ to the horizontal. Driving can be carried out at refusal for long periods of time since the actuator is isolated from the impact zone. Power supply is from a diesel hydraulic power pack.

3.0 DESCRIPTION OF SITE

The tests were carried out at a site at Cowden on the coast of Holderness, north of Kingston-upon-Hull, which has been used extensively by the Building Research Establishment as a test bed for the evaluation of the properties of glacial till and the performance of site investigation equipment, Marsland (Ref. 1, 2). At this site there is a sequence of clay tills which were laid down during a series of glacial advances, having plasticity indices ranging from 12 to 20 per cent and a clay fraction of 30-40 per cent. Layers and lenses of fine sand are contained within the sequence. Fig. 2 gives the undrained shear strength variation over the top 13 metres as obtained by Marsland (Ref. 2) using 865mm diameter plate bearing tests. These tests were performed in an area about 200 metres distance from the test piles, but numerous CPTs have been carried out over the surrounding area and the general uniformity of the clays' properties established. Four records of cone tip resistance and one of sleeve friction from tests performed in the immediate vicinity of the test piles are shown in Fig. 2.

The water table is close to the surface. A plan of the pile test area is given in Fig. 3 showing the position of the instrumented piles A and B, and the adjacent cone penetrometer tests. The ten piles surrounding the test piles were used for reaction when the static tests were performed. These were cast in-situ in holes left by the extraction of a closed ended vibropile.

4.0 PILE INSTRUMENTATION

Details of test piles A and B are shown in Fig. 4. A is identical to B except that the 25mm thick base plate is omitted. The piles were 457mm o.d. and 19mm w.t..

The behaviour of the hammer during driving was observed by high speed photography. Strain gauges and accelerometers were attached to the outside wall of the pile as shown in Fig. 4.

The high speed camera operated at 200 f.p.s. The position of the hammer relative to a fixed scale painted on the guides was observed. This enabled the displacement-time behaviour to be obtained. The impact velocity of the ram could then be calculated.

The strain gauges were Ailtech SG 129, 120 ohm weldable gauges. They were positioned as shown on Fig. 4. At each level there were two gauges attached diametrically opposite each other. The leads from each gauge to the signal conditioning units were individually screened in order to reduce noise to a negligible level. The signals were then stored on magnetic tape. By summing the signals or differencing them at each level axial and bending stresses could be deduced.

At the pile head two PCB 5000g accelerometers were attached diametrically opposite each other. These were screwed into small metal blocks welded to the pile wall. During the driving of Pile B a mains supply was used for the accelerometers. An earth loop was established as the pile penetrated the soil and the noise derived therefrom masked the signals. The difficulty did not arise when an independent power supply was used. Screened cables were used to connect the accelerometers to the signal conditioning units.

All sensors and associated cables were protected by 38x38mm angles welded along the length of the pile, Fig. 4. Details of the driving head are shown in Fig. 5.

The pressure cells used were of the strain gauged diaphragm type and were initially constructed by Transducers (CEL) Ltd especially for fitting to a 1372mm diameter pile which was driven in the BP Forties field. The pore pressure cells differed from the total pressure cells only in the fact that the 28mm diameter active face was covered by a porous disc. The pressure cells operated over a range of 0-4137 kN/m².

5.0 DESCRIPTION OF TESTING

5.1 Driving Tests

Pile A was driven on Tuesday, 2nd May over a period of 45 minutes. There were two breaks in driving to prime the pore pressure cells and one break to paint a continuation of the penetration markings. The following morning the pile was re-driven a further 150mm.

Pile B was driven on the following day to 2.1m and then extracted. It was then driven over a period of approximately 30 minutes with two interruptions.

5.2 Loading Tests

The pile loading tests were carried out using BRE's plate loading test equipment which has a maximum capacity of 500 tonnes. Fig. 6 shows the equipment set up in position over test pile B. Reaction was obtained using two spreader beams which were each fixed to three of the tension piles. Head displacement was measured relative to a reference beam which was supported on two concrete A blocks placed outside the region of influence of the piles. Load was measured using a load cell placed between the hydraulically operated loading head and the pile cap. During testing all quantities were recorded visually and remotely.

Each pile was first tested under various levels of maintained load over a period of around three days, left for one day unloaded, and then loaded under a constant rate of penetration, eventually to failure. The rate of loading was 0.33% of the pile diameter per minute. Tests on pile A were completed before the test equipment was moved to pile B. Both piles were left for a period of around one month between driving and testing to failure.

5.3 Pressure-Measurements

The signals from the twelve pressure cells were recorded on a Solatron Compact data logger with tape cassette at a rate of four channels per second during the driving phase. The interval between successive scans was then increased according to the expected rate of change of the signals. During driving efforts were made to maintain the pore pressure cells in a deaired state. This was done by placing the porous discs in position in a water filled trench once each cell had gone just below ground level.

6.0 DRIVING RESULTS AND INTERPRETATIONS

A large volume of test data was collected. Only a representative selection of this is given herein.

6.1 Hammer Impact Velocity

Nine high speed films were made covering a range of penetrations for piles A and B. From these fifteen hammer displacement-time curves were constructed using frame by frame analysis of the negative. The average impact velocity of the hammer for piles A and B was found to be 4.11 m/s and 4.21 m/s respectively.

6.2 Pile A Test Results

590 blows were required to drive Pile A to a final penetration of 9.14m. At intervals of 50 blows throughout the drive two or three consecutive blows have been analysed. From these the variations on a blow by blow basis and during the course

of the whole drive could be assessed. The scatter on all data was of the order of 10%. This is deemed to be acceptable.

Figs. 7 and 8 show typical sets of data corresponding to penetrations of 4.572 and 9.144m. Several checks based upon the theory for propagation of a solitary wave in a pile can be made on the consistency of this data. To help facilitate these, the stresses at the various levels are plotted below each other at distances apart proportional to the levels of the gauges on the actual pile.

Referring to Fig. 7 if a line ab is drawn through the points corresponding to the initiation of the disturbance at each level, then the slope of this line should be equal to that of the speed of sound in steel. From some 21 tests so analysed an average value of 5200 m/s was obtained.

If the hammer is striking uniformly the peak stress and the peak acceleration at the pile head should be almost independent of the soil conditions. For 21 blows the average values of these was found to be 138 MN/m^2 and 224 g respectively. The initial slope of the acceleration-time curve should also be independent of the soil conditions. An average value of 146 $g\ ms^{-1}$ was found for this. The average scatter in all these results was of the order 10%.

The time to zero acceleration and the time to peak stress at the pile head should be nearly equal. Average values of 2.25 ms and 2.4 ms were obtained for these.

Eighteen hours after completing the main drive a redrive test was carried out. The pile was advanced a further 150 mm by means of 21 blows. This is equivalent to 138 blows/m and should be compared with 131 blows/m at the end of the main drive. The set-up is therefore small.

By subtracting the signals from diametrically opposite pairs of strain gauges the bending stress may be obtained. Typical results are shown in Fig. 9. The peak bending stresses were generally of the order of 28 MN/m^2.

6.3 Pile B Test Results

351 blows were required to drive the pile to a final penetration of 9.144m. The experimental data obtained was checked in the same manner as Pile A. In general the quality of the data matched that of Pile A.

Figs. 10 and 11 show typical sets of data corresponding to penetrations of 4.572 and 9.144m. These results show that a closed end produces noticeably higher toe stresses than an open end. The passage of the peak stress down and up the pile can be more readily seen, i.e. a stronger reflection is observed.

At blow 250 a strain gauge at the toe ceased to function. The reason for this is unknown. Prior to failure it had been operating satisfactorily.

As mentioned earlier the accelerometer data had a noise component superimposed upon it due to an earth loop caused by using a mains power supply. Hence no useable data was obtained.

6.4 Dolley Compression Tests.

On completion of the field work the pile cap was tested in a 6000 kN compression testing machine. Seven cycles of loading and unloading were applied and a typical cycle is shown in Fig. 5.

The stiffness of the cushion during loading and unloading was found by fitting a straight line through the loading and unloading curves respectively. Average values of 1.1×10^6 kN/m and 1.36×10^6 kN/m respectively were obtained. The coefficient of restitution is given by square root of the ratio of these two. An average value of 0.9 was obtained.

6.5 Theoretical Calculations

The driveability studies were carried out by the wave equation method Smith (Ref. 3). For this the damping parameters J' (side) and J (point) and the soil quake Q are needed.

No measurements of J' or J are known for Cowden Clay. Because this material has a clay content of only 45% and has layers and lenses of sand it is thought to be relatively insensitive to rate of deformation. Thus J' will be closer to that of sand than that of pure clay. Accordingly a value of J' = .328 sec/m has been used. For clays the work of Litkouhi (Ref. 4) has shown that J is small, accordingly a value of 0.00328 sec/m has been used.

Qualitative information on Q may be deduced from Fig. 16. This indicates Q is in the region 2.54mm.

For modelling the pile/soil adhesion a constant value of 85 kN/m^2 has been assumed.

At the conclusion of driving Pile A, the plug was found to have risen 40% of the penetration. Hence under all driving conditions a plug height of 40% of penetration has been assumed and the adhesion on the inside wall taken as 85 kN/m^2.

For Pile B the base pressure was taken as 9 C_u x base area.

The results of the driveability calculations are given in Figs. 7, 8 and 10, 11. Agreement between experimental and theoretical curves is on the whole satisfactory.

6.6 Discussion of Driving Results

Prior to entering the field an extensive laboratory investigation to determine the best method of attaching strain gauges and cable leads to the pile wall was undertaken. All instruments and circuits were thoroughly tested in the laboratory.

A rehearsal was held at the main works of BSP to test the procedures which had been devised and modify them in the light of the experience gained.

The importance of this preparatory work in reducing unforeseen difficulties in the main tests cannot be overstated. As a result instrument failure was reduced to an insignificant level, and the time spent on site was minimised.

Throughout the test series the performance of the hammer was found to be very uniform. This was confirmed by the high speed photography and the strain gauge data. A scatter of 10% was observed on the hammer data.

The experimental data and the wave equation calculations confirmed that the mechanism by which the hammer impulse is transmitted down the pile is of the nature of a wave propagation phenomena.

At the completion of each main drive, the residual strains in the pile were observed. They were of the order of 50 microstrain, which indicates a high level of stress relaxation in the soil adjacent to the pile wall. These strains correspond to about 5% of the peak observed during driving.

At final penetration the blow counts of piles A and B were observed to be 131 and 98 bl/m respectively. The ratio of these is 1:.75. The values calculated by the wave equation were 107 and 78 bl/m which give a ratio of 1:.73. Thus although the absolute values of the observed and theoretical blow counts differ, due to lack of measured data on J' and J, the ratios are in close agreement. From this it can be concluded that relative movement takes place between the soil plug and the inside pile wall and the hypothesis that under dynamic conditions the pile does not plug appears to be proved.

7.0 PRESSURE CELL RESULTS AND INTERPRETATION

7.1 Results

Measurements during driving were of limited use because of the slow response of the pressure cells. Several of the cells appear to have suffered from significant zero changes during driving, even before they entered the ground. Additionally it appeared from the results that the majority of the pore pressure cells retained some air. Consequently readings continued to increase after driving.

Fig. 12 shows measurements from pore pressure cell AP2 during driving, together with those from the three levels of total pressure cells on pile A. Fig. 13 shows the readings from the two lower levels of total pressure cells on pile B during driving. The periods during which driving took place are marked on the figures.

Fig. 14 shows the pressure measurements after driving for both piles for a period of around 27 days. Events affecting readings are marked. Pressure changes were also monitored during axial load testing and changes were found to be very small.

7.2 Discussion

It is difficult to draw many quantitative conclusions from the measurements. It can be seen from Fig. 12 that the immediate response to driving is a reduction in both total pressure and pore pressure, followed by a steady increase. The only level at which this was not observed to occur was near the bottom of the closed ended pile (B), where the lateral pressures are likely to be significantly affected by the penetration of the pile tip. It is possible that the pressure drop is caused by the irrecoverable strains in the soil created by the rapid expansion of the pile under the first hammer blow Heerema (Ref. 5). However the magnitude of the change is greater than would be expected.

It is not possible to make a distinction between the values of pressure generated by the two different piles. The summation of the changes that take place during driving gives a maximum total pressure change at the tip of the pile of about 700 kN/m², i.e. around $6 \times C_u$.

It appears from the measurements after driving that pressures have at least stabilised by day 20, although some of the cells are exhibiting inexplicable behaviour during the period 15-23 days. Little change is apparent at the uppermost level. This is probably explained by the proximity of a higher permeability layer between 2 and 3 metres. Fig. 15 shows the plot resulting from the subtraction of the pore pressure measurements from the total pressures, which enables the changes in effective stress to be examined. It is apparent that over the time period shown small increases in radial effective stress have been recorded on pile B, at the lower levels but not on pile A.

8.0 PILE LOAD TEST RESULTS AND INTERPRETATION

8.1 Results

The results of the constant rate of penetration tests on piles A and B are presented in Fig. 15 in the form of a plot of applied load against pile deflection at mudline. Before the piles were loaded to failure both were subjected to a load cycle; pile A to 50 tonnes, pile B to 75 tonnes. In both cases the response to loading was identical to that shown in Fig. 15 over the same portion of the initial loading curve.

The recorded maximum loads were

$$p\ max = 143 \text{ tonnes for the closed ended pile B}$$
$$p\ max = 118 \text{ tonnes for the open ended pile A}$$

Pile B which was unloaded after a penetration of around 40mm had been reached was reloaded to failure and maximum load of 137 tonnes was achieved.

The tests were stopped when a penetration of around 8.5% of the pile diameter had been reached.

8.2 Discussion

The load-deflection curves from the two pile tests show some marked differences. The open ended pile, A, very clearly has a bilinear response before failure is reached where the majority of shaft friction has been mobilised by a deflection of 1% of the diameter, and the increase in load is thereafter taken by the plugged pile base. The difference between this and the curve for pile B is caused by the fact that the response is stiffer due to the stresses locked in under the solid base after driving, Cooke (Ref. 6). This is in contrast to pile A, in which the base resistance can only by mobilised once sufficient movement has occurred between the internal column of soil and the shaft to cause the pile to plug.

The piles behaved identically under initial loading. It is possible to deduce the elastic properties of the soil from the initial loading curve using the simple mathematical model suggested by Randolph (Ref. 7) which models the pile as an axially loaded elastic rod which is restrained by elastic tractions along its length and at the base. For the sake of comparing results from large scale plate tests a secant slope of the load deflection curve is taken between 0 and 50 tonnes load. It can be shown that by making sensible assumptions about the relationship between pile movement and mobilised tractions, a value of the shear modulus (G_s) lying between 14 and 18 MN/m² is obtained. In addition, by analysing the response of pile A during the period that base resistance was being mobilised in the same manner as a plate bearing test result would be treated (eg Marsland, Ref. 1), a value of G_S of 14 MN/m² is obtained. This latter value compares very favourably with the values obtained from plate bearing tests on the site.

The effect of closing the end of the pile was to increase its capacity by approximately 21% under axial compressive loading. There was no apparent significant variation in soil properties within the region where the piles were driven. The only factor differing from one pile to the other apart from geometry was the length of time between driving and testing. However, the observed rate of pore pressure dissipation indicates, and measured permeabilities (1 to 10x10⁻⁹ m/sec) would predict that all excess pore pressures generated during driving had dissipated before either of the piles were tested.

The difference between the ultimate capacities of the two piles can only be explained by looking at the effective stress changes that take place in the soil around the pile, and the effect that these have on the shaft resistance. If pile driving is modelled as a cylinder which is expanded into the low permeability soil, and the soil is then allowed to consolidate, it can be demonstrated theoretically, Wroth et al (Ref. 8) that the effective radial stress on the pile surface after consolidation increases as the ratio

$$p = 1 - (r_i / r_e)^2 \text{ increases}$$

where r_i is the internal radius of the pile
and r_e is the external radius

The effect of an increase in the radial effective stress is to increase both the drained and undrained resistance that can be mobilised.

The ultimate capacities of the two piles have been analysed in terms of some of the common predictive theories.

(i) The α method. If undrained shear strengths (C_u) are taken from the plate bearing test results (Fig. 2), and it is assumed that the base resistance mobilised is given by

$$Q_B = 9 \times C_u \times A_B$$

then $\alpha = (\text{Ultimate load (Qult)} - Q_B)/(C_u)_{ave} A_S$

where C_u is the value at the base
$(C_u)_{ave}$ is the average value down the shaft
A_B is the base area of the pile
A_S is the total shaft area.

For pile A, α = 0.64
pile B, α = 0.8

(ii) The CPT method. Let it be assumed that the total resistance of the pile is given by

$$Q_{ult} = (A_B \times (q_c)_{tip} + A_S \times (f_S)_{ave}) \times \delta$$

where $(q_c)_{tip}$ = Measured cone resistance at pile tip

$(f_S)_{ave}$ = Average measured sleeve friction along the total depth of the pile.

From the results for

pile A   δ = 0.73
pile B   δ = 0.88

(iii) The interpretation of the load test results in terms of the effective stress method enables values of the ratio $K = \sigma_r'/\sigma_v'$ to be calculated, where σ_r' is the radial effective stress on the pile at the time of the test, and σ_v' is the initial in-situ vertical effective stress. The values of K range from 4 to 6, with lower values relating to the open ended pile.

9.0 CONCLUSIONS

The tests confirmed that the hydraulically actuated hammer behaved very uniformly while driving the test piles.

The closed ended pile drove more easily than the open ended pile and had a greater axial capacity. This justifies further investigation.

The tests successfully demonstrated that piles can be instrumented to give reliable field observations of stress, acceleration, total pressure and pore pressure.

Further interpretation of the field data obtained is still being carried out.

10.0 ACKNOWLEDGEMENTS

The Authors wish to thank Messrs R. Elliott and D. Keen of BSP, Mr. J. Petty formerly of BSP, and, Mr. V.J.R. Sutton of BP for useful discussions.

Dr. L. Cuthbert of Queen Mary College advised on the instrumentation and Mr. J. Dolwin of Queen Mary College assisted with the interpretation of data. Mr. T. Freeman of BRE assisted with the field work and the interpretation of the site data.

The work carried out by BRE was funded by the Department of Energy. The Authors are indebted to BP Trading Ltd, BSP International Foundations Ltd, and the Director, Building Research Establishment for permission to publish.

11.0 REFERENCES

1. Marsland, A. "The interpretation of in-situ tests in glacial clays". Paper presented to the S.U.T. International Conference on "Offshore Site Investigation", 5-8 March 1979.

2. Marsland, A. "Evaluating the large scale properties of glacial clays for foundation design". Conference on "Behaviour of Offshore Structures, London, 28-31 August 1979.

3. Smith, E.A.L. "Pile driving analysis by the wave equation", Trans. A.S.C.E., Vol. 127, Part 1, Paper 3306, (1962).

4. Litkouhi, S. "The behaviour of foundation piles during driving", Ph.D. Thesis, London University (1979).

5. Heereman, E.P. "Predicting pile driveability; Heather as an illustration of the friction fatigue theory", Vol. 1, pages 413-423, Proceedings EUROPEC, London, 24-27 October 1978.

6. Cooke, R.W. "The influence of residual installation forces on stress transfer and settlement under working loads of jacked and bored piles in cohesive soils". Presented at the ASTM symposium on "Behaviour of deep foundations", Boston, Massachusetts, June 1978.

7. Randolph, M.F. "A theoretical study of the performance of piles". Ph.D. Thesis, University of Cambridge (1978).

8. Wroth, C.P., Carter, J.P. and Randolph, M.F. "Stress changes around a pile driven into cohesive soil". To be presented at the Conference on "Recent developments in the design and construction of piles". I.C.E. London March 1979.

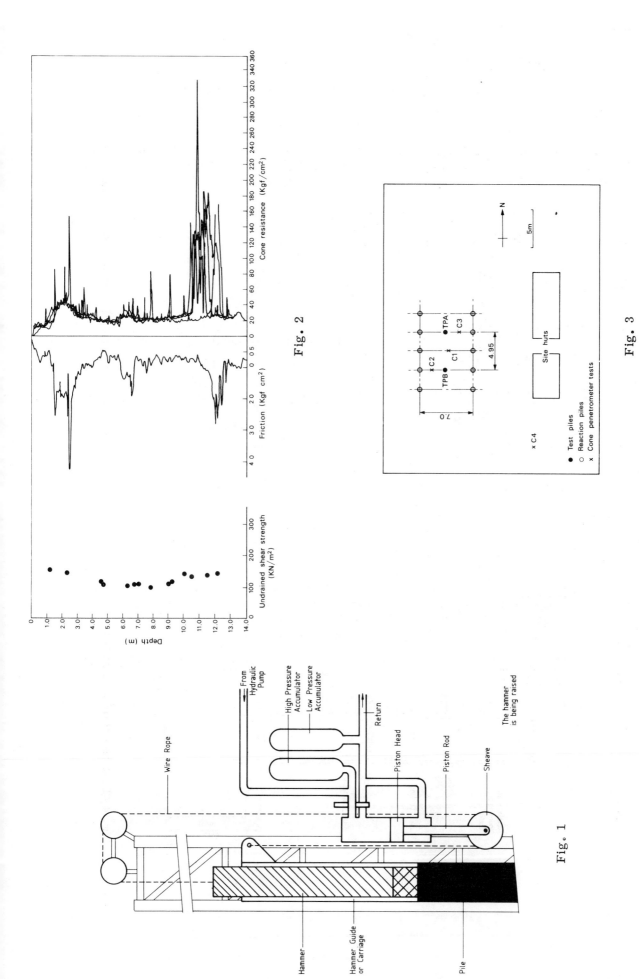

Fig. 1

Fig. 2

Fig. 3

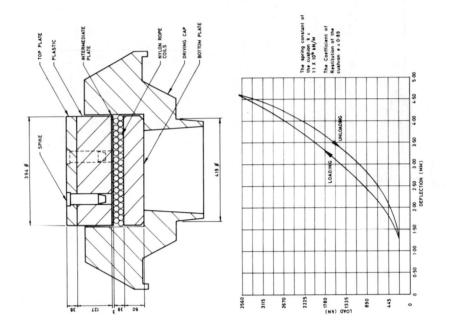

Fig. 5 Typical Dolley compression test

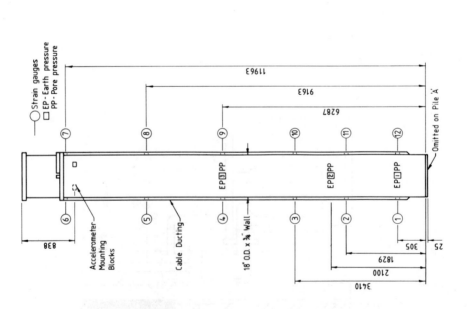

Fig. 4 Test pile B (closed ended)

Fig. 6

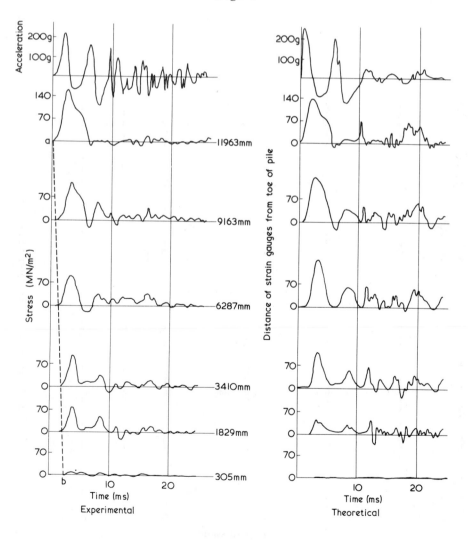

Fig. 7 Pile A penetration 4572 mm

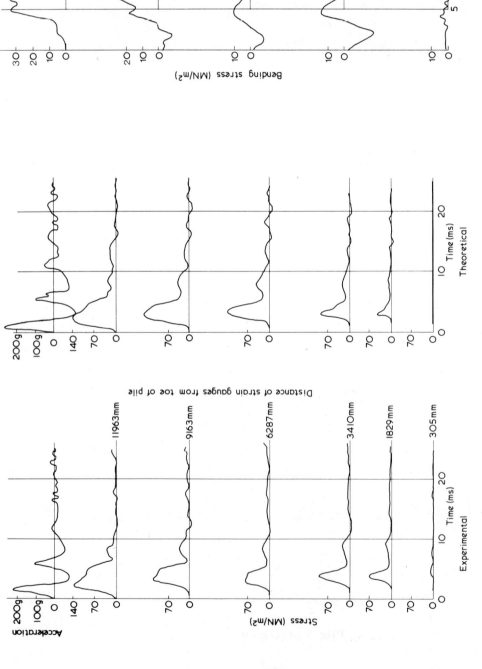

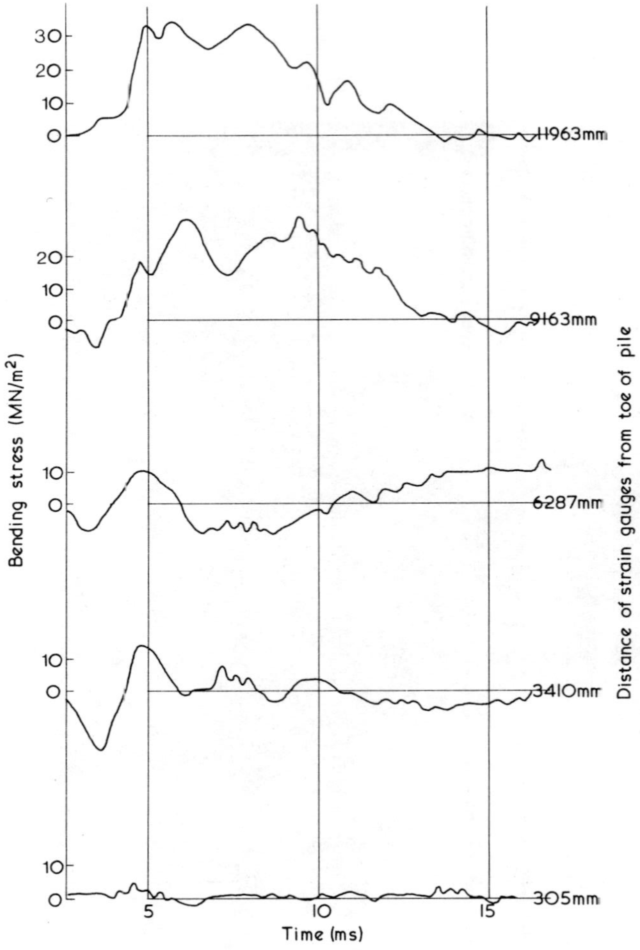

Fig. 9 Measured bending stresses. Pile A penetration 9.144 m

Fig. 8 Pile A penetration 9144 mm

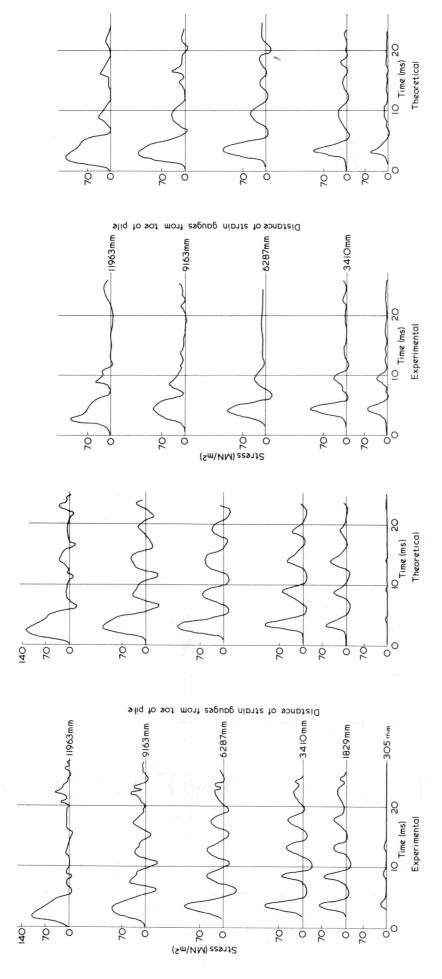

Fig. 10 Pile B penetration 4572 mm

Fig. 11 Pile B penetration 9144 mm

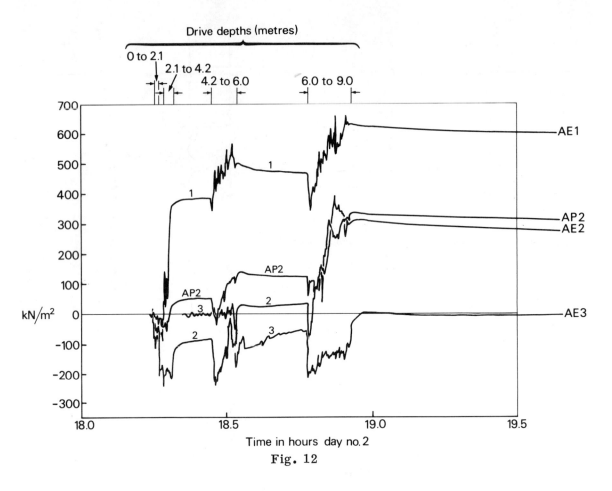

Fig. 12

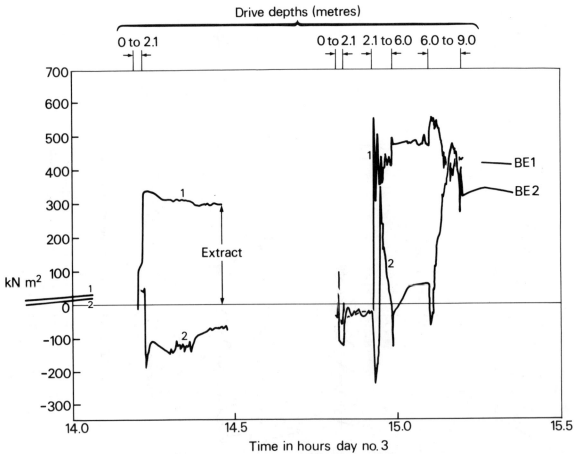

Fig. 13

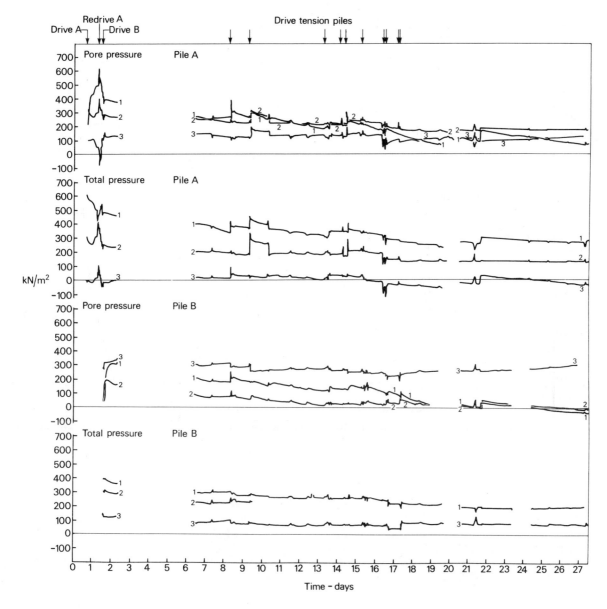

Fig. 14

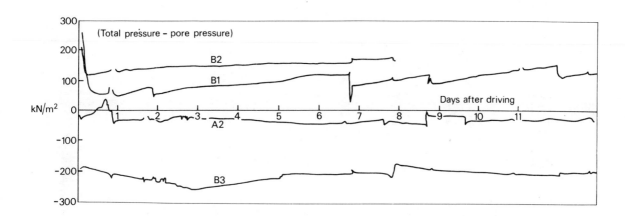

Fig. 15

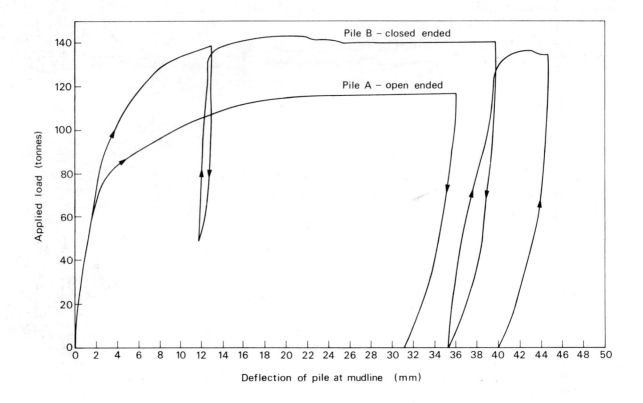

Fig. 16

BOSS'79

Second International Conference
on Behaviour of Off-Shore Structures

Held at: Imperial College, London, England
28 to 31 August 1979

PAPER 69

THE INFLUENCE OF HYDROSTATIC PRESSURE ON THE PROCESS OF UNDERWATER EXCAVATION WITH SPECIAL REFERENCE TO THE UNDERWATER BULLDOZER

V.A. Lobanov

Odessa Institute of Civil Engineering, U.S.S.R.

Summary

In order to develop techniques for the construction of offshore structures it is necessary to establish scientific design principles. With this in mind complex investigations have been carried out in the Laboratory of the Technique of Shelf Exploration of the Odessa Institute of Civil Engineering.

Experiments were performed under natural conditions and also in special pressure chambers which simulate the working conditions for underwater apparatus on the shelf.

Laboratory investigation showed that certain patterns can be determined in the process of underwater excavation. This has made possible the development of a technique of engineering calculation for the design of underwater building and mining machines and the creation of experimental underwater bulldozers and rippers.

For operation and control, special underwater surveying instruments have been designed which can determine reference points on the seabed.

This research has given rise to the development of new types of apparatus that can be used on the seabed for the construction of the foundations of any type of offshore structure.

Sponsored by: Delft University of Technology, The Netherlands
Massachusetts Institute of Technology, U.S.A.
The Norwegian Institute of Technology, Norway
University of London, England

Secretariat provided by: BHRA Fluid Engineering

Copyright: © BHRA Fluid Engineering
Cranfield, Bedford, England

The exploitation of mineral resources from the seabed is only just beginning but already 30% of world oil production comes from such locations.
When wells are drilled underwater and when deep water harbours are made the cost of constructing the foundations has a big influence on the total cost of such structures. Since building machinery which is specifically designed for use underwater has not yet been developed it is necessary to rely on divers to do this work.

The Laboratory of the Technique of Shelf Exploration of the Odessa Institute of Civil Engineering has developed, built and tested experimental models of underwater bulldozers, rippers and cabling machines and machines for the excavation, transportation and levelling of soils under water.

Before proceeding with the building of underwater machines it was necessary to develop general design principles which would be applicable in the conditions which would be encountered.
The first stage involved the investigation of the tools which would interact with the soils of the seabed. In the development of a physical picture of the interaction between the cutting tools and underwater soils the influence of hydrostatic pressure on the cutting resistance may be of major significance.
The first tests showed that the effects of compression produced by only 10 m of water can be compared to the cohesive forces of clay. The influence of hydrostatic pressure varies according to the soil type. In soils of high porosity it is small and unimportant but in hard clays cutting resistance is increased under hydrostatic pressure. In newly exposed layers of the seabed the effects of hydrostatic pressure must be investigated on each particular occasion.
When only a short time elapses between the cutting of different layers, hard clay soils are comparable to paraffin wax and also repel water. For both paraffin wax and hard clay soils which are cut experimentally under hydrostatic pressure the cutting force required increases rapidly with increasing pressure and decreases when pressure is reduced. When, during a period of 30 minutes, paraffin wax and hard clay soils were cut under hydrostatic pressures which were increased at intervals and were then reduced the resulting deformation did not affect the cutting resistance.
Thus it is possible to conclude that when underwater excavation is carried out level by level the condition of the surface layer will differ from that of underlying levels. The geological formation of the uppermost layer is almost complete but the condition of newly exposed levels will change in a way which depends on the type of soil and the duration of exposure. For this reason, in order to prepare a regime of work for underwater machines, it is necessary to study the properties of the soil in situ immediately prior to excavation. It is also necessary to determine the condition of the soil periodically during the process of excavation.

Owing to the difficulties encountered in attempting to study the mechanical cutting of soil in deep water a special test rig, which permits the simulation of the mechanical cutting of soil below water within shelf limits, was developed at the Odessa Institute of Civil Engineering. In this installation hydrostatic pressure can be varied within the limits 0 bar - 30 bar, that is for depths to 300m.

The test rig, Fig.1. consists of a high pressure chamber (14) containing a cutting installation with a blade (9). The blade is mounted in a dynamometer (8) connected by a cable to a measuring system which consists of an amplifier (5) and an oscillograph (4).
Inside the chamber is a cassette (12) in which a number of soil samples are mounted in such a way that they can be cut in rotation.
When the soil samples have been placed in the cassette the chamber is filled with water via the aperture (7) and the chamber is then hermetically sealed. Valve (16) is opened and air from cylinder (17) passes through the pressure reducing valve (15) into the chamber until the pressure required is reached.
The level of the hydrostatic pressure is controlled by a manometer (10) which is installed on the three position valve (11) together with a safety valve.
For reasons of safety hydrostatic pressures were produced by means of a hydraulic pump in further constructions.
The drive of blade (9) is provided by a hydraulic cylinder (6).

The hydraulic system (1) feeds this hydraulic cylinder via a throttle-valve (2) and a reverse flow valve (3) which is also a safety valve. The connection of hydraulic cylinder (6) to the compensating rod (13) makes it possible to keep a constant hyperbaric pressure during tests.

As a result of the construction of this rig research into the cutting of silt, clay, calcareous rocks, paraffin wax and other materials was possible. The first propositions put forward in this paper were supported by this research.
With a low cutting velocity the influence of hydrostatic pressure on soil with a high porosity is negligible. In this case the pressure on both sides of the blade is in equilibrium owing to the porosity of the soil.
The hydrostatic pressure changes the mechanical properties of the soil which is being studied and therefore the resistance to cutting. Further analysis of the experimental data relating to the cutting of soils is necessary.

Fig. 2 illustrates the relationship between cutting force N, hydrostatic pressure P, cutting angle δ and the depth of the cut h. when clay with a coefficient of internal friction tg ϕ = 0.5 and cohesion C = 4.7 N/cm^2 is cut with a blade of width 28 mm the cutting velocity = 0.8m/min..
Values for C were measured by using the shear test on a cylinder with a diameter greater than 70 mm and a height between a half and a third of the diameter. The shearing velocity was .01mm per second. The top part was fixed and the lower part was moved. Similar tests have been conducted by Prof. Maslov and by the Moscow Geological Institute.
Analysis of the curves shows that hydrostatic pressure has a big influence on the cutting force. When hydrostatic pressure increases to 20 bar the cutting force increases in relation to the cutting depth h and cutting angle δ from 2.8 to 5.7 times. With a rise of 1 bar, which corresponds to an increase in depth of 10 m, the cutting force on soils is increased by between 3.75 and 12.5%, depending on the cutting angle. The maximum increase in the cutting force is related to cutting with a cutting depth of more than 3mm. When cutting clay with a cutting depth h = 1 mm the increase is between 3.75 and 7.77% for hydrostatic pressures up to 1 bar.

Fig. 3. has curves showing the relation of cutting force N to cutting angle δ, cutting depth h and hydrostatic pressure for the cutting of soil with the above mentioned parameters. In this case there is a large range of cutting angles, making it possible to compare curves 1 (influence of cutting angle on the cutting force without the action of hydrostatic pressure), with curves 2 and 3 which were obtained under hydrostatic pressure. Analysis of the curves in this figure demonstrates that hydrostatic pressure changes the general characteristics of the relationship between cutting force and cutting angle.
When cutting without hydrostatic pressure (curve 1) the cutting force is minimal for angles between 30° and 35°. If hydrostatic pressure increases up to between 10 and 20 bar, for the minimum energy consumption the cutting angle is 45°-55°.

Fig. 4 shows the relationship between cutting force, cutting depth and hydrostatic pressure when soil with a coefficient of internal friction tg ϕ = 0.35 and cohesion C = 8 N/cm^2 is cut. The width of the blade is 28 mm, the cutting angle δ = 60° and the cutting velocity 0.8 m/min. These experiments confirm that, as anticipated, the relationship between the cutting force and the cutting depth for the range 0-30 bar is linear. At the same time it was established that, as shown in fig. 2., hydrostatic pressure exerts a big influence on the cutting force when clay with lower cohesion is being cut. Further research into the process of cutting soil underwater, using blades with larger geometric parameters confirmed the results of the first research programme. This allowed the formulation of design principles for underwater apparatus moving on the seabed and the testing of different experimental machines was commenced.

The underwater bulldozers which were created to further the development of the design principles for underwater machinery will serve as an example in this paper. Three types of bulldozer were developed for different working depths:
1. With a rigid connection between the caterpillar track and the source of energy.
2. With a hinged connection between the underwater caterpillar track and the floating energy source.
3. With a hose and cable connection between the underwater caterpillar track and the floating energy source (ship), for depths exceeding 12 m.

Fig. 5. is a schematic diagram of an experimental bulldozer for depths not exceeding 12 m. the bulldozer consists of a caterpillar track (19) with a blade (1). The drive of the bulldozer is provided by hydraulic equipment (18) mounted on the caterpillar track. Oil for the hydraulic drive is supplied from energy source (11), which is mounted on pontoon (9), through the hoses (10).
Pontoon (9) is connected to the caterpillar track (19) by the pipe column (14) by hinges (13) and (16). If a rigid connection for the parts above or below water is required this can be obtained by use of the cylinders (8). By means of these cylinders the caterpillar track can be lifted and, with the aid of additional pontoons transported over the seabed.

Preliminary tests of the bulldozer indicate that the mechanical parts function well and that it may provide a highly effective means of excavating underwater soils during the construction of the underwater parts of offshore structures.

Operation of the bulldozer in turbid water has proved difficult. Three systems have been developed to reduce the problems caused by such waters.
1. Operation from pontoon (9)
2. Operation from a hermetically sealed capsule.
3. Operation by a diver using a portable control panel.
In all these cases specially designed underwater surveying instruments (2) which can determine reference points on the seabed have been used. These include depthmeters, barometric levelling instruments, hydrostatic levelling instruments for the transfer of spot levels from a fixed ground level to the seabed and optical and acoustic theodolites.
To control movements in fixed directions the bulldozer may be attached to a pre-tensioned cable by means of the blocks (3).
The high precision of the underwater surveying instruments, (up to 6 mm) makes them particularly useful for the definition of spot levels under water. This permits the controlling of the position of the underwater bulldozers and other apparatus at work on the shelf.
Underwater surveying instruments were also used by divers in the preparation of an underwater experimental area in the White Sea, in which a number of fixed points were established.

The underwater bulldozer has three life-support systems for the operators in capsule (7). The operator can leave the capsule by using ladder (7), situated inside column (14), which is closed by valve (12) to prevent penetration by water. He can also leave the capsule via the security hatch (15) or he can separate the capsule from the caterpillar track by using the separable hinge construction (16).
In addition to the control panel (4) the ventilation system (6) and reserve air (8) are situated in the capsule.

Tests on the underwater bulldozer revealed the principal faults and led to the construction of more advanced machines. In order to provide facilities for the further testing of the working tools for underwater machines a new hyperbaric chamber, of diameter 2m and length 11.5 m was developed. In this chamber depth conditions to 400 m can be simulated. The experience thus obtained gives grounds for the belief that in the near future underwater machinery will play an important role in the construction of offshore structures.

REFERENCES

V. A. Lobanov "Stend dlja podvodnogo resanija gruntov". Transportnoje Stroitelstro, 1975, no. 6.

V. A. Lobanov, A. A. Shatalov and S. K. Nanestnikov "Podvodnij buldozer". Transportnoje Stroitelstro, 1979, no. 2.

V. A. Lobanov "Konstructivnie i technologicheskije schemi podvodnich buldotherov". Transportnoje Stroitelstro, 1979, no. 3.

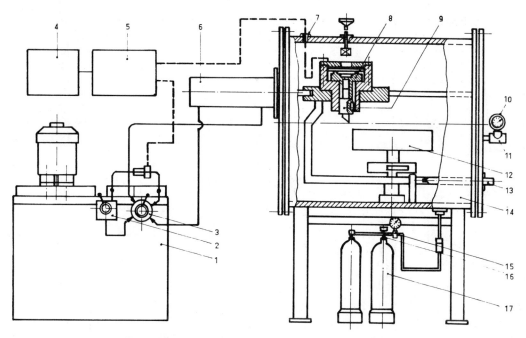

Fig. 1 Diagram of test rig for cutting soil under hydrostatic pressure.

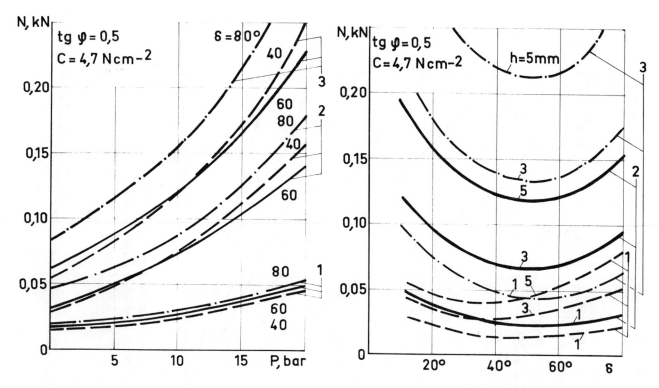

Fig. 2 Relationship between cutting force N and hydrostatic pressure P for cutting angles δ and cutting depths: 1. h = 1 mm, 2. h = 3 mm, 3. h = 5 mm.

Fig. 3 Relationship between cutting force N and cutting angle δ for cutting depths h and hydrostatic pressure: 1. P = 0 bar, 2. P = 10 bar, 3. P = 20 bar

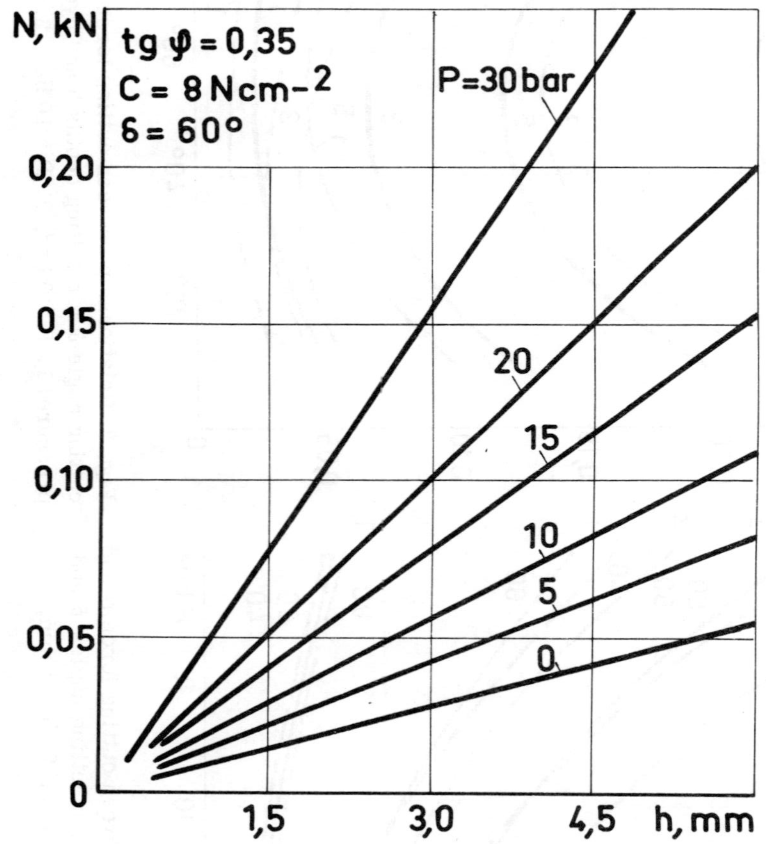

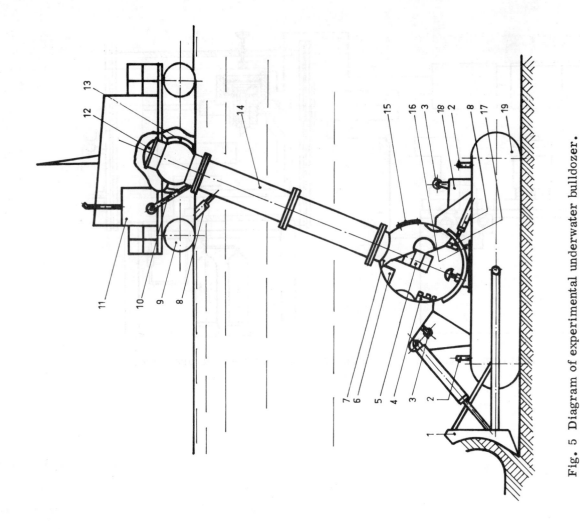

Fig. 5 Diagram of experimental underwater bulldozer.

Fig. 4 Relationship between cutting force N and cutting depth h for hydrostatic pressure P.

BOSS'79

PAPER 70

Second International Conference
on Behaviour of Off-Shore Structures
Held at: Imperial College, London, England
28 to 31 August 1979

BUCKLING STRENGTH AND POST-COLLAPSE BEHAVIOUR OF TUBULAR BRACING MEMBERS INCLUDING DAMAGE EFFECTS

C.S. Smith, W. Kirkwood and J.W. Swan

Admiralty Marine Technology Establishment, U.K.

Summary

Results are presented of a theoretical and experimental study of the bending, buckling and post-collapse behaviour of axially compressed tubular bracing members as employed extensively in the construction of offshore steel production platforms and mobile drilling rigs. A parametric evaluation is made of the influence on compressive strength, pre-collapse stiffness and post-collapse load-carrying capacity of:

a) Lack of straightness of various forms, both within and outside current design-code tolerances;
b) weld-induced residual stresses;
c) residual stresses caused by cold-rolling of plating and cold-bending of completed tubes.

An assessment is made of the performance of bracing members containing various levels of damage (lateral deformation and associated residual stress) caused by supply-boat collisions. Results are described of a series of compression tests on 1/4 scale steel tubes whose objects were:

a) To provide experimental confirmation of theoretical results, referring particularly to post-collapse behaviour and the performance of damaged members;
b) to provide empirical information on the effects of local buckling and damage in the form of local dents on collapse and post-collapse behaviour.

The provisions of current API and DNV design rules for offshore structures are examined critically in the light of results obtained.

Sponsored by: Delft University of Technology, The Netherlands
Massachusetts Institute of Technology, U.S.A.
The Norwegian Institute of Technology, Norway
University of London, England

Secretariat provided by: BHRA Fluid Engineering

Copyright: © BHRA Fluid Engineering
Cranfield, Bedford, England

INTRODUCTION

Long unstiffened cylinders of low diameter/wall thickness (typically D/t < 50) are used extensively to form the primary structural framework of fixed offshore platforms and mobile drilling rigs. As part of the process of ultimate limit state design careful consideration must be given to the strength and stiffness of such members under axial compression combined in some cases with bending and local lateral loads. Buckling failure of tubular bracing members is likely to take place inelastically and to be affected by manufacturing imperfections including initial deformations and misalignments combined possibly with residual stresses caused by welding and cold-forming. The ultimate limit state of an offshore steel platform, which is usually a highly redundant structure, may be influenced not only by the buckling strength but also by pre-collapse loss of stiffness and post-collapse load-carrying capacity of individual tubular elements.

Damage, in the form of lateral elasto-plastic bending and/or local indentations, occurs occasionally in tubular bracing members as a result of minor supply-boat collisions and accidental dropping of weights from platform decks. Because of the difficulty and high cost of repairing such damage at sea, it is clearly important to be able to assess accurately the reduced capability of damaged braces.

The object of the present paper is to review briefly present offshore design practice relating to tubular braces and to outline the results of a parametric study and series of small-scale tests which were carried out with the objects of:

(i) evaluating the effects of boundary conditions and imperfections on the strength and stiffness of tubular members under axial compression, with particular reference to post-collapse behaviour;

(ii) examining illustratively the effects of damage on tube performance.

DESIGN AND CONSTRUCTION: PRESENT OFFSHORE PRACTICE

The API RP2A requirements for design of tubular members under compressive load(1) follow closely the AISC and AISI specifications (2, 3), being based on the American CRC column curve as shown in Figure 1. The corresponding requirement in the 1977 DNV Rules (4) is based on the European column curves (5), which are also shown in Figure 1 together with test data for large-scale fabricated tubular columns (6, 7). Design of tubular beam-columns under combined axial compression and bending is dealt with in both the API and DNV rules by means of interaction formulae which stipulate that total outer-fibre stresses should be less than yield stress by an appropriate margin. According to API RP2A local buckling of the tube wall may be ignored if $D/t < 0.1\ E/\sigma_y$ (ie D/t < 70 for typical steels); the corresponding DNV requirement is $D/t < 0.088\ E/\sigma_y$. For thin-walled tubular columns in which D/t exceeds these limits both API and DNV specify that the local-buckling stress σ_c for thin-walled cylinders under axial compression should be substituted for σ_y in the column-buckling curves (σ_c is given in API RP2A by $\sigma_c/\sigma_y = 0.22\ \alpha - 0.0121\ \alpha^2$, where $\alpha = Et/(\sigma_y D) < 9.1$, and in the DNV Rules by $\sigma_c/\sigma_y = 1.07 - 0.8/\alpha$ for $\alpha < 11.4$).

Tubular chord and bracing members for offshore platforms are normally fabricated by cold-rolling single plates to form "cans" of circular cross-section, each of which is completed with a seam weld; cans, of length up to about 3 m, are then butt-welded together to form tubes of any required length, seam welds being staggered circumferentially to avoid continuous lines of weakness; tubular members are finally welded at their ends to prepared nodal "stubs" in the structural framework. Fabricators normally take great care to maintain straightness of tubular members during assembly of cans although it appears that straightness is not usually checked following erection of completed tubes or installation of platforms at sea. The maximum permissible deviation from straightness as a fraction of length (δ_o/L) is 0.001 in API RP2A and 0.0015 in the DNV Rules.

THEORETICAL ANALYSIS

Methods of Analysis

Provided that local buckling of the tube wall does not occur, buckling and beam-column behaviour of a tubular member may be analysed with reasonable accuracy, allowing for initial deformation, residual stresses, eccentricity of load and any specified boundary conditions, using elasto-plastic beam-column or frame theory. In certain cases it may strictly be necessary to treat a tube as a biaxially loaded column or space frame in order to represent three-dimensional deformations. For most practical purposes, however, it is satisfactory to assume that boundary conditions, imperfections and lateral bending combine to produce deformations in a single plane: a two-dimensional beam-column or plane-frame analysis is then adequate.

Computer-based methods of inelastic beam-column analysis have been developed by many investigators. These methods include:

(i) numerical solutions of the beam-column equilibrium equations using elasto-plastic moment-thrust-curvature relationships computed for the appropriate section geometry with allowance for initial residual stresses (eg 8, 9, 10);

(ii) incremental finite element methods (11, 12, 13, 14).

Results described in the present paper were based on the latter approach following a solution procedure which can be summarized as follows:-

(i) Beam-columns or frames of arbitrary geometry are represented as assemblies of straight beam elements. By treating a beam-column as a plane-frame it is possible to carry analysis well into the post-collapse range where very large displacements may occur. About twenty elements (or ten where a mid-length plane of symmetry exists) are required to represent a typical tubular column, small element lengths being taken in areas where plasticity is expected.

(ii) Frame cross-sections are divided into elemental areas corresponding to element "fibres" as illustrated in Figure 2: residual stresses, which may have any distribution over a cross-section, are represented as initial stresses at fibre centroidal positions.

(iii) Loads (or imposed displacements) are applied incrementally, a linear solution being obtained for each incremental load application of the equation

$$K\delta = (K_O + K_G) \delta = R \qquad [1]$$

where K_O is an incremental stiffness matrix based on the current (tangent) element stiffness coefficients, K_G is a geometric stiffness matrix related to the current (cumulative) destabilizing forces in elements and R represents incremental loads and imposed displacements. Where displacements are imposed and forces corresponding to the same degrees of freedom are unknown quantities, equation [1] is partitioned in the form

$$\begin{bmatrix} K_{11} & K_{12} \\ K_{21} & K_{22} \end{bmatrix} \begin{bmatrix} \delta_1 \\ \delta_2 \end{bmatrix} = \begin{bmatrix} R_1 \\ R_2 \end{bmatrix}$$

where δ_2 are known (imposed) displacements and R_2 are undetermined forces. Incremental solutions are then obtained as

$$\left.\begin{aligned}\delta_1 &= K_{11}^{-1}(R_1 - K_{12}\delta_2)\\ R_2 &= K_{21}\delta_1 + K_{22}\delta_2\end{aligned}\right\} \quad [2]$$

Where imposed displacements and undetermined forces correspond to different degrees of freedom, equation [1] is partitioned in the form

$$\begin{bmatrix}K_{11} & K_{12} & K_{13}\\ K_{21} & K_{22} & K_{23}\\ K_{31} & K_{32} & K_{33}\end{bmatrix}\begin{bmatrix}\delta_1\\ \delta_2\\ \delta_3\end{bmatrix} = \begin{bmatrix}R_1\\ R_2\\ R_3\end{bmatrix}$$

where δ_1 and δ_2 are unknown and δ_3 are known (imposed) displacements; R_1 and R_3 are known and R_2 unknown external forces (δ_3 and R_2 are of equal, usually unit, order). Solutions are found to be

$$\left.\begin{aligned}\delta_2 &= \left[K_{32} - K_{31}K_{11}^{-1}K_{12}\right]^{-1}\left[R_3 - K_{31}K_{11}^{-1}R_1 - (K_{33} - K_{31}K_{11}^{-1}K_{13})\delta_3\right]\\ \delta_1 &= K_{11}^{-1}\left[R_1 - K_{13}\delta_3 - K_{12}\delta_2\right]\\ R_2 &= K_{21}\delta_1 + K_{22}\delta_2 + K_{23}\delta_3\end{aligned}\right\} \quad [3]$$

(iv) Following each incremental solution, cumulative values of nodal displacements, together with stresses, strains and element destabilizing forces, are updated. The state of stress in each fibre of each element is examined and where the total stress (including initial residual stress) exceeds yield, the fibre is taken either

(a) to contribute no stiffness in the next incremental solution, in accordance with an elastic-perfectly plastic stress-strain curve, or

(b) to have a reduced modulus derived from a numerically defined stress-strain curve of any prescribed shape.

Option (b) allows treatment of strain-hardening materials and of beam-columns or frames in which local buckling occurs provided that the effective stress-strain curve for locally buckled parts of a cross-section can be specified. Allowance is made for recovery of elastic stiffness in yielded fibres which experience strain-reversal. The geometry of the structure is re-defined to take account of deformations caused by the preceding incremental load application: a system of moving nodal co-ordinates is employed so that updating of structural geometry is achieved simply by adjusting the angular orientation of elements relative to nodes. Initial deformations are included in the initial definition of structural geometry. An optional, iterative equilibrium correction of the modified Newton-Raphson type is provided following each incremental load application.

The method outlined above has been used to carry out full-range analysis of over 200 typical tubular bracing members, with particular reference to post-collapse load-carrying capacity and to the effects of initial deformations, residual stresses and certain forms of damage.

Post-Collapse Behaviour

By applying incremental end-shortening displacements instead of loads as indicated in equation [2], it is possible in many cases to carry analysis well into the

post-collapse range without numerical difficulty. For some columns, however, the
incremental stiffness matrix K is found to become non-positive definite just after the
peak load is reached and the solution procedure represented in equation [2] breaks
down. In such cases, which are generally associated with a sharply peaked load-
shortening curve, it is usually possible to compute a static post-collapse load-
shortening curve by incrementing lateral displacement, say at mid-length on the column,
with end-shortening displacements unrestrained and with axial force as an undetermined
parameter: this corresponds to the solution procedure of equation [3], which in the
pre-collapse range gives results algebraically identical with those of equations [1]
or [2] but in the post-collapse range maintains a positive-definite incremental stiffness matrix and yields a form of unloading, illustrated by curve PAQ of Figure 3, in
which end-shortening displacements are temporarily reversed. Reversal of end-
shortening is not, of course, likely to be physically possible in real structures.
Columns tested under displacement control in a rigid testing machine would presumably
follow the unloading path PBQ' shown in Figure 3, involving a sharp drop of load at
constant end-shortening. In the practical case of a column tested in an elastic
machine, or of an offshore bracing member axially compressed by the flexible walls of
supporting chord members, the correct theoretical model is that shown in Figure 3 in
which axial load is effectively applied through elastic springs at the ends of the
column: the presence of the springs has negligible influence on pre-buckling
behaviour and collapse load but post-collapse reduction of axial load, resulting in
extension of the springs, leads to an unloading curve for the column of the form PCQ
as indicated in Figure 3. The break-down of static incremental analysis, together
with experimental observations as described below, indicate that unloading behaviour
represented by curves PCQ or PBQ' is essentially dynamic and would require inclusion
of inertia and damping terms in the equilibrium equations [1] to [3] for its correct
evaluation. This phenomenon has been observed previously during strut tests (19) and
analysis of struts as components of truss frameworks (20).

For cases of the type shown in Figure 3, numerical evaluation of strain energy
and potential energy of applied loads indicates that the transition from point P to
point Q is associated with a sharp reduction in total energy. At P, the strain energy
is associated almost entirely with direct elastic compression; at Q much of this
energy has been released and only partly replaced by elasto-plastic bending strain
energy. From the theoretical results described below it is evident that this unstable
post-collapse behaviour is most likely to occur where imperfections are small and the
Euler and squash loads are approximately equal ($\lambda \approx 1$ for simply supported columns,
$\lambda \approx 2$ for clamped columns); in very long and very short columns or where imperfections
are large, post-collapse behaviour is stable and unloading occurs statically. It is
also evident from the experimental results described below that post-collapse behaviour
may be even more unstable and unloading violently dynamic when collapse is precipitated
by local buckling of the tube wall: in this case relatively little bending strain
energy is developed to compensate for the release of compressive strain energy; consequently the jump from P to Q is associated with a greater net energy reduction.

A comparison is made in Figure 4 of lateral deformations computed for tubes of
varying slenderness using incremental elasto-plastic analysis, with simple upper-
bound estimates of lateral deformation assuming elastic pre-buckling and fully-plastic
post-buckling behaviour. It is evident that the simple analysis, represented by
dotted lines in Figure 4, can provide a reasonable approximation to collapse strength
and pre- and post-buckling behaviour, particularly for tubes with low L/r. The
largest errors in estimated collapse load occur for tubes with $\lambda \approx 1$; for slender
tubes the lateral deformation at and following collapse is substantially overestimated.

Effects of Initial Deformation

Using the incremental analysis method described above an examination has been
made of the influence of initial deformations on stiffness and strength of tubular
bracing members. Deviations from straightness of amplitude δ_o were considered, having
the form:

(i) of uniform curvature, as might be caused by end-moments at chord
attachments;

(ii) "dog-leg" distortions involving a mid-span slope discontinuity as might be caused by misalignment of can joints.

Calculations were carried out for a standard tube having D/t = 35 and an elastic-perfectly plastic stress-strain curve with E/σ_y = 638: control calculations for various D/t in the range 20 to 60 and E/σ_y in the range 480 to 960 indicated that results, when related to the column slenderness parameter $\lambda = L\sqrt{\sigma_y}/(\pi r\sqrt{E})$, were virtually independent of D/t and E/σ_y. Tubes with simply supported and clamped ends were considered with slenderness ratios in the range $30 \leq L/r \leq 120$ ($0.38 \leq \lambda \leq 1.51$). Various amplitudes of initial deformation were examined, both above and below the usual design tolerances (0.001 in API RP2A, 0.0015 in the DNV Rules).

The influence of initial deformation on strength of tubular columns is summarized in Figure 5: the curves obtained are similar to those computed as a theoretical basis for the ECCS column curves (15). The effect of initial deformation on pre-collapse stiffness and post-buckling behaviour is indicated in Figure 6, which shows complete load-shortening curves for a range of simply-supported and clamped tubes. Little difference was found between the effects of uniform-curvature and dog-leg forms of deformation: curves shown in Figures 5 and 6 may thus be assumed to apply in either case. As might be expected, the influence of initial deformation on compressive strength was found to be greatest and post-collapse load-reductions were most severe in tubes for which the elastic buckling and squash loads were approximately equal ($\lambda \approx 1.0$ for simply-supported columns). Where δ_o/L exceeds 0.001, significant loss of axial stiffness may precede collapse in slender tubes ($\lambda > 1.0$).

Effects of Residual Stress

Longitudinal residual stresses caused by a seam weld may be assumed to have the form shown in Figure 7a: a block of tensile yield extending over a width ηt in the region of the weld is balanced by compressive and tensile stresses which vary linearly over the depth of the cross-section such that the net force and moment on the section are zero. The plate thickness in an unstiffened tubular member is usually sufficient to require several weld passes, each pass relieving to some extent the stresses induced by previous passes. Application of formulae developed at Cambridge University (16) for single or multi-pass weld-induced stresses suggests values of η of up to about 4 for typical offshore tubulars. Measurements by Chen and Ross (7) on a large-scale test specimen have indicated a value of η of about 5. Calculations carried out as part of the present study show that such stresses will normally affect tubular column strength by less than 10%, which is consistent with the results of large-scale tests (see Figure 1): as illustrated in Figure 8, however, which shows computed load-shortening curves for initially straight tubes with various levels of residual stress, larger values of η, as might result from heavy single-pass welding of thinly plated tubes, would probably cause more serious loss of strength.

Theoretical evaluation of residual stresses caused by cold-rolling of plates to form cans (17, 18) has indicated distributions of circumferential and longitudinal stress through the plate thickness as shown in Figure 7b: cylindrical bending followed by elastic springback is found to produce circumferential residual stresses up to about 15% higher than the uniaxial yield stress σ_y and longitudinal stresses of up to about $0.7\,\sigma_y$. Residual stresses measured in large-scale test specimens (7) have however been much smaller, as shown by the full line in Figure 7b: the reason for this discrepancy may be that residual stresses are to some extent wiped out by reverse bending during the cold-rolling process. For the purposes of the present study, in which uniaxial stressing is assumed, longitudinal residual stresses were represented directly in the analysis as initial fibre stresses while the effect of circumferential stress was represented by increasing or reducing effective fibre yield stresses in accordance with the von Mises yield criterion: on this basis, for an initially straight simply supported tube with $\lambda = 0.88$, theoretical cold-rolling residual stresses were found to reduce the compressive strength drastically to less than $0.5\,\sigma_y$, while residual stresses similar to those observed experimentally caused a strength reduction of only 15%. Theoretical results are thus inconsistent with test data for fabricated tubular columns (Figure 1) which suggest that loss of column strength caused by cold-rolling residual stress is less than 10%.

Longitudinal residual stresses, having the form shown in Figure 7c, may also result from cold-bending of a tube followed by elastic springback, either applied deliberately in order to correct misalignment or lack of straightness or caused inadvertently by rough handling during erection. Theoretical analysis indicates that such residual stresses may cause reductions of up to 20% in the column strength of typical bracing members. While such severe treatment of a tubular member is unlikely to occur normally, designers and fabricators should be aware of the potentially serious consequences of indiscriminate cold-bending. Residual stresses of the form shown in Figure 7c will also occur in bracing members bent laterally by collision loads.

Loss of strength caused by residual stress is generally greatest in tubes for which the elastic buckling and squash loads are approximately equal and is most marked where initial distortions are small: residual stress effects diminish as initial deformations increase. As shown in Figure 8, residual stress may cause significant loss of pre-collapse axial stiffness, particularly in short columns.

Damage Effects

Damage caused by collisions, falling weights etc may take the form of

(i) local indentation, which is likely to occur, particularly in short, thin-walled bracing members, where lateral loading is impulsive or is applied by a sharp, rigid structure (eg a supply boat deck-edge);

(ii) lateral bending without denting, which is likely to occur in longer, thick-walled tubes and where loading is applied slowly or by a "soft", deformable structure (eg a supply boat superstructure).

Bending and denting may of course occur in combination.

Where lateral bending of a bracing member occurs without denting or local buckling, the effects of damage on compressive strength and stiffness can be evaluated using the analysis method outlined above. In order to represent correctly the residual stresses associated with elasto-plastic bending, which will have distributions over tube cross-sections similar to that shown in Figure 7c, analysis may first be carried out simulating the application and removal of lateral load. Load/lateral-deflection curves computed in this way for a typical bracing member are shown in Figure 9, illustrating the possible importance of large-deflection effects and of axial as well as rotational conditions of support at the ends of the brace. Curves such as those shown in Figure 9 may be used to assess the energy absorption characteristics of bracing members for use in collision analysis: it is clear that resistance to collision loads may be substantially enhanced by development of tensile axial forces in bracing members at large deflections if sufficient axial restraint is provided by supporting chords. The analysis may finally be continued to evaluate the stiffness, strength and post-collapse behaviour of the damaged tube under axial compression. Some illustrative results, computed in this way for a representative range of damaged bracing members, are summarized in Table 1: collapse strengths of tubes with various levels of damage are compared with strengths of undamaged tubes having "characteristic" imperfections, ie $\delta_o/L = 0.001$ with weld-induced residual stress ($\eta = 4$). Damage was in each case assumed to occur by application of a concentrated lateral load at mid-span: the level of damage was defined by the permanent lateral displacement δ_p following removal of lateral load.

The analysis described above does not of course include the effects of local buckling and damage in the form of dents: these effects, which become important in thinner-walled tubes, have been examined experimentally as described below.

EXPERIMENTAL INVESTIGATION

Details of Tests

In parallel with the theoretical study outlined above, compression tests were carried out on a series of 16 tubes representing offshore steel bracing members at about 1/4-scale. The aims of the tests were

(i) to check the accuracy of large-deflection elasto-plastic beam-column analysis in representing the collapse and post-collapse behaviour of tubular members, including the performance of damaged members;

(ii) to provide empirical information about the influence of local instability and damage in the form of dents on collapse and post-collapse behaviour.

In view of the prohibitively high cost of manufacturing and testing full-scale bracing members and in view of the virtual impossibility of simulating at 1/4-scale the full-scale rolling and welding processes with correctly scaled distortions and residual stresses, tests were carried out on cold-drawn CDS-24 and CDS-35 steel tubes, stress-relieved in order to eliminate the effects of unknown residual stresses. The process of stress-relief, carried out by heating unrestrained tubes to 600° C for 2 hours followed by slow cooling, led as expected to some reduction in material yield strength.

Following heat-treatment, the thickness, circularity, straightness and material properties of tubes were surveyed. Thickness was measured at 40 points on each tube using an ultrasonic transducer, measurements being checked against micrometer readings taken at the tube ends. Deviations from circularity (ovality), based on diameter measurements, were characterized as $(D_{max} - D_{min})/D_{mean}$. Following assembly of each tube in the test rig, straightness was measured in vertical and horizontal planes relative to stretched wires.

Material properties were evaluated by carrying out a stub-column test on a specimen of length $\simeq 3D$ cut from the end of each tube. The ends of stub-columns were machined to provide intimate contact with the plattens of an ESH 1000 kN test machine; a ball-support arrangement was provided at one end to eliminate end-moments. Compressive load was applied under displacement control at a strain rate of about 10^{-5}/sec, with frequent stops for periods of about 5 min during which stress was allowed to fall at constant strain in order to define both "static" and "dynamic" stress-strain curves. A typical autographic stress-strain curve is shown in Figure 10.

The geometric and material properties of test specimens are summarized in Tables 2 and 3, together with applied load conditions. Specimens A correspond to bracing members with D/t at the lower end of the range normally found in offshore platforms in which local buckling effects would normally be ignored; specimens B correspond to tubes with medium D/t and specimens C to tubes with D/t at the top end of the practical range in which local buckling might be expected to influence post-collapse behaviour. Specimens D correspond to high-strength thin-walled tubes, as might be of interest to designers where weight-saving is important.

As shown in Figure 11a, each tube was fitted at its ends with steel plugs, designed to transmit compressive load as uniformly as possible, to provide some support against premature local buckling at the tube ends caused by any non-uniformity of compressive stress and to provide points of attachment for lateral supports. Tubes were mounted in a test-rig as shown in Figure 11b. Conditions of simple support were simulated at the ends, lateral displacement being prevented by vertical and horizontal spring steel strips of negligible flexural stiffness providing virtually zero resistance to rotations and axial displacement: outer ends of strips were attached to the test rig through a screw arrangement which allowed fine adjustment of the tube position vertically and horizontally.

Axial load was applied under displacement control by means of a Dartec 500 kN servo-hydraulic jack acting through calibrated load transducers and steel balls as shown in Figure 11b. Load was applied at an average strain rate of about 10^{-5}/sec with frequent stops for periods of about 5 min, during which load was allowed to drop at constant jack extension and displacements and strains were recorded when the load had settled down. Axial end-shortening displacements over the length of each tube were derived from pairs of displacement transducers mounted at tube ends: displacement signals were automatically meaned and subtracted to obtain a net end-shortening signal which was fed into an X-Y plotter together with a load signal to provide autographic load-shortening curves. Lateral (vertical and horizontal) displacements were measured at several sections. Longitudinal strains were measured at mid-length and at sections $\approx$ 2D from each end using foil gauges at 90° spacing. Initial alignment of tubes was carried out by reference to strains under small axial loads within the elastic range. During tests, maximum and minimum diameters were measured by micrometer at several sections on each tube in order to monitor the development of any ovality, particularly following collapse.

Lateral loads, where imposed to simulate damage conditions, were applied by a mechanical jack acting through a load transducer. Loads were applied either

(i) through a solid, 60° steel knife edge with a tip radius of about 3 mm, simulating collision with a sharp, rigid structure;

(ii) through a cradle of semi-circular cross-section lined with a rubber pad, simulating collision with a soft structure.

Lateral load was in each case removed before subsequent application of axial load.

Discussion of Results

The estimated static and dynamic collapse strengths of test specimens, defined as the ratios σ_u/σ_y of peak average compressive stresses to corresponding (static and dynamic) yield stresses derived from stub-column tests, are summarized in Table 3. Experimental load-shortening curves, normalized with respect to static yield, are shown as full lines in Figure 12 together where appropriate with theoretical curves shown as dotted lines. Downward spikes on experimental curves correspond to stops in load application: theoretical load-shortening curves (based on static yield) should be compared with lines through the lowest points on the experimental curves. Photographs of collapsed tubes are shown in Figure 13. Collapse and post-collapse behaviour of each specimen is described briefly below.

Specimen A1: Collapse of this tube, which was virtually imperfection-free, occurred sharply at a load within 5% of the theoretical prediction. Post-collapse unloading (points P to Q on the experimental curve) was dynamic: this, together with elasticity of the test rig, accounts for the shift to the right of the experimental unloading curve relative to the theoretical curve. There was no evidence of local buckling following collapse and ovalization was negligible.

Specimen A2: Collapse under eccentric load occurred slowly and in fairly close accordance with theoretical predictions: no local buckling occurred after collapse.

Specimen A3: This tube was first subjected to damage by lateral load acting through a knife-edge: formation of a dent reduced the lateral load-carrying capacity of the tube to 95% of the theoretical fully-plastic value. Collapse under axial load then occurred slowly with growth of the dent depth to about 2.0 t at collapse and a maximum of 3.8 t following collapse.

Specimen A4: Lateral load was applied through a padded cradle causing lateral deformation and slight ovalization without measurable denting; maximum lateral load was 11% higher than the theoretical fully-plastic value. Collapse under axial load occurred slowly and in reasonable agreement with theory; the collapse load was almost identical with that for A3, suggesting that the effect of a dent is secondary to that of lateral deformation in a thick-walled tube.

Specimen B1: Collapse of this virtually imperfection-free tube occurred sharply, almost at the squash load. As in the case of A1, post-collapse unloading was dynamic over the range PQ and therefore differed in form from the computed curve. There was no sign of post-collapse local buckling.

Specimen B2: Collapse under eccentric load occurred slowly, in good agreement with theory and without post-collapse local buckling.

Specimen B3: Damage was first applied by a lateral load acting through a knife edge; lateral load-carrying capacity was 81% of the fully plastic value. Collapse under axial load then occurred slowly with growth of the dent depth to 4.0 t at collapse and a maximum of 7.0 t following collapse.

Specimen B4: Lateral load was applied through a cradle causing lateral deformation with slight indentation. Collapse under axial load took place slowly and in accordance with theory up to end-shortening strains of about 1.7 ε_y, where local buckling occurred at the point of application of lateral load. The collapse load was 17% higher than that for B3.

Specimen C1: Failure occurred very suddenly and was associated with formation of a well-defined, single-lobe local buckle. Post-collapse unloading was dynamic over the range PQ.

Specimen C2: Failure under eccentric axial load occurred gradually and was in reasonably good agreement with theory. No local buckling was observed until ε_{ave} exceeded 2.5 ε_y: the post-collapse load-carrying capacity was, consequently, up to twice that of C1.

Specimen C3: This tube was subjected to damage in the form of a mid-span dent of depth 2.0 t without lateral deformation by applying lateral load through a knife-edge with the back of the tube supported. Failure occurred sharply, with dynamic unloading over PQ, and involved a well-defined single-lobe local buckle at the position of the initial dent.

Specimen C4: Damage was imposed of the same form as in C3 but with a smaller dent of depth 0.9 t. Failure occurred less sharply and at a somewhat higher load than in C3, involving incipient local buckling in the mid-length region of the tube but not at the dent position.

Specimen D1: Collapse occurred explosively at σ_u = 0.78 σ_y and was associated with formation of a 3-lobe local buckle in a quarter-span position. Comparable estimates of collapse loads obtained from the API and DNV formulae for buckling of thin-walled tubes were respectively 0.62 σ_y and 0.63 σ_y. Following collapse the load dropped dynamically to less than 10% of the collapse level.

Specimen D2: Collapse under eccentric axial load occurred very sharply, the load falling dynamically to about 30% of the collapse value. The mid-span outer fibre strain measured just before collapse was 1.2 ε_y.

Specimen D3: Damage in the form of a dent of depth 3.2 t at mid-length was imposed by applying lateral load through a knife-edge with the back of the tube supported. Failure, initiated by local buckling at the dent, occurred sharply with dynamic unloading.

Specimen D4: Damage was imposed of the same form as in D3 but with a smaller dent of depth 1.9 t. Failure, again initiated by local buckling at the dent position, occurred very sharply with dynamic unloading.

CONCLUSIONS

From a review of present design practice and from the theoretical and experimental studies outlined above, the following results have been obtained and conclusions reached.

(i) Data curves have been established defining the effects of initial distortion and weld-induced residual stress, both within and outside normal design tolerances, on the strength of typical bracing members.

(ii) Guidance has been provided on pre-buckling stiffness and post-collapse load-carrying capacity of bracing members as required for evaluation of ultimate limit state behaviour of complete platforms.

(iii) Large-deflection elasto-plastic beam-column theory has been found in most cases to give a satisfactory account of the collapse and post-collapse behaviour of tubular bracing members provided that local buckling does not occur.

(iv) The API and DNV rules give a safe, rather conservative estimate of collapse strength for bracing members with $D/t < 0.1\ E/\sigma_y$ having imperfections within prescribed tolerances: in view of uncertainty about the straightness of bracing members in completed platforms, the conservatism of present design practice seems appropriate.

(v) The compressive strength of high-yield thin-walled tubes ($D/t > 0.2\ E/\sigma_y$) with small imperfections has been found to exceed the strength indicated by current design rules. However, in view of their catastrophic post-collapse behaviour and sensitivity to damage in the form of small dents, it is doubtful whether such tubes should be considered seriously for use in the primary structure of offshore platforms.

(vi) Theoretical and experimental results have been obtained which provide some guidance on the effects of collision damage on performance of bracing members. It is suggested that by judicious application of theoretical analysis combined where necessary with small-scale tests, an effective assessment could be made of the implications of any damage condition arising in service in the bracing members of a North Sea platform.

ACKNOWLEDGEMENT

The theoretical study and tests described in this paper were carried out with financial support from the Petroleum Engineering Division, Department of Energy, by whose permission the paper is published. The authors are indebted to Messrs A Adamson, A Bennett and C Duncan for assistance in carrying out experiments.

REFERENCES

1. American Petroleum Institute: "Recommended Practice for Planning, Design and Constructing Fixed Offshore Platforms". API RP2A (8th Ed), April 1977.

2. AISC Specification for the Design, Fabrication and Erection of Structural Steel for Buildings, American Institute for Steel Construction, Feb 1969.

3. AISI Specification for the Design of Cold-Formed Steel Structural Members: Cold-Formed Steel Design Manual, Part 1. American Iron and Steel Institute, 1968.

4. DNV Rules for the Design, Construction and Inspection of Fixed Offshore Structures. Det Norske Veritas, Oslo, 1977.

5. ECCS 2nd Internat Colloquium on Stability (Introductory Report). European Convention for Structural Steelwork, July 1976.

6. J G Bouwkamp: "Buckling and Post-Buckling Strength of Circular Tubular Sections". Proc Offshore Technology Conference, Houston, 1975.

7. W F Chen, D A Ross: "Tests of Fabricated Tubular Columns". Proc ASCE (Struct Div), Vol 103, March 1977.

8. H Matlock, T P Taylor: "A Computer Program to Analyze Beam-Columns under Movable Loads". Report No 56-4, Center for Highway Research, Univ of Texas, Austin, June 1968.

9. W H Chen: "General Solution of Inelastic Beam-Column Problem". Proc ASCE (EM Div), August 1970.

10. B W Young: "Axially Loaded Steel Columns". Cambridge Univ Report No CUED/C-Struct/TR11, 1971.

11. D A Nethercot, K C Rockey: "Finite Element Solutions for the Buckling of Columns and Beams". Int J Mech Sci, Vol 13, 1971.

12. Y Fujita, K Yoshida, M Takazawa: "Compressive Strength of Columns with Initial Imperfections". Paper presented to ISSC Committee II.2, Hamburg, 1973.

13. S Rajasekaran, D W Murray: "Finite Element Solution of Inelastic Beam Equations". Proc ASCE (Struct Div), Vol 99, June 1973.

14. C S Smith, W Kirkwood: "Influence of Initial Deformations and Residual Stresses on Inelastic Flexural Buckling of Stiffened Plates and Shells". Proc of Internat Symposium on Steel Plated Structures, Imperial College, London, 1976.

15. H Beer, G Shulz: "Theoretical Basis of the European Buckling Curves" (Bases Théoriques des Courbes Européennes de Flambement). Construction Métallique, No 3, September 1970.

16. A G Kamtekar, J D White, J B Dwight: "Shrinkage Stresses in a Thin Plate with a Central Weld". Jour Strain Analysis, Vol 12, No 2, 1977.

17. L Ingvarsson: Contribution to discussion of Ref 14.

18. B Kato: "Column Curves for Cold-Formed and Welded Steel Tubular Members". ECCS 2nd Internat Colloquium on Stability of Steel Structures (Preliminary Report), Liege, April 1977.

19. N J Hoff: "Buckling and Stability". Jour Roy Aero Soc, Vol 58, No 517, January 1954.

20. G Davies, B G Neal: "The Dynamical Behaviour of a Strut in a Truss Framework". Proc Roy Soc, Vol 253, p 542, 1959.

BHRA and Crown Copyright reserved.

TABLE 1: Effect of Damage on Strength of Tubular Columns

$\frac{L}{r}$	Support Condition		Damage Deformation δ_p/L	$\frac{\sigma_u}{\sigma_y}$	% Loss of Strength Caused by Damage
	Flexural	Axial*			
30	Simple Support	–	Undamaged	0.94	–
"	"	Free	0.0031	0.93	1
"	"	"	0.0166	0.67	29
"	"	"	0.0368	0.50	47
"	"	Fixed	0.0031	0.92	2
"	"	"	0.0168	0.64	32
"	"	"	0.0370	0.57	39
"	Clamped	–	Undamaged	0.97	–
"	"	Free	0.0022	0.97	0
"	"	"	0.0197	0.83	14
"	"	"	0.0412	0.68	30
"	"	Fixed	0.0034	0.97	0
"	"	"	0.0210	0.85	12
"	"	"	0.0426	0.75	23
70	Simple Support	–	Undamaged	0.75	–
"	"	Free	0.0044	0.63	16
"	"	"	0.0128	0.43	43
"	"	"	0.0232	0.33	56
"	"	Fixed	0.0058	0.51	32
"	"	"	0.0162	0.30	60
"	"	"	0.0284	0.30	60
"	Clamped	–	Undamaged	0.94	–
"	"	Free	0.0035	0.89	5
"	"	"	0.0126	0.69	27
"	"	"	0.0264	0.53	44
"	"	Fixed	0.0038	0.84	11
"	"	"	0.0136	0.64	32
"	"	"	0.0279	0.51	46
120	Simple Support	–	Undamaged	0.38	–
"	"	Free	0.0005	0.38	0
"	"	"	0.0081	0.28	26
"	"	"	0.0169	0.21	45
"	"	Fixed	0.0062	0.22	42
"	"	"	0.0189	0.21	45
"	Clamped	–	Undamaged	0.85	–
"	"	Free	0.0032	0.74	13
"	"	"	0.0196	0.40	53
"	"	"	0.0331	0.31	64
"	"	Fixed	0.0038	0.65	24
"	"	"	0.0239	0.37	56

*during application of lateral load only

TABLE 2: Geometry and Material Properties of Tubes (dimensions in mm)

Specimen No	Diameter (D) to mid-thickness		Thickness (t)		Length (L) to support points	Initial Ovality $\left(\dfrac{D_{max} - D_{min}}{D_{mean}}\right)$		Max Deviation from Straightness (δ_o/L)	Young's Modulus (E) $N/mm^2 \times 10^5$	Yield Stress N/mm^2	
	mean	cov	mean	cov		mean	cov			Static	Dynamic
A1	61.5	0.001	2.11	0.022	2150	0.0015	0.54	0.0002	2.01	228	261
A2	61.5	0.002	2.12	0.021	"	0.0010	0.37	0.0003	2.08	229	262
A3	61.5	0.001	2.11	0.021	"	0.0023	1.11	0.0002	1.98	226	259
A4	61.5	0.001	2.12	0.023	"	0.0010	0.15	0.0003	1.83	218	244
B1	77.8	0.001	1.74	0.049	"	0.0014	0.25	0.0003	2.04	195	223
B2	77.8	0.001	1.71	0.026	"	0.0025	0.19	0.0001	2.29	199	223
B3	77.8	0.001	1.72	0.030	"	0.0013	0.30	0.0003	2.14	198	230
B4	77.8	0.004	1.70	0.020	"	0.0068	0.81	0.0002	1.95	200	231
C1	100.0	0.001	1.66	0.024	"	0.0043	0.88	0.0005	2.01	211	247
C2	99.9	0.001	1.73	0.022	"	0.0017	0.77	0.0005	2.06	284	304
C3	100.0	0.001	1.72	0.023	"	0.0026	0.23	0.0004	1.97	233	255
C4	100.0	0.001	1.73	0.025	"	0.0026	0.43	0.0005	2.07	252	272
D1	89.0	0.001	1.02	0.023	"	0.0017	0.41	0.0003	2.21	485	496
D2	89.0	0.005	1.01	0.025	"	0.0012	0.63	0.0005	2.55	466	484
D3	88.9	0.002	1.03	0.032	"	0.0039	0.28	0.0003	2.32	469	481
D4	89.0	0.002	1.05	0.028	"	0.0053	0.53	0.0010	2.21	477	485

TABLE 3: Test Conditions and Collapse Loads

Specimen No	$\frac{D}{t}$	$\frac{L}{r}$	$\lambda = \frac{L}{r\pi}\sqrt{\frac{\sigma_y}{E}}$ (static σ_y)	Damage Condition					Max Tube Ovality $(D_{max} - D_{min})/D_{mean}$		Experimental Collapse Load σ_u/σ_y		% Loss of Strength Caused by Damage
				Eccentricity of Axial Load	Lateral Deformation δ_p/L	Depth of Dent	Max Lateral Load $PL/4D^2t\sigma_y$		at Collapse	post-Collapse	Estimated Static	Dynamic	
A1	29.2	98.9	1.06	0	–	–	–		0.002	0.003	0.84	0.78	–
A2	29.0	98.9	1.04	0.16 D	–	–	–		0.003	0.006	0.49	0.45	–
A3	29.2	98.9	1.06	0	0.0055	1.4 t	0.95		*	*	0.48	0.43	43
A4	29.0	98.9	1.09	0	0.005	–	1.11		0.001	0.024	0.50	0.47	40
B1	44.7	78.2	0.77	0	–	–	–		0.001	0.046	1.00	0.93	–
B2	45.5	78.2	0.73	0.13 D	–	–	–		0.003	0.038	0.60	0.58	–
B3	45.2	78.2	0.76	0	0.005	3.7 t	0.81		*	*	0.52	0.47	48
B4	45.8	78.2	0.80	0	0.005	0.5 t	1.17		0.021	*	0.61	0.56	39
C1	60.2	60.9	0.63	0	–	–	–		0.004	*	1.10	0.98	–
C2	57.8	60.9	0.72	0.10 D	–	–	–		0.002	*	0.58	0.58	–
C3	58.1	60.9	0.67	0	–	2.0 t	–		*	*	0.76	0.73	31
C4	57.8	60.9	0.68	0	–	0.9 t	–		*	*	0.84	0.82	24
D1	87.3	68.4	1.02	0	–	–	–		0.002	*	0.75	0.75	–
D2	88.1	68.4	0.93	0.17 D	–	–	–		0.034	*	0.50	0.50	–
D3	86.3	68.4	0.98	0	–	3.2 t	–		*	*	0.53	0.52	29
D4	84.8	68.4	1.01	0	–	1.9 t	–		*	*	0.64	0.63	15

* affected by dent or local buckling

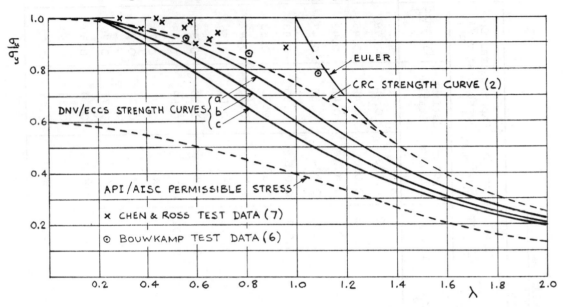

FIG 1   DESIGN CURVES AND TEST DATA FOR FABRICATED TUBULAR COLUMNS

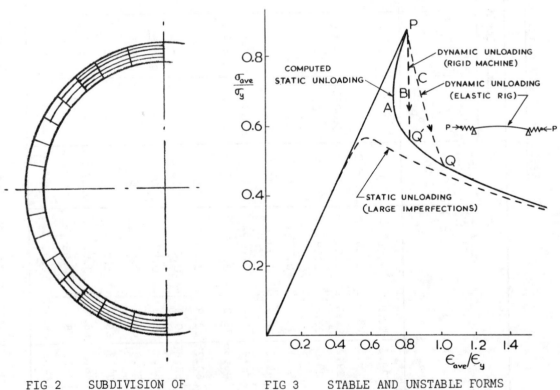

FIG 2   SUBDIVISION OF TUBE SECTION INTO "FIBRES"

FIG 3   STABLE AND UNSTABLE FORMS OF POST-COLLAPSE BEHAVIOUR

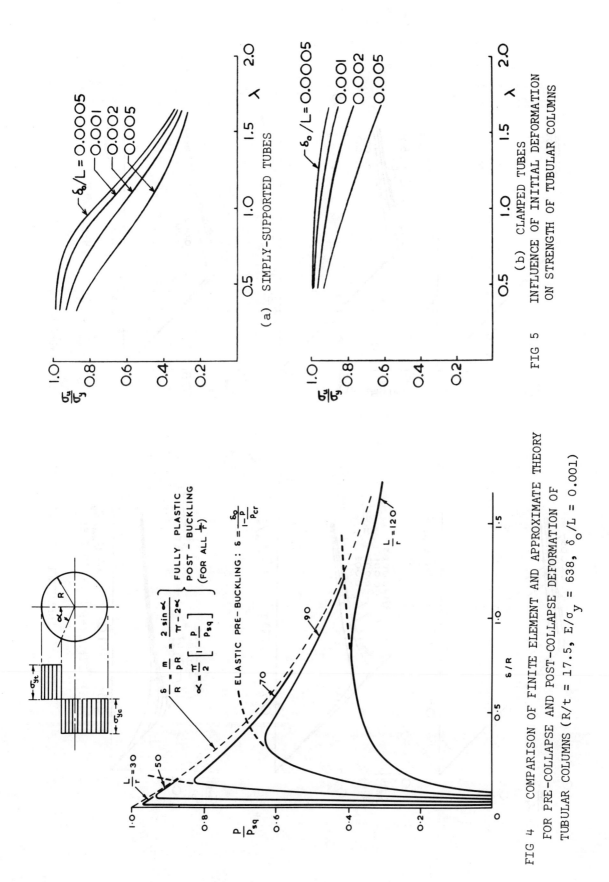

FIG 5 INFLUENCE OF INITIAL DEFORMATION ON STRENGTH OF TUBULAR COLUMNS

FIG 4 COMPARISON OF FINITE ELEMENT AND APPROXIMATE THEORY FOR PRE-COLLAPSE AND POST-COLLAPSE DEFORMATION OF TUBULAR COLUMNS (R/t = 17.5, E/σ_y = 638, δ_o/L = 0.001)

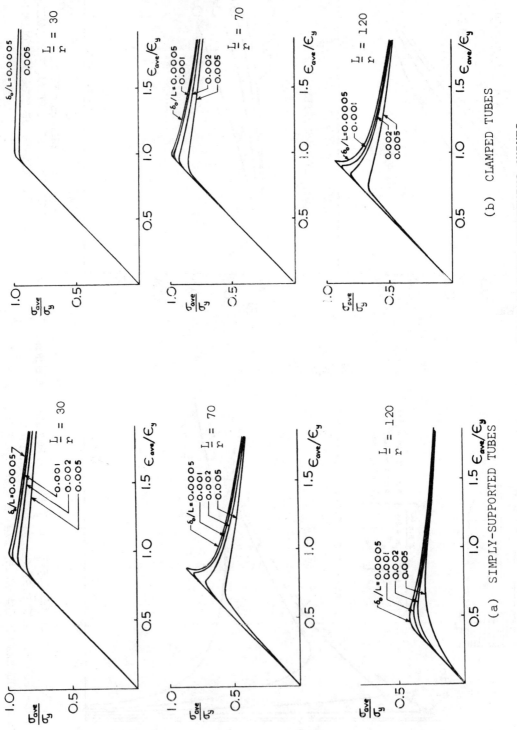

FIG 6   INFLUENCE OF INITIAL DEFORMATION ON LOAD-SHORTENING CURVES

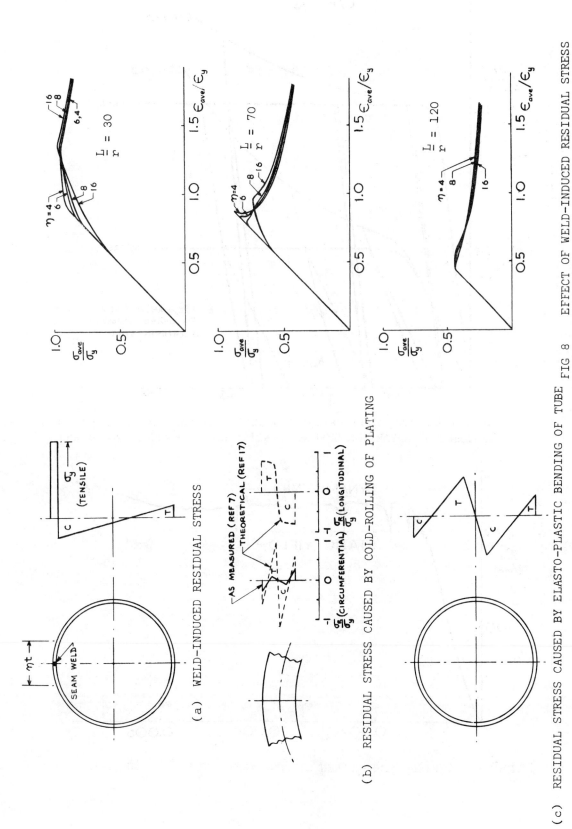

(a) WELD-INDUCED RESIDUAL STRESS

(b) RESIDUAL STRESS CAUSED BY COLD-ROLLING OF PLATING

(c) RESIDUAL STRESS CAUSED BY ELASTO-PLASTIC BENDING OF TUBE

FIG 7 RESIDUAL STRESS DISTRIBUTIONS

FIG 8 EFFECT OF WELD-INDUCED RESIDUAL STRESS ON LOAD-SHORTENING CURVES FOR SIMPLY-SUPPORTED TUBES

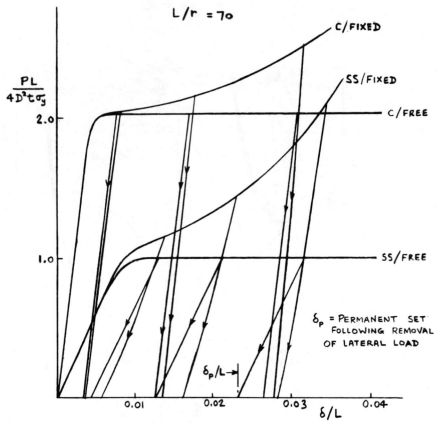

FIG 9   MID-SPAN LATERAL DISPLACEMENT (δ) UNDER LATERAL COLLISION LOAD (P) FOR TUBES WITH VARIOUS SUPPORT CONDITIONS (L/r = 70, E/σ_y = 638)

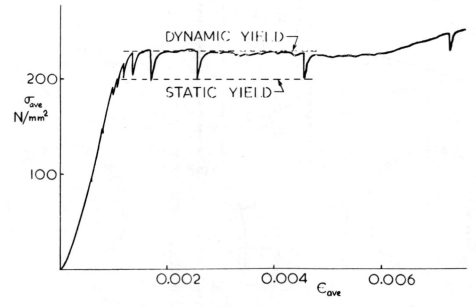

FIG 10   TYPICAL STUB-COLUMN STRESS-STRAIN CURVE (SPECIMEN B2)

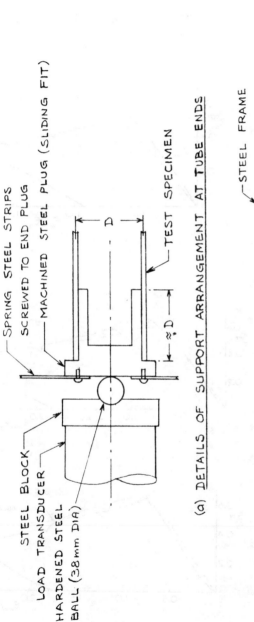
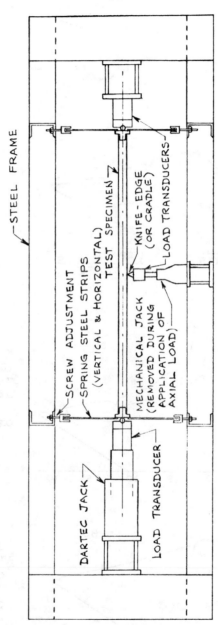

(a) DETAILS OF SUPPORT ARRANGEMENT AT TUBE ENDS

(b) GENERAL ARRANGEMENT OF TEST RIG

FIG 11

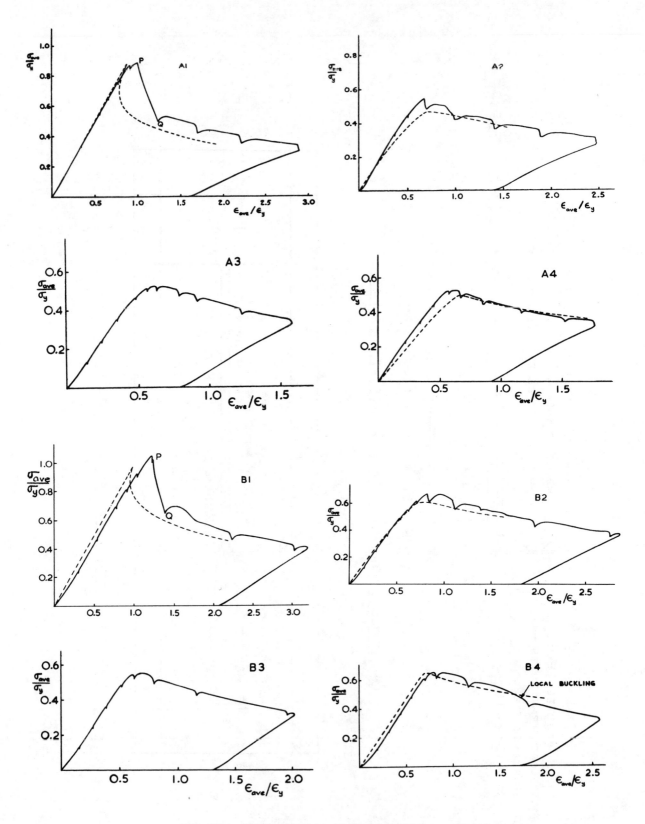

FIG 12   EXPERIMENTAL LOAD-SHORTENING CURVES

(Theoretical Curves shown as Dotted Lines)

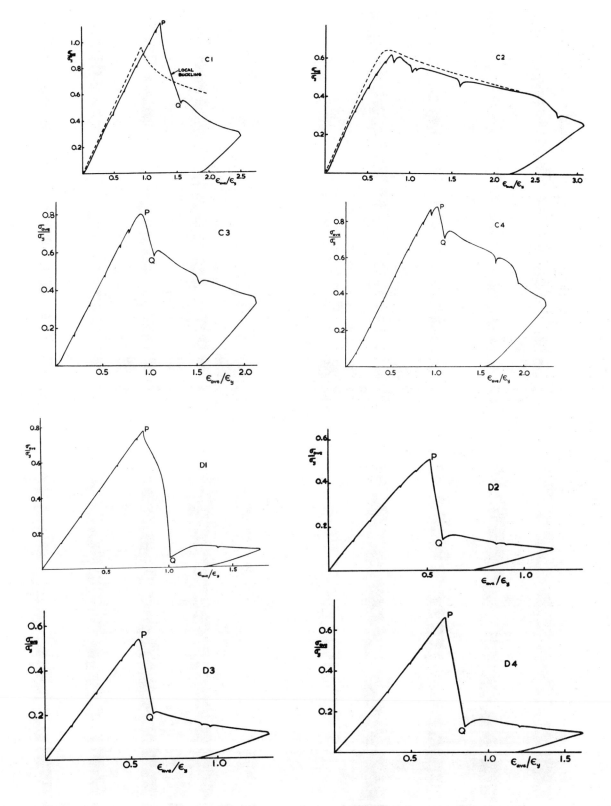

FIG 12 (Continued)

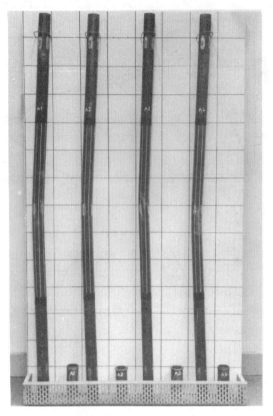

TUBES A1 TO A4

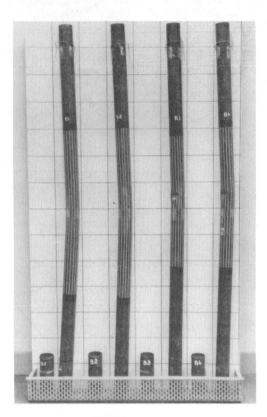

TUBES B1 TO B4

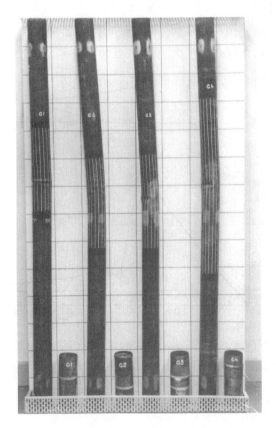

TUBES C1 TO C4

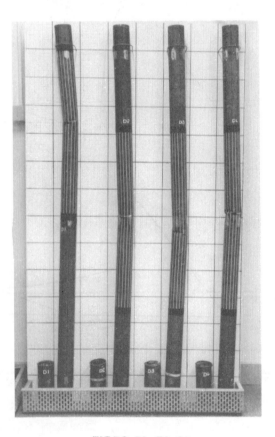

TUBES D1 TO D4

FIG 13   COLLAPSED TEST SPECIMENS

BOSS'79

PAPER 71

Second International Conference on Behaviour of Off-Shore Structures

Held at: Imperial College, London, England
28 to 31 August 1979

CURRENT RESEARCH INTO THE STRENGTH OF CYLINDRICAL SHELLS USED IN STEEL JACKET CONSTRUCTION

P.J. Dowling and J.E. Harding

Imperial College, U.K.

Summary

The paper summarises the aims and accomplishments of the research programme at Imperial College looking at the buckling of structural members typical of those used in the legs and bracing members of steel rigs.

A combination of numerical approaches is being used and cross checked with large and small scale experiments to clarify the situation regarding the design of the rig members, leading eventually to a rationalisation of existing design rules.

The numerical approaches have already achieved some success and sample results will be presented in the paper. The experiments have reached the stage where models have been fabricated and rigs partially built. A discussion is also included in this paper on this aspect of the work.

INTRODUCTION

A series of research centres has been set up in the U.K. by the Science Research Council to look into various aspects of the behaviour of offshore structures. The theme of the London Centre, comprising both Imperial College and University College, is that of marine structures and materials. This paper concentrates on the work at Imperial College on one aspect of this study, the buckling of steel shells of the type used in conventional and future marine structures. In particular, the research concentrates on the behaviour of unstiffened, ring stiffened, or orthogonally stiffened cylinders and cylindrical elements.

Over the last few years considerable progress has been made in updating the steel bridge design code in the light of experimental research(Ref. 1) and computer based numerical analyses(Refs. 2-4). These analyses have been based on finite difference dynamic relaxation solutions of the large deflection plate equations. The solutions take account of plasticity and have been used to obtain ultimate load data for direct corelation with experimental results and subsequently for updating and rationalising the existing design rules.

A similar approach is being used in the field of offshore structures. Small and large scale models, sponsored by the Department of Energy, are being tested, and existing flat plate finite difference programs have and are being modified to include the curvature effects associated with cylindrical elements. In addition, because of the complexity in geometry associated with orthogonally stiffened offshore legs, the foundations are being laid for a treatment of the buckling problem using the more powerful finite element method.

Associated research is being undertaken at University College and at other centres throughout the country. This research will be discussed briefly later in this paper.

NUMERICAL APPROACHES

Finite Difference Elasto-plastic Analyses

The finite difference/dynamic relaxation analytical approach had already been developed successfully to deal with unstiffened and stiffened flat plate structures under various types of in-plane and out-of-plane loading.

Versions of finite difference programs have been used to look at isotropic plate panels in compression (Ref. 3), in compression, tension, in-plane bending and shear (Ref. 2) and in biaxial compression (Ref. 5). The problem of stiffened plating was investigated in the elastic range by Basu (Ref. 6) and subsequently expanded into the plastic range by Djahani (Ref. 7) and Webb (Ref. 8). The final formulation is capable of handling flat plates orthogonally stiffened by flat, angle or tee stiffeners.

The version used as the basis of the finite difference work accomplished so far in the offshore field is that of references 2 and 9. The second author of this paper has modified the flat plate version of the program to include curvature terms and linked two of the plate edges so that a complete unstiffened cylindrical element has been formed. Ho (Ref. 10), using a similar approach has been looking at the problem of a curved panel typical of those used in an orthogonally stiffened cylinder. He has also produced a version of this curved plate program incorporating longitudinal stiffeners following the method used by Webb (Ref. 8). Sample results from these analyses will be presented later.

Finite Element Analysis

In the finite element field Javaharian (Ref. 11) has developed a generally curved shell element to include geometric and material non-linear effects. The element used as the basis for this work was the powerful 'semi-loof' rectangular element developed by Irons. This element is currently being tested against available solutions for cylindrical shells.

Mechanism Approach

This work is concerned with the extension of a rigid plastic mechanism approach developed at Imperial College by Dean (Ref. 12) on the collapse of flat plates under in-plane and lateral loading. The formulation by Dean allowed for normal and shear forces and moments along the mechanism lines. Normality was also taken into account. The resulting analysis produced unloading characteristics for flat plates which compared well with the results of more sophisticated numerical approaches. This approach is being modified to include the effect of curvature so that the unloading characteristics of shells under in-plane and lateral loading can be investigated. The numerical analyses have shown that buckling of shells under in-plane and lateral loading is extemely complex and it is hoped that this method will provide a cheap and simple way of studying the effect of various parameters on post-buckling response.

Other Analytical Approaches

Within the steel structures section involved in the above work, other more fundamental research by Burgoyne and Brown has been going on into the imperfection sensitivity of shell structures. This is hoped to provide additional information on the buckling modes of different shell geometries.

EXPERIMENTAL WORK

Two series of experimental tests are in progress. The first is concerned with small scale cylindrical models at a scale of about 1 in 20. The second is concerned with larger scale cylinder models at a scale of about 1 in 4. Both sets will be subjected to axial compression with, in some cases, a small eccentricity. These tests are being sponsored by the Department of Energy along with other tests at University College, London, the University of Glasgow and the Admiralty Marine Technology Establishment at Dunfermline. Table 1 shows how our own test series fits in with those of the other centres.

Large Scale Models

At present four tests are in the progress of construction or being planned. These are comprised of three tests on ring stiffened cylinders (IC1, IC2 and IC4, Table 1) and one test on an orthogonally stiffened cylinder (IC3). The tests are moderately large with ultimate load capacities up to about 3500 kN. Models IC2 and IC4 are identical cylinders ($R/t = 250$, $L/R = 1.0$), one being loaded axially while the second will be loaded with a small eccentricity. The third ring stiffened cylinder will be smaller ($R/t = 150$, $L/R = 0.2$) subjected to axial loading.

At the time of writing, models 2 and 4 have been fabricated but not yet tested. Model 1 is being fabricated. Figure 1 shows a photograph of Model 2 with the loading plattens under construction.

Residual stress readings are being taken on all models and those of models 2 and 4 are shown in Figs. 2 and 3 respectively. It can be seen that the major residual stresses are in the hoop direction and are compressive. These are caused, in the main, during the rolling process. Very high bending strains are set up in the plate during rolling and in this case the compressive component (which exceeds yield) is the larger, resulting in the average plate strains shown in the figures. It is not thought that these strains are likely to affect behaviour under longitudinal loading to a large degree, but this will be investigated in future analytical work. Longitudinal strains can be seen to be more variable. For cylinder 1 they are of the order of -100 to -200 µ-strain while for cylinder 2 they are between about -50 and $+50$ µ-strain. These appear to alter during both welding and rolling processes.

Imperfections have not yet been measured in detail but examination using a straight edge indicates that DnV tolerance levels have been met on the first two cylinders.

It is hoped that these first-two models will be tested in the near future.

Small Scale Models

The use of large scale models is expensive and time consuming. They have obvious benefits in terms of direct representation of structual steels, welding techniques, normal fabrication imperfections, etc. It was felt that the use of small scale modelling would provide a more economical solution to experimentally cover a sufficient range of parameters to calibrate analytical solution techniques. University College have pioneered the use of small scale stiffened plates and subsequently cylinder models (Ref. 13), and it was decided to generally follow these techniques while introducing some fabrication differences of our own. It was still felt necessary to validate the small testing techniques and for this reason the large scale models at Imperial College are being duplicated by small scale counterparts.

The first small scale models at Imperial College are longitudinally stiffened and these have now been fabricated. Figure 4 shows one of these models (IC16) at the completion of fabrication. The fabrication techniques used on these small scale models are interesting. They are different to those used by University College and will therefore provide an interesting comparison. The models are being made as follows. The cylinder itself is cut in one piece allowing for contractions due to welding distortions. The stiffeners are cut and then welded onto the unrolled cylinder sheet. After welding, this stiffened sheet is formed around a segmented mandrel and a closing tee butt weld is then run over one stiffener. An outer clamping cylinder is fitted and clamped to this assembly and the whole unit is stress relieved, ensuring that the shape is maintained during this process. After stress relieving the ends are turned square and parallel and bound in an araldite and sand resin between heavy steel collars to maintain the rigid end condition required in the testing.

The technique used at University College is rather different. The cylinders are formed from individual plate panels, each T-butt welded to the longitudinal or ring stiffeners.

Figure 5 shows the test rig for the Imperial College models near completion. Basically it is similar to a standard testing machine with a screw jack driven by an electric motor loading the specimens. Load cells are provided between the two bottom plattens to accurately measure the load going into the specimens. Eccentric loading will be provided by the introduction of two spherical bearings between the cylinder and the loading plattens. These will, of course, be eccentric to the centre line of the cylinder.

RESULTS OBTAINED FROM THE NUMERICAL ANALYSES

Finite Difference Results on a Complete Cylindrical Element.

Results have been produced for complete cylindrical elements with end boundary conditions approximating those provided by ring stiffeners which are designed not to fail before local buckling occurs. This is the design philosophy commonly adopted in offshore rig design. The flexural boundary conditions at the ends were assumed to be simple supports, while zero circumferential displacement was assumed around the loaded ends. The loading was applied in the form of a meridional displacement.

Results for cylinders in compression have already been presented in detail in reference 14.

Cases have also been run for cylinders under external pressure and meridional compression, and for cylinders under bending and tension or compression.

Figures 6, 7 and 8 show a summary of these results for the three load types compared with existing design rules.

Figure 6 shows the results of a parametric study on cylinders in compression in the form of lower and upper bound analytical curves. The large difference in these two curves reflects the high degree of imperfection mode sensitivity. The imperfection mode, perhaps even more than the imperfection magnitude, in the practical range for offshore structures, has a considerable effect on the plate buckling strength of cylinders of any slenderness other than the most stocky ones.

The design recommendations of the API, DnV and ECCS are shown plotted on the same figure. It can be seen that the API (and DD55) rules are the most conservative of the three. They might be considered over-conservative in the range of R/t's from 100 to 500. The DnV rules and ECCS recommendations are quite close to the lower bound of the experimental results. The DnV rules seem to be slightly unconservative for the stockier range of cylinders.

Figure 7 shows analytical curves produced by applying axial compressive displacements and an initial external pressure. It can be seen that for pressure heads in excess of 50m, axial strength decreases rapidly. The curves shown are for imperfections similar to those used to obtain the lower bound analytical curve of Fig. 6. The three graphs refer to different ring stiffener spacings. Three cylinder slendernesses have been considered. For these curves a curve with a fixed ratio of applied compressive stress to external pressure has been produced corresponding to the case of a capped cylinder under external pressure. This can then be compared with the BS5500 pressure vessel recommendations shown. The agreement is reasonably satisfactory.

Figure 8 shows interaction curves produced by applying combinations of bending and axial compressive loading to a cylinder. The imperfection mode used is again the lower bound mode of Fig. 6. Also shown are the corresponding API (DD55) and DnV rules and the ECCS recommendations. The API rules are generally conservative especially in the intermediate slenderness range (R/t = 100 to 500). DnV are slightly conservative except in the predominently compressive cases for R/T's from 100 to 200. The ECCS recommendations are again generally slightly conservative apart from the intermediate R/t values (200 - 500). Agreement in general is quite good.

Finite Difference Results on Cylindrical Panels.

Results have been produced for the behaviour of unstiffened cylindrical panels subjected to axial compression using the analysis of reference 10. Various slenderness levels, curvatures, initial distortions and boundary conditions have been considered. Simply supported boundaries have been assumed with three degrees of membrane boundary restraint. Constrained edges in which the edges are free to move in-plane but are held straight are considered to be the most appropriate for design. Figure 9 shows the agreement between the theoretically predicted curves and the DnV rules. For the stocky plates (b/t up to 40) the rules are conservative over the range of R/t's plotted. There is some underestimation of buckling capacity for more slender panels for the level and shape of initial distortion assumed in the analysis. The level used is an inwards imperfection of $W_0/t = 0.87\beta^2$ ($\beta = b/t\sqrt{\sigma_0/E}$). This expression is one adopted for ship structures.

In the absence of specified plate tolerances in the DnV rules it is interesting to take a panel of typical slenderness and curvature and study the level of distortion which is most closely represented by the design approach. Figure 10 shows such a panel. An inward distortion equal to 0.374t (equal to the β level given by the above expression) approximates to the design rule curve for the plate slenderness used.

A recent DnV report (Ref. 14) does, however, indicate the level of tolerances expected to be appropriate for the use of the design rules (Ref. 15).

This states that for curved panels:

- The deviation from a straight generator of length ℓ_1 should not exceed δ_1.
- The deviation from a circular template of length ℓ_2 should not exceed δ_2.

$\delta_1 = 0.01\ell_1$, $\ell_1 = 4\sqrt{rt}$, $\ell_1 \leqslant$ max panel length.

$\delta_2 = 0.01\ell_2$, $\ell_2 = 4\sqrt{rt}$, $\ell_2 \leqslant$ max panel breadth.

This means that panels are allowed a basic tolerance (W_0/t) for b/t equals 40 (R/t > 100) of 0.4, W_0/t for b/t equals 60 (R/t > 225) of 0.6, and W_0/t for b/t equals 80 (R/t > 400) of 0.8. In these cases the maximum panel length dominates.

For lower R/t values the W_0/t value is lower for each b/t and becomes a function of R/t.

Figure 11 shows analytical curves for square panels with the above imperfections and also the corresponding DnV design rules. It can be seen that if the suggested tolerances are applied the DnV rules are significantly non-conservative in the case of the b/t = 40 and 60 results. It may well be that the tolerances specified are incorrect and larger than would normally be accomplished but some clarification is clearly needed.

This conclusion ties in with a similar discussion about ring stiffened cylinders (Ref. 13) which suggests that DnV tolerances are too large and are not compatible with their design curves.

FINITE ELEMENT RESULTS

At present, only a limited number of comparison cases have been run. These do not specifically refer to offshore structures and will not be reproduced here. They are given in the report of reference 11. It is hoped that more results will be available in the near future.

WORK AT OTHER CENTRES

The work at University College, London is concentrating on small scale experimental studies (detailed in table 1) together with elastic critical buckling analyses looking at the interaction between different critical buckling modes.

Work at the Civil Engineering Department of Glasgow University is based on large scale experimental work. Future analytical comparisons are planned.

The work at AMTE is again concerned with smaller scale testing work but is concentrating on the buckling of long stocky cylinders. AMTE is particularly interested in the effect of damage sustained to rig structures. Bracing members incorporating local initial buckles are being tested in compression to see the effect of large initial deformations and eccentricities on their column collapse behaviour.

FUTURE WORK

The main advance on the theoretical front will be in the sophistication of the stiffening used in the analyses. T-section ring stiffeners, for example, will, in the immediate future, be added to the cylindrical elements to accurately represent cross-frame action. The finite element work will also be advanced to include suitable beam elements so that, together with the shell element, stiffened shell components can be analysed using this alternative approach.

By including discrete stiffeners it will be possible to study the elasto-plastic interactive buckling between shell and stiffeners.

On the experimental side, apart from testing additional models, it is hoped that a hyperbaric facility will be designed and constructed to look at reasonably large scale models under external pressure and axial loading. Support has already been obtained for a design study on a potential facility. This is seen as an essential part of the research programme as the numerical results already obtained have highlighted the importance and complexity of the effects of pressure loading on the buckling behaviour of offshore structures. The state of uncertainty in the design of structures for this type of loading is a potentially hazardous situation.

CONCLUSIONS

This paper has presented the overall aims of the offshore research steel buckling program at Imperial College. It has also shown that considerable progress has been made, especially in analytical techniques and it is hoped that the experimental side of the programme will be producing results in the near future.

It can be seen from the results presented so far that more has been achieved on a theoretical level than experimentally. This is partly due to the length of time necessary to prepare experimental tests but is also due to the low financial backing given to experimental work at present in offshore research. Theoretical results, without experimental verification, can not, with certainty, be used to upgrade design rules. An element of risk would always be present in their application. It is essential that more experimental work be done and it is hoped that this will be achieved in the future.

The following preliminary conclusions can be drawn from the analytical work so far undertaken bearing in mind the above comment.

1. Numerical methods of analysis have shown the influence of imperfection magnitude and more particularly of mode on the buckling strength of curved components.

2. The existing API rules for local buckling of cylindrical shells (and the draft UK rules) appear to be conservative for cylinders over the range of practical slenderness.

3. The new (1977) DnV rules are generally less conservative and may be unconservative for the stockier range of cylinders. More information is, however, needed on actual imperfection levels.

4. The existing design rules for cylindrical panel buckling in the DnV rules seem to be satisfactory. This, however, assumes that the levels of initial distortions assumed in the analyses are representative of practical imperfections.

As stated, there seems to be an incompatibility between the levels of tolerances assumed in the DnV rules and the resulting design curves. It is suggested that the tolerance levels may be too high but it is imperative that information on actual fabrication tolerances is obtained before rational design rules can be produced.

ACKNOWLEDGEMENTS

The authors gratefully acknowledge the cooperation of their colleagues forming the shell buckling team at Imperial College. They also thank the Science Research Council for their financial support for the theoretical approaches discussed in the paper and the Department of Energy for their support of the experimental programme.

REFERENCES

1. Dowling, P.J., et al., "Experimental and Predicted Collapse Behaviour of Rectangular Steel Box Girders," Proceedings, International Conference on Steel Box Girder Bridges, The Institution of Civil Engineers, London, England, 1973.

2. Harding, J.E. and Hobbs, R.E. "The Ultimate Load Behaviour of Box Girder Web Panels", to be published in part B of the Structural Engineer.

3. Frieze, P.A., Dowling, P.J. and Hobbs, R.E. "Ultimate Load Behaviour of Plates in Compression", Steel Plated Structures an international symposium, Crosby Lockwood Staples, London, 1977.

4. Harding, J.E. and Dowling, P.J. "The Basis of the Proposed New Design Rules for the Strength of Web Plates and Other Panels Subject to Complex Edge Loading", Stability Problems in Engineering Structures and Components, Cardiff, September, 1978.

5. Coombs, M.L. "Aspects of the Elasto-plastic Behaviour of Biaxially Loaded Plates", Thesis presented in partial fulfilment of the requirements of the degree of Master of Science, Imperial College, London, 1975.

6. Basu, A.K. "The Theory of Orthotropic Plates Under Lateral Pressure and its Application to Corrugated Plates", Thesis presented in partial fulfilment of the requirements of the degree of Doctor of Philosophy, Imperial College, London, 1964.

7. Djahani, P. "Large Deflection Elasto-plastic Analysis of Discretely Stiffened Plates", Thesis presented in partial fulfilment of the requirements of the degree of Doctor of Philosophy, Imperial College, London, 1977.

8. Webb, S.E. "The Behaviour of Stiffened Plates Under Combined In-plane and Lateral Loading", To be submitted in partial fulfilment of the requirement of the degree of Doctor of Philosophy, Imperial College, London, 1979.

9. Harding, J.E., Hobbs, R.E. and Neal, B.G. "The Elasto-plastic Analysis of Imperfect Square Plates Under In-plane Loading", Proc. Instn. Civ. Engrs., Part 2, 1977, 63, March, 159-179.

10. Dowling, P.J. and Ho, T.K. "Effect of Initial Deformations on the Strength of Axially Compressed Cylindrical Panels", Conference on Significance of Deviations from Design Shapes, Institution of Mechanical Engineers, London, March, 1979.

11. Javaharian, H. and Dowling, P.J. "Non-linear Finite Element Analysis of Shell Structures", Engineering Structures Laboratories, Civil Engineering Department, Imperial College, CESLIC Report to be produced.

12. Dean, J.A. "The Collapse Behaviour of Steel Plating Subject to Complex Loading", Thesis presented in partial fulfilment of the requirements for the degree of Doctor of Philosophy, Imperial College, London, 1975.

13. Harding, J.E. "The Elasto-plastic Analysis of Imperfect Cylinders", Proc. Instn. Civ. Engrs, Part II, Vol. 65, London, December, 1978.

14. Odland, J. "Buckling of Slightly Curved Panels Subject to Axial Compression" Report by the Research Division of Det norske Veritas, No. 78-679, November, 1978.

15. Det norske Veritas, Rules for the Design, Construction and Inspection of Offshore Structures. Oslo, 1977.

TABLE 1 U.K. DEPT. OF ENERGY BUCKLING TEST PROGRAM

Type of Cylinder	Test No.	R/t	L/R	d/t_s	t_s/t	N	Scale	Laboratory	Scale
Orthogonally	1	200	0.4	8	1.0	20	1/20	*UC 1	
Stiffened Cylinder	2	"	"	"	"	40	"	UC 2	
OSC	3	"	1.11	16	"	20	"	UC 3	
Axial Compression	4	"	"	"	0.67	30	"	UC 4	GU 3	3/16
	5	"	"	"	1.0	40	"	UC 5	
	6	280	0.78	"	"	20	"	UCB 4	
	7	"	1.11	8	"	40	"	UC 6	
	8	"	"	16	"	20	"	UCB 3	
	9	"	"	"	"	40	"	UC 7	
	10	"	1.56	10	0.67	"	"	UCB 5	
	11	"	"	16	"	20	"	UCB 2	GU 2	1/8
	12	"	"	"	"	40	"	UCB 1	
	13	360	1.11	"	"	20	"	UC 8	
	14	"	"	"	"	40	"	UC 9	
	15	94	1.31	"	"	8	"	UC A	GU 1	3/8
Ring Stiffened	16	150	0.2	8	1.0	–	1/20	UC 10	IC 1	1/4
Cylinder	17	50	1.0	16	"	–	"	UC 11	
RSC	18	150	"	"	"	–	"	UC 12	
Axial Compression	19	250	0.2	8	"	–	"	UC 13	
	20	"	1.0	16	"	–	"	UC 14	IC 2	1/4
OSC	21	200	0.4	8	1.0	20	1/20	IC 15	
Eccentric Loading	22	"	"	"	"	40	"	IC 16	
	23	"	1.11	16	"	"	"	IC 17	IC 3	1/4
	24	360	"	"	"	"	"	IC 18	
RSC	25	250	0.2	8	1.0	–	1/20	UC 19	
Eccentric Loading	26	"	1.0	16	"	–	"	UC 20	IC 4	1/4
Unstiffened Cylinder	27.34	15-50	30-120	–	–	–	1/4	AMTE 1 - 8	

*UC = University College.
IC = Imperial College.
AMTE = Admiralty Marine Technology Establishment.
GU = Glasgow University.

R Cylinder radius
t Cylinder thickness
L Spacing of ring stiffeners
d Stiffener depth
t_s Stiffener thickness
N No. of longitudinal stiffeners

Fig. 1 Large scale cylinder model and loading plattens

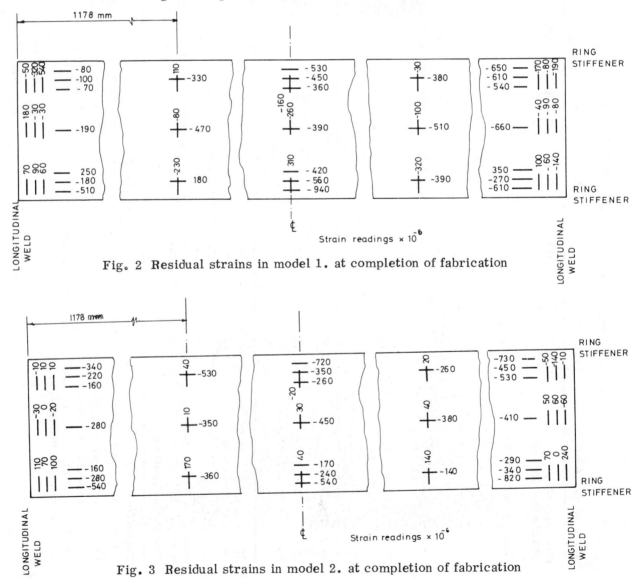

Fig. 2 Residual strains in model 1. at completion of fabrication

Fig. 3 Residual strains in model 2. at completion of fabrication

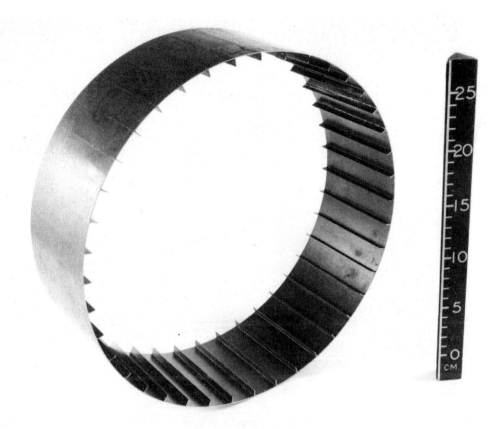

Fig. 4 Small scale cylinder model

Fig. 5 Small scale cylinder test rig

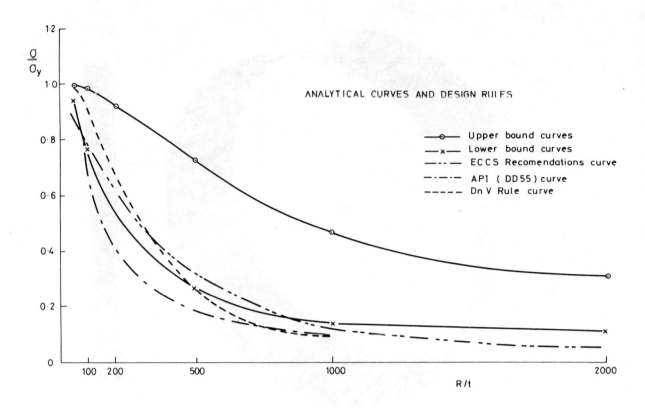

Fig. 6 The behaviour of ring stiffened cylinders under uniaxial compression

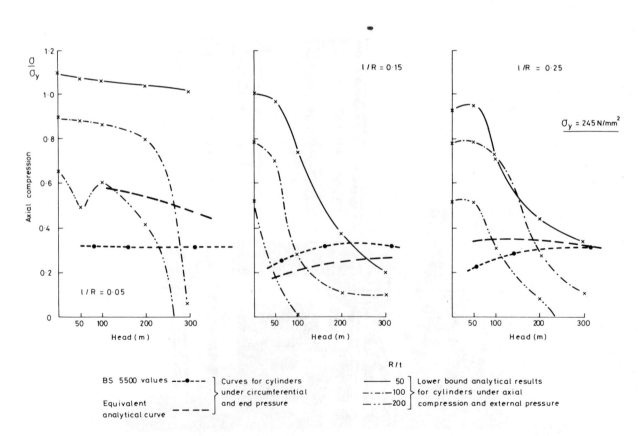

Fig. 7 The behaviour of ring stiffened cylinders under uniaxial compression and external pressure

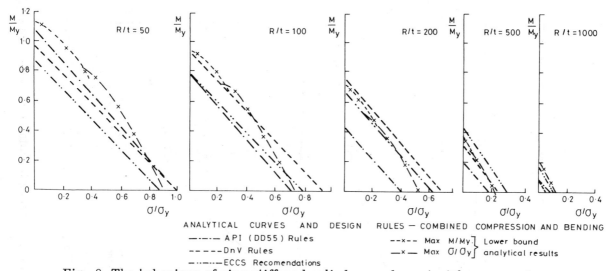

Fig. 8 The behaviour of ring stiffened cylinders under uniaxial compression and bending

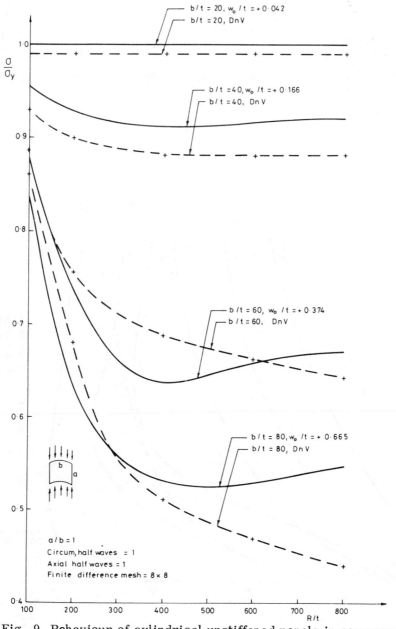

Fig. 9 Behaviour of cylindrical unstiffened panels in compression

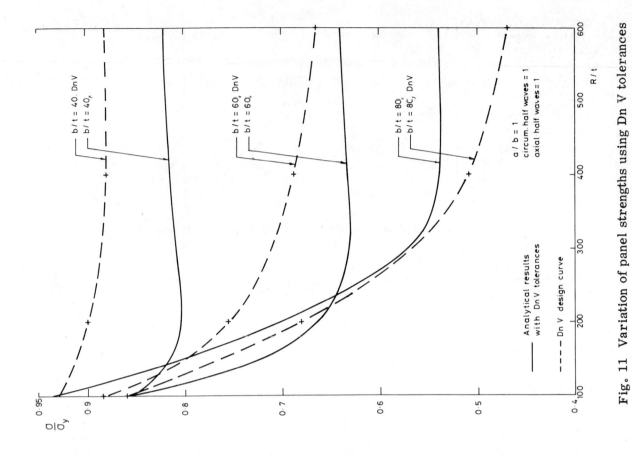

Fig. 11 Variation of panel strengths using Dn V tolerances

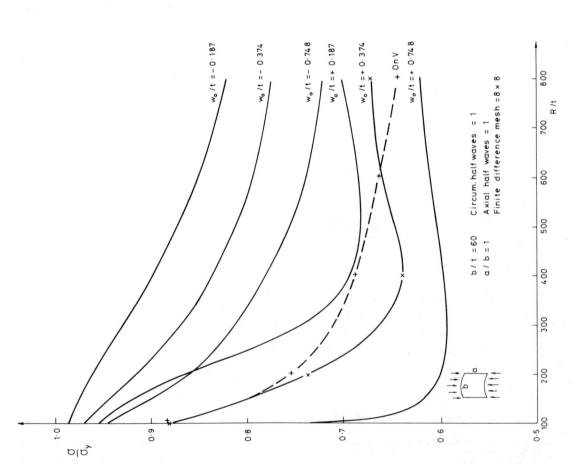

Fig. 10 Variations of panel strength with imperfection level

BOSS'79

Second International Conference on Behaviour of Off-Shore Structures

Held at: Imperial College, London, England
28 to 31 August 1979

PAPER 72

BUCKLING OF COMPRESSED, LONGITUDINALLY STIFFENED CYLINDRICAL SHELLS

A.C. Walker, DSc, PhD, MICE, MIMechE,

and S. Sridharan, MSc, PhD.

University College, London, U.K.

Summary

This paper reports on a series of tests on small scale models of welded steel stringer stiffened cylinders. The specimens were clamped at the ends and subjected to axial concentric loading. A description is given of the method of manufacture, measurement of residual stresses and measurement of initial imperfections. The test results are summarised and outlines are given of simplified methods for predicting the failure stresses.

Sponsored by: Delft University of Technology, The Netherlands
Massachusetts Institute of Technology, U.S.A.
The Norwegian Institute of Technology, Norway
University of London, England

Secretariat provided by: BHRA Fluid Engineering

Copyright: © BHRA Fluid Engineering
Cranfield, Bedford, England

NOMENCLATURE

E	:	Young's Modulus
E_{tan}	:	Tangent Modulus
N	:	Number of Stiffeners
b	:	Width of Panel
d	:	Depth of Stiffener outstand
ℓ	:	Length of shell, clear distance between the end fittings
n	:	An integer
r	:	Radius of the shell
t	:	Thickness of the shell
t_s	:	Thickness of the stiffener outstand
σ_ℓ^*	:	Local buckling stress
σ_ℓ	:	$\sigma_\ell^*/\sigma_{yc}$
σ_m	:	Failure stress of the plate
σ_o^*	:	Modified Overall buckling stress, buckling load ÷ shell area only
σ_o	:	σ_o^*/σ_{yc}
σ_p	:	Ratio of design stress from consideration of panel buckling to yield stress in compression
σ_{st}	:	Ratio of design stress from consideration of stiffener buckling to yield stress in compression
σ_{yc}	:	Yield stress in compression
σ_{yt}	:	Yield stress in tension
σ_{u1}^*	:	Collapse stress
σ_{u1}	:	$\sigma_{u1}^*/\sigma_{yc}$
σ_{u2}^*	:	Stress at which $E_{tan} = E/3$
σ_{u2}	:	$\sigma_{u2}^*/\sigma_{yc}$

1. INTRODUCTION

The frequent use of stiffened cylindrical shells as components of off-shore structures has made the sound understanding of their behaviour essential in marine structural design. The stiffening of the shells takes the form of longitudinal stringers and rings and the loading is a combination of axial compression and hydrostatic pressure. The present paper is concerned with an aspect of the problem, viz. stringer stiffened shells subject to concentric axial compression.

A wealth of literature already exists on stiffened cylindrical shells (1). **This is the result of their extensive use in the aeronautical industry** in the past four decades and more recently in aerospace structures. **The current research in that area is directed towards a prediction of buckling** loads of the cylinders on the basis of measured geometric imperfections (2). Despite the undoubted value of that research it may not be immediately relevant to the design of stiffened shells in off-shore construction, due to following considerations:

1. Its preoccupation with elastic buckling which is the usual mode of failure of very thin aluminium shells used in the aerospace industry; on the other hand the shells of the type used in off-shore construction are so proportioned that plastic yielding often precedes collapse.

2. The differences between the stress-strain characteristic of aluminium and that of steel, the material used in off-shore construction.

3. The differences in the mode of manufacture of the shells; the fabrication of shells by welding high strength steel plates leads to geometric imperfections and residual stress levels quite different from those in the integrally machined aluminium alloys.

4. The differences in the mode of proportioning of stiffening; the stiffeners in off-shore work are much less closely spaced and torsionally less rigid than those in aerospace. This makes an analysis of the local buckling taking into account the mutual interaction of stiffeners and shell panels very necessary (3).

Thus it appears necessary to continue the study of stiffened shells, now in the context of off-shore structural design. The small-scale tests reported in this paper are a contribution to this study and are motivated by an aim of arriving at suitable simple analysis methods to be applied in conjunction with more precise theoretical results for use in design. A first step in this direction has already been taken in the formulation of DnV rules.(4) These will be examined in the light of the present test results later in the paper.

2. FABRICATION OF TEST SPECIMENS

2.1 *The Material* The models were manufactured from thin sheet steel, fairly representative of that used in full-scale structures. Typical tensile stress-strain characteristic as obtained for a coupon of the material is shown in the Fig.1. Since the shells were being loaded in compression it seems logical to compare the collapse loads to the corresponding squash load, i.e the product of the compressive yield stress and the cross-sectional area. The measurement of the compressive properties of a thin strip of the material is complicated by the possibility of buckling masking the true value of the yield stress. The difficulty was overcome by nesting together several small samples of the strip bent through about 10° and then compressing them (5). Typically the compressive yield stress is about 10% greater than the corresponding tensile value as shown in Table 1.

2.2 *The Welding Technique* The cylinders were built up segmentally by rolling and machining the steel sheets into curved panels and welding these together with the stiffeners. The welding technique is described in Ref.6. Briefly it consists of clamping the component panels and stiffener plate between suitably shaped blocks of copper and traversing the junction under a T.I.G. welder electrode to form a fusion weld. The purpose of the copper blocks is to provide heat sinks so that there is little diffusion of heat in the sheets and thus to keep the residual stresses at a low level.

2.3 *Residual Stresses* The residual stresses were inferred from the change in distance due to welding between gauge points 50 mm apart. The measurements before and after welding on a panel were made using a standard Demec gauge in which the usual clock gauge had been replaced by a proximity transducer. This modification increased tenfold the sensitivity of the instrument and permitted the evaluation of the stress levels to a possible accuracy of 0.1% of σ_y. However, in fact the results obtained are probably of a lesser accuracy as small changes in the condition of the sheet surface did affect the readings taken. Figure 2 shows a typical set of results obtained, for one of the narrower panels. The average residual stresses ranged from 0.25 σ_y for the narrowest panel in the current test programme to 0.08 σ_y for the broadest panel.

2.4 *Imperfections* The geometric imperfections were measured by placing the shell, mounted into its end supports, on a turntable which was driven at a suitable speed relative to a row of linear displacement transducers. The output of these is fed into a computer programme to obtain (i) the best fit cylinder (ii) the relative imperfections and the Fourier coefficients which represent them and (iii) pictorial representations of the relative imperfections as exemplified by Figure 3 (a-b). A study of these figures shows that in general two types of imperfections occur most frequently:- the "overall" type which takes the form of one half-wave in the longitudinal direction and produces ovalities in the circumferential; and the extremely localised type of imperfection which occurs in the form of "blips" at each stiffener. The magnitudes of the imperfections were found to be well within the tolerances prescribed by the DnV rules.

2.5 *End-Fittings of the Shell* The ends of the shell were flat and parallel to each other to close tolerances and attached to the end fittings shown in Fig.4. These consisted of (i) flat plates to which the shell was butted at its ends, (ii) segmental steel pieces placed on the inside between the stiffeners and bolted on to the plates (i), and (iii) outer rings to which the inner segments were bolted through the shell. Thus very nearly a clamped end condition, membrane and bending, was achieved.

3. DETAILS OF TESTING PROCEDURE

3.1 *Loading under controlled end compression* An Instron Servo-Hydraulic machine of 2500 kN capacity specially commisssioned for the purpose was employed to test the specimens in axial compression. In order to be able to sudy the post-collapse unloading characteristic, the cylinders were tested under a condition of controlled end-shortening. Three displacement transducers (see Fig.5) were located round the cylinder at equal intervals to measure displacements between the outer rings at the top and bottom fittings of the specimen. The arrangement helped to monitor any lack of concentricity of loading in the initial stages which could then be corrected by a more accurate machining of the surfaces in contact. The machine was programmed such that the average of the displacements recorded by the three transducers was increased at a constant rate during tests. The rate of loading was extremely slow, thus permitting time for observation and recording of radial displacements indicated by clock gauges. Typically the machine was set to impose an end-shortening equivalent to five yield strains during a period of one hour.

3.2 *Bow Transducers* During the numerous trial runs before collapsing the test cylinders it was found that the record of the end-shortening as given by the direct-reading displacement transducers, even though generally reliable, was not always of sufficient accuracy. This was found to be due to the fact that the displacements were being measured between the end fittings of the specimen and thus involved an error due to the displacements and more importantly the small local rotations, of the fittings themselves. The need for a direct measurement of end-shortening between points on the specimen itself led to the fabrication of another kind of transducer shown schematically in Fig.6. The essential component of this was a bent strip of high strength spring steel to which strain-gauges were bonded. This was fitted between V-grooves cut on two horizontal flat bars one located near the top of the specimen and the other near the bottom. Each of the bars was pivoted at one end and in contact at the other with a hardened steel ball bonded to the shell surface. As the shell was compressed, the flexing of the steel strip resulted in a signal from the strain-gauges which could be related to the values of end-shortening by a previous calibration. Note that the values of lateral displacements of the strut are an amplification of its end-shortening and that the amplification-factor decreases with increasing curvature. Thus by suitably adjusting the initial curvature of the strip, it was possible to achieve any desired degree of accuracy. The device had the particular merit that its output is little affected by the local rotation of the shell surface in the vicinity of the steel balls. Three of these 'bow transducers' were arranged round the specimen in location mid-way between the direct-reading transducers round the shell. Thus in all there were six recordings of end shortening at various locations of the shell.

4. RESULTS AND DISCUSSION

A summary of the geometric and material characteristics and test results of the specimens is given in Table 2. The values σ_{u1} and σ_{u2} are stresses, averaged over the entire cross-section and rendered dimensionless by dividing them by σ_{yc}. The former is the collapse stress and the latter is the stress at which the stiffness of the specimen (as given by the tangent modulus) drops to $\frac{1}{3}$ of its initial value (i.e $\simeq E/3$). The ratio $\frac{1}{3}$ was chosen as it represents a sufficiently significant loss of stiffness to cause serious concern in practice and also because the corresponding stress can be read out with a small margin of error in any experimental set-up. Table 2 gives, in addition, the elastic buckling stresses and the DnV design strengths for the specimens. The cylinders UC1-UC9 form part of the current programme while B1-B5 have been reported on earlier (6-7).

4.1 *Characteristic Buckling Modes* As may be seen from Table 2, the number of stiffeners (N) was chosen either to be 20 or 40 in each specimen. In the ranges of r/t and ℓ/r investigated here, this choice resulted in two distinct modes of failure. The broad panelled cylinders (N=20) failed exclusively by local buckling in which the junction of the stiffeners and panels remained essentially straight until after the failure load had been reached. The narrow-panel shells (N=40) failed generally by overall buckling (i.e. a long wave mode with considerable radial movement of the stiffener-panel junctions) with some evidence of local buckling in the larger diameter cylinders (r/t = 280,360). The two buckling patterns are typified by the shells shown in Fig.7(a-b).

4.2 *Broad-Panelled Cylinders* The characteristic feature of these cylinders subject to axial loading is that they buckle locally between the stiffener/panel junctions. The buckling is dynamic and apparently the process is initiated while the shell is elastic. However, in all cases buckling results in short wave-length plastic mechanisms being developed. The mechanisms generally form a complementary pattern round the shell. With the type of control of the machine described earlier the dynamic buckling takes place at a constant value of end-shortening and is observed as a series of "snaps" each taking the cylinder away from one equilibrium path on to another.

This process is well illustrated by Fig.8(b) which shows the average stress-strain relationship of cylinder UC8, having the broadest panels. The tests here showed that the broader the panel, the more pronounced the tendency to violent snap buckling. Thus, in comparison to UC8 the cylinder UC1 which has a relatively narrow panel, though failing by local buckling showed no drop in the load at the onset of local buckling (see Fig.8(a)). The stress σ_{u2} indicates approximately the level at which the first local buckling occurred. This first buckle is probably due to one panel having a higher level of significant imperfection than the others. However, when such an event takes place there is redistribution of the load to the remaining panels and the local buckling process once begun tends to spread to the other panels fairly rapidly till all the panels are buckled, with only a small increase in the load. At that stage the plastic mechanisms continued to carry some load and the shell as a whole does not fail until yielding has spread to the stiffeners.

The theoretical prediction of the load at which the panels will buckle is fairly complex. Analysis developed by the senior author (3) has shown that the broad panels investigated in this programme have unstable post-buckling behaviour. Moreover, taking the torsional rigidity of the stiffeners into account it has been shown that there can be a number of buckling loads with very close values. The critical stress σ_ℓ (which is expressed as a fraction of σ_{yc}) shown in Table 2 is the lowest of these values. This closeness of critical loads, presages the possibility of mode interaction in the post-buckling behaviour and, together with the theoretically unstable behaviour indicates that the panels are highly imperfection sensitive. These results go some way towards explaining the marked reduction of the experimental first local buckling load ($\approx \sigma_{u2}$) from that predicted, σ_ℓ. However, before an accurate assessment of the panel buckling can be obtained further analysis is required, taking into account factors such as the measured imperfections and residual stresses.

Paradoxically, it seems likely, that an assessment of the maximum load for the cylinders, σ_{u1}, is relatively much simpler. In view of the imperfection sensitivity of the broad panels, and the high level of imperfections which are inescapable in off-shore work, it may be realistic to treat the curved panel as a plane one in the limit and thus neglect the additional strength contributed by the curvature. This means that in a conservative design approach the elastic loading curve of the curved panel may be replaced by that of the corresponding plate panel. Careful examination of the plastic mechanisms formed in the shell panels showed that these are remarkably similar to those observed in tests on flat plate panels (8). Thus the plastic unloading characteristics of the shell and plate panels are nearly the same. As a consequence the ultimate loads which are given to a close approximation by the intersection of the respective elastic loading and plastic unloading characteristics (9) can be expected to remain sensibly the same in the cases of curved panel and flat plate. The ultimate stress σ_m of a locally buckled long rectangular plate simply supported and restrained against inplane waving along its longitudinal edges is given by the expression (10)

$$\frac{\sigma_m}{\sigma_{yc}} = 0.43 + 0.72 \frac{\sigma_{cr}}{\sigma_y} - 0.15 \frac{(\sigma_m/\sigma_{yc})^2}{(\sigma_{cr}/\sigma_{yc})} \quad \{\sigma_m/\sigma_{yc} < 1\} \tag{1}$$

where

$$\sigma_{cr} = \frac{\pi^2 E}{3(1-\nu^2)} \cdot \left(\frac{t}{b}\right)^2 \quad \text{and} \quad b = \frac{2\pi r}{N}$$

Assuming that the stiffeners reach the full yield stress, the ultimate stress (in its dimensionless form) is given by

$$\sigma_{u1} = \left(2\pi r \frac{\sigma_m}{\sigma_{yc}} + Nd\right)/(2\pi r + Nd) \tag{2}$$

The good agreement of the values of σ_{u_1} obtained using the above procedure with the test results is illustrated in Fig.10(a). Also shown therein are the values of theoretical and experimental failure stress for a cylinder tested at a larger scale (11).

4.3 Narrow Panelled Cylinders As mentioned above, the collapse of the narrow panelled cylinders (n=40) is characterised by overall buckling. Local buckling, if it occurs at all, is not the primary cause of failure. On the other hand, local buckling can undermine the resistance of the structure to overall buckling and thus encourage premature failure in the overall mode. In the case of the cylinders with the narrowest panels (UC2 and UC5) local buckling was minimal, (see Fig.9(a) for the average stress-strain characteristic of UC5). In the case of cylinders with r/t = 280 (UC6,UC7,B1 and B5) there was some evidence of local buckling. However, this tended to have the form of local plastic mechanism and obviously the level of residual stresses were important in causing this localised yielding. In the case of UC9, which has the widest panel in this category, there was noticeable interaction of local and overall elastic modes of buckling (Fig.9(b)).

The elastic analysis for the local buckling stress showed that the post-buckling behaviour of these narrow-panelled cylinders should be stable. This is partially confirmed by the absence of any dynamic changes of local mode shapes during the tests. The overall elastic buckling stress σ_o (expressed in Table 2 as a fraction of σ_{yc}) has been calculated using a theory which "smears out" the stiffeners and assumes an equivalent orthotropic characteristic for the cylinder. The present analysis (12) takes into account the clamped boundary conditions which are used in this test series. The calculated values of σ_o are all considerably greater than unity which indicates that plasticity plays a large part in the failure of these narrow panelled cylinders.

A particularly simple design analysis approach has been formulated (13) at U.C.L to predict the maximum loads of cylinders which fail by overall buckling. This takes into account (i) initial imperfections in the overall mode of buckling (ii) the reduction in the effectiveness of the panel when it buckles locally due to applied and residual stresses (iii) shell orthotropic action. Preliminary calculations indicate that the narrow panelled shells in this programme will experience failure of the stiffener outstands at loads well below the maximum loads. The predicted stress at which this stiffener tripping should occur coincides fairly closely with σ_{u2}, i.e. the stress at which significant loss of stiffness is observed in the tests. Further development of the approach is in hand and it promises to provide a very simple and realistic design method.

Meanwhile, an alternative line of attack is to consider an interaction formula. If σ_o and σ_ℓ are indeed the proper parameters representing cylinder behaviour, the interaction of the corresponding modes of buckling with material yielding to cause collapse can be typified by an equation of the type

$$\left(\frac{A}{\sigma_o^*}\right)^n + \left(\frac{B}{\sigma_\ell^*}\right)^n + \left(\frac{C}{\sigma_{yc}}\right)^n = \left(\frac{1}{\sigma_{u1}^*}\right)^n \qquad (3)$$

The values of σ_{u1} obtained with

A = 2, B = C = 1 and n = 2

are shown in Fig. 10(b) compared with the corresponding experimental results; the agreement is reasonable.

4.4 *DnV Rules* DnV recommends procedures for the calculation of design strengths for panel buckling (σ_p) and lateral buckling of stiffener with shell, using appropriate partial factors of safety.

The calculation for the lateral buckling strength is based on neglecting the effects of curvature, i.e. treating the shell panel as a plane plate panel. If d/t_s exceeds a certain recommended value, then the characteristic strength to be used in the calculation must be a fraction of the yield stress, the fraction being dependant on the torsional buckling stress of the stiffener outstand. The procedure distinguishes between the plate-induced and stiffener induced failure. The lower of these values (σ_{st}) is the design strength in the lateral buckling of the shell, and is shown along with σ_p in Table 2. An examination of the design strengths together with the **test results reveals some possible lacunae in the present rules;** these are

1. The lower of the design strengths does not necessarily indicate the failure mode. This is illustrated by considering B4,B3 and B2 all of which fail by local buckling with nearly the same collapse load, while the rules predict stiffener failure in all cases with widely varying collapse strengths.

2. There seems to be no recognition of the fact that broad panelled cylinders tend to fail exclusively by local buckling.

3. No attempt has been made to evaluate realistically the interaction of local and overall buckling in narrow panelled cylinders.

4. There is little discussion of the possible loss of stiffness the structure may suffer due to imperfections, residual stresses or plastic yielding of the stiffeners in the course of its loading history. Thus, even though the recommended design strengths appear to be conservative in comparison to test results, the structure would have lost part of its stiffness at the projected design loads.

5. CONCLUSIONS

1. A series of tests on small scale models of welded steel stringer stiffened cylinders has been carried out which show two character--istic buckling modes.

2. The initial buckling of broad panel cylinders occurs suddenly and at loads significantly below the failure load. The process can be violent and theory and experiment indicate that such shells are imperfection sensitive.

3. A simplified model of the shell behaviour leads to an accurate assessment of the failure stress in the local buckling mode. However, the range of parameters covered in this test programme is probably insufficient to fully validate the approach.

4. Narrow panel cylinders fail in an overall buckling mode with significant loss of axial stiffness at loads well below the maximum load. This is explained by the onset of stiffener tripping and mention is made in the paper of a simplified theory which is currently being developed to predict this load and the failure load.

5. A simple interaction formula could meanwhile be used to calculate the failure stress of stiffened cylinders similar to those in this test series, i.e. with clamped ends.

6. The simple analysis methods are shown to give prediction for collapse stress which are in better agreement with the experimental results than those published by DnV. The latter, while being very conservative, give little insight to the actual shell behaviour.

7. More testing is required to enable the results of simple analysis, and those from computer based methods currently being developed elsewhere, to be compared with the actual behaviour of stiffened shells subject to hydrostatic as well as axial loading and with end conditions other than clamped. Particularly, there is need to study the failure modes of shells which have geometries intermediate between the broad-panel and narrow-panel models reported here. These intermediate shells may have significant coupling between the local and overall buckling modes which would result in failure loads which are very sensitive to imperfection levels.

6. ACKNOWLEDGEMENTS

The authors gratefully acknowledge the support of the Department of Energy in funding this programme of testing. Also, we are aware of the tremendous debt we owe the technicians at the Department of Civil Engineering, UCL for for their skillful, imaginative and enthusiastic work in preparing and testing the models. The tests were performed on an Instron machine which was provided by the Science Research Council as part of their support for the London Centre for Marine Structures and Materials.

REFERENCES

1. Singer,J. "Buckling,Vibrations and Post-buckling of stiffened metal cylindrical shells" BOSS, The University of Trondheim,1976.

2. Arbocz,J. and Williams,J.G."Imperfection Surveys on a 10ft Diameter shell structure", AIAA Journal,15, No.1.pp.949-956.(July 1977).

3. Syngellakis,S. and Walker,A.C. "Instability of curved panels",3rd I.U.T.A.M. Symposium on Shell Theory, Tsibilisi(1978). (In press).

4. Rules for the Design and Construction and Inspection of Off-shore structures, Det Norske Veritas (1977).

5. Walker,A.C. and Sridharan,S. "Experimental Investigation of the Buckling Behaviour of Stringer-stiffened Cylinders". Report OT-R7836 Department of Energy,1979.

6. Walker,A.C. and Davies,P.,"The Collapse of Stiffened Cylinders", Steel plated structures, An International Symposium,Ed. P.J.Dowling et.al Crosby Lockwood Staples,London, pp.791-801(1976).

7. Walker,A.C. and Kemp,K.O."Buckling of Stringer stiffened welded steel cylinders", Discussion on paper quoted as ref.1, BOSS'76.

8. Davies,P.,Walker,A.C. and Kemp.K.O. "An analysis of the Failure Mechanism of an Axially Loaded Simply-supported steel plate", Proc.Inst. of Civil Engineering,Part II,59,(1975).

9. Davies,P."The Collapse and Post-Collapse Behaviour of Steel Plating", Ph.D. Thesis, University of London (1977).

10. Walker,A.C. and Murray,N.W. " Analysis for stiffened Panel Buckling" Report 2/1974, University of Monash, Melbourne, Australia.

11. Nelson,H.M.Green.D.R. and Philips,D.V. "Buckling studies of Large Diameter stiffened Tubes", Interim Report No.2 Submitted to the Department of Energy (1978).

12. Syngellakis,S."Orthotropic Buckling of Cylinders with clamped Boundary Conditions", to be published.

13. Walker,A.C. and Sridharan.S. "Simplified methods of calculating the buckling loads of Stringer-stiffened steel shells" to be published.

©Crown Copyright and ©BHRA Copyright reserved.

TABLE 1
Comparison of tensile and compressive yield stresses

Nominal Steel Thickness (mm)	Tensile yield Stress σ_{y_t} (N/mm²)	Compressive Yield stress σ_{y_c} (N/mm²)	Percentage difference $[(\sigma_{y_c}-\sigma_{y_t})/\sigma_{y_t}] \times 100$
0.8	267.8	289.2	8.0
0.8	261.6	284.2	8.7
1.27	312.5	344.7	10.3
1.27	294	339.5	15.5

TABLE 2(a) BROAD PANELLED CYLINDERS
N=20   t = 0.81mm*

Test	r/t	d/t_s	ℓ/r	E kN/mm²	σ_{yc} N/mm²	ELASTIC THEORY		DnV RULES		TEST RESULTS	
						σ_o	σ_ℓ	σ_{st}	σ_p	σ_{u_1}	σ_{u_2}
1 (UC1)	200	8	0.4	210	324	12.63	2.01	0.45	0.44	0.82	0.26
2 (UC3)	200	16	1.11	207	322	11.18	1.78	0.33	0.43	0.76	0.36
3 (B4)	280	16	0.78	210	281	9.34	1.57	0.31	0.33	0.61	0.18
4 (B3)	280	16	1.11	210	284	5.27	1.59	0.25	0.33	0.60	0.42
5 (B2)	280	16	1.56	210	324	3.10	1.36	0.16	0.29	0.54	0.25
6 (UC8)	360	16	1.11	201	309	2.95	1.14	0.15	0.22	0.51	0.47

TABLE 2(b) NARROW-PANELLED CYLINDERS
N = 40   t = 0.81mm*

Test	r/t	d/t_s	ℓ/r	E kN/mm²	σ_{yc} N/mm²	ELASTIC THEORY		DnV RULES		TEST RESULTS	
						σ_o	σ_ℓ	σ_{st}	σ_p	σ_{u_1}	σ_{u_2}
7 (UC2)	200	8	1.11	210	320	4.38	2.90	0.28	0.75	1.05	0.56
8 (UC5)	200	16	1.11	203	338	17.11	1.83	0.39	0.72	1.04	0.84
9 (UC6)	280	8	1.11	211	311	2.47	1.76	0.15	0.57	0.65	0.35
10 (UC7)	280	16	1.11	211	311	8.06	1.52	0.32	0.57	0.86	0.60
11 (B1)	280	16	1.56	210	313	4.43	1.54	0.25	0.56	0.82	0.68
12 (B5)	280	10.7*	1.56	210	318	5.44	1.85	0.35	0.56	0.82	0.58
13 (UC9)	360	16	1.11	211	340	3.98	1.06	0.23	0.40	0.66	0.56

* t_s=t in all cases, except for B5, for which t_s=1.5t.

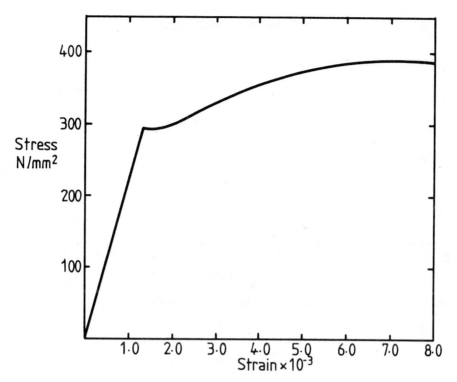

Fig. 1 A typical tensile stress-strain characteristic of the material

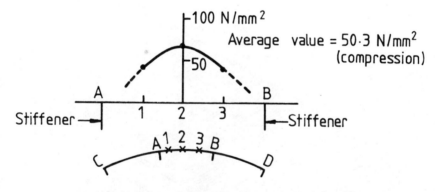

2(a) Residual stresses across a panel of UC7.
width of panel = 36mm

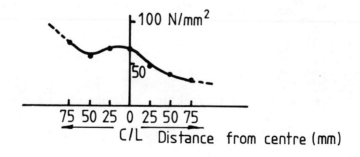

2(b) Peak residual stresses along the panel.

Fig. 2 Residual stress distribution in a panel of UC7

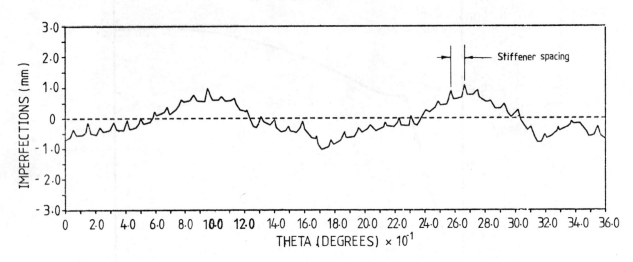

Fig. 3(a) Imperfections around the shell at mid-height of cylinder UC7

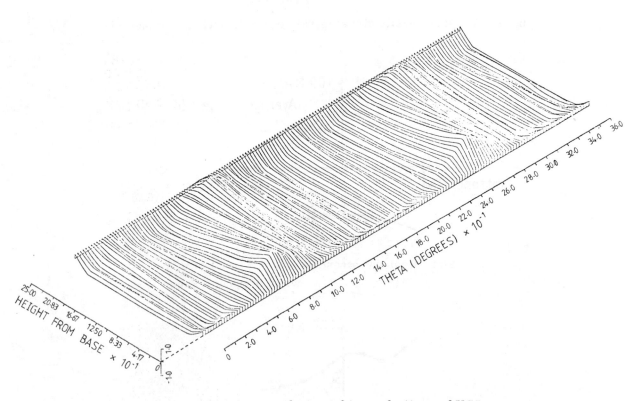

Fig. 3(b) A general view of imperfections of UC7

Fig. 4 A view of the end-fittings

Fig. 5 The test cylinder with the transducers

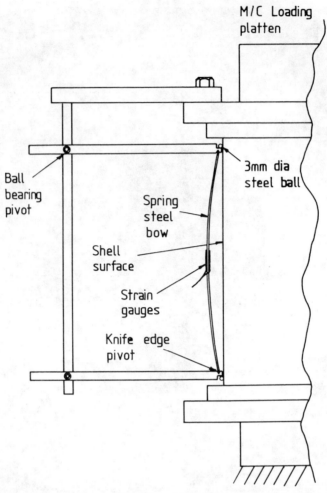

Fig. 6 Schematic diagram of the 'bow transducer'

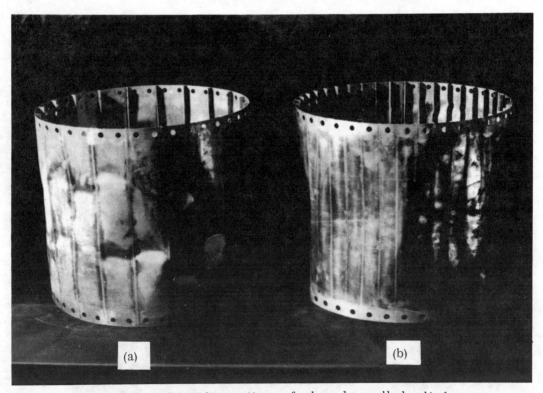

Fig. 7(a) Local buckling pattern of a broad-panelled cylinder

7(b) Overall buckling pattern of a narrow-panelled cylinder

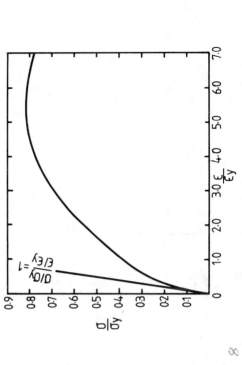

Fig. 8(a) Average stress-strain relationship of cylinder UC1

Fig. 8(b) Average stress-strain relationship of cylinder UC8

NARROW PANELLING

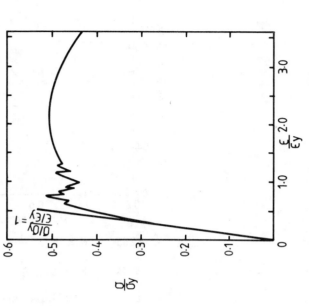

Fig. 9(a) Average stress-strain relationship of cylinder UC5

Fig. 9(b) Average stress-strain relationship of cylinder UC9

BROAD PANELLED

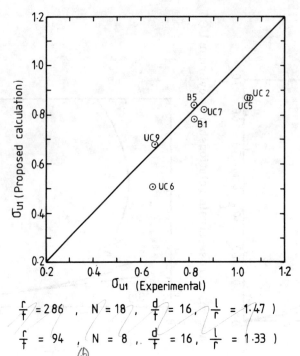

($\frac{r}{t} = 286$, $N = 18$, $\frac{d}{t} = 16$, $\frac{l}{r} = 1.47$)

($\frac{r}{t} = 94$, $N = 8$, $\frac{d}{t} = 16$, $\frac{l}{r} = 1.33$)

Fig. 10 (a) Collapse strengths as given by experiment and theory for broad-panelled cylinders.

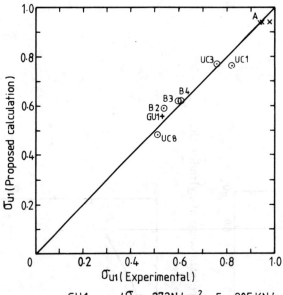

+ GU1 ($\sigma_{yc} = 272 \text{N/mm}^2$, $E = 205 \text{KN/mm}^2$, $N = 18$, $d/t = 16$, $r/t = 286$, $l/r = 1$)

× Specimens A ($\sigma_{yc} = 239 \text{N/mm}^2$, $E = 210 \text{KN/mm}^2$, $N = 8$, $d/t = 16$, $r/t = 94$, $l/r = 1.33$)
(Ref. 6)

Fig. 10 (b) Collapse strengths given by experiment and eqn. (3) for narrow-panelled cylinders.

ID# BOSS'79

Second International Conference
on Behaviour of Off-Shore Structures
Held at: Imperial College, London, England
28 to 31 August 1979

TRANSVERSE IMPACT ON BEAMS AND SLABS

I.C. Brown, BA, CEng, MICE and S.H. Perry, PhD, CEng, MICE, MIStructE

Imperial College, U.K.

Summary

A project is being undertaken to investigate the response of prestressed concrete slabs to impact. A brief survey is made of the methods of analysis which are available for the solution of impact problems. Previous work for the solution of transverse impact on beams, using the method of characteristics, is developed to include contact force phenomena. Using the same method, the governing relationships are found for transverse impact on slabs. After a review of previous experimental investigations of impact on concrete structural members, details are given of the testing programme for the present project.

Nomenclature

A	cross sectional area
$\bar{A}_s$	shear area
c_m	wave velocity in slab
c_o	wave velocity in beam
c_s	wave velocity in beam or slab
D	plate modulus
E	Young's modulus
G	shear modulus
h	thickness of slab
$\bar{h}_s$	shear thickness of slab
I	second moment of area of beam
	second moment of area per unit length of slab
M	bending moment in beam
M_r	bending moment per unit length in radial direction in slab
M_θ	bending moment per unit length in circumferential direction in slab
Q	shear force in beam
Q_r	shear force per unit length in radial direction in slab
q	uniformly distributed load per unit length on beam, uniformly distributed load per unit area on slab
r	radial distance
t	time
v	transverse velocity
w	transverse displacement
x	longitudinal distance
β_o	rotation of cross section
γ_o	shearing strain at neutral axis
κ	curvature of beam
κ_r	curvature in radial direction of slab
κ_θ	curvature in circumferential direction of slab
ν	Poisson's ratio
ρ	density
ω	angular velocity

1. Introduction

The usual conception of an impact is of a sharp, sudden loading such as occurs during the collision of two bodies. The abruptness of the loading may be quantified by comparing its duration with the period of the lowest natural frequency of the colliding bodies. In the case of loading durations which are long compared with the fundamental period it may then be possible to neglect wave propagation effects in calculating the response. Alternatively attention may be confined to the sharpness of the leading edge of the loading pulse. In the case of high loading rates, even though the total duration of the loading pulse may be long, it will then be necessary to include wave propagation effects to determine the initial response.

At sea impact may occur because of collisions between two ships, between a ship and a berthing structure or oil platform, or between waterborne debris or ice and a fixed or floating structure. In addition, on an oil platform, impact may be caused by objects dropped, for example, from a crane, or by a helicopter crashing onto the landing deck.

The progress of an impact may be described as follows. Assume that one body is at rest and is approached by another body with a certain initial momentum and kinetic energy. When the bodies collide, both of them deform locally, either elastically or plastically, and thereby generate a contact force. This force decelerates the contact zone of one body and accelerates the contact zone of the other. At this stage only part of the mass of either body is involved; also, the positions and types of support, if any, of the impacted body have no influence on its response. The effects of the contact force may be so severe that a local failure occurs in one or other body before there has been any significant transfer of momentum or energy. If local failure does not occur, the impinging body continues to decelerate and contact may end. However, if the loading pulse is short, the impacted body will have acquired both strain energy and kinetic energy, and, influenced by its conditions of support and its inertia, will continue to deform. This continued deformation and redistribution of strain energy and kinetic energy may lead to failure. On the other hand, if the loading pulse is of long duration, then a pseudo-static situation occurs and failure is determined by the static strength of either body.

An energy balance for an impact would equate the incident kinetic energy with the sum of the exit kinetic energies, vibrational strain energy, energy transmitted to supports, energy causing irreversible damage to either body and energy lost as heat or sound. It is important to note that for an impact to be successfully withstood it is not always necessary for the impacted body to be capable of absorbing large amounts of energy. Provided that a big enough contact force is generated which does not cause local failure but causes yielding or buckling of the striking body, then the majority of the incident energy will be dissipated in the striking body.

The authors are involved in a continuing investigation into the response of pre-stressed concrete slabs to impact loading. The rest of this paper will be devoted to a brief review of the theoretical and experimental techniques available for tackling impact problems, together with a more detailed description of the methods proposed in the current project. At present only preliminary results are available from the experimental programme. These will be presented when the main series of tests has been completed.

2. Review of Methods of Analysis

2.1 Introduction

An excellent account of most of the available methods of analysis is contained in Goldsmith (Ref.1). Further information is in Abramson, Plass and Ripperger (Ref.2) and Sahlin and Nilsson (Ref.3). In the brief review which follows attention is confined to transverse elastic impact on beams or slabs.

2.2 Exact Solutions

The three-dimensional equations of elasticity may be solved for the special case of an infinite train of sinusoidal waves travelling in a bar of infinite length and uniform cross-section. Suitable periodic displacements are substituted into the equations of elasticity, and the frequency equation is obtained by satisfying the boundary conditions for the stresses on the surface of the bar. The results, the Pochhammer-Chree relationships, are extremely complicated and cannot be applied to finite beams. However, they serve as a standard for comparison with approximate theories. The two-dimensional equations of elasticity for thin plates, either plane stress or plane strain, may also be solved for the special case of an infinite train of waves in an infinite, uniform plate.

2.3 Simple Approximate Theories

In the simplest method only conservation of energy is considered. The beam or slab is assumed to be massless and is replaced by a spring with a stiffness equal to the static stiffness of the system. The maximum deflection caused by an impact is obtained by equating the initial kinetic energy of the striker with the maximum strain energy stored in the spring. An allowance may be made for the loss in potential energy of the striker. This method does not give results in good agreement with experimental observations (Ref.1).

As an improvement, the impact is assumed to be completely inelastic with the striker and a fraction of the beam mass attaining a common velocity immediately after contact. The dynamic displacement curve of the beam or slab is assumed to be similar in shape to the static curve. The appropriate fraction of the beam mass is found by evaluating the kinetic energy of the beam for its assumed displacement. The common velocity after impact can then be evaluated, and the equations of motion are readily solved to give the deflection at the impact point. This method gives reasonable results for the maximum displacement, especially if the impact is not too sharp. However, reliable estimates of stresses and strains are not obtained.

2.4 Solution in terms of Normal Modes

Solutions may be sought to the approximate one-dimensional equations normally used by engineers for the description of free vibrations in beams or slabs. In the case of a beam the displacement may be written as

$$w(x,t) = \sum_1^\infty X_i(x) q_i(t)$$

Substitution into the equation

$$EI \frac{\partial^4 w}{\partial x^4} + \rho A \frac{\partial^2 w}{\partial t^2} = 0$$

leads to the determination of the X_i as a sum of circular and hyperbolic sine and cosine terms. The X_i represent the normal modes of the beam and are dependent on the boundary conditions. As the solution is in terms of the free vibrations of the system, the unknown contact force is replaced by a suitable initial velocity condition which determines the form of the q_i.

The estimates of deflection obtained by this method are good. The infinite series which represents the displacement converges quite quickly. However, the series representing the stress, which is proportional to the second derivative of the displacement, converges very slowly. For this reason, reliable values for the stresses are difficult to obtain.

The equation for free vibration may be modified to include the effects of shear deformation, rotatory inertia and warping. These corrections become more important as the sharpness of the impact increases.

2.5 Method of Characteristics

A completely different approach is to use the method of characteristics to obtain a solution. This method, which is dealt with in more detail in a later section, treats the problem as one of wave propagation. Again the one-dimensional approximate equations of motion are used as the basis for the solution, and corrections for shear, rotatary inertia and warping may be included.

2.6 Contact Phenomena

Further progress can be made if more consideration is given to local deformations in the contact zone. All theories of contact use the concept of the "approach", that is, the relative compression, between the two bodies. The most successful theory, due to Hertz, assumes static and elastic conditions. It results in a law in which contact force is proportional to approach to the power 3/2. If vibrations in the striker can be neglected, the Hertz law, or another suitable relationship, can be used directly to describe the motion of the striker. The equation of motion for the beam or slab may also be written to include the effect of the unknown contact force. From these two relationships it is then possible to obtain a solution for both the beam or slab displacement and the contact force. The results are in closed form and numerical iterative procedures are required for the solution.

2.7 Numerical Methods

Many practical problems are too complicated for solution by the more sophisticated analytic techniques. In these cases, if a detailed description of the response is required, resort must be made to the general numerical methods now available. The two usual methods employed are those of finite elements and finite differences. Sahlin and Nilsson (Ref.3) review finite element methods and details of a finite difference method are given by Carlton and Bedi (Ref.4).

3. Method of Characteristics

3.1 Introduction

A general introduction to the method of characteristics may be found in Abbott (Ref.5). The development of the method for the investigation of both elastic and plastic wave propagation in beams is traced in the review article by Abramson et al (Ref.2).

3.2 Transverse Elastic Impact on Beams

The analysis presented here follows the previous work except that a time varying uniformly distributed load is included in the equations to account for the unknown contact force. An element of the beam is shown in Fig.1 where the positive directions of the variables are indicated. Shearing deformations and rotatory inertia are included in the analysis but warping is neglected. As a result of the shearing strain, the plane of each displaced cross-section is not at right angles to the displacement curve of the neutral axis. The shearing strain at the neutral axis is given by

$$\gamma_o = \frac{\partial w}{\partial x} - \beta_o \qquad (1)$$

The constitutive relationships and the equations of motion and compatibility are expressed as first order partial differential equations in terms of the following six variables: bending moment M, shear force Q, angular velocity of section $\omega (= -\partial \beta_o/\partial t)$, transverse velocity v $(= \partial w/\partial t)$, approximate curvature $\kappa (= -\partial \beta_o/\partial x)$ and shearing strain γ_o.

The resulting equations are

$$\frac{\partial M}{\partial t} - EI \frac{\partial \kappa}{\partial t} = 0 \qquad (2a)$$

$$\frac{\partial Q}{\partial t} - G\bar{A}_s \frac{\partial \gamma_o}{\partial t} = 0 \qquad (2b)$$

$$\frac{\partial M}{\partial x} - \rho I \frac{\partial \omega}{\partial t} = Q \qquad (2c)$$

$$\frac{\partial Q}{\partial x} - \rho A \frac{\partial v}{\partial t} = -q \qquad (2d)$$

$$\frac{\partial \kappa}{\partial t} - \frac{\partial \omega}{\partial x} = 0 \qquad (2e)$$

$$\frac{\partial \gamma_o}{\partial t} - \frac{\partial v}{\partial x} = \omega \qquad (2f)$$

The quantity $\bar{A}_s$ is a proportion of the cross-sectional area and arises from the non-uniform distribution of shear across the section. In addition to these six equations there are a further six equations which express the total derivative of the variables in terms of their partial differentials, of the form

$$dM = \frac{\partial M}{\partial x} dx + \frac{\partial M}{\partial t} dt \tag{3}$$

Taken together, these equations may be considered as a set of twelve simultaneous equations for the determination of the partial derivatives of the variables. A characteristic is a line along which the partial derivatives cannot all be uniquely determined. The characteristics in the x - t plane are found by equating to zero the determinant of the coefficient matrix of the set of simultaneous equations. The characteristics are given by

$$dx = 0 \text{ (twice)}$$

$$\frac{dx}{dt} = \pm\sqrt{\frac{E}{\rho}} = \pm c_0$$

$$\frac{dx}{dt} = \pm\sqrt{\frac{G\bar{A}_s}{\rho A}} = \pm c_s$$

Along the characteristics the partial differential equations for the motion of the beam reduce to ordinary differential relationships as follows:

$EI\, d\kappa - dM = 0$	along $dx = 0$
$G\bar{A}_s d\gamma_0 - dQ = 0$	along $dx = 0$
$dM - \rho I c_0 d\omega = Q c_0 dt$	along $dx = +c_0 dt$
$dM + \rho I c_0 d\omega = -Q c_0 dt$	along $dx = -c_0 dt$
$dQ - \rho A c_s dv = (\rho A c_s \omega - q) c_s dt$	along $dx = +c_s dt$
$dQ + \rho A c_s dv = (\rho A c_s \omega + q) c_s dt$	along $dx = -c_s dt$

Previously these relationships have been used to follow the motion of a beam subjected to an assumed variation of one of the variables, for example, the transverse velocity at the impact point. However, now that uniformly distributed loads are included, the relationship can be combined with an appropriate contact force law, and the solution obtained for the unknown contact force as well as the other variables.

As an illustration, an example quoted in Goldsmith (Ref.6) has been re-worked using the method of characteristics. The example concerns the central impact of 12.7mm diameter steel sphere on a 12.7mm x 12.7mm x 762mm simply supported steel beam. The Hertz theory for contact force has been used and the differential relationships integrated numerically along the characteristics. The step length for the integrations was one microsecond. Unfortunately, a convergence test using other step lengths has not yet been carried out. A comparison of the contact force computed by this method and by 65 harmonics of the normal mode solution is shown in Fig.2. The agreement is reasonable. In Fig.3 are shown the distributions of bending moment, shear force, transverse velocity and angular velocity just after contact ends, calculated by the method of characteristics.

3.3 Transverse Elastic Impact on Slabs

The solution for transverse impact on an isotropic slab may be obtained in the same way as for a beam. Fig.4 shows an element of the slab and the forces acting on it. The response is axisymmetric and the resulting constitutive relationships and equations of motion and compatibility are given below in terms of first order partial differential equations. In these equations bending moments, shears and moments of inertia are evaluated per unit length of slab.

$$\frac{\partial M_r}{\partial t} - D\frac{\partial \kappa_r}{\partial t} - \nu D\frac{\partial \kappa_\theta}{\partial t} = 0 \tag{4a}$$

$$\frac{\partial M_\theta}{\partial t} - \nu D\frac{\partial \kappa_r}{\partial t} - D\frac{\partial \kappa_\theta}{\partial t} = 0 \tag{4b}$$

$$\frac{\partial Q_r}{\partial t} - G\bar{h}_s\frac{\partial \gamma_o}{\partial t} = 0 \tag{4c}$$

$$\frac{r\partial M_r}{\partial r} - \rho I r \frac{\partial \omega}{\partial t} = rQ_r - M_r \tag{4d}$$

$$r\frac{\partial Q_r}{\partial r} - \rho h r \frac{\partial v}{\partial t} = -rq - Q_r \tag{4e}$$

$$\frac{\partial \kappa_r}{\partial t} - \frac{\partial \omega}{\partial r} = 0 \tag{4f}$$

$$\frac{\partial \gamma_o}{\partial t} - \frac{\partial v}{\partial r} = \omega \tag{4g}$$

$$\frac{\partial \gamma_o}{\partial t} - r\frac{\partial \kappa_\theta}{\partial t} - \frac{\partial v}{\partial r} = 0 \tag{4h}$$

The term $\bar{h}_s$ arises from the non-uniform shear distribution across the section.

These eight equations together with the further eight equations expressing the total variation of the variables in terms of their partial differential coefficients may be used to determine the characteristics. In the r-t plane the characteristics are given by

$$dr = 0 \text{ (four times)}$$

$$\frac{dr}{dt} = \pm\sqrt{\frac{D}{\rho I}} = \pm\sqrt{\frac{E}{(1-\nu^2)\rho}} = \pm c_m$$

$$\frac{dr}{dt} = \pm\sqrt{\frac{G\bar{h}_s}{\rho h}} = \pm c_s$$

Along the characteristics the following ordinary differential relationships are found:

$Dd\kappa_r + \nu Dd\kappa_\theta - dM_r = 0$	along $dr = 0$
$\nu Dd\kappa_r + Dd\kappa_\theta - dM_\theta = 0$	along $dr = 0$
$G\bar{h}_s d\gamma_o - dQ_r = 0$	along $dr = 0$
$rd\kappa_\theta = \omega$	along $dr = 0$
$rdM_r - \rho I c_m r d\omega = (\rho I c_m \nu \omega - M_r + rQ_r)c_m dt$	along $dr = +c_m dt$
$rdM_r + \rho I c_m r d\omega = (\rho I c_m \nu \omega + M_r - rQ_r)c_m dt$	along $dr = -c_m dt$
$rdQ_r - \rho h c_s r dv = (\rho h c_s r\omega - Q_r - rq)c_s dt$	along $dr = +c_s dt$
$rdQ_r + \rho h c_s r dv = (\rho h c_s r\omega + Q_r + rq)c_s dt$	along $dr = -c_s dt$

3.4 Summary

For a problem which is essentially one of wave propagation, in which the response is determined at first only by the initial conditions and not by the boundary conditions, it is very attractive to use the method of characteristics. It is possible to include the effects of boundaries to extend the use of the method into the realm where the normal mode method also becomes appropriate.

Although the analysis shown here is for elastic waves, the method of characteristics may be used to investigate the effects of plasticity both in the contact zone and also in the beam or slab itself. Indeed, one of the earliest developments of the method was for the study of plastic waves in a beam of strain rate dependent material.

Other extremely powerful numerical methods, the finite difference and finite element methods, are available for the solution of impact problems. They have the advantage of being applicable to a wider range of problems and are capable of including the effects of steel reinforcement in concrete construction and the development of cracking. Against this must be set their extra complexity and computing cost. Also, the relationships which have been derived by the method of characteristics help to clarify the physical processes involved in an impact. For loading with high rates of strain when material properties are not accurately known this physical understanding is perhaps the best that can be expected.

4. Review of Previous Experimental Work

A review of experimental work relating to impact on reinforced and prestressed concrete members may be found in an article by Skov and Olesen (Ref.7). The most extensive series of the tests in which the impact loading was caused by a dropped mass was carried out by Bate (Ref.8). However, since the review article another important experimental programme has been undertaken by Kavyrchine (Ref.9).

In the earlier series of tests by Bate measurements were made of the height of fall of the dropped mass, allowing the initial velocity to be calculated, and of beam displacements. Also, high speed photography was used to record the impact. In the later tests by Kavyrchine support reactions were measured as well as beam displacements and steel and concrete strains. Again, use was made of high speed photography. In neither series of tests was the contact force measured. Neither were the conditions in the contact zone investigated thoroughly, although both authors recognize the importance of local effects. The possibility of high shear forces leading to premature failure is also noted. Both authors estimate the proportion of the incident energy that is absorbed by the beam after impact. This estimate is compared with the measured energy absorbed by the beam in reaching a similar deflection, or failure as the case may be, in a static loading test. Simple approximate theories are used to estimate the energy absorbed during the impact. However, provided that the impact is not too sharp, that vibrations in the striker can be neglected and that local damage in the contact zone is not too extensive, then the results obtained are reasonable, especially if allowance is made for the increased resistance of concrete and steel at high rates of strain.

5. Experimental Investigation of Impact on Slabs

The present project is concerned with the response of prestressed concrete slabs to impact. In the experimental programme the impact loading is provided by means of a freely falling mass. Only moderate approach velocities, up to about 10 m/s, can be attained by this method because of limitations of space.

The test slabs are approximately 1/4 scale, being 1.5m square on plan; ordinary concrete mixes are used with a maximum aggregate size of 10mm. Prestressing is by means of 5mm or 7mm wires which are post-tensioned and grouted.

It is intended to measure the contact force during the impact. To this end a short cylindrical load cell has been manufactured from high strength steel. This load cell is placed at the front of the impacting mass. Semi-conductor strain gauges are used in the load cell. These gauges have a gauge factor of 100, compared with a gauge factor of 2 for wire or foil strain gauges. Consequently, less amplification is required to obtain a readable signal. There is evidence that the elastic modulus of steel remains unchanged as the rate of loading is increased (Ref.10). For this reason a static calibration of the load cell should be adequate for the dynamic loading.

In addition to recording the contact force it is intended to measure steel strains and slab displacements during the impact. The transient signals from transducers will be transferred through a multiplexer and high speed A-D (analogue to digital) converter into the core store of a microcomputer to await subsequent processing.

Since it is expected that local effects will be more important in the response of slabs to impact than in the response of beams, special attention will be paid to conditions near the contact zone. Also it is hoped to simulate both "hard" and "soft" impacts. A hard impact occurs when the striking mass is rigid: a soft impact occurs when the striking body yields and itself absorbs the majority of the incident energy. Sørensen (Ref.11) points out that most ship collisions are in the second category and that the final response of the impacted body depends on its static resistance. However, the initial response in such a collision also requires investigation. Of course, many impacts will be intermediate in nature with, perhaps, both bodies yielding and absorbing large amounts of energy.

Apart from these special considerations the parameters to be varied during the experimental programme include the approach velocity, the properties of the slab concrete, the layout and amount of prestressing and the influence of secondary reinforcement. After completing the tests on flat slab specimens it is intended to extend the project by investigating the response of concrete shell elements.

6. Conclusions

The design of a structure for impact loading presents many problems. For a particular postulated impact the acceptable extent of damage must be decided. This will be either

    a) superficial,

    b) considerable, providing that the structural member remains intact,

    c) destruction of one part of the structure providing that the integrity of the remainder is maintained, or

    d) total loss.

Furthermore, similar consideration must be given to the impacting body. Bearing in mind the probability of occurrence of a particular impact, the engineer must weigh the possible loss of life, economic loss and political or military consequences against the cost of designing the structure to provide a certain level of protection. The final problem is to estimate, by analysis or model testing, the structural behaviour during the impact.

Previous work has helped to explain the structural effects of impact on prestressed and ordinary reinforced concrete beams. It is hoped that the present project will supplement this previous work and improve the understanding of the response of slabs.

7. Acknowledgements

Funds for this project have been provided by the Science Research Council and are gratefully acknowledged.

8. References

(1) Goldsmith, W: "Impact". Edward Arnold (1960)

(2) Abramson, H.N., Plass, H.J. and Ripperger, E.A.: "Stress wave propagation in rods and beams". Advances in Applied Mechanics, $\underline{5}$, pp.111-194. (1958).

(3) Sahlin, S. and Nilsson, L: "Theoretical analysis of stress and strain propagation during impact". Materiaux et Constructions, 8, No. 44, pp. 88-101. (March-April 1975).

(4) Carlton, D. and Bedi, A: "Theoretical study of aircraft impact on reactor containment structures". Nuclear Engineering and Design, 45, No. 1, pp.197-206. (January 1978).

(5) Abbott, M.B: "An introduction to the method of characteristics". Thames and Hudson, (1966).

(6) Goldsmith, W: op. cit., p.119.

(7) Skov, K. and Olesen, S: "Impact resistance of reinforced and prestressed concrete members". Materiaux et Constructions, 8, No.44, pp.116-125, (March-April, 1975).

(8) Bate, S.C.C: "The effect of impact loading on prestressed and ordinary reinforced concrete beams". National Building Studies, Research Paper 35. H.M.S.O. (1961).

(9) Kavyrchine, M: "The effect of impact on reinforced concrete". (Effets de choc sur le béton armé). Annales de l'Institut Technique du Batiment et des Travaux Publics, Serie: Beton No. 173. Supplement to No. 356, pp.134-179. (December 1977). (In French).

(10) Mainstone, R.J.: "Properties of materials at high rates of straining or loading". Materiaux et Constructions, 8, No. 44. pp. 102-116. (March-April 1975).

(11) Sørensen, K.A.: "Behaviour of reinforced and prestressed concrete tubes under static and impact loading". Proc. 1st International Conf. on Behaviour of Off-Shore Structures (BOSS '76). Paper 4.16. Organised by Delft Univ. of Tech., M.I.T., Univ. of London and Norwegian Inst. of Tech., Trondheim (August 2nd-5th, 1976).

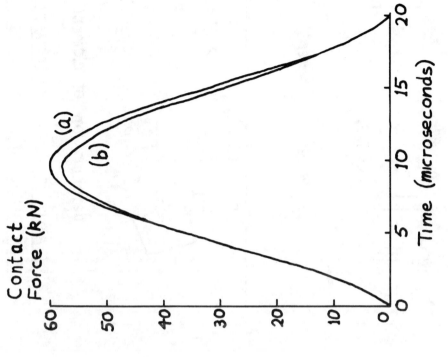

Fig. 2 : Comparison of contact force calculated by different methods.

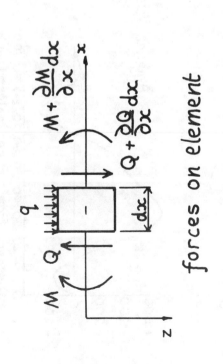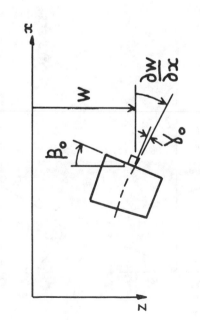

Fig. 1 : Element of beam

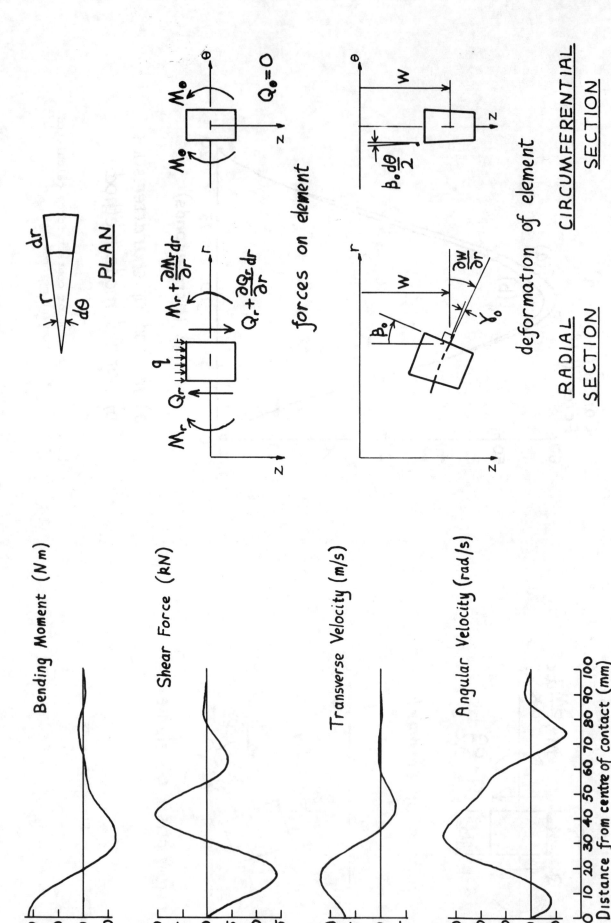

Fig. 3 : Variation of internal forces and velocities along beam just after contact ends.

Fig. 4 : Element of slab.

BOSS'79

Second International Conference on Behaviour of Off-Shore Structures

Held at: Imperial College, London, England
28 to 31 August 1979

PAPER 74

STRATEGY FOR MONITORING, INSPECTION AND REPAIR FOR FIXED OFFSHORE STRUCTURES

Peter W. Marshall

Shell Oil Company, U.S.A.

Summary

The philosophy of making trade-offs between cost and risk permits rational allocation resources in offshore energy development projects, provided the technical and economic considerations are formulated so as to include indirect human and social consequences. Previous applications of this approach to the selection of design criteria for storm loads, earthquakes, and fatigue in the case of new structures are reviewed. The paper then extends its application to rational decision-making in areas of structural monitoring, inspection, and repair, for existing structures. The roles of structural redundancy and secondary protective measures are illustrated. Current inspection practices and monitoring developments are discussed, along with results obtained over the last several years.

Introduction

There are many risks attendant upon a venture into offshore oil activity--dry holes, political uncertainty, blowouts, collisions, storm damage and earthquakes to name a few. Some of these risks are unavoidable and may require difficult tradeoffs. However, the risks related to structural design can be minimized by the expenditure of money and technical effort, and are amenable to rational value analysis.

Ideally, the objective of both the engineering effort and the rational value analysis should be directed towards minimizing C in the following equation (ref. 1), regardless of who pays:

$$C = C_I + C_D + \sum_F pr(F) \cdot C_F \quad \ldots \ldots (eqn. 1)$$

where: C is the total cost of a project or structure which is built to produce a given benefit or return

C_I is the initial cost

C_D represents deferred operating costs, capitalized or reduced to present value

F represents an undesirable occurrence or failure

$pr(F)$ is the corresponding probability or risk of failure event F

C_F is the cost (inclusive of <u>all</u> disadvantages) which follows event F, also capitalized at present dollar value.

All risks are internalized; that is, the cost of failure includes expected oil spill clean up expenses, and liability for damages, injury, and possible deaths. The private business risks--loss of income and reserves; platform salvage and replacement costs-also provide substantial incentive for careful evaluation. Total failure cost ranges from 2 to 5 times the initial cost of an API-level platform-- e.g. $50 million for a $10 million structure, or $500 million for a $250 million structure.

Strength vs. Load

Occasionally, the offshore civil engineer is asked to evaluate situations involving a finite measure of risk. When this evaluation is made on the basis of a conventionally computed safety factor, there is often a tendency to imply that a structure having a safety factor just slightly larger than some arbitrary value is perfectly safe, while one having just slightly less is doomed to failure. Unfortunately, such a black and white answer to the question, "Will this structure fail?" cannot be provided. It becomes necessary to deal with the various shades of grey which are in the realm of probability.

A computed safety factor represents only the ratio of <u>nominal</u> strength to the <u>assumed</u> load. However, structural failure occurs when <u>actual</u> strength is exceeded by the <u>actual</u> load. Both strength and load are random variables, whose distributions <u>include</u> bias and uncertainty, which we shall now examine.

The shaded band in Figure 1 contains various a priori estimates of the relationship between the nominal design strength and the actual failure load of typical Gulf of Mexico drilling platforms subjected to storm forces. The lower tail of the statistical distribution is shown. Among the various models used to evaluate failure modes for offshore structures (ref. 2), the ductile-redundant model, which yields a median failure load of about twice the design load, comes closest to being consistent with survival experience of older platforms subjected to hurricanes.

Of this reserve strength factor of 2.0, some comes from intentional safety factors and safe-side bias in the design criteria. For the example structure, about ten percent was provided by designer's "lagniappe"; however, a major portion, a factor of 1.4, may be attributed to the difference between elastic single element behavior versus ultimate strength system behavior. The indicated range of uncertainty includes that introduced by calculation procedures for applying wave forces to the structure, as well as inherent variability in strength.

The upper tail for random occurrence of extreme storm forces can be found in Figure 2, with the shaded band representing the 95% confidence limits (ref. 2). These forces have been normalized to the present (1978) API reference level design wave for the Gulf of Mexico.

The annual probability of failure, or risk rate, is obtained by evaluating the overlap of these two distributions (ref. 2, 3). The notional computed risk rate for storm losses comes out to 0.1% to 1% for deepwater platforms in the Gulf of Mexico, designed to just barely meet API criteria. This range appears to conservatively overstate the risk when compared to the historical loss rate of 0.3% to 0.6%, or with the lower risk rates for current designs computed by other investigators (e.g. ref. 4). By way of comparison, all-risk marine insurance rates reflect ship casualties of about 2% annually.

Similarly, the lifetime probability of failure is obtained by evaluating the overlap of the strength for loads experienced during the structure's lifetime.

Now we apply equation 1.

The dollar risks, taken over the platform life, are combined with first cost as shown in Figure 3, with the shaded band covering a range of assumed failure costs, or different degrees of risk aversion versus optimism. Note that increasing the design level relative to API decreases the risk and increases the construction cost. Under optimistic assumptions, the API level represents the least total cost (point A). Under more risk-averse assumptions, a higher design force can be justified (point B), and some offshore operators elect to exceed the API guidelines. In so doing, they are making a trade-off which puts more steel into platforms and leaves fewer resources available for finding new energy. However, the optimum point is not sharply defined, and there is room for variation between the design criteria of different companies.

In the Gulf of Mexico, hurricanes are infrequent events for which there is sufficient warning that offshore personnel can be evacuated and the wells secured against leakage in the event of structural collapse. This is done as a practice of long standing. For periods while the platform is manned and the wells are flowing, the maximum winter storm produces less than half the API reference load, and the corresponding risks of collapse due to overload are negligibly small. Such platforms may be termed fail-safe while manned, and the economic optimization of Figure 3 may be done without confronting any socially unacceptable risks.

Figures 1 and 2 also show resistance and intensity distributions for earthquakes. In the case of storm loading, we are accustomed to dealing with direct forces (e.g. due to waves) which, if not resisted, will cause collapse. Earthquakes, on the other hand, do not produce direct forces on the structure, although most simplified design codes treat them this way. Rather, the earthquake ground motions transmitted through the stiffness of the foundation elements and superstructure produce distortions of the members, which are in turn related to forces. If the structure is properly detailed to hang together during inelastic excursions, earthquakes beyond its elastic capacity result in permanent distortion, but not collapse. If such distortion can be tolerated, all that fails is the elastic analysis. It is presumed that the wells and production facilities would be designed to avoid catastrophic fire or leakage for the most extreme motions in any case.

The effect of ductility in reducing the consequences of extreme earthquakes can be seen in Figure 1. Plotted points, in terms of damage, come from inelastic dynamic overload analysis of a typical offshore structure (ref. 4), using several earthquake records at a given nominal intensity. In terms of risk evaluation, we would expect to see a departure from the direct load risk trend at about the design yield level, with the yielding safety-valve effect creating a risk plateau. However, this safety-valve effect has its limits; and when the available ductility is exhausted, the risk curve should resume its upward trend as shown by the dashed lines.

In Figure 2, Housner's intensity distribution for earthquakes at a general site in California is shown. Extrapolation leads to a "maximum credible" event having a site recurrence interval comparable to the next ice age; although on a regional basis, such an event would be more frequent. Specific site studies by geotechnical consultants will generally yield somewhat different results, both in terms of annual probability trends and "maximum credible" events.

Note that both the resistance distribution and intensity distribution for earthquakes fall below the corresponding distributions for storm loading. This implies that more conservative design criteria have been adopted--partly out of concern for the fact that earthquakes occur suddenly, with less time for protective measures than in the case of storms--and partly to compensate for not yet having fully quantified all the uncertainties and risk factors involved in a new technology.

Structural Redundancy

With a few exceptions, fixed offshore structures typically have a multiplicity of load paths such that the sole failure of a single member does not lead immediately to catastrophic failure. While the beneficial effect of redundancy has been recognized for some time, this has largely been treated qualitatively, or in terms of specific examples. We shall define a couple of useful terms which relate to the degree of redundancy--both for simple systems with N_{LP} identical parallel load carrying elements, and for more complex structures in which one effect of member failure must be established by structural analysis of intact vs. damaged structure. The damage being considered here is complete loss of a member, as by brittle fracture or fatigue failure.

The redundancy factor RF ranges from zero for weakest link systems to very large numbers for damage tolerant structures:

$$RF = N_{LP} - 1 = \frac{\text{damaged strength}}{\text{strength loss}} \quad \ldots \ldots (eq.\ 2)$$

Values of RF less than unity imply a high likelihood that initial failure will progress to total collapse in the presence of nominal loads. Very high values of RF require extreme overloads for total failure, assuming the intact design was adequate.

The damaged strength rating DSR is given by

$$DSR = \frac{\text{damaged strength}}{\text{intact strength}} = \frac{N_{LP} - 1}{N_{LP}} \quad \ldots \ldots (eq.\ 3)$$

Figure 4 gives RF and DSR values for a typical Gulf of Mexico 8-pile platform, based on simplified hand analyses for the various damage cases shown. The 10% label on the jacket leg relates its ultimate capacity as a portal frame to that of an adjoining diagonal brace. Note that a plastic mechanism analysis with full load redistribution and portal development, yields higher DSR and RF than elastic analysis. In evaluating the risk of collapse due to overload of a damaged structure, plastic analysis seems most appropriate. Computer programs are now available to perform such analysis to the desired degree of accuracy (ref. 4).

Results of plastic analysis may be applied in terms of a limit state design equation such as that found in the DNV rules (ref. 14):

$$s(\Sigma F_i \gamma_i) \leq \frac{R_k}{k\gamma_m} \psi \quad \cdots \cdots \cdot \text{(eqn. 4)}$$

where $s(\ )$ = effect of applied loads on given element or system

F_i = magnitude of load condition i

γ_i = partial safety factor on load i

R_k = characteristic resistance of element or system being investigated

$k\gamma_m$ = partial safety factor on strength, ranging from 1.15 for elastically designed steel members, to 1.5 for shells in the elastic buckling range; 1.3 applies to ultimate strength (plastic mechanism) calculations

ψ = reserve strength factor, originally intended to reflect unfavorable post buckling behavior with values less than unity. If the post failure resistance is being used, ψ of 1.0 applies.

The factor ψ may also be chosen during routine initial design, as a function of redundancy, to ensure that complete failure of a single member by fracture will not reduce the ultimate strength below the socially unacceptable limit as defined by a consensus code, for periods while the platform is manned. Defining this limit in terms of DNV rules for plastic strength, and taking a typical Gulf of Mexico platform with an intact reserve strength ration of 1.4 x 1.15/1.3 = 1.24, results in the criteria shown in Figure 5. For Gulf of Mexico service, almost any conventional jacket would provide enough redundancy for this fail-safe-while-manned fracture control strategy.

For North Sea or North Atlantic service, structures may be exposed to the design storm while manned, and there is less margin for degradation. What little exists is provided by the system plastic reserve (ψ_p) or by designing for more severe conditions (longer return interval) than that mandated by the regulations. Redundancy factors of 5 or better, or downrating (ψ less than unity) appears necessary to achieve fail-safe-while-manned status.

Additional redundancy factor examples for a North sea type structure are given in Figure 6. Note the jacket braces which also serve as essential supports for deck loads; not only are they in the splash zone and vulnerable to fatigue failure due to direct wave impact, but they also have inadequate redundancy, even with a fully plastic mechanism. Fortunately, such fracture critical designs are the exception rather than the rule.

Fracture Mechanics

Bristoll (ref. 11) indicates that fracture mechanics an be applied to offshore structures in a number of areas, as outlined in Figure 7. The European Offshore Steels Project from which his reference was taken contains vast quantities of significant data on the fracture and fatigue behavior of welded steel in sea water in general, and tubular connections in particular.

Initiation of cracks detectable in the laboratory occurs at N_1, 5% to 50% of the total fatigue life, along the toe of the weld joining the branch member to the chord. Since 50% to 95% of the fatigue life remains, given adequate notch toughness, detection of such tiny cracks hardly indicates imminent doom, while any attempt to do so in practical situations would yield numerous false alarms from built-in weld discontinuities.

Figure 8 shows the general pattern of fatigue crack growth on the punching shear failure surface of a tubular connection. This seems to indicate a strain-controlled regime in which crack growth is not exponential, but rather controlled by the surrounding uncracked material in the connection. Normalized crack growth data from large scale tubular joint specimens, both through the thickness (a) and along the surface (2c), can be combined with fracture mechanics crack growth data from standard specimens in air, to yield estimates of the stress intensity geometry correction term Y(a) for the tubular joint.

Prototype crack growth calculations combine this Y(a) result with fracture mechanics crack growth data in the seawater environment. The effects of cathodic protection, competing process rates, random loads, stress ratio, size effect, and weld quality on the crack initiation and growth may be estimated from the European Offshore Steels Research data, some of which is shown schematically in Figure 9. Crack growth rates for a connection exposed to a wave climate in which various sea states (subscript i) cause hot spot stress range $\sigma_{sig\,i}$ at zerocrossing frequency f_i for various durations HPY_i (hours per year) are estimated as shown in Figure 10. Total fatigue life combines these crack propagation times with a separate estimate of crack initiation time T_i to yield

$$\text{life} = T_i + \Sigma \Delta T_a + \Sigma \Delta T_{c/a}$$
$$\text{or} \quad \text{life} = T_i + \Sigma \Delta T_c \quad \quad \ldots\ldots(\text{eqs. 5})$$

Such extrapolations from laboratory to prototype involve wave climate data and stress analysis, with a large number of factors which must be assumed by the designer, as shown in Figure 11. Equally plausible assumptions give a wide variation in final result. Many designers prefer to use empirical design curves representing 97% survival and safety factors of 5 to 8 on mean life, e.g. AWS curve X. Nevertheless, fracture mechanics provides a useful research tool in evaluating the influence of various parameters on S-N behavior. Also, when one is confronted with a cracked structure in the field, calculations along the line of Figure 10, starting with the known crack size, can be used to evaluate the remaining fatigue life. Even when a through crack is present in a critical member, (N_2), the remaining fatigue life can be usefully long.

Progressive Fatigue Damage

The effect of fatigue failures on collapse risk may be illustrated by reference to a case study (ref. 6) of a North Sea steel structure. As originally analyzed, this structure had the following distribution of fatigue lives among 6 parallel members in the critical elevation within the structure (see Figure 12):

	CALCULATED D=1 PER AWS	MEDIAN TIME TO FAILURE
1 member @	20 yr	100 yr
2 members @	60 yr	300 yr
3 members @	Over 200 yr	Over 1000 yr

Restricting our attention to braces in the critical elevation being studied, there are several members competing to be the first to fail, and the median time for the first failure is 50 years instead of 100. At year 50, such failure is not certain; there could be no failures or more than one failure. However, in the expected case, we have one failure, which because of multiple parallel load paths still leaves the structure with 80% of its original resistance to lateral loads. With the decreased mean resistance and increased scatter, the risk rate has now increased to about twice its original value. With further passage of time the structure gets progressively weaker and both the risk rate and the rate of fatigue damage accelerate (see Figure 13a).

In fail-safe redundant structures, with multiple parallel load paths, the failure of a single brace does not lead immediately to collapse of the structure. Collapse occurs only when an extreme value of the applied load exceeds the remaining strength of the structure, which is reduced by progressive fatigue damage. Over the 20-year life, the added risk of collapse due to fatigue is of the order of half the original risk of collapse due to overload. Nevertheless, structures are often designed so that the calculated fatigue life is a multiple of 2 or 3 times the anticipated service life, to further reduce the fatigue risk.

So far, the risk estimate assumes no inspection; that is, any failures would be allowed to progress to ultimate collapse without any kind of intervention. In practice this is not what happens. For the example structure, the expected interval between first failure and subsequent failures is 11 years, with 95% probability that the interval will be at least one year--that is, long enough to permit the damage (complete failure of one brace) to be found in a periodic inspection and repaired. This interval is represented by a half-normal distribution as shown in Figure 14.

For structures with less redundancy (lower RSR) than the example structure, progressive damage is accelerated in proportion to $(RSR)^{-m}$, where m is the exponent of the S-N curve, as shown in Figure 15. For a redundancy factor of 5, there is a reasonable chance that routine inspection at intervals of 5-10% of the calculated life, and after each occurrence of the design storm, would be able to detect complete loss of one brace before further progressive failure has occurred. More frequent readings of structural integrity would increase the probability of timely detection.

Given inspection opportunity, the lifetime risk of catastrophic total collapse due to fatigue is reduced to less than 0.1%. In most cases of initial failure, detection would permit the structure to be either repaired or abandoned in an orderly fashion. These outcomes are shown in the event tree of Figure 13b.

A large part of this risk reduction is achieved through secondary safety measures such as (1) subsurface safety valves for well control, (2) hurricane shut-in and evacuation procedures, and (3) downrating platforms with known deficiencies. Much of the risk which remains is an economic one on the part of the platform owner, rather than a hazard to human life or the public interest. The strategy of inspection and secondary safety measures is more cost-effective than simply specifying larger safety factors.

Inspection Practices

One Gulf of Mexico operator's inspection practices for fixed offshore platforms during recent years may be summarized as follows:

Extensive underwater inspections are performed each summer on certain platforms selected on basis of age, condition, interval since last inspection, and service history (e.g., unusual loadings, lapses in cathodic protection, etc.). Typically, four to five such inspections are conducted, or about 15% of the major platform population. Platforms which have been subjected to potential overloads in a hurricane are visually inspected under water and above water for major damage immediately after the storm. Above water inspections are performed annually on all platforms, and casual inspections of opportunity are performed in connection with diving for other purposes.

Detailed inspections are basically visual, with documentation in the form of underwater color photographs (sometimes video tape) and of the diver's observations. Techniques do not vary with depth, except that coverage is greatest at shallow depths, with spot checks at deeper levels where diving time becomes much more expensive. Marine growth has been removed by water jet blasting for recent detailed inspections.

Defects found have included the following:

(1) Parted, buckled, and missing braces and legs--always the result of overloads from collisions (boats, ice, etc.) and, for older platforms, storms exceeding design criteria.

(2) Cracks in the chord at tubular joints, usually the result of overload in joints with inadequate wall thickness by today's standards.

(3) Breaks, buckles, punctures from objects dropped overboard, and abrasions/chaffing from cables hanging on braces.

(4) General corrosion pitting and selective accelerated loss of weld metal on platforms with incomplete cathodic protection. In severe cases, members may be penetrated (150 mpy pitting rate) or suffer localized 50 percent loss of net section. Knife-edge corrosion of weld heat affected zones has also been noted.

Physical damage, when it occurs, tends to be concentrated near the waterline due to accessibility to boats, localized wave pressures and slamming effects, and to marine growth increasing the effective diameter.

The current state of the art does not permit reliable and cost-effective detection of minor damage and small cracks. Attempts at using ultrasonics for this purpose have usually resulted in copious reports of "cracks" which turn out to be built-in weld discontinuities. However, by providing fail-safe redundancy in the structure, the more certain detection of major damage will suffice.

Costs for detailed underwater inspections range from $10,000-$50,000 for existing Gulf of Mexico structures, to $130,000 in Cook Inlet to $500,000 estimated for the 5-year inspection of a deepwater North Sea platform. Some improvement might be possible through the use of onboard saturation diving systems and an enclosed diver access tube/elevator; and by concentrating the detailed inspection selectivity on a "critical member list." Critical members may be identified as those with low redundancy factors and short calculated fatigue lives.

However, a major potential source of savings appears to be the use of continuous structural monitoring (when perfected) as a supplement to underwater inspections to reduce the frequency and extent of the latter.

Structural Monitoring

Although measurements of natural periods and damping were obtained on offshore structures in the 1950's, one of the earliest examples of using motion measurements to monitor performance and integrity of manned offshore platforms was on Texas Tower No. 4 shortly before its collapse in 1961. Nicknamed Old Shakey for the noticeable motions which preceded its collapse, the structure lacked the redundancy of typical template-type fixed platforms.

Shell's MGS Platforms A and C in Cook Inlet, Alaska, have been the subject of myriad monitoring schemes since their completion in 1965 and 1967. The first of these involved accelerometer measurements during 1965-66, with peak accelerations of .05 g reported. During the period 1968-72, a second system was deployed which involved velocity transducers instead of accelerometers, with the data recorded on 14-channel FM analog tape, to be later digitized and analyzed using spectral techniques. Initial measurements were taken for verification of calculated natural frequencies from the space frame computer program, and to provide baseline measurements for determining future changes in vibrational characteristics, as might be caused by structural damage. This was accomplished, and in 1971 after complete loss of a horizontal brace, new readings were taken. These indicated insignificant changes in the natural frequencies, which could be explained by changes in deck mass. This does not necessarily condemn the approach for damage detection since analysis showed this was a redundant member whose loss did not affect either the strength or stiffness of the overall structure. While the "no significant damage" indication was essentially correct, those with unrealistic expectations that the frequency signature method could find lesser damage, e.g., cracks, were disappointed.

More recently, Schmidt developed an inertial reference Displacement Meter for such measurements (ref. 7). Accelerometers are wired to a double-integrating, zero-seeking circuit, which provides displacement output for dynamic and quasi-static motions, while eliminating long-term drift. In the Gulf of Mexico, measurements of platform response using the displacement meter described above have yielded valuable feedback for analysis and design procedures, particularly in regard to structural modeling, damping, and overall response level (ref. 8).

A somewhat parallel technology for using topside motion measurements to monitor platform integrity has been developed in academia (refs. 9 and 15). Based on analysis and measurements on a laboratory structure, the developers claimed ability to detect damage (removal of jacket braces) by changes in the higher mode natural frequencies of the structure (which are less sensitive to changes in deck mass and foundation stiffness than the first mode). Field testing of their system in the offshore environment has led to a more realistic assessment of its capabilities. Baseline measurements are now seen as essential since the differences between theoretical and actual frequencies are of the same order of magnitude as the shifts caused by damage. Nonlinearities in real structures and noise in field data prevented the identification of more than the first few modes, so that detection of minor damage (e.g., large cracks or the loss of secondary braces) would require mounting a transducer on each individual member. Changes in the frequency signature due to the loss (or restoration) of a major portion of the structure's strength and stiffness could be detected successfully, however, when such change was anticipated.

Recently, an opportunity was provided to put the structural monitoring approach to the acid test on a Gulf of Mexico structure. Baseline measurements were made on an about-to-be-salvaged quarters platform in 60 ft. of water. The deck mass was then changed and selected underwater braces removed, making it necessary to detect the nature of each change "blind"; i.e., based only on successive motion measurements. The redundancy of the four-pile platform was such that brace removal caused a 40% loss in strength. Although baseline data was impaired by non-linear behavior of ungrouted piles, the test was a qualified success. Out of four subsequent readings, two correctly indentified the condition of the structure, and a third was partially correct.

A probabilistic statement of predicted changes, similar to Weather Bureau "percent chance of showers" format, now seems more appropriate than having to guess at one of several possible interpretations. More formal probabilistic system identification theory should also play a key role in structural monitoring efforts. An initial step is to analytically generate a "failure dictionary" of member removal versus changes in the frequency signature; in highly redundant structures, loss of several braces at a given level may be required to produce a significant change. In connection with the problem of detecting fatigue failures, the designer would specify critical members, which would be targeted for detailed inspection and inclusion in the "failure dictionary." A parameter study on the sensitivity of the calculated platform response to foundation stiffness, the various sources of damping, loading parameters (e.g., directional spreading of waves), deck loads, etc.--together with a prior estimate of the range of variability of these various parameters--is also used in a Bayesian sense to update the analytical model of the structure when baseline platform response data becomes available. Hart et al (ref. 10) have now formalized these procedures.

The objective of current efforts in this area are to develop a cost effective strategy, and the necessary technology, for monitoring the response of operating offshore platorms as required. Such efforts should provide a basis for project-specific applications to structures scheduled for installation in deep water, earthquake zones, and frontier areas; and focus on design verification, damage detection, and realtime data capabilities as outlined below and in Figure 16.

(1) <u>Design verification</u> - comparison of predicted versus measured response given the sea state. Where platforms are sensitive to uncertainties in soil support, damping, or wave load parameters, analysis of response vs. measurements will indicate whether the design assumptions are in the right ball park. Such measurements will provide valuable feedback to subsequent designs. Baseline measurements, system identification, and an updated computer model are also part of most damage detection schemes.

(2) <u>Damage detection</u> - shifts in the frequency signature (or some other aspect of response) may provide an early warning of <u>major</u> deterioration due to corrosion or fatigue failure; or buckling and parting of braces due to hurricanes, earthquakes, or collison. Given that some degradation is permissible and the use of redundant fail-safe structural configura-

tions which can tolerate the loss of one or several main members, significant damage and loss of stiffness can occur prior to imminent collapse--hopefully enough to be readily detectable. With an established baseline and continuous monitoring, the absence of major damage indications could be used to justify a less stringent program of in-situ underwater inspection than would otherwise be necessary.

(3) Real time date - such as sea state and gross deck displacement - is monitored as a check on system operation. While such data may also be useful in daily operation on the platform, this function should be approached very cautiously, since false alarms could in themselves create hazards.

Repair Philosophy

When one goes looking for structural damage or deterioration, one should have established criteria for what to do when it is found. This is particularly important when regulatory agencies are involved, since ad hoc judgments made in the circumstances of a particular situation may be difficult to justify in the dogmatic atmosphere of subsequent review. Yet tradeoffs and judgments are particularly important here, since rigid application to existing structures of criteria adopted for new structures can lead to unconscionable waste.

Ideally, one should be able to evaluate strength and risk for the structure in both the damaged and repaired condition, and then compare the cost of repairs with the risk reduction, making rational trade-offs on a case-by-case basis. For example, the risk rate curves of Figure 17 (ref. 3) were developed for use together with ultimate strength analysis of the platform jacket and piling systems, yielding the risk rate R for each system; with the probability of failure during the remaining life (L years) as given by

$$pr(F) = 1 - e^{-L\Sigma R} \quad \ldots \ldots \ldots (eq.\ 6)$$

This pr(F) is used in a value analysis (equation 1) for each alternative.

Notwithstanding this ideal, however, regulators have a duty to act in the public interest, and may find it difficult to accept private cost-benefit calculations--particularly when there are neither established guidelines on calculation procedures, nor explicit agreement on what constitutes an acceptable trade-off.

Figure 18 is an attempt to provide such a guideline. We implicitly take the design criteria and calculation procedures of the reference standard (e.g. API RP 2A) as a consensus representing acceptable trade-off for new construction. Rated strength is the load at which allowable stress (or permissible limit state) is exceeded, using standard calculation procedures--no fancy footwork allowed. Rated strengths less than 100% of the reference standard represent excess risk over that already found to be acceptable; and reduction of this risk can be used to justify the repair cost involved. Curves A, B, and C correspond to various reference stengths and degrees of risk aversion, as shown in Figure 3, for Gulf of Mexico storm loadings. For other types of load (e.g. drilling and production deck load) quite different (steeper) curves would apply.

Two examples are shown, both for structures having replacement cost of $12 million, and found to be rated at only 80% of the reference strength. If the loss in strength is due to damaged braces costing $1 million to replace, the repair would clearly be justified--except possibly in case B, where the reference strength was very conservative to begin with. If the 80% rating is due to general deterioration, or simply the fact that the platform is an older one designed for 20% lower criteria, the picture changes. The cost of wholesale remedial construction performed offshore and underwater, at $4 million, would not be justified in this example unless the reference criteria were already below optimum (case C). However, the steeply rising curves provide ample incentive for protective measures which would prevent further degradation.

While the shaded band still leaves room for judgment based on the consequences of failure in each particular case, such a guideline--if adopted in advance--would provide a format in which a degree of flexibility and reasonableness could be incorporated into strict and impartial enforcement of regulations. Although Figure 18 is based on total costs and risks as accounted for by the offshore operator, analysis from a wider viewpoint should still lead to a similar result.

In the Gulf of Mexico, there is sufficient warning for hurricanes that for periods while the platform is manned and the wells are flowing, the maximum expected storm produces only 40% of the API reference lateral load. This is shown on Figure 18 as the limit of socially unacceptable risk; that is, we want the platforms to be fail-safe while manned-defined somewhat arbitrarily in terms of traditional safety factors. The shaded band leaves a margin for undetected damage. Gulf of Mexico platforms can suffer considerable degradation and still meet this criterion.

In other areas like the North Sea, evacuation is not always possible, and offshore platforms may be tested with a design storm while manned; further, the expected annual storm may exceed 70% of the specified design force. There is little, if any, margin for degradation, and difficult issues involving human risk will inevitably arise. Here it may be possible to define the socially unacceptable limit by reference to risks which have historically been accepted by the public (e.g. other seafaring occupations), or to adverse effects which would ensue if the offshore reserves were to be prematurely abandoned (e.g. more coal miners at risk working to replace the lost energy supply).

Some of these reference risks are shown in Figure 19. In the lower part of the figure are plotted individual risks, for which the basis of comparison is one's personal mortality. However, in the case of large catastrophes involving multiple deaths, there seems to be a public aversion factor (ref. 13). This is reflected in the broad tolerance bands for voluntary and involuntary risks. Selected reference risks are approximately plotted for comparison. It can be seen, for example, that the risk exposure for a Southern California API offshore structure collapsing in an earthquake (based on the data in Figs. 1 and 2) is an order of magnitude less averse than that of other maritime occupations, and two orders of magnitude less than that of an alternative energy source.

Among the strategy options available to the designer, he may increase the margin for degradation by designing elastically for environmental forces slightly in excess of the legal minimum and make sure his structural configuration provides a high redundancy factor throughout. When such a structure is found to be damaged, it could be checked against the maximum load expected during its remaining life, using ultimate stength analysis with load and resistance factors as suggested by the DNV rules (ref. 14), to determine whether or not a socially unacceptable risk is present. For unmanned platforms which would not cause major pollution upon failure, the question may be resolved in terms of economic trade-offs.

The foregoing issues deserve to be rationally examined, preferably from the broadest possible viewpoint. While this author cannot offer any final answers; and the socially unacceptable limit might end up being more complex than an arbitrary cutoff value, and the economic trade-off curves would no doubt be steeper, a guideline in the format of Figure 18 should still be feasible for these other offshore areas.

Conclusions

In the past, offshore operators have traditionally made trade-offs between benefits (reduced risk) and costs. Calculation procedures used by the author for this purpose appear to conservatively overstate the risk. the extent of platform monitoring and inspection was arrived at as a matter of practical judgment, but consistent with the redundancy of our structures and such secondary safety measures as subsurface safety valves and evacuation procedures. When structural deficiencies are found, the decision whether or not to undertake repairs or strengthening has been couched in terms of economic considerations--attempting to rationally balance the cost of repair against the benefit in terms of risk reduction. Because the cost of underwater repairs is often exorbitant in comparison with what a comparable increment in strength would cost during initial construction, such tradeoffs often indicate the rational choice is to accept a lower safety factor than is specified for new structures. By recognizing that the platform owner is responsible for human casualties, and will bear the full expense of oil spill clean up, as well as such business items as loss of income and loss of investment, making these trade-offs represents responsible behavior and adequately reflects the national interest.

References

1. Abrahamsen, Egil et al, "Safety, Man, and Society", Information from Det Norske Veritas about Safety and Risks, 1977.

2. Marshall, P. W. and Bea, R. G., "Failure Modes for Offshore Structures", Proc. BOSS '76, Vol. II, Trondheim, Norway (1976).

3. Marshall, P. W., "Risk Factors for Offshore Structures", Civil Engineering in the Oceans, ASCE Conference, San Francisco (1967).

4. Gates, W. E., Marshall, P. W., and Mahin, S. A., "Analytical Methods for Determining the Ultimate Earthquake Resistance of Fixed Offshore Structure", OTC 2751 (1977).

5. Housner, G. W., "Strong Ground Motion", Earthquake Engineering (1970).

6. Marshall, P. W., "Failure Modes for Offshore Structures - Part II - Fatigue", Methods of Structural Analysis, ASCE Conf., Madison, WI (1976).

7. Schmidt, T. R., "Offshore Platform Displacement Measurement with Inertial Reference System", OTC 2554 (1976).

8. Ruhl, J. A., "Offshore Platforms: Observed Behavior and Comparisons with Theory", OTC 2553 (1976).

9. Dodds, C. J., et al, "Experiences in Developing and Operating Integrity Monitoring Systems in the North Sea", OTC 2551 (1976).

10. Hart, G. C. and Yao, J. T. P., System Identification in Structural Dynamics", Dynamic Response of Structures, ASCE-EMD Specialty Conf. (March 1976).

11. Bristoll, P., "Review of the Fracture Mechanics Approach to the Problems of Design, Quality Assurance, Maintenance, and Repair of Offshore Structures", European Offshore Steel Research, Preprints Vol. 2, Cambridge, November 1978.

12. Fjeld, S., "Reliability of Offshore Structures", OTC 3027 (1977).

13. Flint, A. R., "Design Objectives for Offshore Structures In Relation to Social Criteria", Proc. BOSS '76 Vol. II, Trondheim, Norway (1976)

14. Det Norske Veritas: Rules for the Design, Construction and Inspection of Offshore Structures, 1977.

15. Vandiver, J. K., "Detection of Structural Failure on Fixed Platforms by Measurement of Dynamic Response," Journal of Petroleum Technology, March 1977.

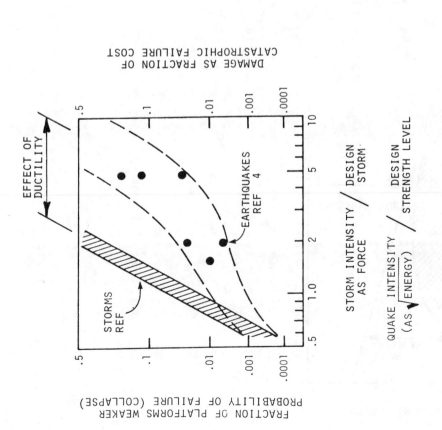

Fig. 1 Resistance characterizations.

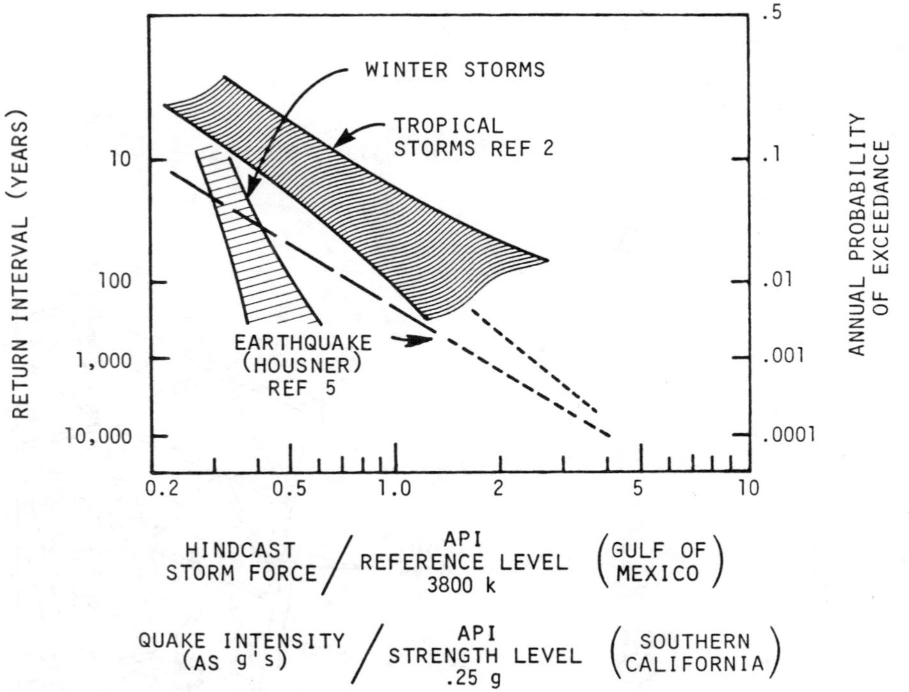

Fig. 2 Applied forces.

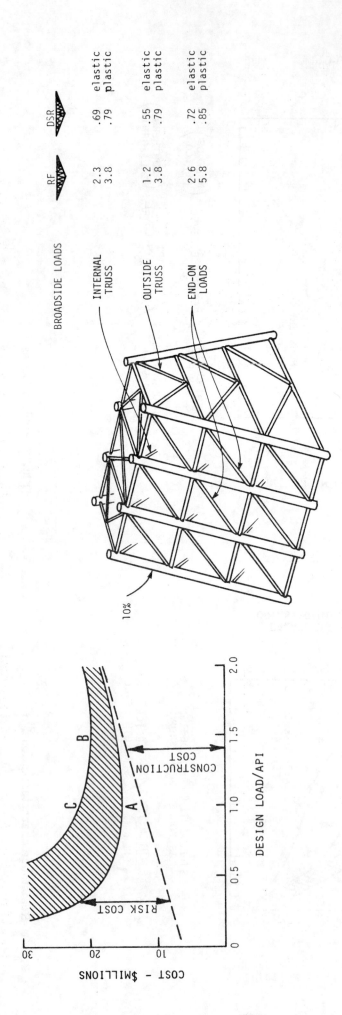

Fig. 3 Economic analysis.

Fig. 4 Redundancy of Gulf of Mexico type 8-legged template.

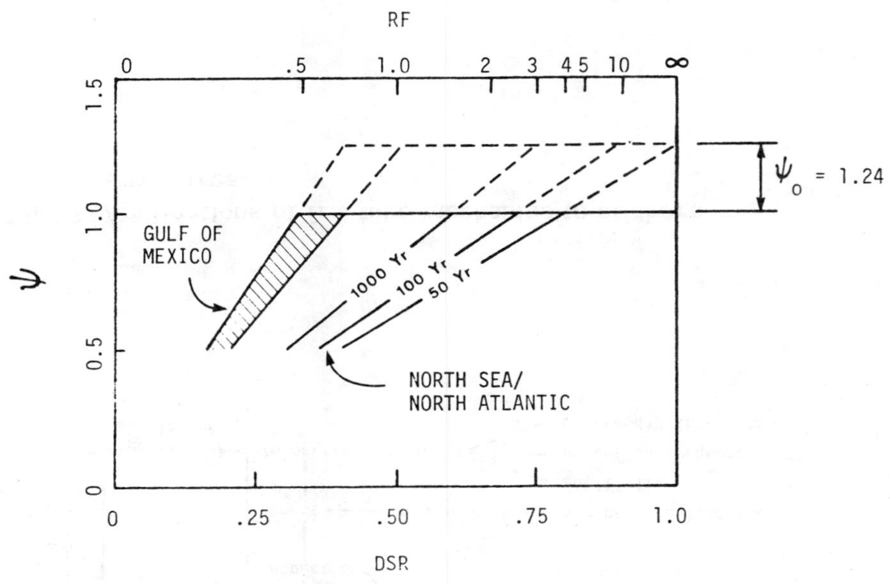

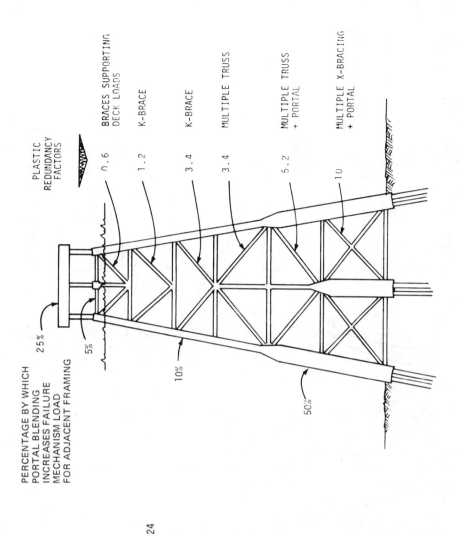

Fig. 5 Fail-safe-while-manned fracture control strategy (see equations 2-4).

Fig. 6 Redundancy of North Sea type 4-legged tower.

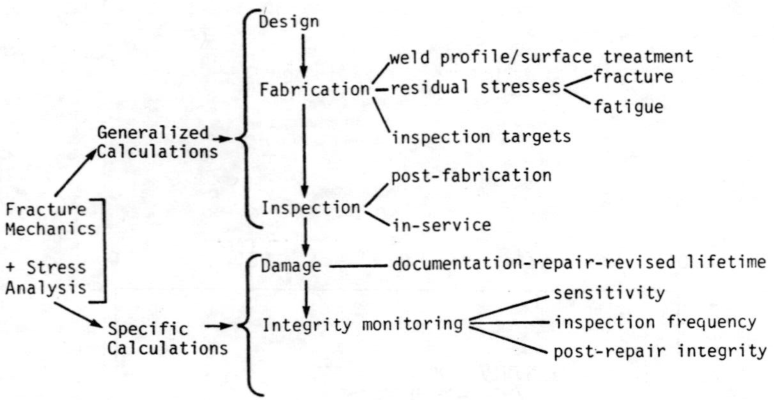

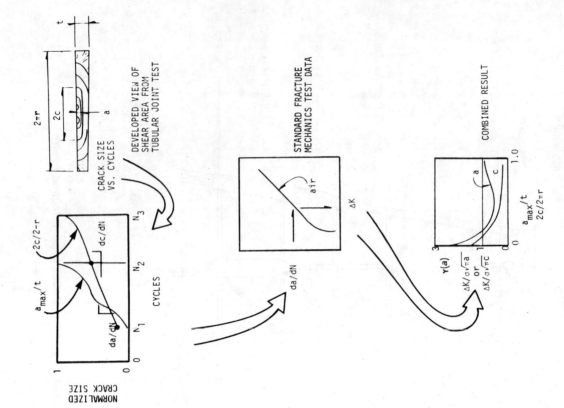

Fig. 8 Laboratory fatigue crack data.

Fig. 7 Applications of fracture mechanics to offshore structures.

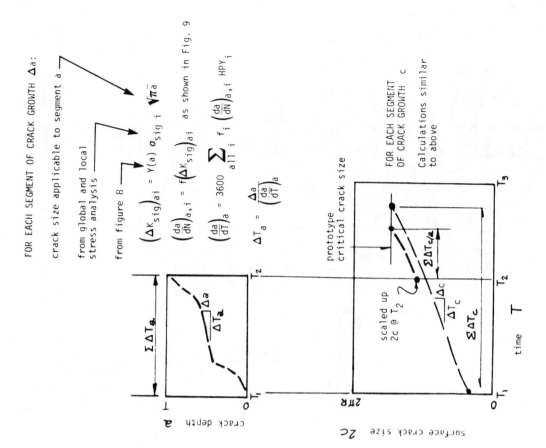

Fig. 10 Prototype crack growth calculations.

Fig. 9 Effect of sea water and cathodic protection on fatigue crack growth.

	MODEL	vs.	PROTOTYPE
Stress concentration factor	2.2		Same or higher.
static overload	proof test 1.25 x design		annual storm 0.7 x design; initial high R partially relieved
restraint	plane stress		thickness transition
environment	air 20°C		sea water, cathodic protection, -2°C
initial flaw	.01 to .1 mm or 2mm = .4t		undercut .25mm or .4t = 8mm
da/dN	Barsom (air)		Vosikovsky (SWCP)
critical flaw	1m - 80 @ 18 ksi		0.5m - 8 @ 40 ksi
life	40,000 cycles		17-400 years

Fig. 11

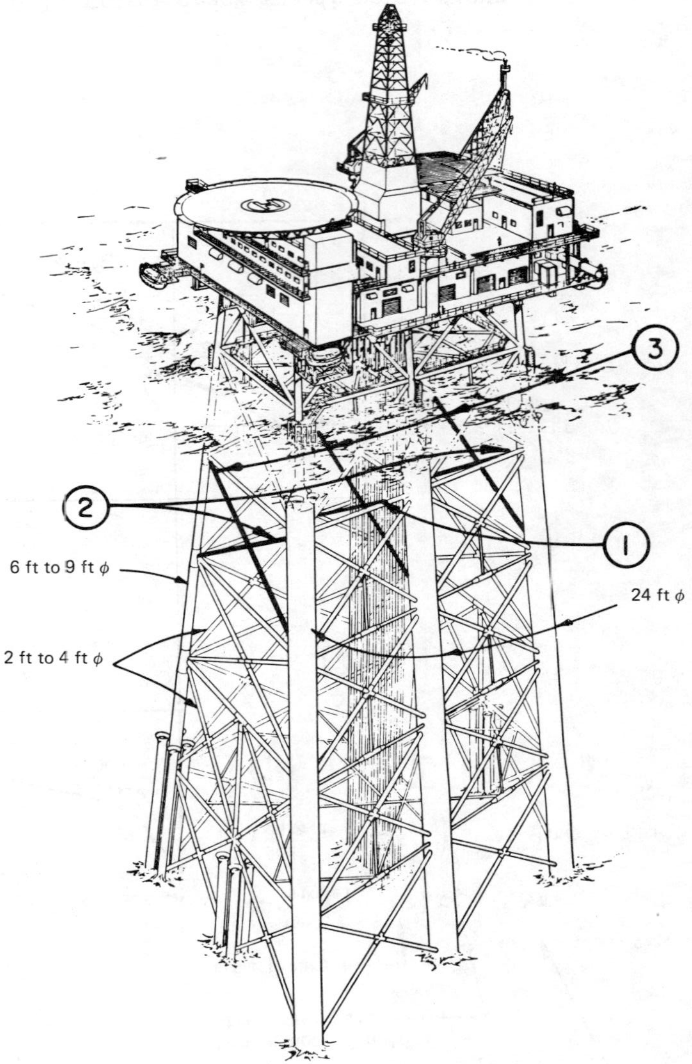

Fig. 12 Example platform for north sea, showing members used in progressive damage study.

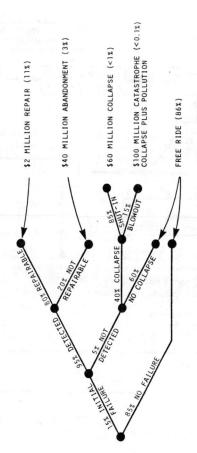

Fig. 14 Interval between initial failure and progressive collapse.

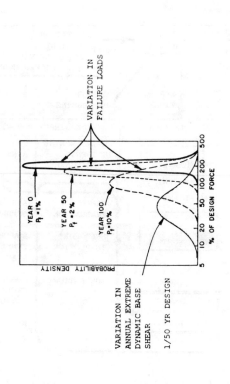

(a) DETERIORATING STRENGTH OF REDUNDANT HYBRID STRUCTURE-BASED ON 20 YEAR CALCULATED LIFE

(b) EFFECT OF INSPECTION

Fig. 13

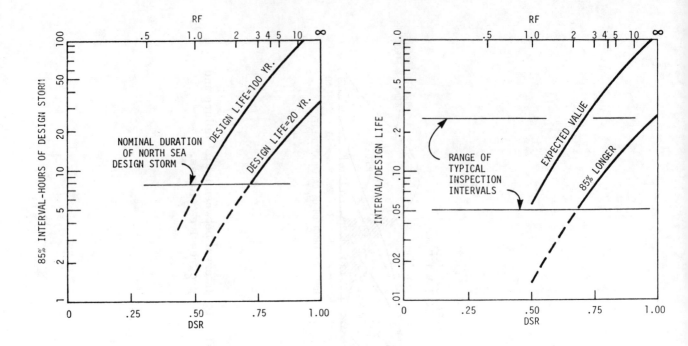

Fig. 15 Effect of redundancy on interval between initial failure and subsequent failures.

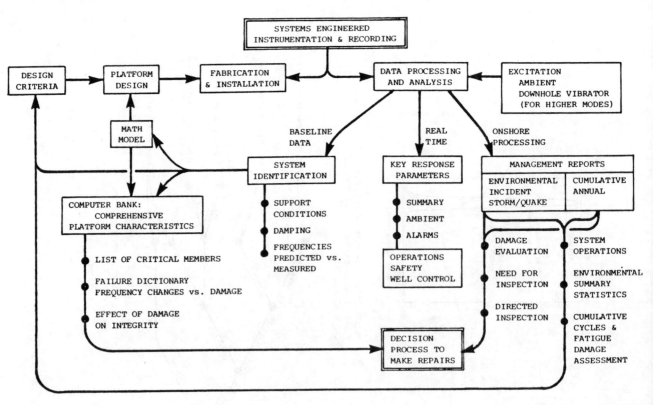

Fig. 16 IEC monitoring system flowchart.

after R. W. Griffiths, Interstate Electronics Corp.

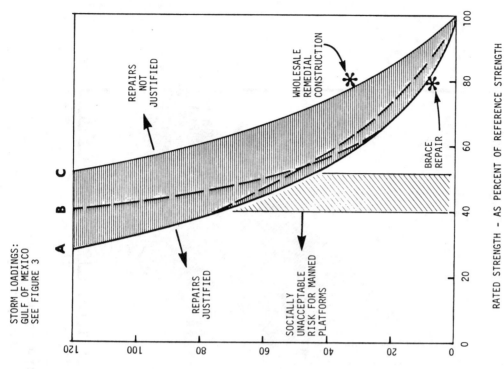

Fig. 18 Repair guideline.

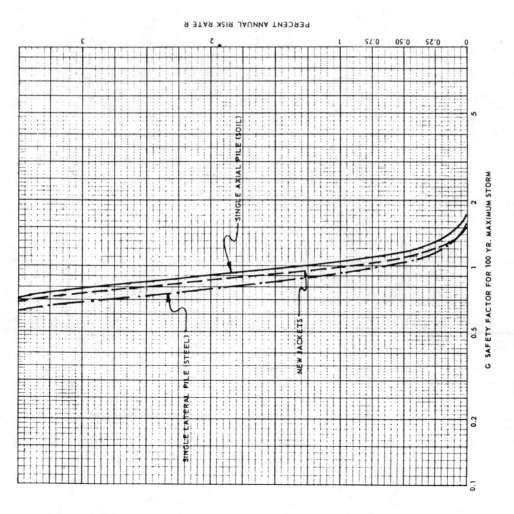

Fig. 17 Annual risk rate.

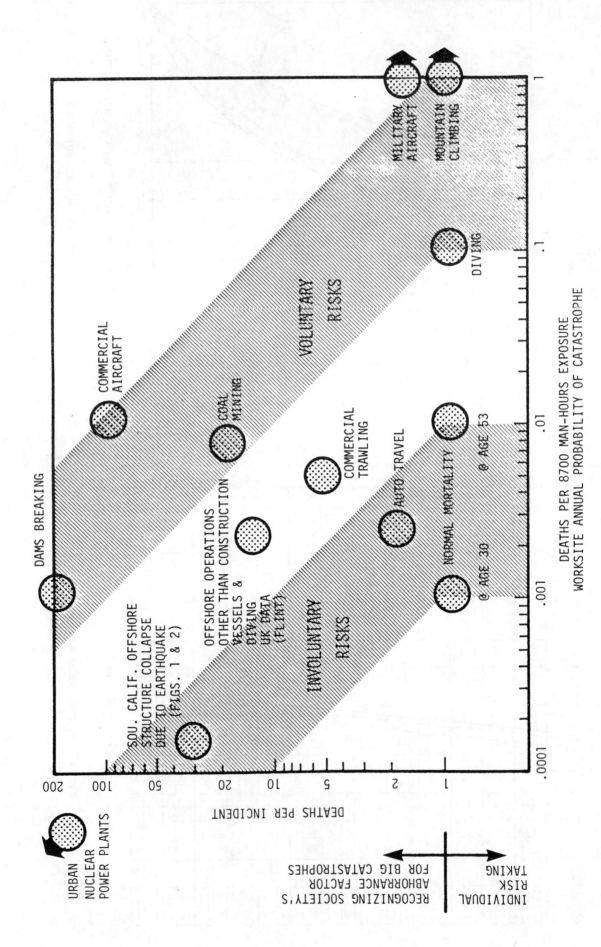

Fig. 19 Subjective risk evaluation.

BOSS'79

PAPER 75

Second International Conference
on Behaviour of Off-Shore Structures

Held at: Imperial College, London, England
28 to 31 August 1979

OFFSHORE OIL PRODUCTION AND DRILLING PLATFORMS.
DESIGN AGAINST ACCIDENTAL LOADS.

Svein Fjeld

Det norske Veritas, Norway

Summary

The requirements to design against accidental loads as laid down in current codes are summed up. A general design philosophy to supplement the code requirements is established. This deals with choice of design loads, safety factors and acceptance criteria. The design procedure in practice is outlined emphasizing the three main principles: event control, indirect design, and direct design..

Direct design methods are outlined for a number of typical accidental loads i.e. ship collision, dropped objects, flying fragments, explosions, earthquake, fire and loss of assumed pressure difference.

Sponsored by: Delft University of Technology, The Netherlands
Massachusetts Institute of Technology, U.S.A.
The Norwegian Institute of Technology, Norway
University of London, England

Secretariat provided by: BHRA Fluid Engineering

Copyright: ©   BHRA Fluid Engineering
Cranfield, Bedford, England

1. INTRODUCTION

In recent years increasing attention has been paid to the safety of North Sea platforms against so called accidental loads i.e. loads from fires, explosions, ship collisions, blowouts, earthquakes, etc. These loads are not assumed to occur under normal use of the platforms. But the majority of structural damage encountered is caused by this type of loads. This is clearly demonstrated by Table 3 recording structural loss and loss of lives depending on cause.

There seems to be a need for a recognized approach to verify the structures to resist this type of loads. Present guidelines are rather short spoken. The U.K. Department of Energy "Offshore installations: Guidance on design and construction" (3) confine themselves to the following considerations (Part 2 § 1.7).

> 1.7 Accidental damage
> Under other than extreme loading conditions the usual types of fixed installation can be expected to have a substantial margin of strength and be capable of withstanding considerable structural damage without collapsing. Pending further investigation into the probability of ship-installation encounters and other sources of serious accident, no current guidance can be given.
>
> Mobile installations must remain afloat and stable with any one compartment flooded. (See Section 4.3).

The present paper is meant as a small contribution to the further investigations asked for by the Department of Energy. The paper is not meant to give basically new results from research, but rather to indicate reasonable approaches which might be used by the designer in practice.

The need for a recognized design approach is accentuated by the fact that design against accidental loads is not just a question of increased dimensions. Appropriate choice of platform layout and static system, sufficient ductility, pressure relief walls, fire protection etc. may be the appropriate design measures.

2. CODE REQUIREMENTS

General structural codes

A majority of modern structural codes mention accidental loads rather briefly. "Common unified rules for different types of construction and materials" (1) (2) states:

> A 2.4 Other Considerations
>
> (a) Overall stability and robustness
> The general arrangement of the structure and the connection of its various members or elements should be such as to give the structure an appropriate stability and robustness.
>
> In particular, there should be a reasonable probability that the structure will not collapse catastrophically under the effect of misuse or accident. No structure can be expected to be resistant to the excessive loads or forces that could arise due to an extreme cause, but it should not be damaged to an extent disproportionate to the original cause.
>
> Such accidental situations may arise in a structure:
> (I) immediately after the occurrence of a very unlikely action causing localized damage;

or

(II) immediately after a very unlikely change in its environment, but changing in a significant manner the action effects in, or resistance of, the structure.

Accidental actions are classified in § A4.2.1

- <u>accidental actions</u>, the occurence of which, in any given structure, and with a significant value, is unlikely during the reference period, but the magnitude of which could be important for certain structures. They are usually assigned nominal values in assessing the resistance of the structure to them.

The accidental loads are required to be included in the ultimate limit state in a so called accidental combination. All load coefficients are taken as 1.0.

For some reason the model codes for steel (1) and concrete (2) prescribe load coefficients somewhat different from 1.0.

Their load coefficients are shown in Table 1.

Codes for offshore structures

Apart from earthquake, which they consider an environmental load, none of the above mentioned accidental loads are mentioned in the traditional structural design codes such as API RP 2A: Planning, Designing and Constructing Fixed Offshore Platforms (8).

Recent European codes are influenced by the Model code and include a few considerations related to accidental loads.

- FIP: Recommendations for the design and construction of concrete sea structures (4).

This document just mentions the safety factors for the ultimate limit states recorded in table 2. Otherwise no guidance is given.

- Norwegian Petroleum Directorate: Regulations for the structural design of fixed structures on the Norwegian Continental Shelf (5).

This document states in § 4.5.5 (see Table 2).

The limit state of progressive collapse shall be checked for accidental loads. A local failure is acceptable provided that the remaining structure satisfies the requirements in the ultimate limit state for the load combinations and load coefficients given in table 4.2. The load combination "Accidental load included" shall only be checked if the accidental load physically can occur after the local failure. Only one accidental load is included in the combination. Loss of intended differential pressure shall be considered as a local failure.

Accidental loads are defined as loads that occur only by accident or under abnormal circumstances, and the magnitude and frequency of which are ill-defined, such as loads from:

- Earthquake

- Explosion

- Collision by aeroplane or ships

- Dropped objects

- Loss of intended differential pressure

- Temperature strains caused by fire

Change in material properties caused by fire shall be accounted for in the assessment of resistance.

Characteristic values of the accidental loads are to be approved by NPD in each case. Associated material coefficients are not specified.

- Det norske Veritas: Rules for the design construction and inspection of offshore structures (6)

DnV defines the accidental loads very similar to the NPD definition. The structural design is to be checked as follows § 4.4.4.5.

> The limit state of progressive collapse is to be checked for the most unfavourable combination of P, L, D, D and A. For accidental loads of impact type e.g. collision loads, dropped objects etc., loads of type E may be ignored. For this check, which may be based on a simplified analysis, all load coefficients may be taken equal to $\gamma_f = 1.0$. Localized damage may be accepted, but is not to be disproportionate to the original cause.

Material coefficients are specified as recorded in Table 2.

As the basis for the choice of accidental loads is not clearly specified a discussion about load coefficients might appear futile. For this reason a further development of the design philosophy seems appropriate.

3. GENERAL DESIGN PHILOSOPHY

Design levels

It seems appropriate to check the safety of structures on three levels see fig. 1.

- Serviceability limit states
  These are related to criteria governing normal use and durability i.e. deflections, vibrations etc. The seriousness of production disturbance should be considered by the owner and the probability level of the design loads chosen accordingly.

- Ultimate limit states
  This corresponds to failure of some structural element. An element failure will normally be considered more serious than some equipment malfunction. Therefore the design load should be chosen so large that this limit state is normally not reached. But even this load cannot be so high as to cover any credible situation. e.g. the design environmental loads are accepted to have an annual probability of exceedance of 1 - 2 %.

- Progressive collapse limit states
  This corresponds to a collapse of major parts of the structure. Consequences such as loss of human lives, pollution and economical losses may be significant. This limit state will therefore have to be checked for such large loads that an exceedance is hardly credible.

Basis for choice of design accidental loads

A progressive collapse limit state design to cover the maximum credible accidental load would result in excessive dimensions and prohibitive costs. The choice of design accidental load will have to be based on cost-benefit considerations. A theoretical economic optimum is obtained if the marginal costs are kept constant for all types of accidents. But this theoretical principle is unfeasible for practical implementation. A simpler strategy is keeping an almost constant residual risk (i.e. risk not covered by the design) for all types of accidental loads. Costs invested to almost eliminate one hazard is wasted if the platform is still threatened by other hazards. An approximately even residual risk is obtained if the accidental loads have a specified low probability of beeing exceeded (14). This basis for choice of design accidental load is compatible with current practice to define design environmental load, certain types of design live load etc.

A reasonable number for the probability may be estimated as follows: The death rate due to accidental collapse must be significantly less than the present number of labour fatalities in similar industrial branches. Considering pertinent statistics this fatality rate is in the range of 4 - 40 per 10^8 hours of exposure (FAFR = 4 - 40). A small minority of these fatalities are associated with major accidental structural collapse. This relation should be maintained for offshore structures.

Assuming:

- Annual probability of an accident not covered by design p

- Number of workers exposed N

- Average number of fatalities in a structural collapse n

- Annual number of hours of exposure for each worker H

the FAFR corresponding to accidents causing structural collapse. is given by:

$p \times n = FAFR \times N \times H \cdot 10^{-8}$

Further assuming H = 2000, $\frac{n}{N} = \frac{1}{5}$ and an acceptable FAFR due to accidents causing collapse FAFR = 1 the appropriate choice of design accident probability exceedance should be:

$p = 1 \times 5 \cdot 2000 \cdot 10^{-8} = 10^{-4}$ accident/platform year

All these estimates are confined to orders of magnitude. As also the magnitude of the accidental load of some given probability level is most uncertain, an annual probability in the order of magnitude 10^{-4} should be an acceptable basis for design. This probability should also be acceptable from an economical point of view i.e. the economic losses due to major accidents will be small compared to other economic losses.

Acceptance criteria

As may be concluded from chapter 2 the present code philosophy is not uniform. Certain codes require the design to stand accidental loads without infringing the ULS-criteria. DOE, NPD and DnV on the other hand accept local structural damage provided a collapse is avoided.

In DnV's opinion the latter philosophy is the only realistic one. A general design requirement to withstand the accidental loads without considering the postfailure strength reserves would be prohibitive. But there are reservations to this:

- cases where thightness is required for seaworthiness or to avoid excessive oil pollution

- cases where restoring the structure after a local damage is hardly possible

In these cases the structure will have to be designed to meet specified serviceability criteria. The first of these two aspects should be considered by the authorities. The latter question should be left to the owner.

Safety factors

The purpose of the design against accidental loads is to verify survival of the structure in case of rare events. As discussed the probability of occurrence of the design accidental load is low and its magnitude uncertain.

Furthermore a check of the collapse resistance against accidental loads is normally based on rough equilibrium checks rather than refined and sensitive analyses.

Thus it might appear too academic to amplify the somewhat uncertain design load effect by the introduction of a load coefficient.

Compliance with criteria normally requires large yielding strain of the materials. Thus also a material factor to ascertain strength margins seems irrelevant.

For these reasons DnV has accepted load coefficients as well as material coefficients to be taken equal to 1.0. The material coefficient of concrete is taken slightly above 1,0 to ensure ductility.

Design load for damaged structure

The requirement explicitly stated by NPD (5) § 4.5.5 that the structure has to stand some environmental load after the local damage has occurred should always be checked. The environmental force to be applied should have a return period at least three times the necessary time to restore the damage. Taking into account delays in connection with decision making, weather delays etc. the return period should not be chosen less than one year.

4. PRACTICAL APPROACH FOR DESIGN AGAINST ACCIDENTAL LOADS

General

As mentioned above design against accidental loads influences the platform layout as well as the choice of static system, choice of materials etc. Certain designers seem to consider the check of safety against accidental loads to be a formality to deal with after completion of the genuine design work.

In fact the accidental load check should be the very first step of the design verification.

If so, safety against accidental loads can often be achieved at negligible extra costs.

The approaches for design against accidental loads may be categorized as:

- event control

- indirect design

- direct design

Event control

The design verification should be initiated by a study of the operations planned in each individual area of the platform. The potential hazards involved should be recorded and the magnitude of accidents of the specified low probability level should be identified. A joint effort involving risk analysts as well as professionals knowing the operations in question is necessary. The effect of the accident should be expressed in engineering terms such as impact impulse, heat flux and duration etc.

The design accidental loads being established, the adequacy of the design may be verified as described in the following chapters. If compliance with criteria is not achieved, the situation may be improved by strengthening the structure or changing the static system.

But there are normally also possibilities to change equipment, working procedures, active protection devices or even change arrangement of the platform to reduce the design loads.

In many cases the adequacy of the static system and platform arrangement has to be verified prior to the detail equipment design. Then the basis to establish design accidental loads is missing. In such cases reasonable design values may simply be chosen by the designer. Afterwards, during the detail design, the platform is to be fitted with equipment and passive protection, so as to ascertain compliance with the definition of the design accidental load. Thus incidents possibly leading to major collapse are avoided.

Indirect design

A majority of the accidental loads are of the impact type. The structure is to absorb a certain amount of impact energy rather than transfer a specific load. Examples of such accidents are ship collision, earthquake, dropped objects, explosions etc. By providing energy absorption capacity, the resistance is improved without directly considering the accidental loads in question.

Energy absorption requires the structure to behave in a ductile manner. Measures to obtain adequate ductility are:

- Connections of primary members to develop a strength in excess of that of the member

- Provide redundancy in the structure so that alternate load redistribution paths may be developed

- Avoid dependence on energy absorption in slender struts with a non-ductile postbuckling behaviour

- Avoid pronounced weak sections and abrupt change in strength or stiffness

- Avoid as far as possible dependence on energy absorption in members mainly acting in bending

- Non brittle materials

For jackets the following API requirements (8) should be fulfilled:

"Tubular joints, members, and piling at locations which are required to maintain their capacity through substantial concentrated, inelastic deformation should be designed to meet the compact section requirements. ($D/T < 9000/F_y$). Portions of tubular members and piling which may be only moderately deformed beyond yield or column buckling need only be sized to preclude premature local buckling ($D/T < 22750/F_y$), provided their limited deformation capacity and degrading post-buckling characteristics are recognized. For tubular members with $9000/F_y < D/t < 15200/F_y$, development of full plastic load and moment capacity, but limited plastic rotation capacity, may be presumed".

(F_y in MPa)

It is prudent to strike a balance between the requirements of strength on the one hand and the need for flexibility on the other to take maximum advantage of ductility.

Tubular joints should be designed for punching shear load 30 % above the yield force in the nominal member cross section.

Joints or nodes should be designed to maintain full plastic section capacity through large ductility ratios.

Ductility of concrete structures is obtained by keeping the amount of reinforcement above the requirements to minimum reinforcement laid down in recognized codes e.g. DnV (6) § 7.9.2.3, but at the same time sufficiently low to ensure all incipient failures to be governed by steel yielding. Excessive use of prestressing should be avoided.

Ductility in shear may be obtained by special transverse reinforcement, but this is only necessary in special cases.

Direct design

Approaches for direct design against the accidental load will be somewhat different from one type of accident to the other as described in the following chapters.

Blowout is not discussed separately. From a structural design point of view the consequences of a blowout are covered by discussions of fires and explosions.

5. SHIP COLLISION

Design accidental load

The design accidental loads are mainly dependent on

- supply ships calling at the field
- tankers to be loaded offshore
- general ship traffic

In practice the general ship traffic can hardly be designed against. Thus it will have to be controlled by active measures, i.e. radar survey, guard ships, sound and light signals etc.

Offshore loading terminals should be designed to stand the impact from the pertinent tankers. In cases of floating or articulated buoys the impact forces should be possible to design for. Certain types of stiff platforms can hardly stand a tanker impact. If such platforms are used the field has to be so arranged that the collision probability is reduced sufficiently to exclude this accident as design basis. DnV have computer programs simulating the ship behaviour in case of malfunction of the power or the helm and which is well suited to determine the collsion probability.

DnV technical note no. A 6/5 deals with ship collisions. Under certain assumptions the note states the velocity of a supply ship out of control drifting against some platform to be in the order of 1,5 - 2 m/sec. The size of future supply ships is estimated to 2500 t displacement. It is important to note that these numbers have never been assumed to be the final answer as to design collision load. Each individual platform should be subject to a special study as to the appropriate design collision impulse, depending on future operation of the platform. When evaluating the impact energy it is important to consider the hydrodynamic mass of the boat. The added mass coefficient may normally be taken as 0,4 when drifting sideways and 0,1 during normal cruise.

The design accidental load is to be expressed in terms of impact energy. Adequacy of the design means that this impact energy may be absorbed by the deformation of the platform and the ship. It is important to note that the strain energy of the ship is to be considered as well. Otherwise stiff structures will be much overdesigned compared to the weaker ones.

Steel jacket structure

The ship can strike the jacket:

a) the bow striking some bracing

b) drifting sideways striking the platform legs

a) The energy absorption in bending of a jacket bracing is negligible compared to the collision impact energy. If the ship is not to penetrate the jacket the bracing will have to develop sufficient deformations to absorb the energy in direct tension.

This action requires the whole length of the bracing to participate. For this reason the bracing or its joints cannot have any weak spots. Otherwise the weaker area would yield when the main part of the bracing is still in the elastic range. Localized parts only will contribute to the plastic energy absorption, and this will normally not be enough to absorb the ship energy. In this context the member joints are to be paid special attention as discussed above under the heading "indirect design".

Thus the procedure to check the jacket resistance against ship collision should be as follows:

1. Identify the bracing members likely to be struck by the ship. Ensure that these members will be deformed to a system which transmits the force in direct tension. Secondary bracing elements (conductor bracing etc.) might confuse the development of the wanted mechanism. They can cause local ruptures or reduced strength at the connections with the brace member.

2. Check that no section of the bracing member including its joints has a lower strength than the yield strength of the plain member. The ultimate punch strength of the joints should exceed this yield strength by 30 %.

3. Verify the energy absorption capacity of the ship and the member to exceed the kinetic ship energy.

4. Assume the impacted member removed and apply forces corresponding to its yield strength in the joints. Check capacity of adjacent element to take these forces.

5. Assume impacted member removed and check the jacket capacity

to stand the load condition specified by NPD in case "accidental load not included" (see table 2).

b) The energy absorption may take into account the following contributions, which in practice may be evaluated one after the other, the sequence to be aborted when an energy balance is verified.

1. Energy absorption capacity of possible fenders or bumpers

2. Elastic strain energy absorbed by the jacket and soil prior to any yielding.

3. Elastic energy absorption of the ship exposed to forces corresponding to first yielding of any jacket member.

4. Inelastic strain energy from local deformation of leg at area of collision.

5. Inelastic strain energy developed in weakest member of the jacket. The designer should check that the weakest member has an adequate energy absorption without excessive damage to the structure.

6. Inelastic strain energy developed in deformation of the ship. This deformation will include both buckling of deck and bottom plating as well as frames. Tensioning of the hull-side contributes significantly to the energy absorption before a rupture occurs.

Concrete structures

Concrete structures fitted with a perforated break water wall of sufficient height have a large extra capacity to take ship collision impact. The present considerations are confined to concrete shafts of diameter less than 20 m.

As for steel structures the design verification is based on the principle of energy absorption.

Punching strength of shaft

Local spalling and chipping at the point of contact can hardly be avoided in an unfendered zone, but the punching shear strength of the column should certainly be checked.

Different designers have estimated the punching strength of concrete shafts on the basis of different assumptions using CEB - FIP recommendation (2) or scaling experimental results. A number of these methods have been used for a typical column (12.0 m internal diameter, 0.6 m wall, typical prestress and reinforcement with stirrups). The results are plotted in fig. 2 (with $\gamma_f = 1.0$, but using the ULS values of γ_m). In the same figure the force – contact zone curves for a representative supply boat with 2500 tonnes displacement are given. It is seen that in the early stages of a collision, for a given equivalent radius of contact zone the strength of the concrete in punching shear is apparently greater than the force which the supply boat can develop.

These calculations point to the conclusion that, because of the limited strength of the ship, punching shear failure of the concrete is unlikely to occur, provided the concrete is appropriately designed.

A possible punching shear failure will not necessarily limit the capacity of a concrete column to absorb energy in a collision, because in a sideways collision the structure of the ship will bridge across the hole and apply force to the undamaged part of the shaft.

Inelastic behaviour of the column

If a punching shear failure can be prevented in the early stages of a collision it seems unlikely that such a failure will occur later. During deformation of the ship (fig. 2) the contact zone and simultaneously the punching shear strength rapidly increases.

However, elastic theory calculations may indicate the reinforcement stress in the contact zone to reach yield or the concrete to crush. Sufficient energy absorption requires inelastic deformations of the concrete. In the absence of a non-linear computer programs to analyse a concrete cylinder under radial load, estimates of the load required to cause a local bending failure of the column have been made using plastic theory. Where possible, conservative assumptions have been incorporated in the plastic theory, and a material factor $\gamma_m = 1.5$ has been included. The corresponding failure loads and contact chord lengths are plotted in Fig. 2.

It seems that the typical supply boat is not strong enough to cause a local failure of a typical concrete column in a sideways collision. But it can cause local yielding and cracking.

The force which the ship can apply is limited in two ways: firstly the ship's inertia can cause a plastic flexural hinge in the ship in a heavy collision, limiting the sideways force on the column, and secondly for heavy indentations the force from the ship's side will drop when the plating reaches its ultimate strain and begins to tear. Both these limits appear to be below the force causing an overall failure of the shaft.

6. DROPPED OBJECTS

Design accidental load

Dropped objects is a rather frequent type of accident. Their ability to demolish parts of the platform has been demonstrated on several occasions.

Objects should be assumed dropped anywhere within the reach of the cranes. The magnitude of the design object should be chosen according to the lifting capacity of the crane. It should be assumed dropped from the maximum lifting height.

On the basis of these assumptions the design energy transferred to the structure may be estimated. Considering objects dropped into the water surface impact and drag forces reduce the velocity. In practice the impact velocity will be near the terminal velocity at a depth of 10 - 20 m. The fall direction in water should be assumed anywhere within a $20°$ cone.

Several experimental and analytical investigations have been carried out to identify the impact velocity on the caisson roof of a gravity platform.

	Weight	Velocity	Energy
Drill collar	2,8 t	23 m/s	76 tm
Hydril	5 t	16 m/s	64 tm
Winch	25 t	6-12 m/s	45-180 t m
Mud pump	33 t	7 m/s	81 tm

As the larger objects seemingly have a shape reducing its velocity, a design energy in the order of 100 tm seems reasonable (hydrodynamic mass added).

A crashing helicopter is a special type of dropped object. Depending on the magnitude of the platform, in the order of 1000 helicopters may be calling a year. The crushing strength of the helicopter will normally govern the design accidental load.

Steel structures

The verification of the energy absorption is very similar to the one described for ship collision. The main difference is the impact velocity calling for attention to

- brittle rupture

- increased yield strength

Concrete structures

For present types of structures the concrete elements above water and at small water depths are approximately vertical. Thus a possible falling object will swerve aside causing no impact damage above the caisson roof. If there is no oil storage, damage to the caisson roof can very well be accepted at some distance from the shafts. In the shaft area and in case of an oil storage the impact energy should be demonstrated to be absorbed by elastic strain energy unless pollution or costly repairs is accepted. In cases where this can hardly be achieved by structural concrete, a gravel or lightweight concrete cover can be used to reduce the impact.

A possible extended base slab might also be struck by the dropped object. In this case local damage necessary to absorb the kinetic energy will normally be acceptable.

7. FLYING FRAGMENTS

Design accidental load

Flying fragment can be originated by:

- desintegration of rotating machinery such as gas turbine blades, helicopter rotors etc.

- fragments of a pressure vessel when bursting

- loose objects thrown by some explosion impulse

The two latter types of fragments will occur in conjunction with explosions.

Reasonable design loads will have to be decided on the basis of the equipment at hand in each individual case. A pressure vessel rupture may produce fragments in the order of 100 kg, but small fragments with high velocity may have a higher damage potential.

Design measures

Fragments alone are not anticipated to knock down main parts of the structure. But they are very likely to occur in combination with other accidents, as a source or as a consequence. If so, they can aggravate the consequences by penetrating fire walls, perforating pressure vessels etc. For this reason possible fragment sources should either be encased or located so as to avoid harmful consequences.

Most of the penetration or perforation formulas for concrete and steel are derived empirically by military designers for bullets, artillery shells, bombs, etc. Large rough fragments, from failed turbines, equipment and from aircraft accidents are outside the limits of the formulas which therefore give conservative results.

A rough estimate of the penetration depth can be obtained by the modified Petry formula (18).

8. EXPLOSIONS

Design basis accident

Explosions can be divided in

- detonations (i.e. a supersonic process)
- deflagrations (i.e. a subsonic combustion)
- shock waves due to sudden energy release e.g. from a pressure vessel rupture

Detonations are not likely to occur in the gas-air mixtures normally encountered on offshore oil and gas platforms and consequently not a design accidental load.

The design accidental loads due to deflagrations are related to whether the explosion is confined, partly confined or an unconfined vapour cloud explosion.

Guidance how to determine the deflagration pressure in confined and partly confined conditions is given by NFPA-68 (16). A revised edition is expected to be issued in the near future.

For unconfined explosions the pressure depends on the shock wave produced. For one platform with a very low degree of confinement 27 kN/m^2 has been accepted as design basis for directly exposed walls and 10 kN/m^2 for indirectly exposed walls.

DnV have computer programs for more accurate determination of explosion pressures in a given case.

Loads due to pressure vessel bursts are given f.inst. by Baker et.al. (21)

Design

Having estimated a reliable pressure-time curve for an explosion the design may proceed by one of the following methods.

1. Energy solutions based on principle of absorption of impact energy as discussed in other chapters of this paper.

2. Methods based on equations of dynamic equilibrium.

The latter methods would mainly be of interest in the cases where the load duration T is in the order of magnitude of the response time of the structure T_N. For internal deflagration static solutions are normally adequate. The dynamic method is of main interest for external shock pressures.

The procedure steps in analysis are as follows:

1. Determine the resistance - deflection function of the equivalent system in elastic, elasto-plastic and plastic ranges

   Real and idealized resistance - deflection curves are shown in Fig. 3 - 4 for a reinforced concrete element.

   Using the unit resistance r (kN/m^2) plastic total resistance is r_u · Area and elastic total resistance = r_e · Area with corresponding stiffnesses in elasto-plastic and elastic ranges. Normally the equivalent unit stiffness K_E and corresponding deflection X_E are simplified as shown in fig. 4. The plastic resistance r_u can be found in appropriate handbooks e.g. (17).

   The above considerations apply to flexural resistance. Normally shear capacity corresponding to r_u is provided for the element in order to preclude the brittle mode of shear failure.

2. Determine natural period of vibration of equivalent one mass system

   The effective natural period of vibration for the single mass-spring system:

   $$T_N = 2\pi \sqrt{\frac{m_e}{K_E}} = 2\pi \sqrt{\frac{K_{LM} m}{K_E}}$$

   m = mass pr. unit area

   m_e = effective mass pr. unit area

   K_E = equivalent unit stiffness of the system (see Fig. 4)

   K_{LM} = "loss-mass" factor determined from consideration of the load, resistance and mass factors necessary to transform the actual dynamic system to the equivalent single mass system.

   Values for K_{LM} for various ranges of behaviour are shown in Fig. 5.

   The values used for m_e and K_E for a particular element depend upon the allowable maximum deflections permitted. When designing for completely elastic behaviour, the elastic values of m and K are used. For elasto-plastic action m and k is taken as average in the elastic and elasto-plastic regions. For design in the plastic range a weighted value of the effective unit mass is used with the equivalent unit stiffness K_E (Fig. 4). When plastic deformations are limited e.g. plastic hinge rotations < 5°, the use of the average value of the average elastic and elasto-plastic effective unit masses and the plastic effective unit mass will normally suffice.

3. Use charts to determine response

In practice the design check is to control that for a given explosion with peak pressure B and duration T, the plastic strain shall be kept within an acceptable value X_m.

Response charts (19,20) are prepared on non-dimensional basis of the types shown in Fig. 6 (Gas Explosion load characteristics) and Fig. 7 (Detonation load characteristics).

For a given structural element, the idealized resistance - deflection function defined by r_u, X_E and K_E can be determined along with the natural period T_N. Then knowing ratios B/r_u and T/T_N, X_m can be readily obtained from the graphs.

9. FIRE

Design accidental load

Governmental regulations and guidance pay a lot of attention to active protection measures. Considering structures they lay down requirements to the fire protection of certain separation walls (mainly A60 and B15). But the requirements to fire protection of the load bearing structure is rather brief. It is simply stated that it should be protected.

The magnitude and duration of the "design fire" should be decided with due regard to all types of active protective measures. The main criteria should be that the active and passive protection in cooperation should almost eliminate the probability of a main structural collapse due to fire.

Steel structures

Unprotected steel structures are vulnerable to fires. Even a fire duration of a few minutes only, can increase the temperatures above 500°C and result in a structural collapse. Thus all steel surfaces of the main structure which can be directly exposed to fire or to heat flux from fires within a distance up to some 50 - 100 m are to be protected. It is important to note that if the structure is intended to resist some impact utilizing plastic strain energy, the fire protection coating must be intact and perform its function under a possible subsequent fire. As an alternative to surface protection, hollow water filled sections may be used.

Concrete structures

Concrete structures of dimensions as applied for offshore structures have normally sufficient fire resistance without further protective measures. Criteria compliance should be verified in each individual case on the basis of published data (9) (10).

The concrete durability is associated with large heat capacity and low thermal conductivity.

10. EARTHQUAKE

Design accidental load

In areas of high seismicity earthquake should be considered an environmental load. It is to be considered in the normal design verification in the ultimate limit state. Irrespective seismic activity an accidental earthquake should be considered. This should be the maximum credible earthquake which could be expected to occur on the site.

The ground motions used as basis for design shall adequately represent the expected conditions at the site, both in terms of frequency and energy distribution.

The effects of local soil conditions in amplifying or attenuating the ground motion and in altering the frequency content are to be studied in order to determine appropriate horizontal and vertical characteristic values of the ground motion.

The ground motion may be described either in terms of time histories or in terms of response spectra. Standard spectra generally recognized as being valid for the region and the site conditions considered, may be considered when describing the ground motion.

The ground motion is normally to consist of three components which are to be applied simultaneously, i.e. in the two horizontal directions and in the vertical direction.

When the response spectra method is used the minimum values of ground motion in the three directions are:

- 100 % in the horizontal direction (i.e. the principal axis) considered most unfavourable to the structure

- 70 % in the orthogonal horizontal direction

- 50 % in the vertical direction

The maximum earthquakes observed in the North Sea the last 100 years is of magnitude about 6. A shallow earthquake of this magnitude might result in severe bedrock accellerations at the epicentre.

Assuming a reasonable attenuation from epicenter to a distant platform site an effective ground accelleration of 0,15 g with a spectrum as recognized for design in California have been accepted for the northern North Sea.

Design measures

A main strategy to obtain survivability of structures exposed to earthquakes is to ensure a ductile performance. Thus a strict compliance with the above mentioned indirect design requirements is indispensable.

Certain authors (11) claim no further analysis to be required for traditional steel jacket structures. API (8) lay down comprehensive provisions for design of steel jackets to resist earthquake. A practical approach to meet the requirements is described in (12). Even if the survivability of the structure as such might be ensured by indirect design measures alone, some dynamic analysis of the platform is strongly recommended. This is to ensure the deck motions to be sufficiently small to avoid damage to pipes and equipment.

Concrete structures should be subject to a dynamic structural analysis.
Ideally the non-linear behaviour should be taken care of by the introduction of constitutive laws incorporating the progressive degradation
characteristics of reinforced concrete and soil. As such analyses are
far from being standard techniques, simplified linear analyses will have
to be accepted. In case of overstressing the amount of yielding should
be estimated from case to case.

Analyses carried out for current North Sea platforms indicate no overstressing to be likely to occur. The more vulnerable elements seems to
be the transition piece of the deck/tower joint.

The main earthquake danger is associated with its impact on piping
equipment and appurtenances essential to operation and safety. In
certain cases these effects may be investigated on the basis of conventional static analysis by introduction of appropriate inertia forces.
The adequacy of these approaches, neglecting dynamic interaction between
components and their supporting structure, should be verified in each
case. More refined methods are discussed in (13).

11. LOSS OF INTENDED DIFFERENTIAL PRESSURE

Design accidental load

The accidents discussed are exclusive for the concrete platforms. They
may have different sources such as:

- accidental filling of shaft intended to remain empty

- loss of intended underpressure in the caisson. This means that a
  favourable prestressing due to the pressure difference is lost. In
  case of an oil storage the external pressure may be turned to a
  significant internal pressure of up to 80 - 100 kN/m^2.

- leakage of pipe guided through a closed caisson compartment resulting
  in excessive internal pressures

All these loads have a specific upper physical limit. Thus the load to
be designed for is easily defined. A common feature of the accidents is
that they must be initiated by some malfunction of equipment or piping.
By improved, foolproof equipment with sufficient redundance the
probability of these accidents might be reduced so much that they may
be disregarded in design.

Design measures

The load cases corresponding to changed pressure differences may
easily be incorporated in the normal design process. The acceptance
criteria should be chosen with some care. A certain local damage might
be difficult to repair, result in reduced durability and render the
oil storage unfit for its purpose because of excessive leakage. The
Owner should decide whether he wants to limit the reinforcement
stresses so as to avoid such calamities.

12. ACKNOWLEDGEMENTS

The author wants to express his gratitude to his colleges in Det norske
Veritas for valuable assistance and advice during the preparation of
the present paper. Above all mr. E. Borse, B. Carlin, G. Foss and
B. Røland should be mentioned.

13. REFERENCES

(1) ECCS: European recommendations for steel construction 1978.

(2) CEB-FIP: Model code for concrete structures. 1978.

(3) Department of Energy: Offshore installations Guidance on design and construction. London 1977.

(4) FIP: Recommendations for the design and construction of sea structures. 3 ed. 1977.

(5) Norwegian Petroleum Directorate: Regulations for the structural design of fixed structures on the Norwegian Continental Shelf 1977.

(6) Det norske Veritas: Rules for the design construction and inspection of offshore structures. 1977.

(7) Heldor: Accidents occurred to structures engaged in offshore oil and gas drilling operations in the periode 1970 - 1976. Det norske Veritas report no. 78-103. Oslo 1978.

(8) API PP2a: Planning, Designing and Constructing Fixed Offshore Platforms. 1977.

(9) FIP/CEB: Recommendations for the design of reinforced and prestressed concrete structural members for fire resistance. London 1975.

(10) FIP/CEB: Report on methods of assessment of the fire resistance of concrete structural members. London 1978.

(11) Kallaby: Considerations for analysis and design of piled offshore structures in severe earthquake environment. OTC 2748 Houston 1977.

(12) Gates, Marshall Mahin: Analytical methods for determining the ultimate earthquake resistance of fixed offshore structures. OTC 2751 Houston 1977.

(13) Kost et.al.: Seismic resistant Design of Piping Equipment and Appurtenances for Offshore Structures. OTC 2750 Houston 1977.

(14) Fjeld et.al.: Risk analysis of offshore production and drilling platforms. OTC 3152 Houston 1978.

(15) Norwegian Petroleum Directorate: Regulations for Production and Auxiliary Systems on Production Installations etc. 1978.

(16) NFPA No. 68 - 1974: Guide for Explosion Venting

(17) K.W. Johansen Slab formulae (Pladeformler) Copenhagen 1949 (In danish)

(18) Doyle et.al.: Design of missile resistant concrete panels. 2 nd international conference on Structural Mechanics in Reactor Technology Vol. 4 part J Berlin 1973 CID Publ. Luxembourg 1973.

(19) R.J. Mainstone: The effect of explosions in buildings. Proceedings of a symposium at Building Research Establishment (1974).

(20) Dept. of the Army, Navy Air Force (US): Structures to resist the effect of accidental explosions 1969.

(21) Baker, W.E. et.al.: Workbook for predicting pressure wave and fragment effects of exploding propellant tanks and gas storage vessels. NASA CR-134906 1975.

TABLE 1

LOAD AND MATERIAL COEFFICIENTS TO BE USED FOR ACCIDENTAL COMBINATIONS
GENERAL STRUCTURAL CODES

Load coefficients	ECCS Steel structures	CEB-FIP Concrete structures	CP 110 Concrete structures		
Dead load	1,1 - 0,9	1,1 - 0,9	1,4	0,9	1,2
Live load	$1,0^x$	$1,0^x$	$1,6^x$	0	$1,2^x$
Environmental load	$1,0^x$	$1,0^x$	0	$1,4^x$	$1,2^x$
Accidental load	1,0	1,0	1,05	1,05	1,05
Material coefficients					
Structural steel	1,0 - 1,12	-	-		
Reinforcement steel	-	1,0	1,0		
Prestressing steel	-	1,0	1,0		
Concrete	-	1,3	1,3		

^{x}Characteristic values reduced by factors taking into account the probability of simultaneous occurrence.

TABLE 2

LOAD AND MATERIAL COEFFICIENTS TO BE USED FOR ACCIDENTAL COMBINATIONS
CODES FOR OFFSHORE STRUCTURES

Load coefficients	Norwegian Petroleum Directorate xxx		Det norske Veritas xxx	FIP Concrete structures		
	Accidental load included	Accidental load not included				
Dead load	1,0	1,0	1,0	1,2	1,1	0,9
Live load	1,0	1,0	1,0	1,6	1,3	0,9
Environmental load	0	1,0	1,0	1,4	1,3	1,3
Deformation load	0^x-$1,0^{xx}$	0-$1,0^{xx}$	0^x-$1,0^{xx}$	1,1	1,1	1,1
Accidental load	1,0	0	1,0	1,05	1,05	1,05
Material coefficients						
Structural steel	1,0	–	1,0	–		
Reinforcement steel	1,0	–	1,0	1,0		
Prestressing steel	1,0	–	1,0	1,0		
Concrete	1,1	–	1,1	1,3		

x Indirect effects

xx Direct effects

xxx Local damage accepted

Material coefficients are not explicitly stated in the NPD regulations.
The numbers are implemented in certification practice. Only those loads
likely to be acting simultaneously to be considered.

	ALL UNITS						LOSS OF LIVES
TYPE OF ACCIDENT	STRUCTURAL LOSS						
	TOTAL	SEVERE	DAMAGE	MINOR	NO	SUM	
Structural	0	5.0	12.8	19.5	3.8	11.4	7
Weather	10.8	25.0	30.8	16.7	13.2	20.5	13
Collision	8.1	5.0	8.5	13.0	39.6	14.4	22
Blow out	29.8	25.0	11.7	11.1	17.0	15.9	35
Fire	8.1	7.5	12.8	15.7	0	10.5	12
Explosion	5.4	5.0	9.6	7.4	0	6.3	24
Capsizing / Grounding / Leakage / Machine / Foundering	37.8	27.5	13.8	16.6	26.4	21.0	32
SUM	100	100	100	100	100	100	145

Table 3. Percentage of the type of cause for each degree of structural loss and number of lives lost. All fixed and floating structures engaged in oil and gas drilling and production the periode: 01.01. 1970 - 31.12. 1977. "Structures" means damage not associated with weather or accidents.

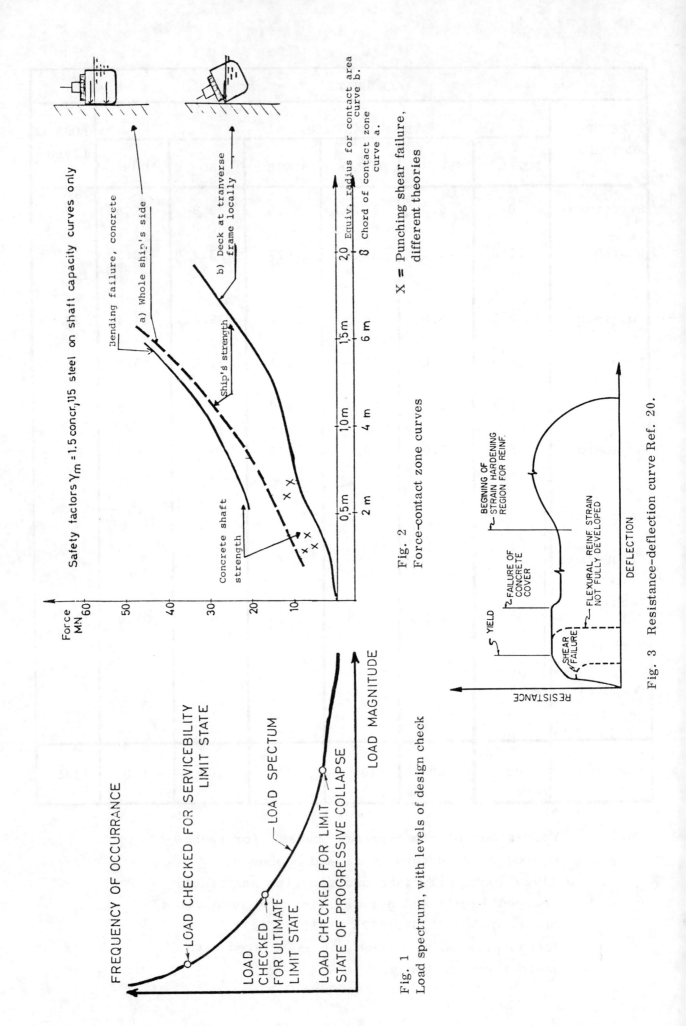

Fig. 1
Load spectrum, with levels of design check

Fig. 2
Force-contact zone curves

X = Punching shear failure, different theories

Fig. 3 Resistance-deflection curve Ref. 20.

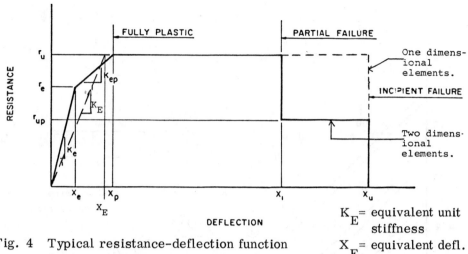

Fig. 4 Typical resistance-deflection function for two-way element   Ref. 20

K_E = equivalent unit stiffness
X_E = equivalent defl. at first yield

One way elements

	Edge conditions	Range of behavior		
		Elastic	Elasto-Plastic	Plastic
Cantilever		0.65	—	0.66
Simple supports		0.79	—	0.66
Fixed supports		0.77	0.79	0.66
Fixed-simple supports		0.78	0.78	0.66

Two way elements

Edge conditions		Values of L/H	Elastic and elasto-plastic ranges (support conditions)					Plastic range	
			All supports fixed	One support simple, other supports fixed	Two supports simple, other supports fixed	Three supports simple, other supports fixed	All supports simple	Before partial failure	After partial failure
Two adjacent edges supported and two edges free		All	0.65	0.66	—	—	0.66		0.66
Three edges supported and one edge free		$L/H < 0.5$	0.77	0.77	0.79	—	0.79	See figure 6-5	0.66
		$0.5 \leq L/H \leq 2$	$0.65 - 0.16\left(\frac{L}{2H}-1\right)$	$0.66 - 0.144\left(\frac{L}{2H}-1\right)$	$0.65 - 0.186\left(\frac{L}{2H}-1\right)$	—	$0.66 - 0.175\left(\frac{L}{2H}-1\right)$		
		$L/H \geq 2$	0.65	0.66	0.65	—	0.66		
Four edges supported		$L/H = 1$	0.61	0.61	0.62	0.63	0.63		0.66
		$1 \leq L/H \leq 2$	$0.61 + 0.16\left(\frac{L}{H}-1\right)$	$0.61 + 0.16\left(\frac{L}{H}-1\right)$	$0.62 + 0.16\left(\frac{L}{N}-1\right)$	$0.63 + 0.16\left(\frac{L}{H}-1\right)$	$0.63 + 0.16\left(\frac{L}{H}-1\right)$		
		$L/H \geq 2$	0.77	0.77	0.78	0.79	0.79		

Fig. 5 K_{LM} - values   Ref. (20)

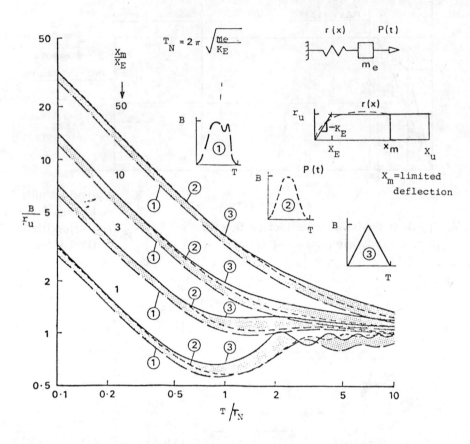

Fig. 6 Further calculated elasto-plastic responses of a simple sprung mass to three types of dynamic load (sketched in the centre) such as might be imposed in internal gas explosions. Ref. (19)

BOSS'79

Second International Conference on Behaviour of Off-Shore Structures

Held at: Imperial College, London, England
28 to 31 August 1979

PAPER 76

SOME ASPECTS OF DOUBLE-SKIN COMPOSITE CONSTRUCTION FOR SUB-SEA PRESSURE CHAMBERS

P. Montague, BSc, PhD, CEng, MICE, MIMechE, MRAeS.
and C.D. Goode, BSc, PhD, CEng, MICE.

University of Manchester, U.K.

Summary

A double-skin, composite construction for vessels subjected to external pressure is described. It has been innovated and is being developed at the University of Manchester and is being included by Sir Robert McAlpine and Sons Ltd in their EEC feasibility study of 12 m internal diameter production chambers for their Subseamac Project.

The construction consists of two concentrically placed cylindrical thin steel shells with the intervening annular space filled with concrete. It is designed to overcome the imperfection sensitivity problems associated with ring-stiffened shells and, by retaining a stable structure, to make use of the full yield strength of the steel skins. Operational depths up to 1000 m are envisaged.

The theoretical basis of the composite form is described briefly, the resulting important stress patterns are described and strength failure criteria are postulated. Examples of experimental tests are used to illustrate the relationship between theoretical and laboratory performance and this is found to be reasonably good. Consideration is given to the interaction between strength and stability and the importance of initial geometric imperfections in determining failure of the shells.

Nomenclature

a	mean radius of shell
D_1	outside diameter of shell
D_2	inside diameter of shell
E_c	tangent modulus of concrete
E_{co}	E_c at zero concrete strain (initial tangent modulus)
E_f	elastic modulus for resin-glass filler
E_s	elastic modulus for steel
f_{cu}	characteristic cube strength of concrete
f_{max}	uniaxial strength of concrete
h	total wall thickness
p	external pressure applied to shell
p_1	radial stress at filler: outside skin interface
p_2	radial stress at filler: inside skin interface
p_f	experimental failure pressure
p_p	p at first skin yield
p_{pp}	p at second skin yield
p_{th}	theoretical strength failure pressure
t_1	outside skin thickness
t_2	inside skin thickness
α	D_1/h
Δ_{max}	maximum deviation from mean circle
γ	$(t_1 + t_2)100/h$ = percentage steel in wall
ε_c	concrete strain
σ_c	concrete stress
σ_{fy}	yield stress of resin-glass filler
σ_{sy}	yield stress of steel skins
σ_{x1}	axial stress in outside skin
σ_{x2}	axial stress in inside skin
σ_{xf}	axial stress in filler
σ_{y1}	circumferential stress in outside skin
σ_{y2}	circumferential stress in inside skin
σ_{yf1}	circumferential stress at outside of filler
σ_{yf2}	circumferential stress at inside of filler

Introduction

The type of vessel considered by this paper is shown in Fig. 1a. Particular consideration is given to the structural behaviour of the circular, cylindrical section between the hemispherical ends when the chamber is subjected to external pressure. It is the shape envisaged by Sir Robert McAlpine and Sons Ltd. for their sub-sea production module (Ref. 1). The double-skin construction (two steel skins separated by a poured and cured filler such as concrete) has the objective of producing a shell wall which is very stiff in bending and therefore insensitive to initial shape imperfections (non-circularity), enabling the shell to remain stable up to and beyond the external pressure which causes the steel skins to develop their full yield strength. The stresses in the wall are defined in Fig. 1b. In the theory (Ref. 2 and 3) it is assumed that the stress distribution is uniform across the thin skins, but the filler is considered as a thick cylinder with the stresses varying across its thickness.

Displacements and stresses

The behaviour of this form of construction is illustrated by the change in radial displacement with increasing external pressure. The theoretical form of this relationship is shown on Fig. 2a for two types of filler allied with two strengths of steel skins. The concrete filler is assumed to follow a parabolic-rectangular stress-strain relationship and the resin-glass filler a linear stress-strain relationship. The resin-glass filler has a low elastic modulus compared with the concrete up to the steel yield. Both types of filler have been used in experimental tests (Ref. 4, 5 and 6). For any of the curves on Fig. 2a, the reduction in radial stiffness occurs when the steel skins reach yield (first, the inside skin at $p = p_p$ followed by the outside skin at $p = p_{pp}$). For a given steel and shell geometry, p_p and p_{pp} will occur at the same two values of circumferential strain (radial displacement) regardless of the filler stiffness. Thus, in Fig. 2a, the stiffer concrete enables the shell to sustain a higher pressure before steel yield than when the low modulus resin-glass filler is used. In all cases, failure of the shell occurs when the filler reaches stress breakdown (see below). The failure pressure does not depend on the stiffness of the filler but only on its breakdown criterion.

Each type of structural behaviour, illustrated in Fig. 2a and dependent upon the type of filler used, has its merits. The low-modulus filler gives a large post-steel-yield pressure range before breakdown takes place. However, during this period, due to the low bending stiffness of the shell wall (with both skins at yield), the initial geometric imperfections of the shell become exaggerated, which may cause premature failure due to the presence of bending stresses in the filler. The concrete-filled wall retains its full elastic stiffness to a higher pressure and the growth of initial imperfections is consequently inhibited. The two types of steel illustrated in Fig. 2a demonstrate the advantage, at both first yield and failure, of using a high strength steel provided the filler is sufficiently strong not to break down before the steel yields.

As previously demonstrated (Ref. 2 and 3), the most important stresses in the circular, cylindrical shell away from the influence of the end restraints are the circumferential stresses σ_{y1}, σ_{y2}, σ_{yf2} and the radial interface stress between the inside skin and the filler, p_2. The increases of σ_{y1} and σ_{y2} with pressure are shown in Fig. 2b for two geometrically similar shells, one with the low-modulus, linear resin-glass filler (the full lines) and the other with the concrete filler (the dashed lines). In each case, the first discontinuity in behaviour occurs when σ_{y2} reaches the steel yield value (i.e. $\sigma_{y2} = \sigma_{sy}$ at $p = p_p$). The slower rise of σ_{y2} in the concrete-filled shell, due to the greater proportion of the circumferential load taken by the concrete filler than by the resin-glass filler, is clearly demonstrated. **For the resin-glass filled shell, yield of the outer skin follows quickly (i.e. $\sigma_{y1} = \sigma_{sy}$ at $p = p_{pp}$). However, for this shell, with steel skins for which $\sigma_{sy} = 355$ Nmm^{-2}, the concrete filler breaks down before the outside skin reaches yield.** A steel skin with a yield stress of about 320 Nmm^{-2} would cause p_{pp} to occur before the concrete breakdown.

Strength failure of the shell

Failure of the shell occurs when the filler breaks down; the different criteria which have been assumed for the two filler materials are illustrated on Fig. 2c. For the resin-glass filler, the maximum shear stress (Tresca) criterion is considered appropriate (Ref. 2 and 4) leading to breakdown when $(\sigma_{yf2} - p_2) = \sigma_{fy}$. For the concrete, the criterion suggested by Hobbs et al (Ref. 7) has been adopted viz. $(\sigma_{yf2} - 3p_2) = 0.67 f_{cu}$. In spite of the less-demanding breakdown criterion used for the concrete than for the resin-glass, the concrete-filled shell fails first. This is because of the lower strength of the concrete viz $0.67 f_{cu} = 46.9$ Nmm^{-2} compared with $\sigma_{fy} = 62.5$ Nmm^{-2} for the resin-glass mixture.

Experimental examples

In Fig. 3 the theoretical predictions are compared with some of the experimental results previously reported (Ref. 4 and 6). The only difference between the two specimens with the concrete filler was the concrete strength. With the weaker mix (Specimen 6.1), the concrete was close to its failure criterion when p reached p_p (yield of the inside skin) and failure took place soon after this. The theoretical curves on Fig. 3 are based on an experimentally derived stress-strain curve given by $\sigma_c = E_{co} \varepsilon_c - E_{co}^2 \varepsilon_c^2 / 4f_{max}$ with $E_{co} = 4000\sqrt{f_{cu}}$ and $f_{max} = 0.67 f_{cu}$. This value of E_{co} was obtained from test measurements on concrete cylinders and was therefore preferred to the CP110 (Ref. 8) value of $5500\sqrt{f_{cu}}$ used on Fig. 2. The cylinder with the resin-glass filler on Fig. 3 is not directly comparable with the concrete-filled shells because, although the geometries of the three specimens were similar, the skin thickness on the resin-glass shell was 1.2 mm instead of 1.96 mm. However, the comparison does show the long post-steel-yield capacity typical of the low modulus filler, which was 4400 Nmm^{-2} for this cylinder.

In order to deduce the true circumferential shape of the cylinder at any stage of loading, it is necessary to relate the radial measurements to a calculated true mean circle (Ref. 4). Fig. 4 shows shape profiles for Specimen 2.3 which had a concrete filler. It is typical of the behaviour of the cylinders tested and shows that little change in profile shape occurred until the steel yield pressure p_p was reached (4.2 Nmm^{-2} for this specimen) and that the initial geometry determined the pattern of subsequent displacements. The points of 'contraflexure' hardly change their positions throughout the loading. The initial lack of circularity is clearly shown in Fig. 4 (the profile at p = 0) and the maximum deviation from the mean circle was 1.3% of the mean radius. This is $2\frac{1}{2}$ times the design figure of 0.5% recommended in BS5500 (Ref. 9) yet the failure pressure for this shell was 20% above the theoretical pressure required to cause both steel skins to yield.

Variation of α and γ for concrete-filled shells

The cross-section of the double-skin shell is completely defined for analytical purposes by the two independent geometric parameters α and γ, α being the ratio of the outside diameter of the shell to its total wall thickness and γ the percentage of steel in the wall. Fig. 5a and 5b show the variation of p_p for two strengths of steel skins filled with concrete of $f_{cu} = 70$ Nmm^{-2}, a range of α-values from 10 to 26 and γ-values between 8 and 20%. Throughout these curves $t_1 = t_2$.

For any value of α, p_p varies linearly with γ; an increase of 50% in p_p requires approximately a 150% increase in the volume of steel. At the end of each line on Fig. 5 is shown the value of $(\sigma_{yf2} - 3p_2)$ at $p = p_p$. Thus, for α = 10 with $\sigma_{sy} = 355$ Nmm^{-2}, $(\sigma_{yf2} - 3p_2)$ varies from 39.8 Nmm^{-2} to 24.6 Nmm^{-2} as the percentage of steel is increased from 7.7 to 20. Moreover, the variation of $(\sigma_{yf2} - 3p_2)$ at p_p is linear with γ, so that paired values of these two quantities can be readily interpolated. On Fig. 5, two values of $(\sigma_{yf2} - 3p_2)$ have been marked; $0.67 f_{cu}$ and $0.45 f_{cu}$. The former is recommended by Hobbs et al (Ref. 7) as the limit of the concrete strength for the purpose of comparison with test results, whereas $0.45 f_{cu}$ would be a prudent value for design. Recent experimental results (Ref. 6) have suggested that these values may be over-cautious for the particular case of concrete totally confined between the steel skins of the composite shell. Whatever criterion is chosen, families of curves such as those shown in Fig. 5 can be readily produced to define the corresponding p_p-values.

The influence of instability on strength

The influence of initial imperfections on failure and the interaction between strength and instability are important factors affecting the performance of shells subjected to external pressure. Recent tests have shown (Ref. 6) that for steel-concrete-steel shells with α-values less than about 18 the stiffness of the wall inhibited the growth of initial imperfections and the full theoretical failure pressure based upon strength alone, p_{th} was achieved and often exceeded.

With thinner walls and less stiff fillers, instability may begin to interact with strength to reduce the failure pressure below p_{th}. The elastic instability pressure of all the double-skin cylinders tested was six times the value of p_{th}, but once a skin has reached yield the stiffness of the shell will decrease rapidly. It has been argued (Ref. 5) that the partially plastic sandwich shell may be analysed for instability by regarding one skin to be operative as an elastic component, totally ignoring the other skin and then considering an equivalent homogeneous shell of thickness $\bar{h}$ and elastic modulus $\bar{E}$. When concrete is used as the filler the E_c used in calculating $\bar{E}$ is that associated with the maximum stress in the concrete at the pressure when the first skin yields. The thinness ratio, λ, defined as

$$\lambda = \sqrt[4]{\frac{(\ell/2a)^2}{(\bar{h}/2a)^3}} \sqrt{\frac{\sigma_y}{\bar{E}}}$$ is a useful measure of the 'stiffness' of the shell.

Fig. 6 shows the experimental results plotted against λ. There is clearly a trend for the failure pressure ratio (p_f/p_{th}) to decrease as the cylinders become less stiff (increasing λ). The results of tests on conventional ring-stiffened shells show the enveloping lower bound curve flattening out asymptotically to $p_f/p_{th} = 0$, whereas the curve in Fig. 6 is flattening out at $p_f/p_{th} = p_p/p_{th}$. The implication is that this double skin construction will achieve at least a pressure equal to that to cause the first skin to yield. Also shown on this figure is the effect of out of roundness (Δ_{max}/a) indicating that the greater the imperfections the lower the strength but that much larger geometric imperfections can be tolerated than the 0.5% recommended in BS5500 (Ref. 8).

Distribution of steel between the two skins

In all the experimental tests completed to date, the steel has been equally distributed between the inside skin and the outside skin i.e. $t_1/(t_1 + t_2) = 0.5$. It is a matter of considerable practical interest whether this is the best arrangement or whether some benefit might be gained by changing this ratio. Fig. 7 shows an example of the theoretical answer for a particular value of α (= 14) and 355 Nmm^{-2} and 280 mm^{-2} steel skins filled with concrete having a cube strength of 70 Nmm^{-2}. The values of p_p, p_{pp} and p_{th} at $t_1/(t_1 + t_2) = 0.5$ on Fig. 7 correspond with those on Fig. 2. The first obvious conclusion from Fig. 7 is that all of these quantities increase as the steel is shifted from the outside of the shell wall to the inside, the clear implication being that, from the point of view of strength, the best performance, for any given percentage of steel, is achieved by putting it all into the inside skin and dispensing with the outside skin. In considering the merits of no outside skin, the previous discussion about strength-stability interaction is important. If the shell circularity was perfect and remained so until filler strength breakdown caused failure, then the best arrangement would be with no outside skin. But the presence of initial out-of-roundness imperfections requires that the bending stiffness of the wall is made as high as possible. This applies before yielding of the steel skins as well as after. The absence of an outside skin would not only reduce the bending stiffness dramatically but would thereby introduce the danger of concrete cracks on the outside surface of the shell due to unfavourable curvature changes. Although the effect of varying the distribution of steel between the inside and outside of the wall has not yet been experimentally quantified and related to the degree of initial imperfection, it does seem that the best disposition might be with a slightly greater proportion of steel allocated to the inside skin.

Conclusions

Some fifty experimental tests on double-skin composite shells have shown a reasonable correlation between performance and theoretical prediction. However, these tests have been carried out at a relatively small scale, the biggest outside diameter being 500 mm, and much larger scale tests are required to supplement the present evidence.

It has been demonstrated that the composite shell is insensitive to initial non-circularity imperfections up to an external pressure which causes the steel skins to yield. Up to this pressure, the radial displacements were nearly axisymmetric. The thinness ratio, λ, is a useful measure for indicating the influence of stiffness on strength.

A vessel with a concrete filler will sustain a higher pressure before the steel skins start to yield than will a similar vessel with a resin-glass filler because the concrete is stiffer. Resin-glass fillers, though more expensive, are much lighter in weight and may have advantages for submersibles.

Acknowledgement

The research described in this paper is being carried out within the North West Universities Marine Technology Consortium supported by the Marine Technology Directorate of the Science Research Council.

References

1. Derrington,J.A. and Barrack,J.W: 'Sub-sea production system, Subseamac One'. Oceonology International '78, Offshore Structures, Brighton 1978.

2. Montague,P: 'A simple composite construction for cylindrical shells subjected to external pressure'.J.Mech.Eng.Sci.,17, 2 pp.105-113 (February 1975).

3. Montague,P: 'The theoretical behaviour of steel-concrete circular cylindrical shells subjected to external pressure'.Proc.I.C.E., Pt II (June 1979).

4. Montague,P: 'The experimental behaviour of double-skinned composite, circular, cylindrical shells under external pressure'.J.Mech.Eng.Sci.,20, 1 pp.21-34 (January 1978).

5. Montague,P: 'The failure of double-skinned composite, circular, cylindrical shells under external pressure'.J.Mech.Eng.Sci.,20, 1 pp.35-48 (January 1978).

6. Goode,C.D. and Fatheldin,Y.T: 'Sandwich cylinders (steel-concrete-steel) subjected to external pressure'. ACI Convention, Milwawkee, March 1979. (Submitted to ACI Journal for publication.)

7. Hobbs,D.N., Newman, J.B. and Pomeroy, C.D: 'Design stresses for concrete structures subjected to multiaxial stresses'. The Structural Engineer, 55, 4 pp.151-164 (April 1977).

8. British Standards Institution: 'The structural use of concrete'.CP110:1972.

9. British Standards Institution: 'Specification for unfired fusion welded pressure vessels'. BS5500:1976.

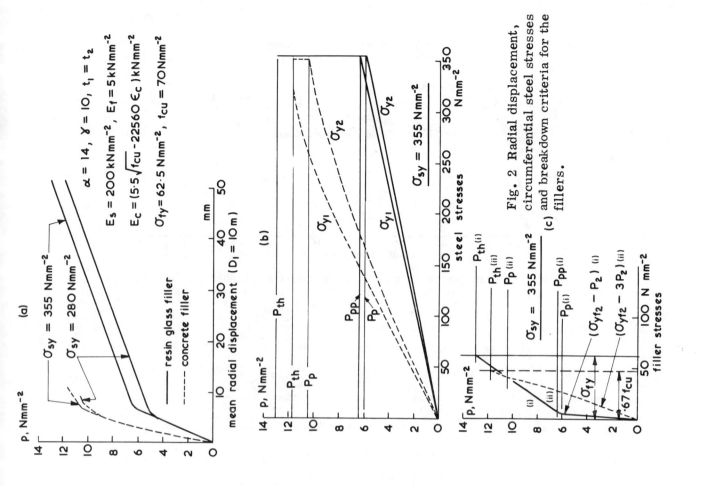

Fig. 2 Radial displacement, circumferential steel stresses (c) and breakdown criteria for the fillers.

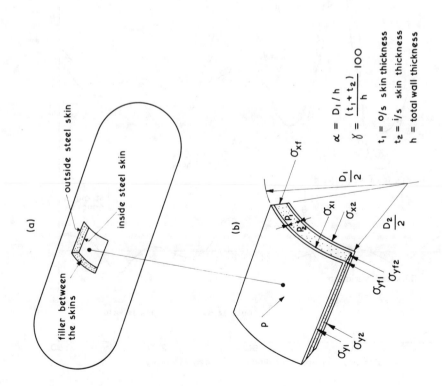

Fig. 1 Stresses in the double-skin shell wall.

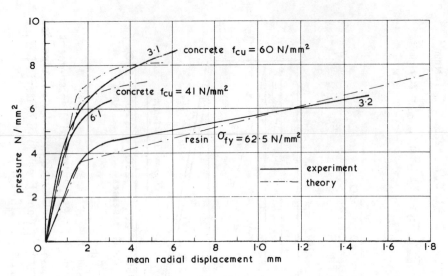

Fig. 3 Pressure-displacement. Experiment and theory compared

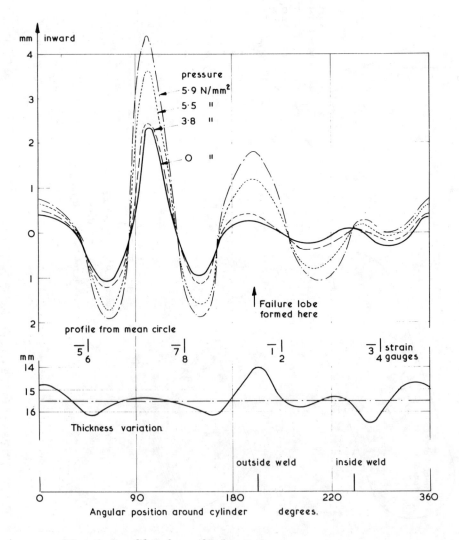

Fig. 4 Profiles for cylinder 2.3

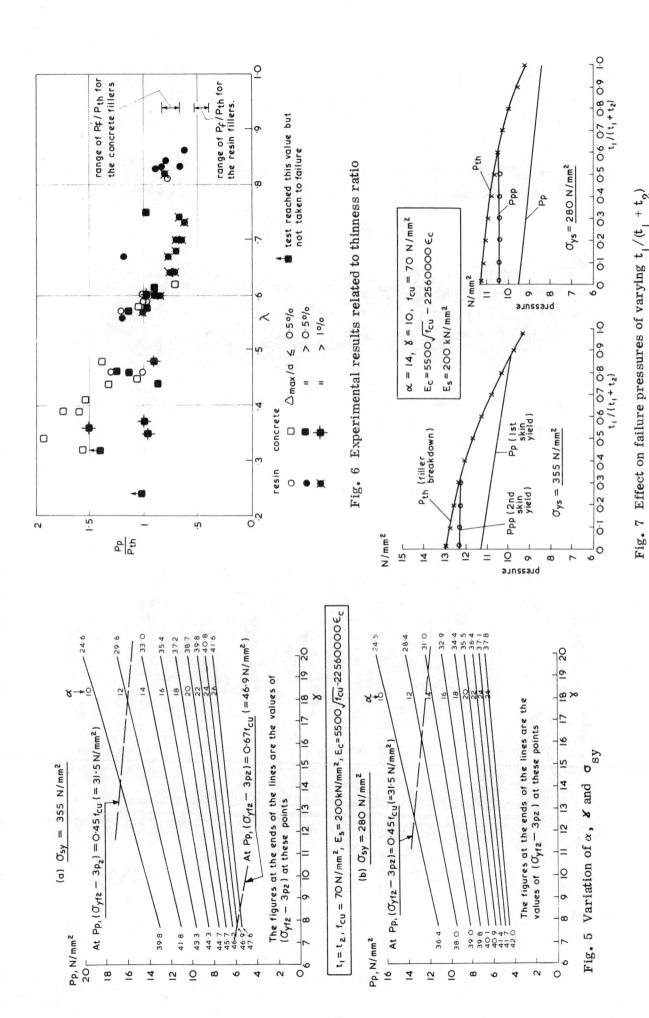

Fig. 6 Experimental results related to thinness ratio

Fig. 7 Effect on failure pressures of varying $t_1/(t_1+t_2)$

Fig. 5 Variation of α, γ and σ_{sy}

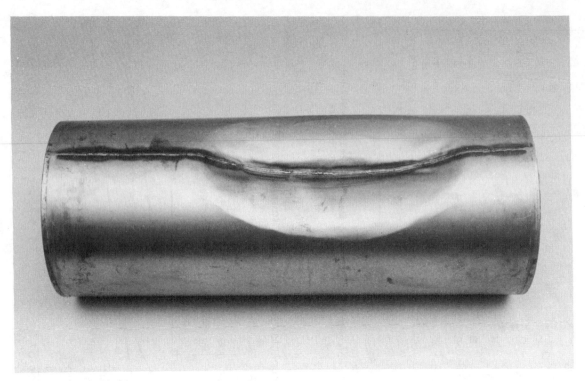

a) after failure.

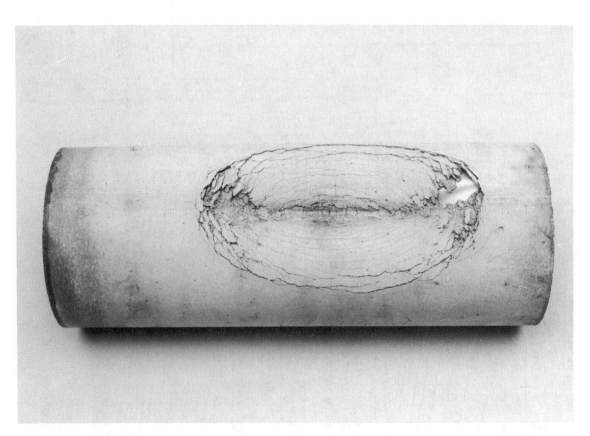

b) with outer steel skin removed

Figure 8. Typical concrete filled cylinder (cylinder 2.3).

BOSS'79

PAPER 77

Second International Conference
on Behaviour of Off-Shore Structures

Held at: Imperial College, London, England
28 to 31 August 1979

ENVIRONMENTAL AND STRUCTURAL INSTRUMENTATION OF PLATFORMS ON THE NORWEGIAN CONTINENTAL SHELF.

I. Holand

The Norwegian Institute of Technology, Norway

and

S. Berg and G. Beck

Offshore Technology Testing and Research Group (OTTER), Norway.

Summary

The Norwegian Petroleum Directorate have issued instrumentation regulations for fixed offshore platforms. According to these regulations, structural data (performance data or P-data) describing the dynamic platform behaviour as well as stresses and strains in the structures and the foundations, are collected on a regular basis. The purpose of these data is to evaluate the safety of the structures. In these evaluations environmental data (E-data) are also needed. A similar recording scheme is therefore specified for waves, current, wind, temperatures etc. These data will also enter a special environmental data bank at the Norwegian Meteorological Institute. This institution will be responsible for mapping the environmental conditions on the Norwegian Continental Shelf on a long term basis. Up to now a total of five different structures (two gravity structures, two jackets and one single point mooring platform) have been instrumented according to these regulations.

For the work in connection with the regulations, storage and analysis of the data as well as the safety evaluations, the Norwegian Petroleum Directorate have chosen the **Otter Group** (Offshore Technology Testing and Research Group) as their consultant.

In this paper the regulations and the instrumentation systems are described.

Sponsored by: Delft University of Technology, The Netherlands
Massachusetts Institute of Technology, U.S.A.
The Norwegian Institute of Technology, Norway
University of London, England

Secretariat provided by: BHRA Fluid Engineering

Copyright: © BHRA Fluid Engineering
Cranfield, Bedford, England

1. INTRODUCTION

Although the oil production industry in Norway is still quite young, it has become a significant part of our industry. Much pioneering work has been done in the North Sea during the last fifteen years and there are still great challenges to be faced.

In the areas where oil and gas have been found in sufficient quantities, production plants and living quarters are built on fixed platforms of steel and concrete. In this context the question of safety is vital. Regarding water depths and environmental conditions, experience of such large structures in areas comparable to the North Sea is very limited. Long term experience in particular is lacking. Therefore, it has been determined necessary to require instrumentation of the platforms in order to assess the safety level of the installations.

2. THE INSTRUMENTATION REGULATIONS

"Regulations for instrumentation, recording and processing of E (environmental) - and P (platform) - data" were issued by the Norwegian Petroleum Directorate on 1978-08-01. These regulations replaced provisional regulations issued 1975-12-22. The purpose of the regulations is twofold:

- Platform data and environmental data in the vicinity of the platforms must be collected in order to assess the safety of the load carrying structure of the platforms and their foundation. This is done by checking the design assumptions and by a semi-continuous registration of the behaviour of the platforms.

- Environmental data must be collected for a systematic statistical mapping of the physical environment.

From this it may be deducted that included in the approval of a design is a requirement that appropriate instrumentation be installed such that e.g. the safety factor applied are verified as sufficient.

According to the regulations it is generally the Norwegian Petroleum Directorate that decide what instrumentation will be required on each platform. The following types of platform data are mentioned specifically in the regulations:

- Strain
- Vibration
- Acceleration
- Water pressure
- Hydrodynamic pressure
- Shock pressure
- Pore pressure
- Earth pressure
- Settlement
- Corrosion data

Furthermore, the following environmental data may be required:

- Wind speed and direction
- visibility
- Weather - present and past
- Atmospheric pressure
- Air temperature
- Sea temperature
- Atmospheric humidity
- Clouds
- Precipitation
- **Waves - height, period and direction**
- Ice formation on structures
- Ice
- Current speed and direction
- Water level

Generally, the E-data sensors will be positioned such that the recordings are unaffected by the structures and the various activities on and around the platforms. In addition, wave measurements very close to the platforms may also be required in order to find the correlation between the waves and the platform response.

It is required that the sea temperature be measured at two depths, one being within 5 metres of the sea surface. Further, at least two current meters must be installed. The instrument should preferably be positioned in such a way that one measures the current speed and direction close to the bottom (within 5 metres of the bottom) and another measures the current speed and direction as close to the sea surface as possible. Additional current meters at other depths may be required.

According to the regulations it is also the Norwegian Petroleum Directorate that decide to what extent data should be recorded on each installation. Generally, the E- and P-data shall be recorded for a period of 20 minutes at 3 hourly intervals throughout the day. The number of P-data recording periods may be reduced during times of calm weather. On the contrary, when the environmental conditions are severe, i.e. when some predetermined threshold value for wave height and/or wind speed has been exceeded, more frequent registrations may be required. Measured response characteristics of the platform and/or foundation may, in addition, be used to trigger more frequent recordings and to set off alarms.

The data shall in general be recorded on magnetic tapes in a format agreed upon between the platform operator and the directorate, the tapes being forwarded to the directorate each month.

The P-data forwarded to the directorate are treated confidentially for a period of two years after the results of the initial data processing are available. After this period the P-data will be made generally available if there is no special reason for further confidentiality (§7 of the regulations).

The E-data are defined as open data. These data will be stored at the Norwegian Environmental Data Centre which is closely related to the Norwegian Meteorological Institute. This institution, in conjunction with other institutions, is responsible for mapping the environmental conditions on the whole of the Norwegian Continental Shelf.

In addition to the regular E- and P-data recorded on tapes, meteorological observations shall be logged manually and weather telegrams for the regular weather forecast sent to the Norwegian Meteorological Institute.

All the practical work in connection with the design, fabrication and installation of the instrumentation systems as well as recording, maintenance etc. is the responsibility of the operator and the costs are borne by the licencee of each field. In addition, the licencee must also cover the costs of the data processing and the safety evaluations done on behalf of the Norwegian Petroleum Directorate. This includes the advisory (consulting) services concerning the instrumentation as well. To perform this work the directorate have engaged the OTTER GROUP (Offshore Technology Testing and Research Group) in Trondheim.

3. PLATFORMS INSTRUMENTED - STATUS AT THE BEGINNING OF 1979

Instrumentation projects according to the regulations just described are being undertaken on the following fields and platforms:

STATFJORD FIELD (operator Mobil Exploration Norway Inc.):
STATFJORD A (production, storage and quarters platform, concrete (CONDEEP)).
STATFJORD ALP (articulated loading platform, steel structure).
STATFJORD B (production, storage and quarters platform, concrete (CONDEEP)).

FRIGG FIELD (operator Elf Aquitaine Norge A/S):
FRIGG TCP-2 (treatment and compression platform, concrete (CONDEEP).
FRIGG DP-2 (drilling platform, steel jacket).
FRIGG E-data (located on FRIGG QP-quarters platform, steel jacket).

EKOFISK FIELD (operator Phillips Petroleum Company Norway):
EKOFISK 2/4 H (hotel platform, steel jacket).

VALHALL/HOD FIELD (operator Amoco Norway Oil Company):
VALHALL QP (quarters platform, steel jacket).

A full E- and M-data collection programme has been required for the three fields, EKOFISK, FRIGG and STATFJORD only. The measurements will be representative of the respective areas. For the individual platforms (except for STATFJORD), additional wave measurements are carried out close to the structure.

The following is a brief description of the instrumentation systems for two different platforms. The instrumentation on these two platforms is representative of what is required for a steel and a concrete platform, respectively. Some particulars of the instrumentation installed on the other platforms will be mentioned as well.

3.1 *Platform data STATFJORD A*

The P-data instrumentation (SIS - Structural Instrumental System) of the platform is shown schematically in Fig.1. The instrumentation (excluding redording system) is designed by the Norwegian Geotechnical Institute and consists of 60 measurement channels:

- linear accelerations in three directions a_x, a_y and a_z, at the bottom of the caisson,
- angular accelerations, $\ddot{\theta}_x$ and $\ddot{\theta}_y$, around two horizontal axes at the bottom of the caisson,
- linear accelerations in two horizontal directions, a_x and a_y, at the deck level,
- angular accelerations around a vertical axis, $\ddot{\theta}_z$, at the deck level,
- settlement (measured on a steel rod fixed to the ground 60 m below the foundation),
- water level ("draught" - water pressure sensor),
- 8 vibrating wire strain sensors around a section at the top of one shaft,
- 2 vibrating wire strain sensors at the lower domes 5 and 7 of the caisson,
- 12 earth pressure cells (vibrating wire) at the lower domes of the caisson,
- 4 vibrating wire strain sensors at one steel dowel below the platform,
- 8 skirt water pressure sensors measuring the pressure at different compartments underneath the platform,
- 15 pore pressure sensors at three locations below the bottom slab. At each location there are 5 sensors at various depths from 3.75 m to 19.95 m below the mudline,
- optical measurement of the platform inclination.

The recording system consists of a digital computer with two magnetic tape stations, a visual display and a hard copy unit. The system has a capacity of 65 data channels. Back-up batteries are used in order to have an uninterruptable power supply.

The sampling frequency ranges from 1 every 30 second to 8 per second. For storm conditions raw data are recorded for 20 min every three hours. Furthermore, reduced data consisting of mean, maximum and minimum value, linear drift, standard deviation etc., based on a 20 minute time series are computed every hour and recorded on one of the tapes. For less severe environmental conditions, the recording is reduced to one raw data sequence every 24 hours and one reduced data sequence every 3 hours.

In addition to the platform data, also data for wave height, wind speed and direction, obtained from the environmental data system (see next section), are recorded simultaneously on the P-data tapes. Also the timing of the recording sequences for the P-data system is automatically initiated from the E-data system. The time for each recording is written on the tape ahead of the data.

The recording system is designed and delivered by the Central Institute for Industrial Research, Oslo.

3.2 *Environmental data - STATFJORD Field*

The environmental data system EMMS (Environmental and Meteorological Monitoring System) is connected to the STATFJORD A platform except for the current meters and water level gauge which all have a self-contained recording system. It consists of

- **2 waverider buoys at approximately 1 km from the platform with radio transmission to the on board recording system,**
- 2 mooring strings at approximately 1 km from the platform measuring the following
    at level -30 m :
    - current speed (impeller/potentiometer via magnetic coupling, or acoustic current meter for two orthogonal directions)
    - current direction (magnetic compass and vane)
    - temperature (thermistor)
    - water pressure
    at approximately 3 m above mudline and
    at midwater level :
    - the same as for level -30 m except that current speed is measured with a Savonius Rotor,
    at approximately 3 m above mudline :
    - tide (pressure sensitive quarts crystal resonator)
    - temperature
    - time
    There is magnetic tape recording within all the instruments suspended in the mooring strings.
- wind speed and direction (propeller, vane, potentiometer) at two locations above the helicopter deck
- air pressure
- humidity
- air temperature

The recording system consists of a digital microcomputer with a floppy disk, two magnetic tape stations, a paper tape punch and a visual display unit.

Data are recorded 20 min every three hours except for current which is stored as 10 min averages every 10 minutes.

In addition to the digital recording there is also analogue recording of the data before digitization. The analogue and digital data may be visualized on a display.

3.3 *Platform and Environmental data - EKOFISK 2/4 H*

The instrumentation of the platform is shown schematically in Fig. 2. The instrumentation (sensors) is designed and delivered by Det norske Veritas (DnV), Oslo. The P-data system consists of 82 channels :

- linear accelerometers in two horizontal directions at 5 different locations (2 locations at the mudline (C1 and C5), 2 locations at the deck level (C3 and C4) and one at the level +6.3 m (C2))
- 6 strain gauges (resistive, welded to the structure) in each section at 13 different locations on the structure (A1.1, A1.2, A1.3, A2.1, A2.2, A2.3, A3.3, A4, A5, A6, A7 and A8 in Fig. 3). For each section only 4 gauges are sampled, i.e. 2 backups.
- 6 strain gauges (vibrating wire) at each of 4 different levels in one of the piles (P1 - P4 in Fig. 2). For each section only 4 gauges are sampled, i.e. 2 backups.
- movable accelerometers for measuring accelerations in one pile at different levels. Linear and angular accelerations are measured along and around two axes orthogonal to the pile axis. These accelerometers have their own

recording system (floppy disk) and are not syncronized exactly with the
permanent instrumentation system.

There is a complete cable system backup to all the sensors on the jacket.

In addition to the P-data described above E-data are also measured. These data are more or less comparable to those described for the STATFJORD field in the previous section, however, all sensors are here hard-wired to the recording system, except for a wave buoy, which transmits via radio. Since both E- and P-data are being recorded by the same recording system (based on a PDP 11/34 minicomputer), all the sensors are time correlated. In addition to the wave buoy, a radar type wave monitor is mounted on the platform, looking straight downwards. It serves as a backup for the wave buoy, but will also provide wave data for time correlation with the platform response analysis. The radar wave monitor will be used for the recording of water level, as well. There are three current meters, all of the electromagnetic type. Water temperature is measured within five metres of the still water level. The recording system is designed and delivered by MAREX, UK. Its capability is more or less comparable to the STATFJORD A P-data recording system, and a further description is omitted.

3.4 *Instrumentation on other platforms*

DP2 at FRIGG has instrumentation similar to EKOFISK 2/4 H. There is no pile instrumentation and also fewer accelerometer stations.

The slamming instrumentation is, however, more comprehensive, both as regards the sensors and the data acquisition. 12 sensors (vibrating wire) are installed on two horizontal braces in the slamming zone, i.e. at three sections with four gauges distributed around the brace. To improve the frequency response, analogue recording of the slamming signal is used. The recording is initiated whenever wave conditions with a high probability of slamming are detected. The recorder will then stay on continuously for 20 minutes. Digitization of the analogue tape will take place on shore.

TCP2 at FRIGG has 32 shock pressure sensors installed above the mean water line on the western half of the utility shaft. These sensors require a very high frequency response recording system and again analogue type recording was chosen. Here the system is triggered by an output from peak pressure detectors which will switch on the analogue recorder for 20 minutes. Because of the high tape speed the recording capacity is limited.

STATFJORD ALP is special because of its flexible support. To obtain sensible platform movements from accelerometer recordings, platform inclination is measured as well. The supporting system on the base section is also instrumented to measure the total force transmitted through the universal joint.

4. DATA PROCESSING

The instrumentation systems described in section 3 will give large data volumes. These data must be easily available for a fairly long period after the initial data processing. The data on the magnetic tapes received from the platforms are multiplexed according to format specified for each installation. As soon as the tapes are received by OTTER, the data are demultiplexed. After that the data pass through quality checking routines before they are stored in a specially designed data base system. When this is completed, the P-data tape is returned to the platform operator. From the data base, time series for specified fields/platforms/channels and for specified time or time periods can easily be retrieved. Since the data are confidential, only authorized persons have access to the data base and only to the particular data for which authorization is given. The data base is shown in principle in Fig. 3.

A specified amount of the data further pass through a routine analysis where simple quantities such as mean-, minimum- and maximum value, standard deviation etc. are computed for the time series. Cycle counting for fatigue evaluations are also performed for the channels selected for this purpose.

The results from this primary routine analysis are input into another computer programme generating plots and statistical values on a monthly basis. Further, short and long term distributions of environmental conditions and selected response quantities are generated.

All the results from the routine analysis are also stored in the data base. Thus, they may easily be retrieved and presented at any time upon request from authorized persons.

After the routine processing, a demultiplexed tape with the environmental data is forwarded to the Environmental Data Centre together with the original E-data tape.

Further data processing for the safety evaluations is performed only for selected periods. Here, stormy periods are of primary interest even if different sea states are to be considered. The system identification is primarily based on spectral analysis. A computer programme, STARTIMES, has been developed in order to find the eigenfrequencies, the correlation between waves and measured response quantities or between different measured response quantities, damping ratios etc.

Even if there are a large number of sensors on each platform, the amount of information is too incomplete to allow a comprehensive depiction of the behaviour of the entire platform. Analytical models have therefore been found to be a necessary tool for the interpretation work in connection with the data analyses. Models for stochastic dynamic analyses of the platforms have been developed. The computer programme DYNOGS [1,2] is tailored primarily to analyse gravity type structures while FEDRA [3] is designed for jackets. The input to these programmes is directional wave spectra. By using these models in connection with the measured data, more reliable conclusions concerning wave forces, ground stiffness, damping ratios etc. may be found.

For special critical regions, fatigue evaluations are performed. These evaluations are based on the cycle counting performed in the routine analysis and on detailed finite element analyses of the stress concentrations.

The primary use of the results from the measurements and the data analyses, is to find out if the design assumptions are reasonably accurate. Furthermore, applied safety margins will be evaluated.

5. EXPERIENCE WITH THE PROJECTS

The instrumentation systems described in this paper are rather complex. A large number of sensors are connected to computer-based recording systems through long transmission lines (cables, radio, troposcatter etc). The sensors and the transmission lines are subjected to a very rough environment. For many of the sensors and the cabling there is no access. Therefore, there are no possibilities of repairing damages, for replacements or recalibration. This must be kept in mind when designing the systems. The experience so far shows, however, that this may not be the most critical factor. There is a very small percentage of malfunctioning sensors and cabling to which there is no access. Most of the damages so far are caused by human activities and can be repaired.

The main problems up till now have been with the substantial delays in getting the systems developed and installed on the platforms. Valuable data from the first winter season after the platforms have been installed is lost. The newness of the instrumentation regulations and the lack of experience with procedures etc must take part of the blame. However, for future platforms it should be required that the entire instrumentation system be ready and the testing finished before the platforms are installed and that the instrumentation system is put into operation as soon as at all practicable after installation. The Norwegian Petroleum Directorate is considering project management routines which are expected to improve the progress of the instrumentation projects.

6. CONCLUSIONS

Large instrumentation systems have been installed on several platforms and there are also such systems being designed and installed on structures which are presently under construction. For all the platforms placed at the fields there have been large delays in the instrumentation projects. However, in 1978 three of the systems have been put into operation.

The potential of the data is very great both with respect to safety evaluations and future design as well as further research. It is only through such data that the design assumptions can be verified or improved.

As the oil activities steadily move into deeper waters and more hostile environmentals, such instrumentation should be a very important part of the total safety system. The lack of experience with these new types of structures operating under such expected severe environmental conditions, can only be compensated for through instrumentation.

7. REFERENCES

BELL,K.: DYNOGS - A computer program for dynamic analysis of offshore gravity platforms. Theoretical Manual. SINTEF report STF71 A78009, Trondheim Apr. 1978.

BELL,K.: DYNOGS - A computer program for dynamic analysis of offshore gravity platforms. User's Manual. SINTEF, Trondheim, May 1978.

MOE,G.: FEDRA - Stochastic dynamic analysis of offshore structures. User's Manual. SINTEF, Trondheim, Nov. 1978.

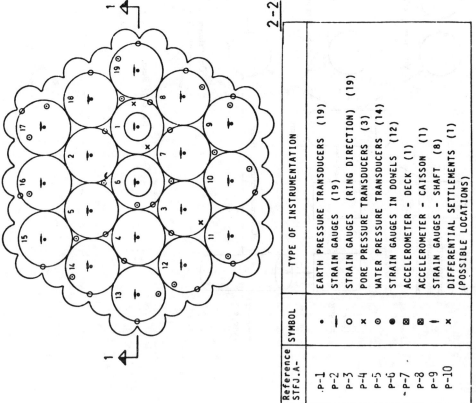
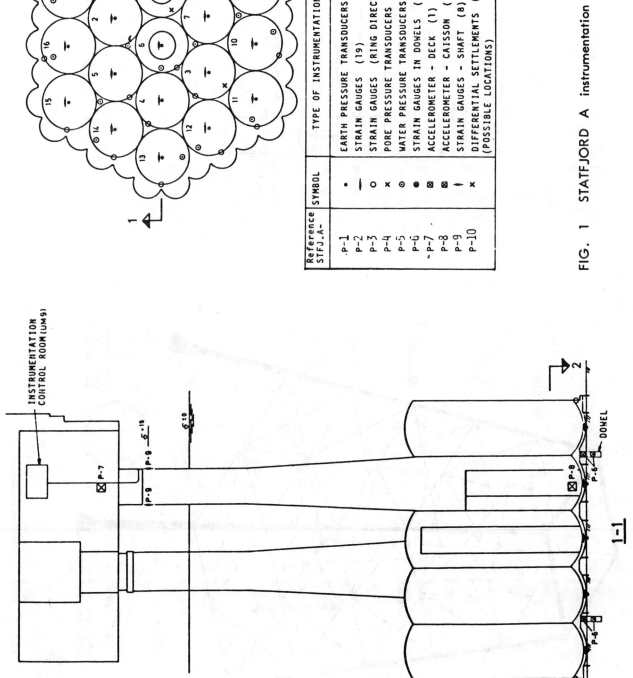

FIG. 1 STATFJORD A instrumentation

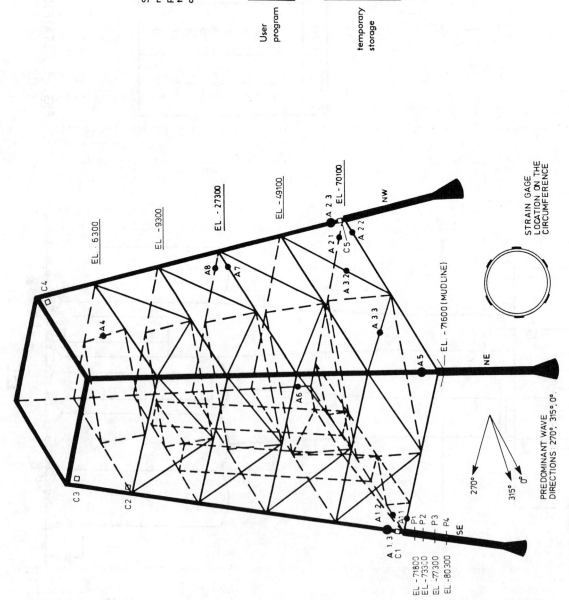

FIG. 3 The Data Base

FIG. 2 EKOFISK 2/4H instrumentation

BOSS'79

PAPER 78

**Second International Conference
on Behaviour of Off-Shore Structures**

Held at: Imperial College, London, England
28 to 31 August 1979

OBSERVED FOUNDATION BEHAVIOUR OF CONCRETE
GRAVITY PLATFORMS INSTALLED IN THE NORTH SEA 1973-1978

O. Eide, K.H. Andersen and T. Lunne

Norwegian Geotechnical Institute, Norway

Summary

Available measured foundation behaviour of the 13 concrete gravity platforms installed in the North Sea in the years 1973-1978 is presented. The measurements cover both the installation phase and the early part of the operation period.

For the installation phase, measurements of dowel and skirt penetration resistance, base contact stresses, tilt, piping and erosion, are presented. For the operation period, the presentation includes measurements of settlements, pore water pressures, base contact stresses, permanent lateral displacements, cyclic displacements, dynamic behaviour and erosion.

The measurements are used to back calculate soil properties and to evaluate the adequacy of the design procedures which have been used. The observations give valuable experience with respect to the foundation performance of offshore gravity platforms.

Sponsored by: Delft University of Technology, The Netherlands
Massachusetts Institute of Technology, U.S.A.
The Norwegian Institute of Technology, Norway
University of London, England

Secretariat provided by: BHRA Fluid Engineering

Copyright: © BHRA Fluid Engineering
Cranfield, Bedford, England

INTRODUCTION

From 1973 to 1978 13 gravity platforms were installed in the North Sea in water depths between 70 and 153 m. Prior to 1973 experience with offshore gravity platforms was limited, and in order to verify design assumptions and make sure that the platforms perform as anticipated, they were extensively instrumented. The platforms were also instrumented to control installation, and in several cases the instrumentation contributed to a successful installation.

This paper summarizes the available observed foundation behaviour of the 13 platforms both during installation and the early part of the operation period.

BRIEF PLATFORM DESCRIPTIONS

The platform locations are shown in Fig. 1. Some key data for the platforms are given in Table I. Of the 13 platforms, there are 5 Condeeps, 4 Howard Doris, 3 Sea Tank and 1 Andoc. Typical examples of the 4 different types are shown in Fig. 2.

All but three of the platforms are equipped with steel or concrete skirts. The purposes of these skirts are to:

- transfer the loads deeper where the soil is stronger
- provide closed compartments to facilitate grouting of open spaces beneath the base after installation
- provide scour protection for platforms on sand.

In the case of steel skirts the platforms have also been equipped with dowels. These dowels will penetrate into the soil and prevent the platform from skidding along the sea bed when the base approaches the sea bed. They thus facilitate positioning of the platform and prevent damage of the skirts from horizontal movements during installation.

In the case of concrete skirts, as well as of steel skirts which were designed strongly enough to take the horizontal forces during installation, dowels have not been necessary. However, for several platforms without dowels, skidding just prior to touch-down has been reported.

The types and dimensions of skirts and the number of dowels for the various platforms are listed in Table I. The geometry of the base, skirts and dowels is illustrated in Fig. 3.

Fig. 3 shows that the platform bases are designed in different ways. The base of the Condeep platforms consists of a number of spherical domes (see also Fig. 2A), whereas the other platform types are designed with flat bases. The platforms installed are all located at sites where the soil is very stiff, and relatively small unevennesses, of the order of a few decimeters, may cause high local contact stresses against the base. The sea bed topography must therefore be mapped, and the base must either be made strong enough to withstand the maximum expected base contact stress or the installation procedure must secure against high local contact stresses. For the Condeeps, the base contact stresses were measured, and ballasting and penetration stopped before approaching critical stresses. Open spaces between the base and the sea bed were then grouted. When ballasting was continued, the additional load was distributed over the entire base area. For the platforms with flat bases and skirts, the penetration was in most cases stopped and grouting performed before applying full ballast weight.

SEA BED TOPOGRAPHY AND SOIL DATA

As mentioned, the sea bed topography is of importance with respect to the danger of developing high local stresses against the base. The average sea bed slope is also of interest since it will influence the platform inclination. At all locations, except for the Ekofisk Tank, the sea bed topography was mapped by a minisubmarine which surveyed the sea bed prior to installation. Relative elevations were obtained with an accuracy of 50 to 100 mm utilizing differential pressure meters.

The results from the survey prior to installation and inspections of the sea bed around the platforms after installation have shown that the sea bed is nearly flat with only gentle slopes at all locations. The maximum observed height difference across the base is 1.0 m. The average sea bed slopes are listed in Table II. The maximum average slope of $0.6°$ is measured at the Frigg field. As an example Fig. 4 shows the results from the survey at the Condeep Brent B field.

At some locations boulders were observed on the sea bed. In most cases they were fully exposed, lying freely on the sea bed. Probably they have been dropped from floating ice. Maximum boulder sizes have been of the order of 1 to 2 m^3. In order to avoid concentrated loads on the platform base or damage to the skirts, boulders greater than 0.1 m^3 were removed by special trawling equipment.

A summary of the soil conditions is given in Table II. The soils are generally Pleistocene dense sands and preconsolidated clays except for recent sand and silt deposits in the upper 0.2 - 3 m. Seven of the platforms are founded on sites with stiff silty clay below 0.2 - 3 m. For the other six, there are dense silty fine sand layers to depths greater than 3 m. Typical boring profiles are shown in Fig. 9.

The clay at all locations is preconsolidated to varying degrees ranging from very stiff to hard with a low water content close to the plastic limit. The upper part of the clay may be softer due to unloading and swelling. Quite often weaker clay layers have been found below stiffer layers (e.g. location C, Fig. 9c), a phenomenon which may be attributed to permafrost (Løken, Ref. 14). At some locations the stratification is quite erratic with layers or lenses of clay and sand.

INSTALLATION

Most of the platforms have been instrumented to measure and control the performance during installation (e.g. DiBiagio, Myrvoll and B. Hansen (Ref. 4)). This has proved very useful and has in several cases been the key to a successful installation. The data are also extremely valuable for checking and improving installation design calculation procedures.

Performance during installation has been reported by several authors, e.g. Eide, Larsen and Mo (Ref. 6); Kjekstad and Stub (Ref. 11); Lunne and St. John (Ref. 13); Kvalstad and Dahlberg (Ref. 12); Foss and Warming (Ref. 8) and Eide, Kjekstad and Brylawski (Ref. 5).

In the present paper a summary of the main conclusions from the available installation performance observations are given.

DOWEL AND SKIRT PENETRATION RESISTANCE

The resistance against penetration of dowels and skirts has to be evaluated in order to make sure that the skirts will penetrate to the depth presumed in the stability analysis.

The measured dowel penetration resistances for six of the platforms are given in Fig. 5. The measured skirt penetration resistances for nine of the platforms are given in Fig. 6. The thickness of the steel skirts have been in the range 15 - 28 mm and for the wedge shaped concrete skirts 0.2 - 1.5 m. From the information in Figs. 5 and 6 it has been possible to improve procedures for calculating penetration resistances for skirts and dowels.

For *dowels and steel skirts* a backcalculation based on a correlation to the tip resistance measured by a standard cone penetrometer with a diameter of 36 mm was carried out by Lunne and St. John (Ref. 13). The unit skirt tip resistance and wall friction are defined as:

$$q = K_t \cdot q_c \quad \text{and} \quad f = K_f \cdot q_c$$

where: q = unit skirt tip resistance
         f = unit wall friction
         q_c = cone tip penetration resistance
         K_t = tip correlation factor
         K_f = friction correlation factor

Lunne and St. John (Ref. 13) found the correlation factors to vary within the following limits:

                 dense, silty, fine sand: K_t = 0.4 - 0.6
                                             K_f = 0.001 - 0.003
                 very stiff, silty clay : K_t = 0.4 - 0.6
                                             K_f = 0.03 - 0.045
                 mixed sand/clay layers : K_t = 0.5
                                             K_f = 0.006 - 0.014

The above values are valid for soil conditions similar to those in the North Sea and 15 - 28 mm thick steel skirts and dowels penetrating with a rate of the order 0.1 to 1 m/hour. The lower values for the correlation factors were found in the upper meter where some piping occurred.

For the wedge shaped *concrete skirts* the backcalculation is based on bearing capacity formulas and wall skin friction from earth pressure theory. A detailed outline of the theory is given by Foss and Warming (Ref. 8).

For sands the backcalculated friction angle is $42°$ - $45°$ for penetration of the 2 m deep wedge shaped concrete skirts of the Frigg TP-1 platform and in excess of $42°$ for penetration of the 0.4 m high concrete ribs beneath the Ekofisk Tank (Clausen, DiBiagio, Duncan and Andersen (Ref. 3)). These friction angles agree with friction angles determined in other ways. The average cone penetration resistance from 6 sites with sand give a friction angle of $42°$ - $44°$ and a relative density of more than 90% (Mitchell and Lunne (Ref. 15)). The variation from one site to another is relatively small. Friction angles backcalculated from two plate loading tests also give a friction angle of $42°$ (Kjekstad and Lunne (Ref. 9)). For the dense sands encountered in the North Sea it thus seems like bearing capacity theory with a friction angle of $42°$ gives a good estimate of penetration resistance for concrete skirts. The friction angle is reasonably well predicted by cone penetration tests.

For clays the backcalculated undrained shear strength from penetration of the 2 - 3 m long wedge shaped concrete skirts on Brent C and Cormorant A are in good agreement with an undrained shear strength determined as the average cone tip penetration resistance, $q_c - \gamma \cdot z$ divided by a cone factor, N_k, of 15 (Foss and Warming (Ref. 8)). Kjekstad, Lunne and Clausen (Ref. 10) report that N_k = 15 to 19 gives the best correlation between cone resistance and undrained triaxial compression strength of non-fissured clays. For other types of tests, like simple shear and triaxial extension, it is expected that the shear strength will be lower, corresponding to a higher N_k. The N_k-values recommended by Kjekstad et al. (Ref. 10) therefore seem to be slightly on the high side for calculating penetration resistance of concrete skirts. However, the available performance data are too scarce to draw definite conclusions. In order not to underestimate the penetration resistance of concrete skirts in stiff clays, it is recommended to apply a correlation factor, N_k, of 15 when utilizing cone penetration test results.

BASE CONTACT STRESSES

The base contact stresses depend upon the geometry of the base, the sea bed topography and the soil properties. The maximum contact stress against the base is limited by the bearing capacity of the soil. The bearing capacity of dense sand is considerably higher than for stiff clays. A layer of dense sand on top of stiff clays may also greatly increase the local base contact stress.

The difference in contact stresses for sand and clay is illustrated in Fig. 7a which shows the measured maximum contact stresses against the tip of the 19 domes of the **Brent B platform at the end of penetration. The reason for the high stress** (160 kN/m^2) against dome 15 was a local dense sand pocket. This sand pocket was not detected by the soil investigation, but the penetration resistance of dowel no. 1 located close to dome 15, indicated a 2 m thick layer of dense sand. The other domes penetrated mainly through clay, and the contact stresses against these domes are on average only of the order of 20% of the stress against dome 15.

The observed contact stresses for the Brent B platform illustrates the importance of performance observations during installation. The high contact stresses against dome 15 was not expected from the initial soil investigation, but turned out to be the governing factor for terminating penetration and starting grouting under the base. If ballasting had continued further before grouting, contact stresses higher than the maximum allowable design values might have developed.

The measured contact stresses against the base of the Frigg TCP-2 platform have been interpreted by Kjekstad and Stub (Ref. 11). Fig. 8 shows that the measured stresses agree with stresses calculated from bearing capacity theory and a friction angle of 42°. The backcalculated friction angle depends upon the assumed distribution of stresses over the contact area. If the stresses against the dome are assumed to be evenly distributed, the backcalculated friction angle will be 44°. The backcalculated friction angle of 42° - 44° agrees very well with the one backcalculated from skirt penetration resistance, cone penetrometer results and plate loading tests described in the previous chapter. It thus indicates that the maximum base contact stress against a dome penetrating into sand can be calculated by bearing capacity theory.

For domes penetrating into clay, the interpretation is somewhat more uncertain mainly due to uncertainties regarding the undrained shear strength of the upper clay layers. For Brent B the average contact stress will be 300 kN/m^2 when excluding the two cells with the highest contact stresses. The high stresses against these two domes are caused by local sand layers. For Brent D, Fig. 8b, the average value is the same, 300 kN/m^2, when excluding five domes which have markedly higher stresses than the others and probably are in contact with sand. The average value of 300 kN/m^2 corresponds to a backcalculated undrained shear strength of approximately 50 kN/m^2, which is within the range of undrained shear strengths determined from laboratory tests and in-situ cone penetration tests.

TILT

Tilt may develop during installation due to uneven skirt penetration resistance or sloping sea floor. It is desirable to keep the platform level, and for most platforms this has been attempted by applying eccentric ballasting during installation. The resulting maximum tilt for the different platforms after installation is shown in Table III.

For many of the platforms eccentric ballasting played an important role in keeping the platform level, and ballasting moments of more than 2000 MNm have been applied. For the Frigg TCP-2 a ballasting moment of 3000 MNm corresponding to an eccentricity as high as 25 m had to be applied to avoid a greater tilt. The conditions at this site are more unfavourable than at most other locations since the sea floor has an average slope of 0.6^o.

PIPING AND EROSION

During skirt penetration, the water entrapped within the skirt compartments flows out through outlet valves. The acceptable rate of penetration is therefore governed by the capacity of the outlet valves. If the penetration is too fast, an excess water pressure will build up within the skirt compartments, and piping may occur, causing the soil beneath the skirts to be washed away.

Differential skirt water pressures may also develop due to overturning moments from the waves. A tendency for cyclic rocking motions leads to alternating water pressures in the skirt compartments.

If piping or erosion occur during installation, it will be difficult to grout beneath the platform afterwards because the grout may just flow out through the eroded channels. Unless successful grouting is obtained and all channels are filled with grout, water may flow in and out through the channels during storm loading and cause further erosion. If open channels remain, it will also be difficult to create a reduced pore water pressure in the foundation by means of the anti-liquefaction drainage systems used under some platforms.

Submarine inspections after installation have revealed local washouts along the skirts of several platforms. These washouts have usually been short trenches not deeper than 0.5 m and have not caused serious problems. Only in a few cases, erosion channels under the skirts are believed to be the main reason for grout leaking to the sea floor outside the platform.

OPERATION

All the platforms are instrumented to some extent for measuring the foundation behaviour during operation of the platform. The measurements may include:

- settlement
- tilt
- pore pressures
- base contact stresses
- long-term lateral displacement
- cyclic displacements
- dynamic behaviour
- differential skirt water pressure
- stresses in shafts
- flow of water from drainage system
- scour
- sea state and environmental data

The number of parameters measured varies from one platform to another. The Brent B platform was selected for an industry-sponsored research project and is more extensively instrumented than the other platforms. The parameters measured at the different platforms are tabulated by Foss (Ref. 7). Instrumentation systems and types of sensors used on some of the platforms are described by DiBiagio, Myrvoll and Borg Hansen (Ref. 4).

Even if most platforms have been in place only for a short time, valuable information has been obtained. In this paper emphasis is placed on presenting settlement records for 6 platforms. In addition conclusions from observations of pore pressures, base contact stresses, permanent lateral displacements, cyclic displacements, dynamic behaviour and erosion are presented.

SETTLEMENTS

Figs. 9a-g present settlement histories for 6 platforms. Typical boring profiles for each platform site as well as load histories are included. Measurements of tilt are available for four of the platforms.

The measurements of settlements and tilt are carried out using a variety of instrumentation methods which are discussed by DiBiagio et al. (Ref. 4). In some cases there are periods when no measurements are taken, and the settlements have been estimated as shown on the presented settlement records. There are uncertainties with most of the load histories as indicated on the figures.

For platforms A and F the settlement observations start during the penetration phase and initial settlements are included in the measurements. However, it is difficult to distinguish settlements from actual penetration, and the initial settlements are somewhat uncertain. For the platforms E, D, B and C the settlement observations started 2 days, 1, 2 and 4 months after platform installation, and initial settlements and the first part of the consolidation settlements are thus not included.

The platforms may experience change in tilt during the operational phase. This tilt may be due to non-symmetrical horizontal forces due to predominant wind, current and wave directions, lateral variations in soil conditions and eccentric additional weight. The measured changes in tilt occurring after installation are shown in Table IV.

PLATFORM	TILT		REMARKS
	mm	DEGREES	
A	13	0.01°	Change from 13 to 20 months after installation
B	-	-	Not measured
C	45	0.03°	Change from 4 to 19 months after platform installation
D	-	-	Not measured
E	50	0.03°	Change from 0 - 14 months (absolute value 0.18°)
F	74	0.05°	Change from 0 - 13 months (absolute value 0.09°)

Table IV   Changes in tilt occurring after installation for the platforms in Fig. 9a-g

The settlement is usually not measured at the center, and the measured settlement is therefore influenced by the tilt. For those platforms where tilt is measured, maximum and minimum settlements are presented in the figures, and the settlement in the center is the average between the maximum and the minimum. The points at which the settlements are measured, are indicated on the figures.

The time required to complete consolidation of the foundation soils may be estimated from the settlement records as the time when the settlement curve levels out. However, this point is not always well defined because the foundation soils contain layers of both sand, silt and clay which consolidate at different rates. Changes in submerged weight which occur after installation cause changes in settlements and complicate the interpretation. With these reservations, estimates of the time for consolidation for the various platforms are given in Table V. For platforms on mainly sand, the estimated consolidation time varies from almost instantaneous consolidation for platform F with a 26 m thick upper sand layer, to 8 - 10 months for platforms B and E with 10 m thick upper sand layers. For platform C founded on an upper layer of clay interbedded with sand layers and sand beneath 45 m, the consolidation time is 10 - 12 months. For this platform the interpretation is confirmed by the pore water pressure observations shown together with the settlements in Fig. 9c. For foundations with mainly clay, platforms A, D and G, consolidation is not completed within the observation period, and exact times for consolidation can not be determined. However, the consolidation times are in these cases exceeding 16 months.

After consolidation, the platforms may continue to settle. The secondary settlements as interpreted from Fig. 9 are given in Table V. The data are very limited, but the platforms on sand, B, E and F have experienced only small secondary settlements. However, platform F did settle about 50 - 70 mm during a period with severe storms which occurred less than one month after final ballasting. The platform on clay interbedded with sand layers has secondary settlements of the order of 15 mm/year.

PLATFORM	CONSOLIDATION TIME (MONTHS AFTER FINAL BALLASTING)	SECONDARY SETTLEMENTS mm/YEAR	REMARKS
A	> 20	-	Observation period too short
B	8 - 10	7	
C	10 - 12	15	
D	> 16	-	Observation period too short
E	~ 5	~ 0	Uncertain reference level
F	~ 0	~ 0	Some uncertainties with bench mark. About 50-70 mm of settlements occurred during periods with heavy storms
G	> 11		Observation period too short

Table V Time for consolidation and secondary settlements evaluated from the settlement record in Fig. 9a-g

Experience from buildings on land indicates that the rate of secondary settlements may be constant for several decades when a building is subjected to cyclic loads (Bjerrum (Ref. 1)). However, the settlement records in Fig. 9 cover only a limited time interval, and it is too early to conclude regarding the long-term secondary settlements.

For four platforms the settlements computed in the design stage (i.e. prior to platform installation) are shown on the settlement records. The settlements are calculated using theory of elasticity and values of Young's modulus and Poisson's ratio as outlined by Kjekstad and Lunne (Ref. 9). In those cases where measurements are started some time after installation, the rate of measured and predicted settlements are compared instead of absolute values.

For platform A it can be observed that the computed immediate settlement is less than the observed. However, as noted earlier there are some uncertainties with the observed initial settlements because it is difficult to separate initial settlements from penetration of skirts and domes. The computed consolidation settlements seem to be in fair agreement with the observed settlements, but there are indications that the actual consolidation occurs faster than assumed in the design calculations.

For platform B the computed settlements seem to be in reasonably good agreement with observed settlements, but the consolidation is completed earlier than anticipated. The increase in settlements occurring at about $4\frac{1}{2}$ months after installation reflects an accidental inwash of sand into the platform during installation of conductors.

For platform C the observed and computed long term settlements are again in fairly good agreement. It was anticipated that consolidation settlements should be completed within 14 months, which is close to observed consolidation time of 10 - 12 months.

For platform D the long term settlements computed in the design stage by far exceed the actually measured settlements.

PORE WATER PRESSURE

The excess pore water pressure in the soil is measured to determine the degree of consolidation. This gives information about how the effective stresses, the soil strength and the stability of the platform increase with time. Secondly, the pore water pressure reflects the effect of cyclic storm loading on the soil, and thirdly, it gives information about the effect of the anti-liquefaction drainage system.

Reliable pore water pressure observations are available on three platforms, two on clay and one on sand, and the results are shown in Fig. 9c and g and in Fig. 10. The time for completion of consolidation can be determined from pore water pressure measurements. For platform C on clay the settlement record indicates a time for complete consolidation of about 10 - 12 months. At this time the pore water pressure has reached equilibrium and thus confirms the interpretation of the settlement record. The equilibrium pore water pressures are slightly below hydrostatic pressure due to the effect of the anti-liquefaction drainage system.

The main reason for the relatively short consolidation time for platform C, is that the clay drains towards sand layers interbedded in the clay. For platform G, the boring profile indicates that the clay is more homogeneous, and it can be seen that the pore pressure dissipation will take much longer time.

For the platform F on sand, no significant excess pore water pressures due to platform weight were observed during ballasting. When the first 1470 MN were applied, only piezometers at the base were in operation. There is a possibility that the base had not achieved a perfect contact with the sea floor at this stage, and in that case these observations will not be quite reliable. The readings of the deep piezometers taken 21 days after ballasting to 1470 MN show that at this time no excess pore pressure due to ballasting remain in the soil. The readings during the first part of the ballasting from 1470 MN to 1860 MN, corresponding to a vertical stress increase of about 60 kN/m^2, show essentially no excess pore pressure, meaning that the pore pressures drain away very rapidly for this platform on dense sand.

The platform F on sand was subjected to severe storms 4.5 months after installation. Figs. 11 and 12 show that during the first of these storms the significant wave height was about 11 m and the pore pressure increased by 10 - 20 kN/m^2. For the two platforms on clay, similar effects have been observed with pore pressure increases of up to 20 kN/m^2 in storms with maximum wave forces less than 45% of the design wave forces. These observations indicate that storm loading generates excess pore pressures in both sand and clay and that the effect of cyclic loading on soil behaviour must be taken into account in design calculations of displacements and stability of offshore platforms.

BASE CONTACT STRESSES

Measurement of base contact stresses after installation is of interest to ensure that load redistribution does not cause stresses that exceed the design values.

Measurements on 4 of the Condeep-platforms have been available. They show that stress changes occur due to special operations such as changes in platform weight (changes in deck load and oil storage), grouting, operation of the drainage system and installation of conductors.

Some stress changes occurred on Brent B and Brent D during conductor installations. At both platforms the effective stress on one of the two domes through which conductors are installed, dropped to zero when conductors were installed through it. This is most likely explained by a crater forming around the conductor during drilling. According to the conductor plans, the distance between the pressure cells and the conductors was only one meter.

No tendencies for long-term increases in stresses which may overstress the base have been observed, but high local stresses which developed during installation seem to remain.

PERMANENT LATERAL DISPLACEMENTS

Long-term lateral displacement may be caused by non-symmetrical horizontal loads due to preferred wind, current and wave directions. Lateral displacements are measured on two of the platforms. The measurements do not cover the very first storms, but do include storms with wave forces up to 45% of the design forces. The measured permanent displacements have been small and less than 2 cm for these platforms.

To the authors knowledge, no excessive lateral displacements causing problems with for instance bridge connections between platforms, have occurred for any of the platforms.

CYCLIC DISPLACEMENTS AND DYNAMIC BEHAVIOUR

Cyclic displacements introduce cyclic stresses in conductors, risers and pipelines connected to the platform and must therefore be below certain limits. From a foundation point of view, resonance frequencies and dynamic amplification factors are of importance for the determination of the wave forces on the platform.

The cyclic displacements and the dynamic behaviour of a platform can be measured by means of accelerometers. Such accelerometers are installed on most of the 13 gravity platforms. Until now, however, interpretation of accelerometer data has been carried out only for a few of the platforms.

The general conclusions for the platforms where some interpretation is performed, are that the resonance frequencies lie above the frequency band where the bulk of the wave energy occurs, and that the measured maximum cyclic platform displacement amplitudes at sea floor elevation have been less than 5 mm. These conclusions are valid for storms with significant wave heights up to 10 m and wave forces up to half the 100-year design forces, and only for platform types for which interpretation has been made. The dynamic soil shear stiffness backcalculated from the measurements is considerably higher than the stiffness used in design for calculating maximum displacements due to the 100-year wave. There are mainly two reasons for this apparent discrepancy. Firstly, the cyclic shear stresses in the storms covered by the measurements were less than half the shear stresses expected in the 100-year design storm. Since non-linearity and effect of cyclic loading increase significantly with increasing cyclic shear stress, the dynamic soil shear stiffness is expected to be considerably lower for the design storm. Secondly, the measurements were taken after consolidation was completed, whereas in design the 100-year storm is assumed to occur before full consolidation is achieved. Theoretical extrapolations of the backcalculated stiffness to design storm conditions taking the effect of stress level and degree of consolidation into account, have shown reasonably good agreement with the stiffness predicted for the 100-year storm conditions at the design stage.

EROSION

The cyclic wave forces cause cyclic pore pressure changes in the soil. These pore pressures vary from point to point in the soil, and there will be pore pressure gradients causing a tendency for flow of water in the soil and along the interface between soil and platform.

At the uplift side, at the heel of the platform, the gradients give a tendency for flow of water from outside, and on the compression side, at the toe, the gradients will be oriented to give flow of water from underneath the platform. If channels of free water exist, piping may occur, causing erosion below the base.

The danger for piping underneath the platform is limited as long as there are no pockets of free water underneath the platform. However, the pore pressures reduce the effective stresses in the soil outside the platform on the compression side. This increases the danger for erosion on the sea floor due to currents. Whether erosion will occur or not, depends upon the type of soil and the currents close to the platform base.

The majority of the gravity platforms in the North Sea are equipped with skirts (Table 1) which to some extent protect the soil beneath the platform against erosion. Other platforms are equipped with an anti-scour perforated wall of Farlan type (Zaleski (Ref. 16)). In some cases the soil has been protected against erosion by placing coarse gravel on the sea bed around the platforms.
In some cases the soil has been protected against erosion by placing coarse gravel on the sea bed around the platforms.

The Frigg TP-1 platform is situated on sand and has a square base and 2 m long concrete skirts. Gravel protection was not placed around it, and it suffered significant erosion to a depth of approximately 2 m locally at two corners during the first winter. In the spring coarse gravel and gravel bags were placed around the platform, and later surveys have shown that this has prevented further erosion.

The experience from the Christchurch Bay research tower founded on sand is interesting (Burland, Penman and Gallagher (Ref. 2)). The tower was neither equipped with skirts nor with scour protection. It was located on uniform, fine, medium dense sand. Measured base contact stresses indicated that the base was in contact with the sea bed only at a discrete number of points, meaning that there were voids with free water between the base and the soil. It is reported that rocking motion during storms was pumping material out from beneath the base causing severe erosion and a complete failure of the foundation. This case history shows that danger for erosion must be carefully studied. It has to be realized, however, that the situation for the tower with no skirts, free water beneath the base, medium dense sand and a low safety against bearing capacity failure, is much more critical with respect to erosion than for the North Sea gravity platforms.

The Ekofisk Tank which is also placed on sand and equipped with 40 cm high ribs beneath the periphery, was provided with scour protection on the sea bed around the tank. The scour protection consisted of a 1 to 1.5 m thick 10 m wide layer of well graded gravel with a maximum grain size of 80 mm. Submarine inspections have shown that the protection worked satisfactorily even if the fine grains had been washed out from the upper part of the gravel layer (Clausen et al. (Ref. 3)).

Platforms on clay have not experienced erosion even if they are not provided with scour protection.

CONCLUSIONS

The observations of the performance of the 13 gravity platforms have given valuable information about their foundation behaviour.

The instrumentation contributed to a successful installation for several of the platforms. Data from the installation phase has also been used to check and improve procedures for calculating base contact stresses and penetration resistance of skirts, dowels and domes.

The conclusions regarding the operation phase are limited by the relatively short observation period and by the fact that except for one platform, the storm loading has been less than 45% of the design wave forces. The existing data do, however, indicate a foundation performance as anticipated in design.

ACKNOWLEDGEMENTS

The authors wish to acknowledge and to express their thanks to a great number of their colleagues in the many companies and organizations being involved in the projects, for agreeing to make this information available.

The financial support to research and development from the Royal Norwegian Council for Scientific and Industrial Research is greatly appreciated.

The authors also wish to thank their colleagues at NGI who have contributed with measurements and interpretations.

The views expressed in this paper are solely the views of the Norwegian Geotechnical Institute.

REFERENCES

1. Bjerrum, L. (1964)
    Secondary settlements of structures subjected to large variations in live load.
    International Union of Theoretical and Applied Mechanics. Symposium on Geology and Soil Mechanics, Grenoble, 1964. Proceedings, pp. 460 - 471.
    Also publ. in: Norwegian Geotechnical Institute. Publication, 73.

2. Burland, J.B., A.C.M. Penman and K.A. Gallagher (1978)
    Behaviour of a gravity foundation under working and failure conditions.
    European Offshore Petroleum Conference & Exhibition. London 1978. Proceedings, Vol. 1, pp. 111 - 120.

3. Clausen, C.J.F., E. DiBiagio, J.M. Duncan and K.H. Andersen (1975)
    Observed behaviour of the Ekofisk oil storage tank foundation.
    Offshore Technology Conference, 7. Houston, Texas, 1975. Proceedings, Vol. 3, pp. 399 - 413.
    Also publ. in: Norwegian Geotechnical Institute. Publication, 108; and in: Journal of Petroleum Technology, 1976, March, pp. 329 - 336.

4. DiBiagio, E., F. Myrvoll and S.B. Hansen (1976)
    Instrumentation of gravity platforms for performance observations.
    International Conference on the Behaviour of Off-Shore Structures, 1. BOSS'76. Trondheim 1976. Proceedings, Vol. 1, pp. 516 - 527.
    Also publ. in: Norwegian Geotechnical Institute. Publication, 114.

5. Eide, O., O. Kjekstad and E. Brylawski (1978)
    Installation of concrete gravity structures in the North Sea.
    To be publ. in: Marine Geotechnology. North Sea State-of-the-art Volume.

6. Eide, O., L.G. Larsen and O. Mo (1976)
    Installation of the Shell/Esso Brent B Condeep production platform.
    Offshore Technology Conference, 8. Houston, Texas, 1976. Proceedings, Vol. 1, pp. 101 - 114.
    Also publ. in: Norwegian Geotechnical Institute. Publication, 113; and in: Journal of Petroleum Technology, Vol. 29, 1977, March, pp. 231 - 238.

7. Foss, I. (1976)
    Instrumentation for operation surveillance of gravity structures.
    International Conference on the Behaviour of Off-Shore Structures, 1. BOSS'76, Trondheim, 1976. Proceedings, Vol. I, pp. 545 - 556.

8. Foss, I. and J. Warming (1979)
    The performance of three gravity structure foundations.
    To be publ. in: International Conference on the Behaviour of Off-Shore Structures, 2. London 1979. Proceedings.

9. Kjekstad, O. and T. Lunne (1979)
    Soil parameters used for design of gravity platforms in the North Sea.
    To be publ. in: International Conference on the Behaviour of Off-Shore Structures, 2. London 1979. Proceedings.

10. Kjekstad, O., T. Lunne and C.J.F. Clausen (1978)
    Comparison between in situ cone resistance and laboratory strength for over-consolidated North Sea clays.
    Marine Geotechnology, Vol. 3, No. 1, pp. 23 - 36.
    Also publ. in: Norwegian Geotechnical Institute. Publication, 124.

11. Kjekstad, O. and F. Stub (1978)
    Installation of the Elf TCP-2 Condeep platform at the Frigg field.
    European Offshore Petroleum Conference & Exhibition. London, 1978. Proceedings, Vol. 1, pp. 121 - 130.
    Also publ. in: Norwegian Geotechnical Institute. Publication, 124.

12. Kvalstad, T.J. and R. Dahlberg (1979)
    Soil reaction stresses on the base structure during installation of gravity platforms.
    To be publ. in: European Conference on Soil Mechanics and Foundation Engineering, 7. Brighton, 1979. Proceedings.

13. Lunne, T. and H. St. John (1979)
    The use of cone penetration tests to compute penetration resistance of steel skirts underneath North Sea gravity platforms.
    To be publ. in: European Conference on Soil Mechanics and Foundation Engineering, 7. Brighton, 1979. Proceedings.

14. Løken, T. (1976)
    Geology of superficial sediments in the northern North Sea.
    International Conference on the Behaviour of Off-Shore Structures, 1. BOSS'76. Trondheim, 1976. Proceedings, Vol. 1, pp. 501-515.
    Also publ. in: Norwegian Geotechnical Institute. Publication, 114.

15. Mitchell, J.K. and T.A. Lunne (1978)
    Cone resistance as measure of sand strength.
    American Society of Civil Engineers. Proceedings, Vol. 104, No. GT 7, pp. 995-1012.
    Also publ. in: Norwegian Geotechnical Institute. Publication, 123.

16. Zaleski, L. (1976)
    Anti-scour protection by means of perforated wall.
    International conference on the Behaviour of Offshore Structures, 1. BOSS'76 Trondheim 1976. Proceedings, Vol. II, pp. 553-555.

TYPE	NAME	OPERATOR	INSTALLATION DATE	SUB-MERGED WEIGHT, 10^6 kN	FOUNDATION AREA, m^2	SKIRTS	DOWELS
Doris	Ekofisk tank	Phillips	June 1973	1.9	7 400	0.4 m concrete ribs	None
Condeep	Beryl A	Mobil	July 1975	1.7	6 200	3.0 m steel 0.5 m concrete	3
Condeep	Brent B	Shell	August 1975	1.7	6 200	3.5 m steel 0.5 m concrete	3
Howard-Doris	Frigg CDP-1	Elf	September 1975	1.8	5 600	None	None
Sea Tank	Frigg TP-1	Elf	June 1976	1.1	5 600	2.0 m concrete wedge	None
Howard-Doris	Frigg - Scotland Manifold	Total	June 1976	1.8	5 600	None	None
Condeep	Brent D	Shell	July 1976	1.8	6 300	4.5 m steel 0.5 m concrete	3
Condeep	Statfjord A	Mobil	May 1977	2.0	7 800	3.0 m steel 0.5 m concrete	3
Andoc	Dunlin A	Shell	June 1977	2.0	10 600	4.0 m steel	4
Condeep	Frigg TCP-2	Elf	June 1977	1.6	9 300	1.2 m steel 0.5 m concrete	3
Howard-Doris	Ninian Central	Chevron	May 1978	3.2	15 400	3.8 m steel	None
Sea Tank	Cormorant A	Shell	May 1978	2.3	9 700	3.0 m concrete wedge	None
Sea Tank	Brent C	Shell	June 1978	1.9	10 100	3.0 m concrete wedge	None

Table I   Summary of platform data

NAME	WATER DEPTH m	ELEVATION DIFFERENCE OVER 100 m	AVERAGE SEA BED SLOPE degrees	SOIL CONDITIONS
Ekofisk Tank	70	0.1 m	0.05	Dense silty fine sand (25 m) 2 m layer of stiff silty clay at 17 m depth
Beryl A	120	0.4 m	0.2	Dense silty fine sand (10 m) overlying very stiff silty clay
Brent B	140	0.2 m	0.0	Stiff silty clay with interbedded sand layers
Frigg CDP-1	98	< 0.5 m	0.0	Dense silty fine sand (8 m) overlying stiff silty clay
Frigg TP-1	104	1.0 m	0.6	Dense silty fine sand (3 - 7 m) overlying stiff silty clay
Frigg Scotland Manifold	94	-	0.0	Dense silty fine sand
Brent D	140	0.2 m	0.0	Stiff silty clay with interbedded sand layers
Statfjord A	145	0.3 m	0.0	Stiff silty clay (coversand 2 - 10 cm)
Dunlin A	153	0.2 m	0.0	Stiff silty clay with interbedded sand layers
Frigg TCP-2	102	1.0 m	0.6	Dense silty fine sand (3 - 6 m) overlying stiff silty clay
Ninian Central	136	-	0.0	Stiff silty clay with interbedded sand layers
Cormorant A	150	0.6 m	0.4	Stiff silty clay with interbedded sand layers
Brent C	140	0.3 m	0.2	Stiff silty clay with interbedded sand layers

Table II   Summary of sea bed and soil conditions

NAME	MAX. TILT AFTER INSTALLATION	
Ekofisk Tank	0.05°	
Beryl A	0.08°	1)
Brent B	0.04°	1)
Frigg CDP-1	0.17°	
Frigg TP-1	0.44°	2)
Frigg Scotland Manifold	-	
Brent D	0.04°	2)
Statfjord A	0.06°	1)
Dunlin A	0.11°	2)
Frigg TCP-2	0.08°	1)
Ninian Central	0.08°	2)
Cormorant A	0.04°	2)
Brent C	0.01°	2)

Table III   Maximum platform tilt at the end of grouting

1) After grouting

2) Before grouting
   Just after ballasting

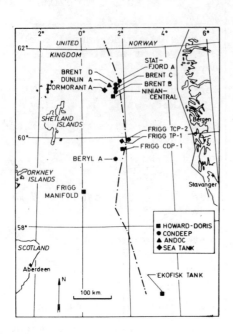

Fig. 1   Platform locations

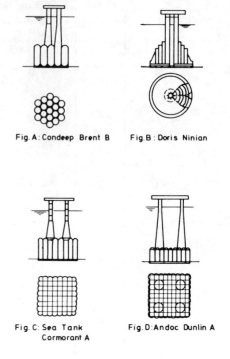

Fig. 2   Typical platform types

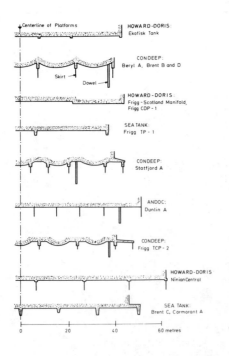

Fig. 3   Base designs (From Eide, Kjekstad and Brylawski, Ref. 5)

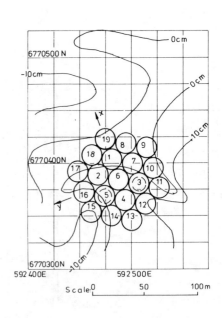

Fig. 4   Sea bed topography of the Condeep Brent B site

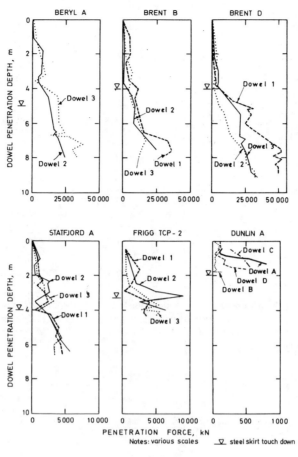

Fig. 5 Soil resistance to dowel penetration for structures equipped with dowels (From Eide, Kjekstad and Brylawski, Ref. 5)

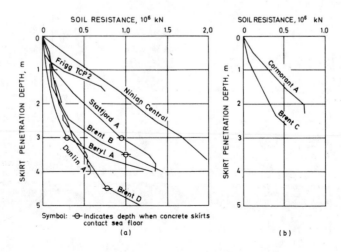

Fig. 6 Soil resistance to skirt penetration
a) steel skirts, b) concrete skirts
(From Eide, Kjekstad and Brylawski, Ref. 5)

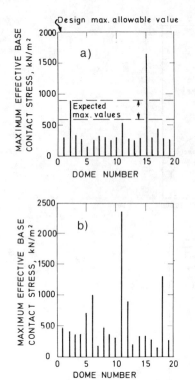

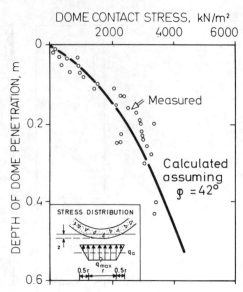

Fig. 8

Comparison between measured and calculated effective maximum base contact stresses (After Kjekstad and Stub, Ref. 11)

Fig. 7

Measured maximum base contact stresses
a) Brent B (From Eide, Larsen and Mo, Ref. 6)
b) Brent D

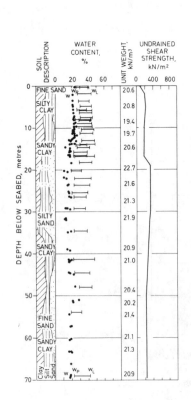

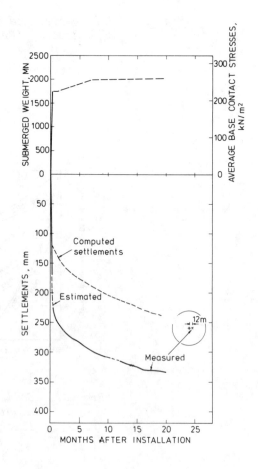

Fig. 9a   Settlements, load history and soil profile for platform A

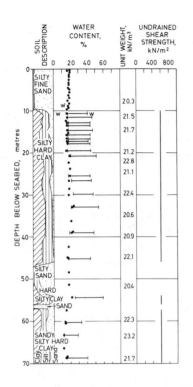
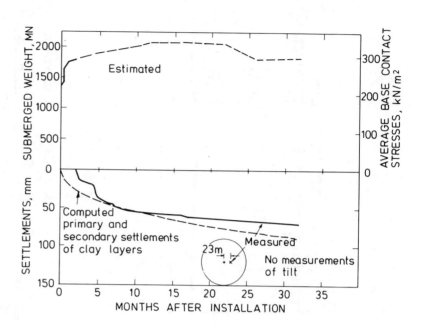

Fig. 9b   Settlements, load history and soil profile for platform B

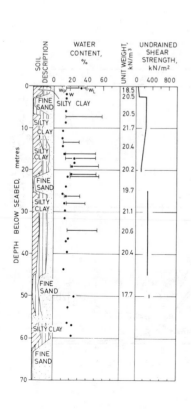
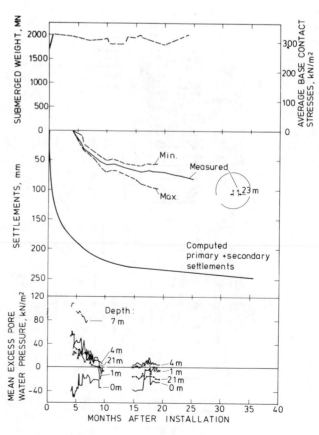

Fig. 9c   Settlements, pore pressures, load history and soil profile for platform C

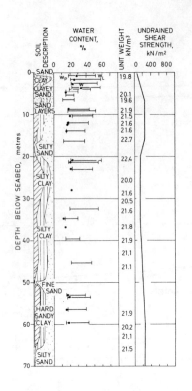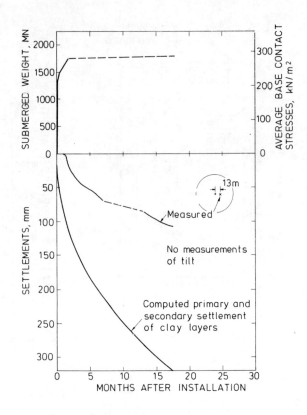

Fig. 9d   Settlements, load history and soil profile for platform D.

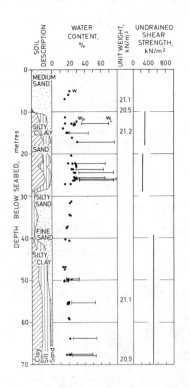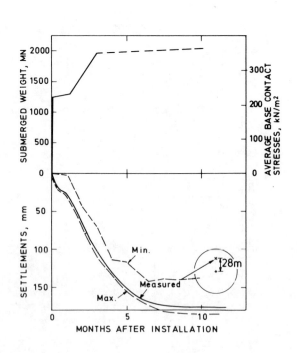

Fig. 9e   Settlements, load history and soil profile for platform E

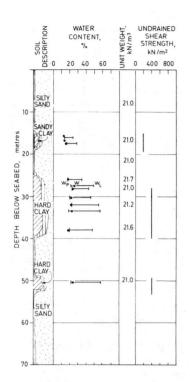
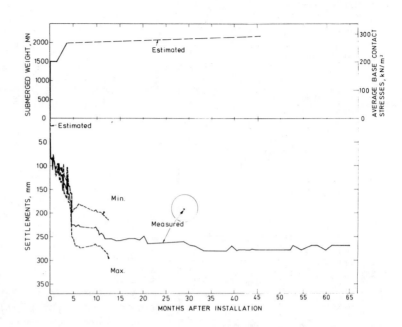

Fig. 9f   Settlements, load history and soil profile for platform F

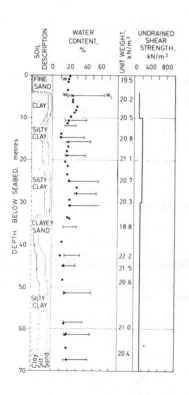
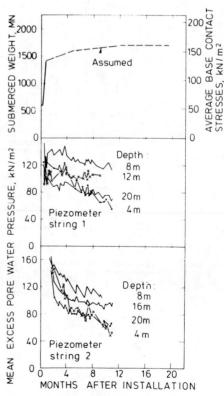

Fig. 9g   Pore pressures, load history and soil profile for platform G
(Pore pressure measurements by the Delft Soil Mechanics Laboratory)

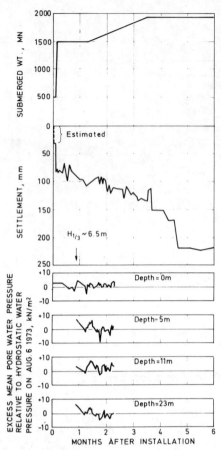

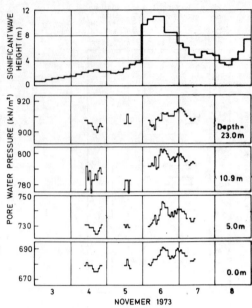

Fig. 11 Pore pressure observations before and during the first severe storm for platform F
(From Clausen, DiBiagio, Duncan and Andersen, Ref. 3)

Fig. 10 Variations in mean pore pressure, settlements and submerged weight during the first months after installation for platform F

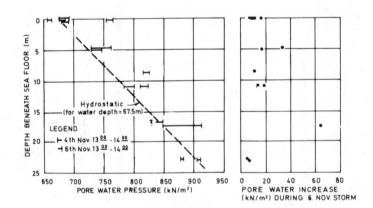

Fig. 12 Pore pressure increase from 4th November at 1300 - 1400 to 6th November at 1300 - 1400
(From Clausen, DiBiagio, Duncan and Andersen, Ref. 3)

BOSS'79

PAPER 80

Second International Conference
on Behaviour of Off-Shore Structures

Held at: Imperial College, London, England
28 to 31 August 1979

RECENT SIGNAL PROCESSING ADVANCES IN SPECTRAL AND FREQUENCY WAVENUMBER FUNCTION ESTIMATION AND THEIR APPLICATION TO OFFSHORE STRUCTURES

A.B. Baggeroer

Massachusetts Institute of Technology, USA

Summary

In recent years several significant advances in spectral and frequency wavenumber function estimation have appeared in the signal processing literature. Many of the algorithms developed have found considerable use primarily because of flexibility of today's minicomputers. In comparison to the correlation function (Blackman/Tukey), periodogram (FFT) and beamforming based methods, these algorithms have usually demonstrated much higher resolution and better controls of sidelobes.

In the spectral analysis of time series, techniques based upon autoregressive modelling, often labelled as the Maximum Entropy Method (MEM), are now frequently employed in speech analysis, automatic control, communications and geophysical exploration. Several adaptive array processing algorithms for estimating frequency wavenumber functions have been developed in the last decade. Two extensively used techniques are the Maximum Likelihood Method, developed simultaneously during research on earthquakes and sonar systems, and the Least Mean Square (LMS) algorithm, developed for adaptive antennas. Both methods produce good estimates primarily because of their capability of adapting to the noise environment.

While there has been some use of these advances in oceanographic and structures research, they have not been exploited to their full potential. In this paper a survey of signal processing research on high resolution spectral and frequency wavenumber function analysis will be presented. Examples of various applications in offshore structures and directional wave spectra are used to indicate the power of the algorithms. They suggest that the adaptive methods can identify both mode frequencies and directional structure with a quarter resolution than previously possible.

Sponsored by: Delft University of Technology, The Netherlands
Massachusetts Institute of Technology, U.S.A.
The Norwegian Institute of Technology, Norway
University of London, England

Secretariat provided by: BHRA Fluid Engineering

Copyright: ©   BHRA Fluid Engineering
Cranfield, Bedford, England

1. Introduction

In the last decade the advent of digital recording and signal processing techniques has led to significant improvements in the spectral analysis of data in both the time and spatial domains. It is now possible to record multichannel data over very wide dynamic ranges and bandwidths which probably far exceed most applications in offshore structures. Moreover, the computational "horsepower" required to process this data has expanded dramatically, especially with the use of array processors with minicomputers. These systems and techniques have had a dominant impact upon such fields as geophysics, transmission of speech, acoustics in both the ocean and the earth where their greater resolution and statistical stability have led to a much better understanding of the physics of these fields.

The potential impact of these methods in the analysis of ocean waves and offshore structures has yet to be assessed. If the success of past applications in other fields is indicative, then these techniques will become the commonly used methods of analysis. In this paper we present a survey of some of the recent, often termed high resolution, advances in the spectral analysis of time series and directional wave spectra estimation. We introduce the methods, but defer to the references for the mathematical details of the signal processing. Several examples related to ocean wave analysis and offshore structures are used to illustrate the capabilities of these techniques. In the brief space of this paper it is hoped that the merits of these methods will be apparent so that future studies in offshore structures will be stimulated to exploit their potential.

2. Critique of Blackman-Tukey and FFT Methods of Spectral and Beamforming for Directional Wave Spectral Estimation

To prompt the discussion of the new methods we first present a very brief review of the "classical" methods of spectral analysis and directional wave spectra estimation primarily to indicate their deficiencies. Blackman and Tukey were the first to introduce a computational method of time series analysis[1-3]. The correlation function is first estimated in the time domain; it is then multiplied by a spectral window; and finally Fourier transformed to obtain the spectral estimate. By a judicious choice of the spectral window and length of the correlation function estimate an experimenter could in effect trade off the two criteria in the performance of a spectral estimate - the bias, or resolution, and the variance, or statistical stability. In practice, one also needed to prewhiten the data, usually by trial and error, if it had intense regions, e.g., strong resonance regions.

The major deficiencies of the Blackman-Tukey method are i) it implicitly assumed properties for the data and its correlation function that were not justifiable physically, e.g., that it vanished outside the observed region, and ii) the estimate at one frequency was contaminated by adjacent frequencies by leakage of power through side-lobes in the spectral window.

A major advance in spectral estimation was made possible by the Cooley-Tukey, of Fast Fourier Transform (FFT), algorithm[4]. This made frequency domain techniques competitive, often even preferred, since a computationally efficient method of going between the time and frequency domains was available. Operating in the frequency domain has a number of advantages, but the most important one is the fundamental theorem of harmonic analysis - disjoint frequency bands are statistically uncorrelated. The basic method of the FFT techniques is to average the uncorrelated Fourier coefficients. The time series was transformed to the frequency domain, then the Fourier coefficients were magnitude squared, and finally averaged across a frequency band for statistical stability. In many applications the data was segmented, primarily because of computer memory limits, and the spectral estimates of the segments were averaged. The averaging operation reflected the tradeoff between the variance and resolution - averaging over a narrowband had high resolution but poor stability and vice versa.

The major definiencies of the FFT techniques are i) they also make implicit as- sumptions about the structure of the data that cannot be justified physically, and ii) the finite data length compromises the content of the fundamental theorem of harmonic analysis by power leaking through the sidelobes of the finite bandwidth of the transformation window.

Estimating directional wave spectra is appreciably more difficult than time series analysis, so it is understandable that there have been far fewer applications. Theoretically, most, but not all, the principles of time series analysis can be extended to spatial processes. Experimentally, the cost of multisensor systems has been large, so applications have been few.

Directional spectra have usually been estimated by "beamforming". This consists of steering, or delaying, the signals from an array of sensors such that those in the direction of interest add constructively and those from other directions interfere destructively. The beamformer output is then squared and averaged to produce the wave spectra estimate in the steered direction. By doing this in all directions the entire directional wave spectra can be estimated. In practice, the beamforming can be done in either the time domain by delay operations or the frequency domain by phase shifting[5]. The most efficient procedure is a rather complicated function of the specific experiment parameters, e.g. the number of sensors, number of beams or wave directions, and bandwidths.

Sensor arrays for directional wave spectra estimation have two limitations - typically they are not large enough and they have too few sensors. The former limits the beamwidth, or resolution, while the latter introduces spurious sidelobes because of spatial aliasing. In most applications the size of the array is constrained either economically or physically, so one has an inherent beamwidth restriction. The second limitation, however, only is in part a limitation, for the aliasing introduced by sparse arrays can be controlled if their geometry is well designed and the signal processing is effective.

3. Spectral Analysis of Time-Series

The spectral analysis methods that have evolved in digital signal processing recently are in many respects extensions of techniques developed in the literature of mathematical statistics[6]. The major difference is that the signal processing literature is very much centered in the frequency domain because of its utility in engineering applications whereas the mathematical literature has a much stronger emphasis on the time domain. While one may argue that the methods are not new fundamentally, the methods have both matured significantly and several completely novel algorithms have been developed. More importantly, the large number of applications made possible of the digital computer has sharpened our insight considerably.

3.1. Autoregressive (Maximum Entropy) Spectral Analysis

The autoregressive (AR) methods of spectral analysis are a subclass of autoregressive, moving average (ARMA) linear difference equation models long studied in statistics. At the current time they are by far the most extensively used.

3.1.1. Autoregression Models

The AR approach to spectral estimation consists of representing the time series of interest as the output of a system with an all pole difference equation structure of the form indicated in Fig. 1, or

$$Y_{n+1} = \sum_{k=1}^{N} A_k Y_{n-k+1} + U_n \tag{1}$$

where U_n is a discrete time white noise process,

A_k are termed the AR coefficients,

Y_n is the represented process,

N is the order of the process.

The power density for such a process is given by

$$S_y(t) = \sigma_u^2 / \left| 1 - \sum_{k=1}^{N} A_k e^{-j2\pi f k \Delta T} \right|^2 \tag{2}$$

where ΔT is the sampling interval. This formula for the power density spectrum forms the basis for all of the AR spectral estimations.

In general the AR representation cannot represent all time series; however, it can be demonstrated that all processes with rational spectra can be represented by (1) if the order is allowed to become sufficiently large.

One can interpret (1) as a shaping operation which takes the white noise process to a correlated one. An alternative approach to AR spectral estimation is based upon the inverse of this, whitening a correlated process. The basis of this is the prediction-error filter indicated in Fig. 2. Here, one wants to use M previous values of a time series to predict the next value. If a minimum mean square error criterion is adopted, one can demonstrate the following$^{(7,8)}$

i) the coefficients for the prediction filter $\{h_i\}_{i=1}^{M}$ are given by

$$\begin{bmatrix} R_y(0) & R_y(1) & \cdots & R_y(M) \\ \cdot & & & \\ \cdot & & & \\ \cdot & & & \\ R_y(M) & & & R_y(0) \end{bmatrix} \begin{bmatrix} 1 \\ -h_M^1 \\ \cdot \\ \cdot \\ \cdot \\ -h_M^m \end{bmatrix} = \begin{bmatrix} P_E(M) \\ 0 \\ \cdot \\ \cdot \\ \cdot \\ 0 \end{bmatrix} \tag{3}$$

(normal equations)

where $R_y(n)$ is the autocorrelation function of the process.

ii) the error in the prediction is given by

$$P_E(M) = R_y(0) - \sum_{m=1}^{M} R(m) h_M^m \qquad \text{(prediction error)} \qquad (4)$$

iii) the error in the prediction is approximately white, i.e. uncorrelated among samples, if the prediction error in (4) has reached its asympotic value as a function of M.

In the special case when the time series is truly autoregressive the coefficients $\{h_i\}$ in (3) will equal those $\{A_i\}$ in (1) and the prediction error will be a constant for M>N. In general this is only satisfied approximately, and one of the real issues in AR spectral analysis is determining an appropriate value of M.

3.1.2. Methods of Autoregressive Spectral Estimation

In the current literature there are essentially three formulations of methods for autoregressive spectral estimation. In the first the correlation function approach, one simply estimates the correlation function of the process, and applies the normal equations. In the second, or maximum entropy method, one adopts a statistical criterion of maximizing the "uncertainty" of the process which then leads back to the normal equations. In the third, one whitens the data with filters of increasing order based upon a "reflection coefficient" concept. We now briefly discuss each of these methods.

In the correlation function approach to AR spectral estimation one simply estimates the correlation function of the time series using the lagged product formula

$$\hat{R}_y(m) = \sum_{\ell=m+1}^{L} Y(\ell) Y(\ell - m) \, L \qquad (5)$$

where L is the data length, then solves for AR coefficients, and evaluates the estimated spectrum using (2). This is illustrated in Fig. 3. The diagonal symmetry of the normal equations, termed a Toeplitz form, leads to a computationally efficient method, the Wiener-Levinson algorithm, for determining the AR coefficients. It can be demonstrated by direct substitution that the normal equations can be solved recursively for increasing order as follows

$$\gamma(M+1) = - \left(\hat{R}_y(M+1) - \sum_{m=1}^{M} \hat{R}_y(M+1-m) h_M^m \right) \Big/ P_E(M) \qquad (6a)$$
reflection coefficient eq.

$$P_E(M+1) = \left(1 - \gamma^2(M+1)\right) P_E(M) \qquad (6b)$$
prediction error eq.

$$\begin{bmatrix} 1 \\ -h_{M+1}^1 \\ \vdots \\ -h_{M+1}^{M+1} \end{bmatrix} = \begin{bmatrix} 1 \\ -h_M^1 \\ \vdots \\ -h_M^M \\ 0 \end{bmatrix} + \gamma(M+1) \begin{bmatrix} 0 \\ -h_M^M \\ \vdots \\ -h_M^1 \\ 1 \end{bmatrix} \qquad (6c)$$
update eq.

It can be demonstrated that the number of computations is proportional to M^2 rather than M^3 in the general case of matrix inversion. This leads to numerical results that are usually far stabler, simply because of the fewer computations performed.

The reflection coefficients, $\gamma(M)$, and the prediction error, $P_E(M)$, are useful in the numerical aspects of AR spectral estimation. In the statistical literature the reflection coefficients are called partial correlations. They have some very useful properties with interesting physical interpretations in propagation in layered media. They are always bounded in magnitude by unity, and they can be specified independently of each other and still correspond to a well behaved time series, which is not true of the correlation function because of its positive semidefinite constraint.

In practice one of the most difficult issues is to determine the appropriate value of M, the estimated order of the autoregression model for the time series. One has the following tradeoff: the larger the order, M, the better the resolution; however, the statistical stability decreases with increasing order. This is the same tradeoff that appears in the "classical" literature on spectral estimation, but it is far more difficult to quantify because of the nonlinear aspects of the autoregressive methods. The most commonly used method is due to Akaike, which measures the goodness of fit versus the decrease in degrees of freedom in the estimation of each coefficient as the order increases for a fixed length, L, of the time series. Alternative criteria have been advocated based upon the asymptotic probability density of the spectral estimate. It is not apparant however, that these conditions are satisfied in a number of practical cases of interest where the data length is short. The literature in this area is extensive and the question is an important practical one, so further progress on this issue may be anticipated.

The second method of AR spectral analysis, popularly termed the <u>m</u>aximum <u>e</u>ntropy <u>m</u>ethod (MEM) involves consideration of the structure of the correlation function of the process. Essentially one extends the correlation to increase the resolution in a manner which imposed no implicit assumptions beyond its positive semidefinite constraint. This led to the maximum entropy criterion, which is the source of the popular name for the AR methods[10].

The entropy of a discrete time Gaussian random process is given by

$$S = C_0 + C_1 \int_{-1/2\Delta T}^{-1/2\Delta T} \ln [S_y(f)] df \qquad (7)$$

where $S_y(f)$ is the power density spectrum and C_0 and C_1 are constants and S is process entropy. The MEM consists of finding the power density spectrum, $S_y(f)$, that maximizes (7) subject to the constraints that its inverse transform agree with the measured correlations, or

$$\hat{R}_y(m) = \int_{-1/2\Delta T}^{-1/2\Delta T} S_y(f) \, e^{j2\pi m f \Delta T} \, df \qquad (8)$$

The solution to this problem is very rich mathematically, and the most remarkable result is that the solution is found via the normal equations (3). One uses the constraints of the measured correlations in the normal equations, determines the AR coefficients, and computes the spectra in the same manner as in the correlation method.

The third method of AR spectral estimation is based upon an algorithm introduced by Burg for calculating the reflection coefficients, $\gamma(M)$, by recursively operating

upon the time series itself for successively larger filter orders. The algorithm
is a stochastic realization of the Wiener-Levinson algorithm (6). Burg's insight was
that one should choose the reflection coefficient in Eq. (6c) in a manner that minimized the prediction error of the updated filter of the next higher order. An
efficient computer program for the algorithm is indicated in Ref. 11.

The Burg algorithm has recently become quite popular. It has not been resolved
whether the correlation method of the Burg algorithm is superior. The experimental
evidence suggests that the Burg algorithm has higher resolution, but is prone to tone
"splitting", i.e. making one peak appear as two[12]. This is certainly a drawback if
one is analyzing a structure with closely spaced oscillation modes.

3.2 Maximum Likelihood Methods of Spectral Estimation

The second class of high resolution methods of spectral analysis is based upon
the concept of a $\underline{m}$inimum $\underline{v}$ariance, $\underline{d}$istortionless $\underline{f}$ilter (MVDF), or $\underline{m}$aximum $\underline{l}$ikeli-
$\underline{h}$ood (ML) filter, when Gaussian statistics are assumed. The model for this filter is
illustrated in Fig. 4. Essentially one wants to determine the impluse response
$(g_1(f_0),\ldots g_n(f_0)) = (\underline{g}^T(f_0))$ of a finite length filter which has minimum output
power when its input is a random process with the correlation function $R_y(n)$. In
addition, the filter must be distortionless at the frequency of interest, f_0, i.e.
its transfer function must have unit gain at this frequency. Mathematically, the
problem is formulated as

$$\min_{\{g_i(f_0)\}} \sigma^2(f_0) = \min_{\{g_i(f_0)\}} \left[\sum_{i,j=1}^{M} g_i^*(f_0) R_y(i-j) g_j(f_0) \right] \quad \text{(minimum variance)} \quad (9)$$

$$\sum_{i=1}^{M} g_i(f_0) e^{-j2\pi f_0 (1-i)\Delta T} = 1 \quad \text{(distortionless at } f_0\text{)}$$

The solution to this problem is easily obtained by the calculus of variations. The
important result vis a vis the $\underline{m}$aximum $\underline{l}$ikelihood $\underline{m}$ethod (MLM) of spectral estimation
is that the minimum output variance is given by

$$\sigma^2(f_0) = (\underline{E}^+(f_0) \underline{R}_y^{-1} \underline{E}(f_0))^{-1} \quad (10)$$

where $\underline{E}^+(f_0) = [1, e^{-j2\pi f_0 \Delta T}, \ldots, e^{-j2\pi f_0 (M-1)\Delta T}]$ is a steering vector of phase
delays and

$$\underline{R}_y = \begin{bmatrix} R_y(0) & R_y(1) & R_y(M-1) \\ & & R_y(0) \end{bmatrix}$$

is a matrix formed from values of the correlation function. The $\underline{m}$aximum $\underline{l}$ikelihood
$\underline{m}$ethod (MLM) of spectral estimation follows from the assertion that the minimum output
power at f_0 is a measure of spectral power, $S_y(f_0)$. Since all the signal power at f_0
is passed undistorted while the power from all the other frequencies has been
attenuated in an optimum manner within the constraint of the finite length of the filter to obtain the entire spectral density estimate one simply evaluates (10) versus
f_0. The remaining issue is that in practice one does not have the ensemble correlation
function, R_y. In the implementation of the MLM this function is estimated usually

according to the sum of lagged products of formula (5). This finally leads to the following spectral estimation formula.

$$S_y(f_0) = [\underline{E}^+(f_0) \underline{R}_y^{-1} \underline{E}(f_0)] \tag{11}$$

The MLM has not been used in time series analysis as extensively as the AR methods because it has lower resolution[13]. One can in fact relate the MLM and AR formulae when ensemble quantities are used in the spectral estimation formulae[14]. The relationship demonstrates that the MLM is an "average" of the AR estimates of lower order. While the MLM does have lower resolution, it is a more stable estimate with less sensitivity to spurious sidelobes.

3.3 Examples of AR and MLM Spectral Estimation in Offshore Structure and Ocean Wave Applications

In this section we use field data to demonstrate the performance of the spectral estimation methods described in the previous sections. Two applications are discussed 1) the modes of oscillation in an offshore structure, the Airforce Combat Manuevering Range (ACMR) platform† and ii) tidal motion in San Francisco Bay[15]. These spectral estimates are compared to the Blackman-Tukey and FFT methods to indicate the improved interpretations that one can obtain.

3.3.1 Analysis of the Airforce Combat Maneuvering Range (ACMR) Platform

Data from an unmanned offshore structure off the coast of North Carolina was recorded and digitized. In the following analysis 100 secs of this data digitized at 51.2 Hz were used. In Fig. 5a the correlation function was measured over a lag of 50 samples or $\simeq$ 1 sec, while in Fig. 5b a lag of 100 samples, or $\simeq$ 2 secs, was used. Three spectral estimation methods are used i) the Blackman-Tukey with a triangular spectral window, ii) the AR, or MEM, implemented via the correlation function method and iii) the MLM. (The scales have been adjusted to facilitate the comparison of the estimates.)

At the shorter lags in Fig. 5a one dominant frequency of oscillation is indicated at approximately 2.2 Hz and several higher modes are suggested at 3.4 and 4.8 Hz. In the Blackman-Tukey estimate one must be cautious of interpreting the higher order modes because they may correspond to sidelobe leakage from the dominant frequency. The MEM indicates that the higher order modes are indeed real. The MLM suggests that there is significant spectral power at frequencies larger than the dominant one, but their location cannot be identified. At the longer lag in Fig. 5b the modes of oscillation are well identified by both the Blackman-Tukey and MLM with the MLM having a much lower sidelobe level. The AR method, or MEM, has further resolved that the dominant frequency is actually composed of two closely coupled modes of oscillation.

---

† This data and analysis was provided by Mr. Brad Campbell, doctoral candidate, and Prof. J. K. Vandiver, Dept. of Ocean Engineering, MIT.

The frequencies for these modes have been subsequently confirmed by numerical models. The power near 2.2 Hz correspond to bending and torsional modes. Further analysis with the MEM has also revealed that the power near 3.4Hz and 4.8Hz also can be resolved into bending and torsional modes. The important point regarding our discussion is the significantly improved interpretations that can be obtained with the high resolution methods.

3.3.2 Analysis of Tidal Motion in San Francisco Bay

In our second example data from a tide gauge in San Franciso Bay was sampled at 1/2 hour intervals and recorded over an 8 day period. The data was analyzed via the three methods used in the previous example and the results are compared in Fig. 6. With the conventional methods one can only identify the two strong components at .04 cph (period ≃ 25 hrs) and at .08 cph (period ≃ 12.5 hrs). The MLM and MEM demonstrate more structure to these peaks separating the various frequencies associated with the solar, lunar and sideral motions as indicated by the designatores on the figure[16]. In order to obtain this resolution with the conventional methods it was necessary to use data over 50 days instead of 8.

These examples are illustrative of the performance of the high resolution algoriths. (There is virtually no additional computational burden in using the methods.) They should be particularly applicable to offshore situations where the length of the data recording is limited either by a changing environment or the cost of recording the data.

3.4 Comments on Current Research on High Resolution Spectral Analysis and Its Relevance to Offshore Structures

At the current time researchers have a good understanding of virtually all aspects of spectral analysis for data from a single sensor. Investigations into some parametric techniques, the more general ARMA models and order selection are popular topics in the literature. Significant questions still remain, however, when one studies systems with more than one sensor i.e. vector time series. All the AR, or MEM, techniques have been extended to vector processes[17-19]. This should be very useful in a number of offshore structure applications. For example, the cross spectra can reveal the various modes of oscillation directly instead of inferring them by matching the frequencies, and the impulse response of a system can be determined by first exciting the structure with a shaker and then determining the cross spectra between in input and the output. To this author's knowledge there have been no experiments where the data has been analyzed using the vector versions of the high resolution algorithms, use in their analysis of offshore structure motion seems particularly appropriate.

4. Data Adaptive Estimation of Directional Wave Spectra

The spatial extension of the spectral analysis of time series is the estimation of frequency wave number function. This function indicates the mean square distribution of plane wave power versus the temporal and spatial frequencies, f,ν. In this section we provide a brief introduction to this function and its relationship to directional wave spectra. We then discuss some of the recently developed methods for estimating this function from an array of sensors such as could be used in offshore structure research.

4.1 Spectral Correlations, Frequency Wavenumber Functions and Directional Wave Spectra

Consider a signal, $y(t,z)$, which is a function of both space, z, and time, t, e.g. the wave height on the sea surface or an acoustic pressure wave in the ocean. If the signal is modelled as a random process, the correlation between samples at two times and locations is given by

$$E[y(t_1,z_1)y(t_2,z_2)] = R_y(t_1,t_2,z_1,z_2) \qquad (12)$$

The process is temporally stationary and spatially stationary, or homogeneous, if the correlation function depends only upon the separation of the samples in time and space, or

$$R_y(t_1,t_2,z_1,z_2) \implies R_y(\tau, z_1 - z_2)$$
$$\tau = t_1 - t_2$$

This assumption is made in most offshore applications.

If we now use Fourier transform with respect to the temporal variable, we obtain the spectral correlation function

$$S_y(f,z_1, - z_2) = \int R_y(\tau, z_1 - z_2) e^{-j2\pi f\tau} d\tau \qquad (13)$$

This function is important in many applications because it is the quantity that is estimated from the data, and it can be identified in the same role as the correlation function in time series analysis. If we next transform with respect to the spatial function, or variable $\underline{\Delta z} = z_1 - z_2$, we obtain the frequency wavenumber function as

$$P_y(f,\underline{\nu}) = \int S_y(f,\underline{\Delta z}) e^{j2\pi \underline{\nu} \cdot \underline{\Delta z}} d\underline{\Delta z} \qquad (14)$$

This function is the power density spectrum for a spatial random process; its most fundamental property is it describes the distribution of power in the process when it is represented as a superposition of uncorrelated plane waves with temporal frequency, f, and spatial frequency, $\underline{\nu}$.

In applications involving wave propagation the structure of the frequency wavenumber function is restricted by the dispersion equation relating f, with the magnitude of the spatial frequency, $|\underline{\nu}|$. For example, in surface wave problems this function is nonzero only on a ring in the spatial frequency domain whose radius is fixed by f. In a number of other applications, e.g acoustics, we observe signals that propagate in three dimensions with a linear or planar arrays of sensors. Here, the signal processing uses the projection of $\underline{\nu}$ on the plane of the array, and we are led to a spatially bandlimited wavenumber function. The important issue is to incorporate the restrictions that the dispersion relation imposes upon the structure of $P_y(f,\underline{\nu})$.

In most offshore applications we analyze the data over narrowbands of frequency. This fixes the magnitude of $\underline{\nu}$ so one analyzes the process according to the angle of the vector $\underline{\nu}$. In acoustics and radar, this is termed the angular, or bearing, power, while in oceanographic work it is the directional wave spectrum.

4.2 Methods of Directional Wave Spectra Estimation

In the following we indicate some of the new data adaptive techniques for estimating directional wave spectra. Before doing this, however, we point out some considerations about the array geometry that influences this estimation.

The array geometry is usually limited by two factors, the number of sensors and the allowable dimensions. To obtain high angular resolution we want to have an array geometry that is large since the resolution depends inversely upon the array extent measured in wavelengths. Conversely, we want to have closely spaced sensors to avoid spatial aliasing. With a limited number of sensors it becomes difficult to satisfy both of these criteria simultaneously. Moreover, design of arrays with irregularly spaced sensors is not easily done. Experimentally, we have found that it is better to avoid equally spaced geometries and to use both closely spaced sensors and some widely separated ones. Fortunately, most of the data adaptive methods have flexibility to incorporate the irregular geometries.

4.2.1 Estimation of Directional Wave Spectra with the Maximum Likelihood Method (MLM)

The most extensively used data adaptive method of directional wave spectra estimation is the spatial version of the maximum likelihood method. (In fact this method was first developed for wave spectra estimation of teleseismic signals[21]. It was subsequently applied to time series analysis[13]. The algorithm has the same formulation as introduced earlier, for it is based upon a <u>m</u>inimum <u>v</u>ariance <u>d</u>istortionless <u>b</u>eamformer. Assume that we have an array with N sensors as indicated in Fig. 7. We want to construct a spatial filter, a beamformer, whose output power is a minimized when the input signal field has the spectral correlation $S_y(f, \underline{p}_i, \underline{p}_j)$. In addition we require that all signals with spatial frequency $\underline{\nu}_0$ pass the beamformer without distraction. We can mathematically formulate this as follows:

$$\underset{\{G_i(\underline{\nu}_0)\}}{\text{Min}} \sigma^2(\underline{\nu}_0) = \underset{\{G_i(\underline{\nu}_0)\}}{\text{Min}} \left[\sum_{i,j=1}^{N} G_i^*(\underline{\nu}_0) S_y(f, \underline{p}_i, \underline{p}_j) G_j(\underline{\nu}_0) \right] \quad \text{minimum variance} \quad (15)$$

$$\sum_{i=1}^{N} G_i(\underline{\nu}_0) e^{-j 2\pi \underline{\nu}_0 \cdot \underline{p}_i} = 1 \quad \text{distortionless} \quad (15b)$$

The solution is similar to the time series result. For the purposes of directional wave spectra estimation the important result is the minimum output power

$$\sigma^2(\underline{\nu}_0) = [\underline{E}^+(\underline{\nu}_0) \underline{S}_y(f)^{-1} \underline{E}(\underline{\nu}_0)]^{-1} \quad (16)$$

where $\underline{E}^+(\underline{\nu}_0) = [e^{-j2\pi \underline{\nu}_0 \cdot \underline{p}_1}, \ldots\ldots e^{-j2\pi \underline{\nu}_0 \cdot \underline{p}_N}]$
is a vector of steering delays and

$$\underline{S}_y(f) = \begin{bmatrix} S_y(f, \underline{p}_1, \underline{p}_1) & \cdots\cdots & S_y(f, \underline{p}_1, \underline{p}_N) \\ \vdots & & \\ & & S_y(f, \underline{p}_N, \underline{p}_N) \end{bmatrix}$$

The argument now follows that the minimum output power is a measure of the frequency wavenumber function at spatial frequency $\underline{\nu}_0$ since this power passes through the beamformer without attenuation, and all other spatial frequencies are attenuated in an optimal manner within the constraints of the array geometry.

The remaining issue is that the spectral correlation in Eqs. 14 and 15 is an ensemble quantity. In the application of the MLM to array processing this is replaced by an estimate based upon recordings of the signals observed at the sensors. Usually one forms an estimate of this by applying the FFT to segments of the sensor data and averaging over the segments, or

$$[S_y(f)]_{ij} = \frac{1}{L} \sum_{\ell=1}^{L} Y_i^\ell(f) \, Y_i^\ell(f) \tag{17}$$

where the $Y_i^\ell(f)$ are Fourier coefficients of the transform of the ℓth data segments.

This leads to the following definition of the MLM frequency wavenumber estimate and the algorithm indicated in Fig. 8.

$$\hat{P}_y(f,\underline{\nu}_0) = [\underline{E}^+(\underline{\nu}_0) \, \hat{\underline{S}}_y^{-1}(f) E(\underline{\nu}_0)]^{-1} \tag{18}$$

The use of this algorithm has often led to dramatic improvements in array processing performance. It has been used extensively in sonar, radar and geophysical exploration.

4.2.2 The Data Adaptive Spectral Estimation Method

A variation of the MLM formulation leads to a <u>d</u>ata <u>a</u>daptive <u>s</u>pectral <u>e</u>stimation (DASE)[22]. The distortionless constraint at a single spatial frequency is replaced by an average one over a band, or

$$\frac{1}{A} \int_{D(\underline{\nu}_0)} G_i(\underline{\nu}_0) e^{-j2\pi \underline{\nu} \cdot p_i} \, d\underline{\nu} = 1$$

where $D(\underline{\nu}_0)$ is a local region spatial frequency about $\underline{\nu}_0$ and A is a normalization constant. It can be demonstrated that the solution to this is a largest eigenvalue problem. Applying this algorithm has led to results that are an improvement over the MLM when estimating continuous directional spectra but are degraded for directional spectra.

4.2.3 Extension of the Maximum Entropy Method (MEM) to Estimating Directional Wave Spectra

Although the problem of extending the AR and MEM techniques to spatial processes has received considerable attention, it has not been solved satisfactorily at the current time. The primary theoretical difficulty is that a general mathematical model for spatial autoregression has not yet been devised. At the current time theoretical results are available only for signals with a time variable and one spatial variable sampled uniformly in space and time.

The entropy concept associated with the AR processes can be extended to directional wave spectra estimation. One is led to an algorithm with nonlinear equations for the estimate which must be solved iteratively. The maximum entropy method for a spatial process can be formulated in a manner parallel to Eq. 7 as follows. The entropy for a spatial process is given by

$$S = C_0' + C_1' + \int d\underline{\nu} \, \ln\left[P_y(f,\underline{\nu})\right] \tag{19a}$$

One has measurements of the cross spectral correlation which constrain the estimate so that the inverse transform must satisfy the

$$\hat{S}_y(f,\underline{p}_i,\underline{p}_j) = \int P_y(f,\underline{\nu}) \, e^{-j2\pi\underline{\nu}\cdot(\underline{p}_i-\underline{p}_j)} \, d\underline{\nu} \tag{19b}$$

At first it was suggested that the computer time required to optimize (19a) subject to the constraints of (19b) variations[19] was excessive. Recently, the hard constraints of (19b) have been relaxed in recognition that the cross spectral correlation values are but estimates, i.e. the equality constraints of Eq. (19b) have been replaced by the inequalities. This has apparently led to successful results when simulated correlation functions were used[23].

In summarizing the state of directional wave spectra estimation it must be said that many issues remain unresolved. The only high resolution algorithm that has been employed successfully in several applications is the MLM. There are no satisfactory AR representations, so this aspect of the spectral estimation is completely open. In addition the additional information that the dispersion equations provide have not been exploited.

4.3 Examples of Directional Wave Spectra Estimation

Experiments in measuring directional wave spectra are often difficult to perform, and those relevant to offshore structures are no exceptions. Two major difficulties are encountered. First the array must be established. In work in the ocean this can lead to some awkward structural requirements since there is no convenient way to preserve the array geometry. At the wavelengths of interest this is particularly difficult to do in the ocean unless the sensors are mounted on some taut wire system. The data adaptive methods of directional wave spectra estimation are very intolerant about imprecision in the array geometry. A useful criterion is that the geometry must be known to within $\lambda/8$. The second problem is recording the data over wide dynamic ranges with synchronous time bases. Multichannel digital recorders for 500 sensors, 150 db dynamic range and 1 kHz bandwidth now exist commercially - the major difficulty is connecting the sensor to the recorder. If the array is near shore or a platform, then cables can be run to the sensors. If they are remote one has a major problem in either data telemetry or maintaining synchronous time bases among several recorders. Again, the uncertainty in the time basis must translate to a spatial error of less than $\lambda/8$.

In the following we present two examples of directional wave spectra estimation. The first is for ocean surface waves in San Francisco Bay[24]. It is indication of the current result in measuring surface wave spectra (see 22b) also. The second is an acoustic experiment in ocean reverbaration in the Arctic. It is indicative of results that can be obtained if one uses a good recording system and exploits the data adaptive estimation methods to their full potential.

4.3.1 MLM Estimation of Directional Wave Spectra in San Francisco Bay

A five element wave gauge array was deployed in San Francisco Bay. The data was recorded on an analog recorder, digitized and processed with the MLM algorithm. Fig. 9 is an example of the results that one obtains where the structures of a cross sea is clearly observed. This structure is not evident using conventional array processing techniques.

4.3.2 MLM Estimation of Acoustic Reverbaration in the Arctic Ocean

In an experiment conducted in the Arctic Ocean explosive charges were set of at the location indicated in Fig. 10a. The acoustic signal propagated throughout the Arctic Ocean basin reflecting off the continental margins and were observed with an array located near the explosion site. The array had an L shaped configuration as indicated in Fig. 10b. The signals were recorded on a digital multichannel recorder and processed with the MLM on a minicomputer.

Fig. 10c illustrates the results of the application of this method. The horizontal axis is time after the explosion while the vertical axis is the direction towards which waves are propagating. The contours indicate the intensity of the return. Some of the locations of important bathemetric relief have been indicated on the figure. One can observe that the MLM has detected virtually all of the boundaries of the Arctic Ocean, and that it is indicating wave power as much as 1 hr, 20 min. after the explosive source which we have tentatively identified as reflection from the continental margin on the opposite side of the Arctic Ocean basin from the explosion location. In essence one is doing sonar ranging with the MLM over two way propagation distances of 7000 km! Conventional methods of array processing simply do not have the capability of resolving directional wave spectra in this much detail.

Conclusions

We have summarized some of the recent techniques that can be used for spectral analysis in offshore structures research. In both time series analysis and directional wave spectra estimation that application of these methods have produced significantly improved results. In an "ideal" experiment one would measure both the environmental directional wave spectra and deploy an extensive instrumentation of the structure. The simultaneous application of these techniques for such an experiment would enhance the understanding of offshore structure response to directional seas significantly.

References

1) Blackman, R.B. and Tukey, J.W., *The Measurement of Power Spectra*, Dover Publications, New York, 1958.

2) Jenkins, G.M. and Watts, D.G., *Spectral Analysis*, Holden-Day, San Franciso,(1968)

3) Bendat, J.S. and Piersol, A.G., *Random Data: Analysis and Measurement Procedures*, Wiley-Interscience, New York, (1971)

4) Bingham, C., Godfrey, M.D. and Tukey, J.W., "Modern Techniques of Power Spectrum Estimation", *Trans. of IEEE on Audio and Electroacoustics*, AV-15, June 1967

5) Baggeroer, A.B., "Sonar Signal Processing", Chapter 6 on *Applications of Digital Signal Processing*, ed A.V. Oppenheim, Prentice-Hall, Englewood, Cliffs, New Jersey, (1978)

6) Anderson, T.W., *The Statistical Analysis of Time Series*, J. Wiley, New York, (1971)

7) Makoul, J., "Linear prediction: A tutorial review", *Proc. of the IEEE*, Vol. 63, pp. 561-580, (1975)

8) Robinson, E.A., *Statistical Communication and Detection*, Hafner Publishing Co., New York, (1967)

9) Akaike, H., "A New Look at the Statistical Model Identification", *IEEE Trans. on Automatic Control*, Vol. AC-19, No. 6, pp. 716-723, (1974)

10) Burg, J.P., "New Concepts in Power Spectra Estimation", presented at the 40th Annual Int. SEG Meeting, New Orleans, Nov. 11, 1970

11) Claerbout, J.F., *Fundamentals of Geophysical Data Processing*, Mc-Graw-Hill, New York, (1976)

12) Fougere, P.F., "A Solution to Spontaneous Line Splitting in Maximum Entropy Spectral Analysis", *J. Geophysical Research*, Vol. 82, pp. 1051-1954, (1977)

13) Lacoss, R.T., "Data Adaptive Spectral Analysis Methods", *Geophysics*, Vol. 36, pp. 661-675, (1971)

14) Burg, J.P., "The Relationship Between Maximum Entropy Spectra and Maximum Likelihood Spectra", *Geophysics*, Vol. 37, pp. 375-376, (1972)

15) Porter, D.L., Baggeroer, A.B., and Briscoe, M.G., "High-Resolution Spectral Analysis of Oceanographic Time Series", Woods Hole Oceanographic Institution, Contribution 3541

16) Neuman, G. and Pierson, W.J. Jr., *Principles of Physical Oceanography*, Prentice-Hall, Englewood Cliffs, New Jersey (1966)

17) Whittle, P., "On Stationary Processes in the Plane", *Biometrika*, Vol. 41, pp. 434-449.

18) Strand, O.N., "Multichannel Complex Maximum Entropy (autoregressive) Spectral Analysis", *IEEE Trans. on Automatic Control*, Vol. AC-22, No. 4, August, 1977, pp. 634-640

19) Baggeroer, A.B., "Maximum Likelihood Estimation of Cross Spectra", Lincoln Laboratory, Massachusetts Institute of Technology, Lexington, Mass. (1974)

20) Steinbug, B.D., *Principles of Specture and Array System Design*, J. Wiley, New York, (1976)

21) Capon, J., "High Resolution Frequency-Wavenumber Spectrum Analysis", *Proc. of the IEEE*, Vol. 57, pp. 1408-1418

22a) Reiger, L.A. and Davis, R.E., "Observations of Power and Directional Spectrum of Ocean Surface Waves", *Journal of Marine Research*, Vol. 35, No. 3, pp. 433-452

22b) Davis, R.E. and Reiger, L.A., "Methods for Estimating Directional Wave Spectra from Multielement arrays", *Journal of Marine Research*, Vol. 35, No. 3, pp. 453-477

23) Weinecke, S.J. and D'Addario, L.R., "Maximim Entropy Image Reconstruction", *IEEE Trans. on Computers*, Vol. C-26, No. 4, pp. 351-364

24) Oakley, O.H. Jr., and Lozow, J.B., "Directional Spectra Measurements by Small Arrays", Paper No. 2745, *Proceedings of the Offshore Technology Conference*, 1977

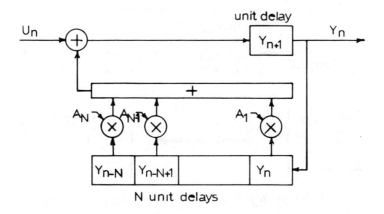

Fig. 1 Autoregressive model for processes

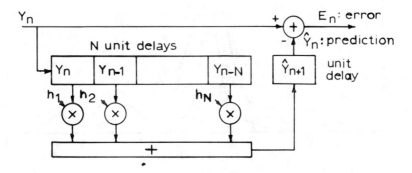

Fig. 2 Prediction error model for whitening filter

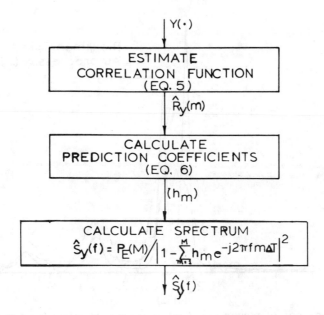

Fig. 3 Autoregressive, or maximum entropy, method of spectral estimation

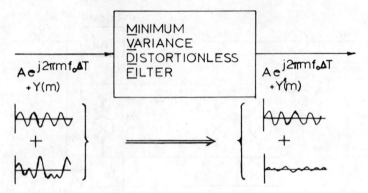

Fig. 4   Model for Minimum Variance Distortionless Filter

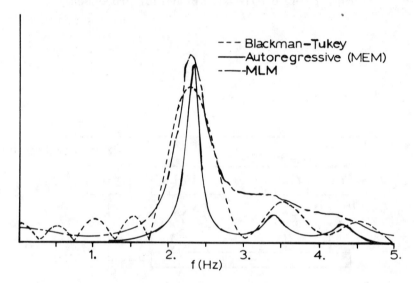

Fig. 5a   Comparison of power spectral density estimates for the ACMR structure, 100 secs. of data, 1 sec lag

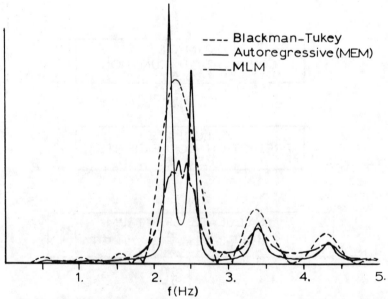

Fig. 5b   Comparison of power spectral density estimates for the ACMR structure 100 secs. of data, 2 sec lag

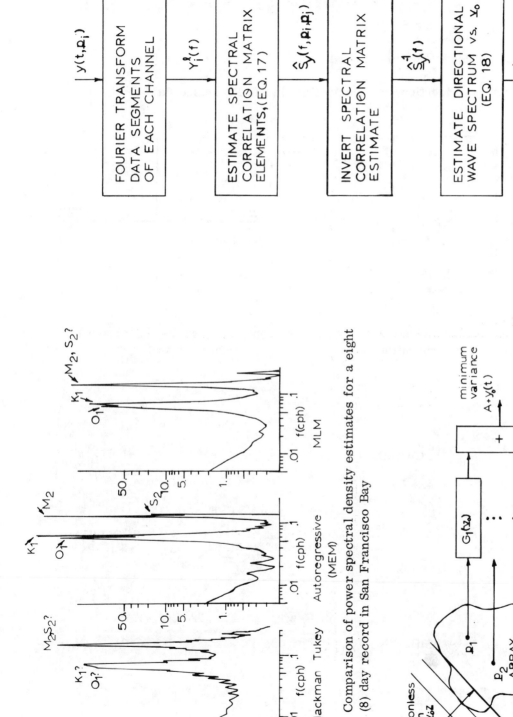

Fig. 6 Comparison of power spectral density estimates for a eight (8) day record in San Francisco Bay

Fig. 7 Model of minimum variance distortionless beamformer

Fig. 8 Algorithm for maximum likelihood method of estimating directional wave spectra

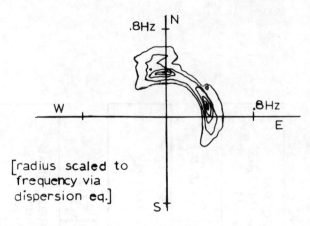

Fig. 9 Directional wave spectrum in San Francisco Bay

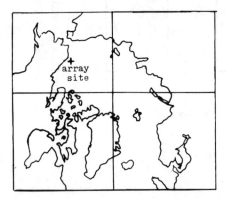

Fig. 10a Map of Arctic Ocean indicating array site for acoustic reverbaration expt.

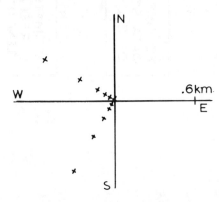

Fig. 10b Array geometry for acoustic reverbaration expt.

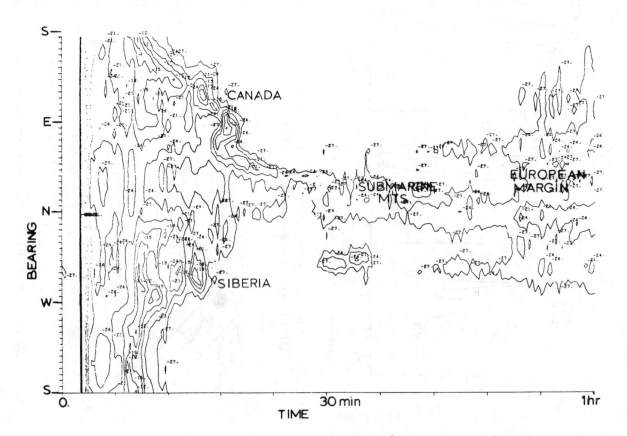

Fig. 10c Directional wave spectrum vs. time for Arctic reverbaration expt.

BOSS'79

PAPER 81

Second International Conference
on Behaviour of Off-Shore Structures

Held at: Imperial College, London, England
28 to 31 August 1979

PRELIMINARY RESULTS OF THE GEOTECHNICAL INSTRUMENTATION INSTALLED BELOW THE TCP 2 GRAVITY STRUCTURE (FRIGG FIELD) DURING ITS FIRST WINTER (1977-78) PERIOD

J.P. Mizikos

SNEA(P) - France

Summary

Some of the geotechnical gauges of the TCP 2 instrumentation were operated during the first winter season endured by the platform, but without simultaneous structural or environmental instrumentation records.

The records consist in pore pressure and lateral earth pressure measurements, in addition to settlements and lateral displacements.

Sponsored by: Delft University of Technology, The Netherlands
Massachusetts Institute of Technology, U.S.A.
The Norwegian Institute of Technology, Norway
University of London, England

Secretariat provided by: BHRA Fluid Engineering

Copyright: ©   BHRA Fluid Engineering
Cranfield, Bedford, England

1 - THE STRUCTURE

The TCP 2 structure, standing on FRIGG FIELD (North Sea), is a CONDEEP concrete platform, with 3 columns and 19 caissons.

The hexagonal slab is cantilevered from the cells. The sides are 60 m long and the total foundation area is 9 340 m^2. Concrete skirts are extended downwards by steel skirts. Their penetration after installation is estimated to lie between 1.3 and 2.3 m according to the sea-bottom slope of about 1 per cent oriented N 150°E.

Other data are :

- platform buoyant weight : 1 620 MN
- expected 100 year significant wave : 17 m, with a period of 15 s.
- maximum load on sea-bottom : 0.174 MN/m^2 (1.74 bar)
- maximum horizontal force : 463 MN
- maximum overturning moment : 10.075 GN x m

2 - SUB-BOTTOM CONDITIONS

2.1 - The sea-bottom consists of a dense ($\phi' = 39°$), uniform, medium to fine sand, with an irregular thickness of 3-4 m. The lower part of the layer contains some silt lenses.

2.2 - Lying below is a heterogeneous silt or silty clay, (up to approx. 8 m) which becomes a medium plastic or plastic clay at the base, i.e. up to 11 to 12 m.

2.3 - The deeper layers consist of alternating sand and silty clay, which are themselves interbedded with seams of sand, silt or clay.

2.4 - All layers had been overconsolidated by past glaciers, at least between 0.5 to 1 MPa of effective preconsolidation.

3 - INSTRUMENTATION EQUIPMENT

3.1 - Data acquisition

The automatic system was not ready during the first winter (1977-78) endured by the structure. Only two paper recorders were available, providing 9 paper tracks that were used as follows :

- 5 for deep piezometry measurement (abrev. : DPPs), operated from 23/11/77 to 08/04/78 ;
- 4 for lateral earth pressure (passive cell of dyprometers or PCD), two for total stress, two for corresponding pore water pressure ; this system was continuously recorded from 23/11/77 to 15/03/78, and temporarily used out of this period.

Simultaneously, we measured :

- total settlement by classical topography ;
- layer per layer settlement, using in situ radioactive markers located down to - 51 m in a bore-hole below the slab ;
- horizontal displacements all along the same bore-hole ;
- significant wave height, estimated from the platform deck and from a stand-by boat ;
- influence of the anti-liquefaction system (abrev. : ALS).

3.2 - DPP (a Syminex device)

The piezometers are MAHIAK transducers operating on the principle of continuously vibrating wire. There are two transducers (one as a spare) per level. They are fixed on a steel block (90 mm O.D.) into which two ceramic filters are inserted. Above and below the steel block are fitted two tubes (0.65 m long) each equipped with a rubber packer.

The level to be instrumented is identified by gamma-ray logging. The borehole tubing (uncemented on the walls here) is perforated by bullets fired from a logging gun. Once the gauges are located in front of such perforation the packers are inflated.

The non-correlation between all kinds of simultaneous pore pressure measurements and water tightness tests confirm the absence of leakage or channeling between each gauge location. Calibration was controlled and considered to be "perfect", i.e. within the range of experimental error : $\pm$ 0.015 bars.

3.3 - PCD (a Syminex device)

A lateral earth pressure cell is an elastic membrane inflated at a pressure slightly higher than the expected local earth pressure, and after that maintained at constant volume and pressure in order to act as a passive gauge (details in section 9.1). A pore pressure sensor is installed at the edge of the membrane in order to compute both total and effective stresses.

3.4 - Settlement

Optical settlement measurements refer to a mark on QP jacket and are accurate to within $\pm$ 1 cm.

In situ settlement measurements result from the detection of 14 radioactive bullets previously fired along the walls of a bore-hole (P4) 51 m deep (Syminex device).
Experimental accuracy is $\pm$ 2 mm.

3.5 - Horizontal displacements

Measurements consist in lowering two inclinometers along the P4 bore-hole under the slab. Accuracy is estimated to be $\pm$ 5 mm.

3.6 - Antiliquefaction system (ALS)

This is not an instrumentation device but a system of drains installed in the sea-bottom sand only, either inside the skirt compartments or along the dowells, (i.e. below the level of the bases of the skirts). ALS was requested by the constructor as a protection against the liquefaction risks. The drains are connected through a system of tubes to eight sets of valves opening inside a dry column at 5, 10, 15 or 20 metres below the mean sea level.

3.7 - All the other instrumentation systems (accelerometers, swell radar, shock gauges on columns, shallow piezometers distributed just below the slab, etc. making a total of 56 permanent and 32 intermittent) worked perfectly during the first part of this winter (1978-79) and are presently being processed using the tapes of the automatic digital recording system which has been operating since September 1978.

4 - EXCESS DEEP PORE PRESSURES

The DPP bore-hole P1 is situated approximately half-way between the centre and the edge of the slab in the NE direction. The location of the five pairs of gauges is presented on figure 1 together with the main properties of the lithology.

4.1 - Results of the upper DPP

It is important to note that the gauge n° 5 is not situated in the fine sea-bottom sand but just at its base inside a silty layer. This situation is assumed to exaggerate the hydrodynamic behaviour of the sand.

Only once was an excess pore pressure observed. This occurred during the severe storm from Dec. 7th to 12th, 1977 (85 % of the 100 year storm) while H 1/3 increased continuously from 2 m on the morning of the 7th to about 13 m one day later. Only the upper DPP5 showed an increase of the average pore pressure of 0.15 b. that developed within 3 hrs during the afternoon of Dec. 7th.

This excess remained until the afternoon of Dec. 15th, that is, more than 3 days after the end of the storm. The same delay was observed on a piezometer installed in the sea-bottom sand below another gravity structure on FRIGG FIELD.

Further storms with significant wave attaining as much as H 1/3 = 12 or 13 m, but lasting only 24 to 36 hrs no longer produced any excess pore pressure. This result is interpreted in terms of <u>pre-shearing by the first and most severe storm.</u>

Note : Inside the lower silty - clay, a pre-shearing or work-hardening, under cyclic loading, had been demonstrated further by in situ cyclic and static testing using active cells of dyprometers (results still unpublished). The sea-bottom sand was not instrumented in the same way.

4.2 - Lower sensors 4, 3, 2 and 1 (refer to fig. 1)

During the same storm, a lull (H 1/3 = 10 m) occurred on Dec. 9th. As the waves again increased, a slight excess pore pressure of about 0.04 b and 0.09 b seemed to develop around gauges 3 and 2 respectively. The former ended with the storm while the latter remained until Dec. 15th like DPP n°5. It is remarkable that during the highest H 1/3 peak of about 15 m, which occurred around Dec. 11th at noon, all the DPPs showed a simultaneous increase in their average pore pressure which attained 0.15 b (only 0.12 b for the deepest sensor n° 1).

4.3 - Interpretations

The pre-shearing has already been mentioned.

The only observed excess pore pressure of 0.3 bar corresponding to the storm of Dec. 7th - 12th which represents 85 % of the 100 year storm, reaches 15 per cent of the static load at this depth.

We believe that this range of excess pore pressure <u>is coherent with the cyclic mobility</u> of this dense sand, as induced by horizontal shear displacements of the slab, but is inconsistent with some predictions of liquefaction risk that were presented to us.

During this storm, the ALS was open and is assumed to produce an under pressure of about 0.3 bars around the DPP n° 5. The excess observed of 0.3 b above this level agrees with the method of computation of Dr SMITS (Delft University), reducing his prediction for a 100 year storm by 15 % .

The absence of excess pore pressure for DPP n° 4, but not for DPPs n° 3 and 2, thus independently of the sediment nature, may be interpreted in terms of drainage capability. It seems that the amplitudes of cycles for DPP n° 1 are too low to generate an excess pore pressure.

5 - CYCLIC VARIATIONS OF DEEP PORE PRESSURE

This is one of the most interesting features. Only the most representative example will be presented here.

5.1 - The example

Figure 2 presents the peak to peak amplitudes, A, recorded by each DPP gauge (selected as the higher value among sinusoidal cycles on DPP records) to be compared to the estimated H 1/3 value and to an estimated wave period (same selection as for the amplitude of cycles).

Referring to figure 2, we see that all DPP cycles were low on Dec. 13th as a consequence of the storm ending on Dec. 12th.

Although the sea-state is calm, during Dec. 14th and 15th a rise in the amplitude appears on all sensors. This rise correlates on the other hand with a sharp rise of the (estimated) wave period from 8 or 10 s, up to 20 s. The amplitude of cycles then reach about one half of the values recorded during the previous severe storm.

Later on, when the period returned to its normal value of 11 or 12 s. , the cycles stopped except for the upper gauge n° 5. Remembering that its excess pore pressure dissipates during Dec. 15th, we find no correlation between the excess pressure in the sea-bottom dense sand and the pore pressure cycles recorded in front of sensor n° 5.

A second general rise of the amplitude of cycles, starting early in the day of Dec. 19th, develops progressively in relation with a new increase of the wave period. The cycles accentuate again when a moderate storm (H 1/3 $\neq$ 6 m) combines with the period elevation (11 to 15 s.) , but the cycles decrease rapidly with the reduction of the periods whereas the storm continued.

5.2 - Generalization

The multiplication of such examples during the observation period leads to the following preliminary conclusions :

- <u>During calm weather</u>, that is H 1/3 lower than 3 m, a wave period of 10 s. appears necessary to induce noticeable cyclic variations of the pore pressure in the top fine dense sand (piezometer 5). The threshold for the deeper sensors seems to be 12.5 or 13 s. Their induced cyclic variations are lower than those of sensor 5.

- <u>During light storms</u> such as :

$$3 \text{ m} < \text{H } 1/3 < 8 \text{ m}$$

the wave amplitude combines with the period which is rarely lower than 10 s. in this type of storm. It results in a modification of the previous threshold.

For DPP n° 5, the threshold varies between 10 and 13 s. while the modification for the deeper piezometers is not clear.

- <u>During heavy storms</u> i.e. with H 1/3 greater than 8 m, the wave amplitude influence cancels that of the wave period. The amplitude of the cycles correlates with H 1/3 rather than with the period.

5.3 - Preliminary interpretations

5.3.1 - We have no proof that the periods considered herein represent correctly the wave spectra but it is doubtful that periods as high as 14 to 20 s. correspond to a vibration mode of the structure, especially as the vibration mode is excited by a calm sea or light storms. An interpretation at the present stage can only be preliminary because no wave spectra or accelerograms were recorded.

Consequently, the more probable explanation is that the transfer of kinetic energy from waves to columns is better under long period swell than under waves of short period (less than 10s.). This kind of coupling becomes progressively overlapped by the shock impact of the waves when their amplitude rises.

5.3.2 - The maximum amplitudes of cycles were recorded during the storm of February 23rd - 24th, 1978 with H 1/3 of 12 to 13 m

(in KPa) $\dfrac{A5}{32} > \dfrac{A2}{19} > \dfrac{A3}{18} > \dfrac{A4}{13} > \dfrac{A1}{4}$

This pattern is very common during storms or high periods swell. A5 predominates. The other amplitudes of cycles increase as a function of depth from sensor 4 to sensor 2, independently of the nature or permeability of sediments. This is a consequence of the drainage conditions. The amplitudes for the sensor n° 1 are always low.

5.3.3 - The interpretation is completed with the general conclusion in chapter 9.

6 - INFLUENCE OF THE ANTILIQUEFACTION SYSTEM (Remark)

The distribution of drains and/or the distribution of connections to the evacuation valves is not coherent with the delimitation of the skirts compartments so that the few existing measurements of underpressures and corresponding flows can only exceptionally be interpreted in terms of Soil Mechanics.

The system was opened at the level - 10 m below sea-level on August 24th and produced underpressures ranging between - 0.3 to - 0.9 bars per valve and flow ranging between 0.1 and 7 m^3/h per valve. The flows decreased by about 90 % within 15 days and became negligible one month after the opening ; simultaneously, the range of underpressures reduced from - 0.1 to - 0.7 b. From that time, the system became progressively choked up and can now be considered as ruined, except for one valve, number 12.

This valve 12, as it happens, drains the sea-bottom sand around the DPPs borehole and normally produces a water flow exceeding 3 m^3/h (it is normally closed).

The permeability of the sea-bottom sand is about 10^{-2} m.s.

Considering the influence of the ALS on the DPPs reading, we observed an influence estimated at - 0.5 bars at the end of November, - 0.33 bars on January 11th (the valve n° 12 contributing with 0.23 b). As valve n° 12 was definitively closed at the end of January 1978, the influence of the ALS on DPP n° 5 (no longer controlled) could be considered as increasingly negligible.

7 - SETTLEMENT

7.1 - Observations

The structure was completely laid on the sea-floor on June 25th at midnight. At this instant, the optical measurement of the vertical displacement began using a fixed point on QP jacket as a reference. One part of the immediate settlement was probably missed.

From June 25th, an immediate total settlement of about 3.5 cm was observed which ended exactly with the main ballasting on July 17th. This demonstrates a purely elastic deformation of the sea-bottom (consequence of the preconsolidation). After this date and until the first storms in November 77, the structure no longer moved vertically.

Fluctuations of the observations can be explained by the thermic variations of the position of the reference level. They prevent any correlation (accuracy being $\pm$ 1 cm) with later occasional ballasting operations or ALS manipulations.

From September 1977, the settlement was also measured layer by layer owing to the radioactive markers system (SYMINEX).
The displacements of each marker are presented on figure 3. The assumed fixed point is the lower marker n° 0.

The distance between the reference marker n° 0 and the next marker n° 1 is about 1.5 m ; their relative movement at 50 m depth and inside a dense overconsolidated sand can thus be considered as negligible, so that a measured vertical displacement of marker n° 1 in fact indicates the actual movement of the reference marker. Till now, it remains within the limits of error : 2 mm and no corrections appears necessary.

The markers device indicates the following main features :

- the upper sand layer (markers 11 and 12) settles again from November, in correlation with the occurrence of the first storms endured by the structure. This vertical displacement increases in December in connection with the severe storm of Dec. 7th - 12th and thus remains quasi-constant.

- layers of the sub-bottom settle from Dec. 77 until Feb. or March. 78. This deep settlement is noticeable between markers 6 and 7, 8 and 9, and possibly, 2 and 3, which are pairs of markers limiting clay layers (see fig. 3).

Consequently, an additional deep settlement is generated by the storms, and propagates downwards from the upper sea-bottom sand layer affecting only the clay layers down to 50 m depth.

The total settlement induced by storms during the winter of 1977-78 is 3.5 cm according to markers systems. This tallies exactly with the value indicated by the optical measurement.

7.2 - Interpretations

An immediate elastic settlement, as a consequence of the preconsolidation, is followed by a storm induced settlement limited to sea-bottom sand and deeper clay layers. Numerical interpretation of the observed settlement values leads to a value of 400 MPa for the elastic modulus whereas a conservative value of 50 Mpa was recommended at the design stage and a realistic value of 100 MPa should have been decuced from triaxial testing. Correlatively, the total settlement was overestimated with a ratio 1 to 6.

8 - HORIZONTAL DISPLACEMENT

Measurement began in September 1977, but deals only with the permanent monthly deviations.

The variations of the horizontal displacement at slab level are summarized on figure 4. It seems that the orientation became eastward during the stormy period and westward during summer, but one year of observation is not sufficient to provide a reliable conclusion.

9 - LATERAL EARTH PRESSURE (from PCD)

9.1 - Installation procedure and operation conditions

The PCD or passive cell represents one of the three cells of the dyprometric probe (SYMINEX) (see figure 5). The two others are active and will not be examined here.

The passive cell has 68 mm OD (2"11/16) and 0.34 m of sensitive length. The probe is lowered into the soil in a pre-drilled 2 3/8" (60.3 mm) hole, by jacking. The lower part of the probe consists of a cutting head equipped with a water nozzle. When the tip of the probe penetrates into the 2 3/8" hole, the water jetting starts with a pressure of 25 to 50 bars and a flow rate of 5 l/mn. Liquified sediment is drained off by a central hose. This penetration procedure is supposed to minimize the remoulding of the sediments in contact with the cell. Nevertheless, just after installation, the lateral pressure of the passive cell of P3 was 1.56 MPa whereas the static horizontal pressure is assumed to be : 1.36 x 0.8 $\neq$ 1.09 MPa.

Once introduced, the passive cell is inflated at a value slightly higher than the expected hydrostatic pressure, in order to ensure a good contact with the hole wall but not to overload the sediment.

Two dyprometric probes were installed this way below the slab, in bore-holes P2 and P3.

9.2 - Discussion of results

9.2.1 - Absolute pressures

We shall discuss the results of the probe installed in bore-hole P3, situated at mid-distance from centre and edge of the slab, like DPP bore-hole P2, but in the S.E. direction (direction of most storms). Each so-called absolute value represents the average of a 10 or 20 minute paper record values, each 20 min sequence being repeated every 3 hrs or every hour during severe storms.

a.a - One week after installation, the (assumed) total lateral pressure stabilized at around 1.195 MPa, whereas the theoretical value of total vertical stress could be :

- mean sea level   : 102.85 m x 1.027 x 9.82 = 1.037 MPa
- overburden pressure :   7.55 m x 1.97 x 9.82 = 0.147 MPa
- platform load   :   0.177 MPa
                                                 ─────────
                                                  1.36 MPa

This leads to σ_H/σ_V ratio of 1.195/1.36 = 0.88

A piezometer is installed between the connection of the cell and the upper extension pipe ; it is protected by a ceramic filter saturated before installation. At the same date, this gauge indicated 1.175 MPa, then an effective lateral stress of 0.02 MPa, whereas the theoretical value (in agreement with DPPs results) is 1.113 MPa.

Note that during the lowering of the probe along the installation tubing, the passive cell and the piezometers indicated the same pressure reading at a given level. Nevertheless, no accurate calibration had been possible.

The results became even more questionable from the moment when we observed a slow continuous decrease of about 0.01 MPa per month, leading in August 78 to the equalization of the measured total lateral stress with the corresponding pore pressure, despite a new inflation of the passive cell. Later pore pressure even exceeds the total pressure !

The same trend was observed for the other PCD (bore-hole P2), but more accentuated.

We conclude that the absolute measurements of (supposed) lateral earth pressure by PCD are not valid at least concerning the long term variation.

b.b - The same pressure had been deduced from the active cells of dyprometers when they were used for the study of clay fatigue. The results seem more reliable because the readings are observed in very different tests (during one week).

In bore-hole P3 where the ACDs were located at - 8.2 and - 8.95 m, the following results were found in November 1978 :

- a total horizontal earth pressure of :
  1.299 ± 0.006 MPa

- a corresponding pore pressure of :
  1.090 ± 0.007 MPa

(the theoretical value, in agreement with DPPs being 1.124 the difference presumably resulting from the lack of real calibration).

- Thus : $\sigma H/\sigma V = 1.299 / 1.38 = 0.94$

$$\sigma'H/\sigma'V = \frac{1.299 - 1.090}{1.38 - 1.090} = 0.72$$

c.c - A simple calculation supposing an elastic expansion of the clay can be done for comparison.
Stating :

$\sigma'Vm$: maximum effective vertical stress supported by the sub-bottom during its stress history.

$\sigma'Hm$: corresponding horizontal stresses.

The present effective horizontal stress should be :

$$\sigma'Ha = Ko \cdot \sigma'Vm - \frac{V}{1-V} (\sigma'Vm - \sigma'Va)$$

At a depth of - 8.6 m where $\sigma'Vm \neq 0.09 + 0.18 = 0.27$ MPa including the structure load, we can state that :

$$\sigma'Ha = 0.5 \times 1 - \frac{0.3}{1-0.3} (1 - 0.27) = 0.19$$

which gives a ratio $\sigma'H/\sigma'V = 0.66$. Taking into account the different uncertainties we can conclude on a fair agreement with the measured values and to results compatible with the preconsolidation.

9.2.2 - Short term variations of average values (P3 bore-hole)

The variations of lateral earth pressure as a function of time which are higher than the rate of drift can be considered as reliable in the short term. In such conditions, the values averaged per sequence (thus eliminating wave influence, but keeping tide influence) correlates with the similar DPP observations :

- Only the severest storm of Dec. 7th - 12th produced a variation of this averaged value, here a decrease in total lateral earth pressure with values of :

  1.180 MPa before Dec. 8th

  1.171 MPa from this date to Dec. 13th

  1.164 MPa later on

- Influence of the manipulation of ALS on Jan. 28th and 31st, 1978. Note that the P3 bore-hole is influenced by the efficient valve 12 (ref. chap. 6) ;

$$\begin{cases} \text{opening of valve 12 on Jan. 28th : } -0.012 \text{ MPa max.} \\ \text{decreasing to zero on Jan. 31st} \end{cases}$$

- No more influence of storms after Dec. 7 - 12th, 1977, the severest storm of the recording period, except for some increases of the tide amplitude without any modification of the mean lateral pressure value.

9.2.3 - Amplitude of cycles

The peak to peak amplitude of the cycles (recorded by the passive cell) always exceeds the corresponding pore pressure record, but also the corresponding DPP records. This is not surprising for a total pressure measurement. It is also possible that the PCD acts as a pore pressure sensor rather than a total earth pressure sensor. In this case, the excess in amplitudes would result from the flexibility of the elastic membrane.

Consequently, we cannot conclude that the PCD really represents the total lateral earth pressure cycles.

Nevertheless, they exhibit the same features as the DPPs amplitudes of cycles, i.e :

- correlation with H 1/3.

- intense amplitude of cycles during small swell of long period ; an example is presented in figure 6.

- maximum recorded amplitudes (here $\pm$ 0.4 bars = $\pm$ 0.04 MPa, during the storm of February 23rd, 1978.

9.3 - Interpretation

Due to its flexibility, compared to the surrounding overconsolidated rigid silty clay, the membrane of the passive cells probably acts more as a pore pressure sensor than as a total pressure sensor. This interpretation combined with the similar distribution of DPPs and PCD P3 bore-holes may explain the strong correlation between DPPs and PCD results.

I present the PCD results more as an experiment which could be useful for further improvements of the technique for in situ lateral earth pressure measurement. Similar equipment had worked perfectly in underconsolidated or normally consolidated fine subsea sands.

CONCLUSIONS

From the geotechnical data obtained during the winter of 1977-78, without available records of environmental conditions and no record of structural gauges, we are able to analyse only the shear response of soils to the slab displacements, but not the slab overturning influence on soils (punching, etc) nor clay fatigue. These relatively limited observations nevertheless lead to fundamental conclusions:

1. The results of vertical displacements as well as the current lack of pore-over pressure in deeper layers indicate an elastic behaviour of the sub-bottom, which is a consequence of the preconsolidation of the layers.

2. The location of the analysed geotechnical instrumentation at mid-distance between the centre and the edges of the slab imply that observed soil responses reflect mainly horizontal shear movements below the foundation.

During storms but also when the sea is calm with long wave periods, intense cycles of pore pressure develop in the sea-bottom sand (where the skirts are installed). Cycles became negligible below 30 m depth. In between, the amplitude of cycles depends on the drainage conditions but not on the sediment nature.

3. The only non-linearity involving an excess pore pressure was observed temporarily in the sea-bottom dense sand, during the first and severest storm of this period. The fact that this did not reoccur can be interpreted as proof of pre-shearing, confirmed by the active dyprometer tests (not presented here). Correlatively, layer per layer settlement begins in this sand. When this upper sand becomes pre-sheared, this settlement is relayed by the deep clay strata.

All results seem to indicate that the structure foundation is safe in respect of the phenomena analysed here. Only the operation of all the instrumentation will, of course, indicate if the structure has been overestimated or not. This is the second objective of the instrumentation after the control of its safety (some results may be presented in August).

To summarize, the first interesting result pointed out by this partial instrumentation is the possibility of pre-shearing of the sub-bottom preconsolidated layers. The second interesting point, observed also below another structure of the FRIGG FIELD is the possibility of intense cyclic loading of soils by calm swell with long period.

Reference for SYMINEX devices :

BOURGEOIS (D) and LEROY (J.P), 1976.

Geophysical and mechanical platform Instrumentation

Symp. Design. Constr. Offshore Struct. (Inst. Civ. Engrs, Lond.)

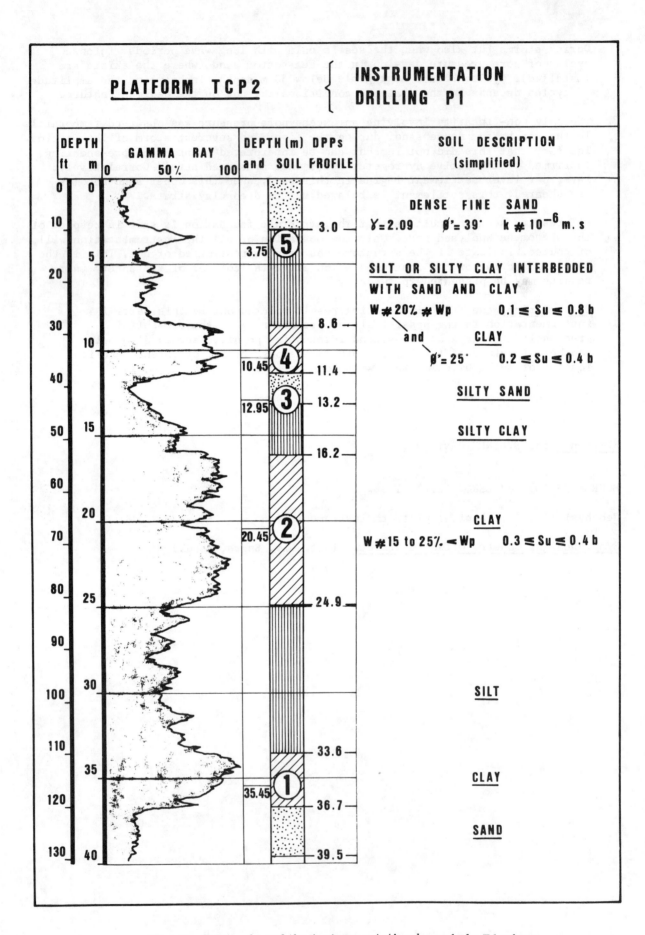

Fig. 1 Boring log of the instrumentation bore-hole P1 where the Deep Pore Pressures gauges or DPPs are located.

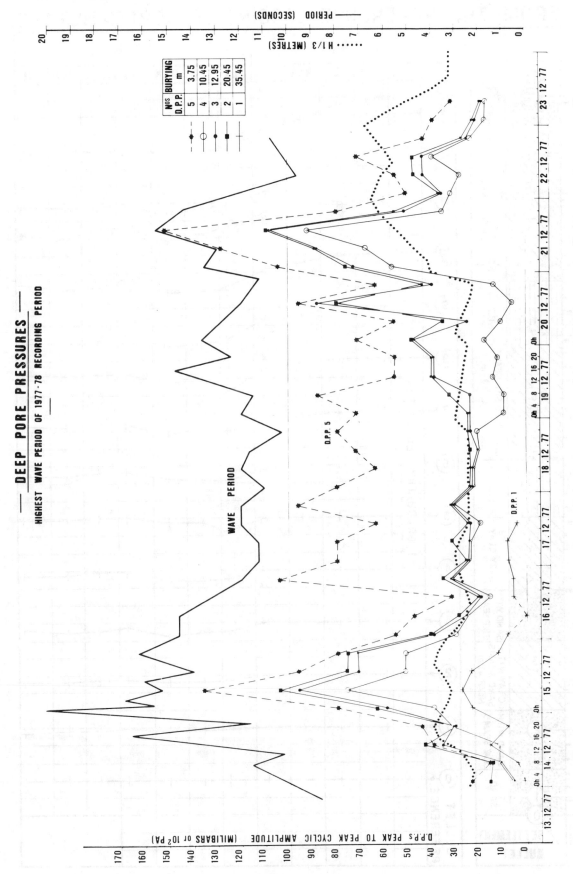

Fig. 2 Example of the variation of the peak to peak amplitude of the deep pore pressure cycles recorded by the five sensors as a function of (supposed) sea wave main period and of the estimated significant wave. This figure points out the development of cycles under calm sea conditions but long period swell.

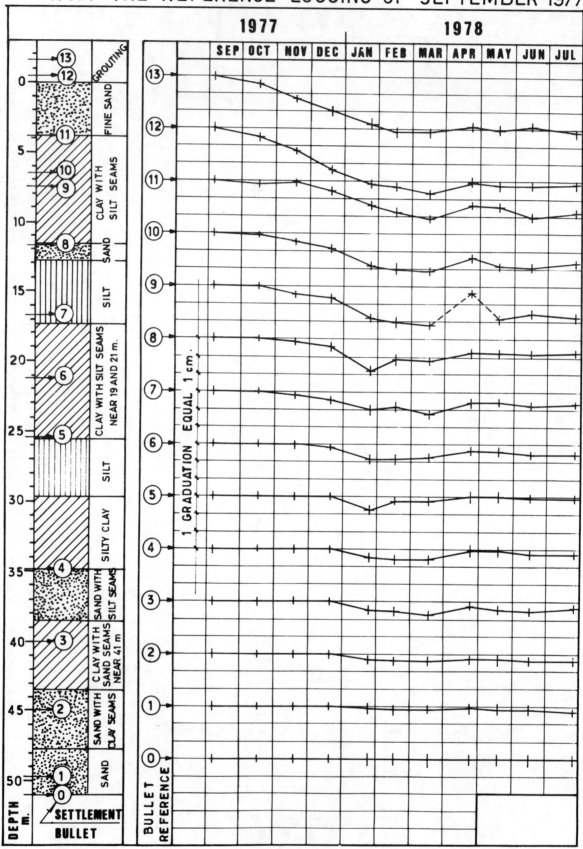

Fig. 3 Displacements of the radioactive markers of bore-hole P4 from September 1977 to July 1978, with indication of the lithology.

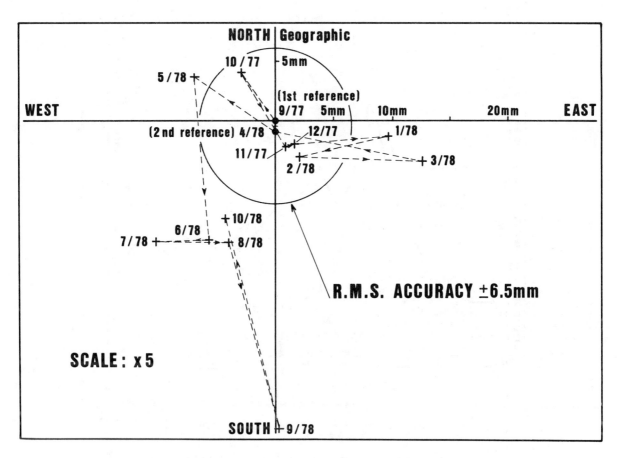

Fig. 4 Horizontal displacement of the slab at P4 bore-hole location from September 1977 (reference) to June 1978.

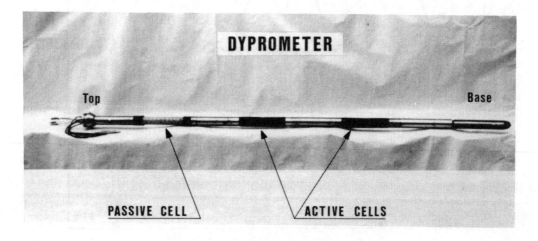

Fig. 5 Description of the dyprometer.

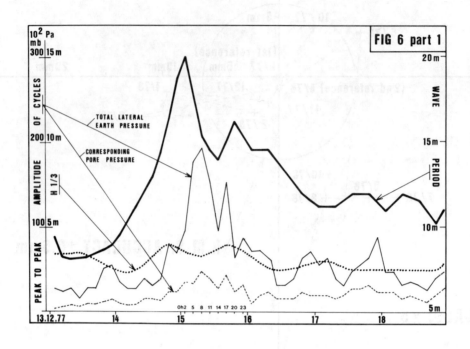

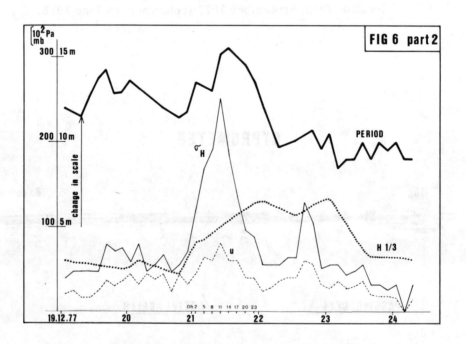

Fig. 6 Variation of the peak to peak amplitude of cycles of the passive cell which should show the total lateral earth pressure and of the adjacent piezometer, as a function of sea wave period and significant wave height.

This example corresponds to dates (DPPs results) in Fig. 2.

BOSS'79

PAPER 83

**Second International Conference
on Behaviour of Off-Shore Structures**

Held at: Imperial College, London, England
28 to 31 August 1979

MEASURING EQUIPMENT FOR FIELD INVESTIGATIONS ON NEAR SURFACE WAVE FORCES

F. Busching, E. Martini and U. Sparboom

Techn. Univ. Braunschweig, Federal Republic of Germany

Summary

Two field investigation programmes on simultaneous wave force and water particle velocity measurements are described with reference to tubular members subjected to offshore and near shore wave kinematics.

Sponsored by: Delft University of Technology, The Netherlands
Massachusetts Institute of Technology, U.S.A.
The Norwegian Institute of Technology, Norway
University of London, England

Secretariat provided by: BHRA Fluid Engineering

Copyright: © BHRA Fluid Engineering
Cranfield, Bedford, England

INTRODUCTION

For want of anything better MORISON's equation is still used for the calculation of wave forces on circular cylindrical structural members. Hence, until now many experiments in the laboratory or in the field are based on it, however, in such a way that only water level deflexions and wave forces on a test section are measured, whereas velocities and accelerations are determined using some suitable wave theory. As is well known this procedure turned out to be one of the reasons for the wide range of scatter in the reported force coefficients.

DEAN 1 has pointed out the necessity of also measuring undisturbed flow characteristics down in the fluid. To the author's knowledge this was recently done only by KIM and HIBBARD 2 measuring the local water particle velocities in a full scale experiment. In a laboratory high enough REYNOLDS's numbers were only obtained applying special model techniques (for instance SARPKAYA [3], HOGBEN [4], YAMAMOTO and NATH [5]). It remains, however, still a question how well the respective results apply to real wave motion and especially to irregular waves with varying directions of propagation.

Being also aware of some more uncertainties arising from:

      different roughness characteristics due to marine fouling,

      the coincident presence of wave and (tidal and wind induced) currents,

      different shapes of test sections (vertical or inclined),

      different wave kinematics (deep versus shallow water) etc.

the authors started two field investigation programmes which are sponsored by the GERMAN MINISTRY OF RESEARCH AND TECHNOLOGY (MTK 0053) and the GERMAN RESEARCH FOUNDATION (Sonderforschungsbereich 79, C5) respectively.

Because of the substantially differing kinematics of deep and shallow water waves, one project (MTK) is being carried out on the GERMAN RESEARCH PLATFORM "Nordsee" about 100 km offshore in a water depth of approx. 30 m, and the other one (C5) on the island of NORDERNEY on the GERMAN NORTH SEA COAST. Both investigation programmes are at first restricted to the measurement of near surface wave forces (derived from the measured pressure distributions on the circumference of tubular members), and water level deflexions (waves), and the measurement and analysis of the ambient flow characteristics including tidal, wind and wave induced currents.

In the future these research programmes will be extended. As regards the offshore programme (MTK), this will be completed by the measurement of directional spectra (from an array of 3 sonar devices), whereas the near shore measuring configuration (C5) will be combined with measurements of additional forces exerted by wave spray loadings, see FÜHRBÖTER [6].

At present the MTK-project deals with the wave loadings exerted on an inclined member of a platform leg, whereas the near shore measuring configuration (C5) consists of a vertical pile structure for the force measurement and a satellite measuring station for the measurement of water level deflexions and particle velocities.

Because of the lack of space here, only the test structures and measuring devices subjected to the waves are described in the following, whereas the synchronous signal transfer and data processing is similar to that reported by FUHRBÖTER and BUSCHING [7]. It need only be mentioned here that different analyzing techniques will be used in the time domain as well as in the frequency domain. In particular the cross power spectra, transfer functions and coherence functions between any two signals will be calculated.

OFFSHORE MEASURING CONFIGURATION

As may be seen from the sketch of the measuring devices (Fig. 1) the test section consists of a packing ring clamped on a member of a platform leg which is inclined $30.225°$ with reference to the vertical axis. This tubular structure (5 m long, 1.92 m diameter) contains 24 KISTLER-pressure transducers on its circumference centered about 5m below mean low water spring (MLWS).

In the same depth of water there are 3 two-component electromagnetic COLNBROOK-current meters (No. 1 - 3) oriented in such a way (angular spaced $22.5°$ and 2.5 m distant from the test section) that the particle velocities in certain vertical planes containing the respective main wave propagation direction can be determined from the measurements to a high degree of certainty. For the direct measurement of the wave propagation direction and the phase velocities the current meters No. 4 and 4' are used, each measuring two velocity components in a horizontal plane 3.5 m below MLWS.

Finally, only 2 m below MLWS there is a third horizontal measuring plane again containing 3 current meters (No. 1', 2' and 3') for another measurement of the water particle kinematics in vertical planes with reference to the above mentioned current meter positions No. 1, 2 and 3.

The corresponding water level deflexions are measured at a certain distance from the test section by a BAYLOR wave staff, see Fig. 2, whereas additional wave data can be received from another set of 3 sonar devices fixed to different members of the platform structure, see [8].

4. NEAR SHORE MEASURING CONFIGURATION

Fig. 4 shows the measuring facilities located about 120 m distant from the shoreline on a slightly inclined beach $\gamma = 1 : 35$.

The vertical pile structure (diameter 0.70 m) consists of 3 parts connected by flanged coupling joints.

The bearing member driven some 7 m into the sea bed also contains two cable inlet structures, whereas the actual test section is installed in the center of the middle part at a water depth NSL + 0.70 m. For the measurement of the pressure distribution here there are 16 KISTLER transducers distributed on the circumference of a tubular member which is fixed to a cantilever supporting structure. The total force (magnitude and direction) can also thus be measured directly by some strain gauges attached to the cantilever.

The upper part of the pile extends to a height of NSL + 10 m which is more than 5 m beyond the highest measured storm tide still water level.

On top of the pile there is a working platform and a junction box where the submarine cables are connected to the cables of the measuring devices.

At a distance of 5.0 meters from the measuring pile there is a pile supported satellite measuring station provided with 4 electromagnetic 2-component NSW-current meters and 2 KISTLER pressure devices for the measurement of the water particle kinematics (in horizontal and vertical planes) and water level deflexions respectively.

In the future this programme will be extended to the measurement of the total behaviour of the structure (total wave forces and pile oscillations) especially under storm surge conditions.

REFERENCES:

[1] DEAN, R. G. — Methodology for Evaluating Suitability of Wave and Wave Force Data for Determining Drag and Inertia Coefficients
Proceedings of BOSS'76, Norwegian Institute of Technology, Trondheim, 1976

[2] KIM, Y. Y.
HIBBARD, H. C. — Analysis of Simultaneous Wave Force and Water Particle Velocity Measurements, Paper 2192, OTC 1975

[3] SARPKAYA, T. — Vortex Shedding and Resistance in Harmonic Flow about Smooth and Rough Circular Cylinders
Proceedings of BOSS'76, Norwegian Institute of Technology, Trondheim, 1976

[4] HOGBEN, N. — The Wavedozer: A Travelling Beam Wave-maker, Proc. 11th ONR Symposium on Naval Hydrodynamics, London, April 1976

[5] YAMAMOTO, T.
NATH, J. H. — Hydrodynamic Forces on Groups of Cylinders
OTC 2499, 1976

[6] FÜHRBÖTER, A. — Remark on the Influence of Wave Spray on Wind Load, Safety of Structures under Dynamic Loading
Vol. 2, Tapir; Norwegian Inst. of Technology, Trondheim 1977

[7] FÜHRBÖTER, A.
BÜSCHING, F. — Wave measuring Instrumentation for Field Investigations on Breakers, Ocean Wave Measuring and Analysis, Vol I, New Orleans, 1974

[8] LONGREE, W.-D. — Aspects of the Instrumentation and Measurement Performances of the Research Platform NORDSEE
Proc. BOSS'76, Norwegian Inst. of Technology, Trondheim, 1976

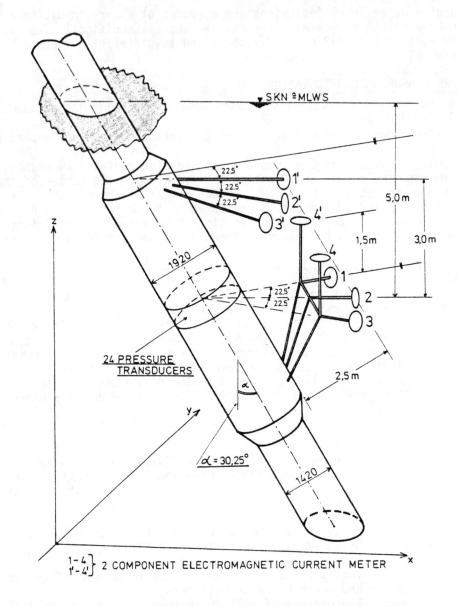

Fig. 1: SKETCH OF THE OFFSHORE MEASURING CONFIGURATION

Fig. 2: TEST SECTION CLAMPED ON AN INCLINED MEMBER OF THE PLATFORM LEG TO BE SEEN AT AN EXTREMELY LOW TIDE WATER LEVEL AT STORM CONDITIONS.
BAYLOR WAVE STAFF ON THE LEFT HAND SIDE.

Fig. 3: LOWERING PROCESS AND INSTALLATION OF THE TEST SECTION. 1.92 m-DIAMETER PACKING RING CONTAINING 24 PRESSURE TRANSDUCERS AND 8 TWO-COMPONENT CURRENT METERS FIXED TO CANTILEVERS

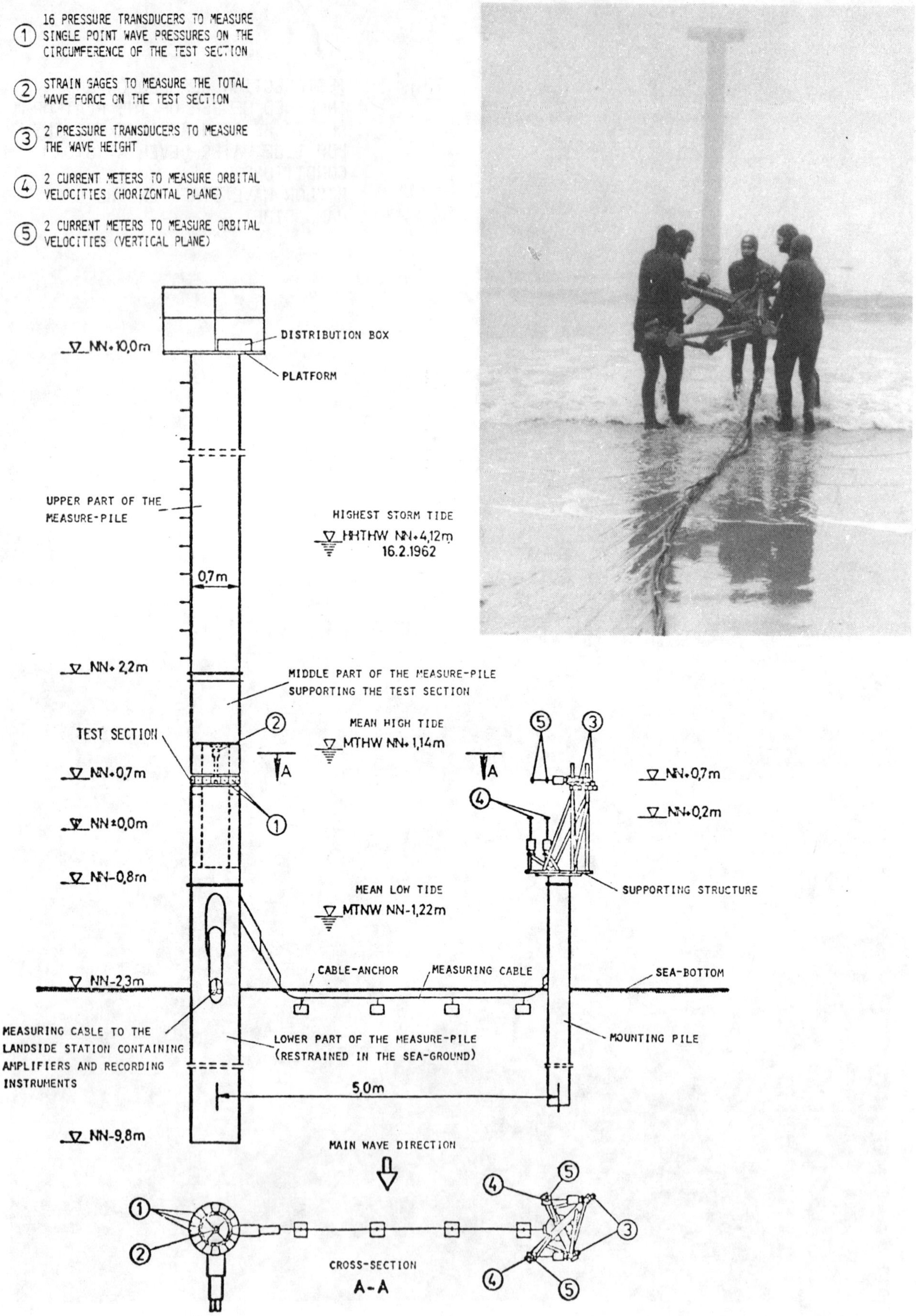

① 16 PRESSURE TRANSDUCERS TO MEASURE SINGLE POINT WAVE PRESSURES ON THE CIRCUMFERENCE OF THE TEST SECTION
② STRAIN GAGES TO MEASURE THE TOTAL WAVE FORCE ON THE TEST SECTION
③ 2 PRESSURE TRANSDUCERS TO MEASURE THE WAVE HEIGHT
④ 2 CURRENT METERS TO MEASURE ORBITAL VELOCITIES (HORIZONTAL PLANE)
⑤ 2 CURRENT METERS TO MEASURE ORBITAL VELOCITIES (VERTICAL PLANE)
DISTRIBUTION BOX
NN+10,0m
PLATFORM
UPPER PART OF THE MEASURE-PILE
HIGHEST STORM TIDE
HHTHW NN+4,12m
16.2.1962
0,7m
NN+2,2m
MIDDLE PART OF THE MEASURE-PILE SUPPORTING THE TEST SECTION
TEST SECTION
MEAN HIGH TIDE
MTHW NN+1,14m
A
A
NN+0,7m
NN+0,7m
NN+0,2m
NN±0,0m
NN-0,8m
SUPPORTING STRUCTURE
MEAN LOW TIDE
MTNW NN-1,22m
CABLE-ANCHOR
MEASURING CABLE
SEA-BOTTOM
NN-2,3m
MEASURING CABLE TO THE LANDSIDE STATION CONTAINING AMPLIFIERS AND RECORDING INSTRUMENTS
LOWER PART OF THE MEASURE-PILE (RESTRAINED IN THE SEA-GROUND)
MOUNTING PILE
5,0m
NN-9,8m
MAIN WAVE DIRECTION
CROSS-SECTION
A-A

FIG. 4: NEAR SHORE MEASURING CONFIGURATION

BOSS'79

PAPER 84

Second International Conference
on Behaviour of Off-Shore Structures

Held at: Imperial College, London, England
28 to 31 August 1979

METHODS OF RELIABILITY ANALYSIS FOR JACKET PLATFORMS

M. J. Baker and T. A. Wyatt

Imperial College of Science and Technology, U.K.

Summary

Structural reliability analysis is playing an increasingly important role in the development of practical structural codes and there is considerable scope for its application in the offshore industry. The nature of the subject is such that misunderstandings between specialists and practising designers are not infrequent and this opportunity is therefore taken to re-examine some of the fundamental assumptions concerning the interpretation of reliability statements. This leads to an examination of the effects of gross errors on optimum design safety levels and some new results are presented based on a simplified structural model.

Various computational techniques for reliability assessment are discussed and the errors resulting from different approaches and approximations are considered. The so-called advanced Level 2 techniques are shown to be extremely powerful.

Finally, the problems of applying these techniques to complex jacket structures are discussed and some typical results are presented.

Sponsored by: Delft University of Technology, The Netherlands
Massachusetts Institute of Technology, U.S.A.
The Norwegian Institute of Technology, Norway
University of London, England

Secretariat provided by: BHRA Fluid Engineering

Copyright: © BHRA Fluid Engineering
Cranfield, Bedford, England

1. INTRODUCTION

The need to exploit oil and gas reserves in deeper offshore waters is leading to greatly increased structural costs and to many new technical problems which require solution. Taking a global view, perhaps the most important of these is the need to ensure the correct balance between the conflicting requirements of economy of initial construction and of subsequent reliability, including personnel safety. The problem is made even more difficult when design decisions have to be made in the absence of complete information about true structural behaviour or when data on environmental loads or foundation conditions are limited. This is nearly always the case.

Structural reliability analysis is playing an increasingly important role in the solution of such practical design problems, particularly in selection of partial safety factors for use in Codes (1,2). There are many difficulties in applying the techniques and in particular in choosing appropriate standards of reliability for different structures and components; nevertheless, no rational alternative exists. The aim of this paper is to review the available methods of reliability analysis and to discuss their application to offshore engineering and in particular to jacket-type platforms. Because such methods are now beginning to be used in earnest and are being recommended to the profession (3), it is important that some of the fundamental assumptions concerning reliability analysis should be re-examined to provide a firmer basis for discussion between practising engineers, research workers and Code writers. In particular, misunderstandings often occur even over the interpretation of the term reliability. The first part of this paper is therefore concerned with some of these fundamental questions.

Evidence (4) suggests that most structural failures occur because of accidental overloading or because of the occurrence of gross errors during design or construction. A proper assessment of structural reliability cannot therefore neglect the possibility or the effects of gross errors. What would be useful to know in practice is the extent to which designs should be modified to allow for gross errors and the optimum level of expenditure on gross error control. Using a simple structural model, the principles of how this can be achieved are demonstrated.

The remainder of the paper is concerned with the calculation errors that are likely to occur as a result of using different methods of reliability analysis and the problems associated with applying these techniques to offshore structures.

2. STRUCTURAL RELIABILITY ANALYSIS

2.1 The meaning of reliability

The essence of the structural reliability problem can be stated as follows:

All quantities (except physical and mathematical constants) that currently enter into engineering calculations are in reality associated with some uncertainty. If this were not the case, a "safety factor" only slightly in excess of unity would suffice in all circumstances. The attainment of appropriate standards of safety and serviceability requires the quantification of these uncertainties by some appropriate means and a study of their interaction for the type of structure under consideration.

This statement is indisputable. What is debatable is the best way in which these uncertainties can be processed in order to optimise design decisions.

In a general sense it is clear what is meant by reliability - the reliability of a structural system is its ability to fulfil its design purpose for some specified time. However, two questions arise immediately from this statement. Is it possible to quantify this ability? If so, what use can be made of this information? The answers to these questions are not straightforward and will first be examined in the context of an elementary example.

Assume that an offshore platform is idealised* as a uniform vertical cantilever rigidly connected to the sea bed (Fig.1). The structure will fail in flexure when the moment S induced at the root of the cantilever exceeds the flexural strength R. Assume further that R and S are random variables whose statistical distributions are known as a result of a very long series of measurements. R is the distribution of strength representing the variations between nominally identical structures, whereas S is the distribution of the maximum load effects in successive T year periods. The distributions of R and S are both assumed to be stationary with time and each realisation of resistance r is assumed to be constant with time. Under these assumptions, the probability that the structure will collapse during any reference period of duration T years is given by the well known relationship (5)

$$P_F = \Pr(Z < 0) = \int_{-\infty}^{+\infty} F_R(x) \, f_S(x) \, dx \qquad \ldots(1)$$

where Z is the failure indicator given by

$$Z = R - S \qquad \ldots(2)$$

and where $F_R(r)$ is the cumulative probability of R and $f_S(s)$ is the probability density of S. Exact closed form solutions exist for Eqn.(1) for various combinations of types of statistical distributions for R and S, but, in general, solution by some numerical or simulation procedure is required. In practice, however, it is rarely possible to simplify a structural problem to the extent that Eqn.(1) is applicable and other more complex formulations are required.

Because of the definition of R and S in terms of frequentist probabilities, the probability determined from Eqn.(1) may be interpreted as a long-run failure frequency. Similarly the reliability Q, defined as

$$Q = 1 - P_F \qquad \ldots(3)$$

may be interpreted as a long-run survival frequency or long-run reliability. Q may therefore be called a <u>frequentist reliability</u>. If, however, we are forced to focus our attention on one particular structure (and this is generally the case for "one-off" civil engineering structures), Q may also be interpreted as a measure of the reliability of that particular structure. This interpretation of reliability is fundamentally different from that given above, because, although the structure may be sampled at random from the theoretically infinite population described by the random variable R, once the particular structure has been selected (and, in practice, constructed) the reliability becomes the probability that the fixed, but unknown, resistance r will be exceeded by the as yet "un-sampled" reference period extreme load effect S. The numerical value of the failure probability remains the same but is now dependent upon two radically different types of uncertainty - firstly, the physical variability of the extreme load effect, and, secondly, lack of knowledge about the true value of the fixed but unknown resistance. This type of probability does not have a relative frequency interpretation and is commonly called a subjective probability. The associated reliability can be called a subjective or <u>Bayesian reliability</u>. For a particular structure, the numerical value of this reliability changes as the state of knowledge about the structure changes - for example, if non-destructive tests were to be carried out on the structure to estimate the magnitude or r. In the limit when r becomes known exactly, the probability of failure given by Eqn.(1) changes to

$$P_F = \Pr(r - S < 0) = 1 - F_S(r) \qquad \ldots(4)$$

This special case may also be interpreted as a <u>conditional failure probability</u> with a relative frequency interpretation.

i.e. $P_F = \Pr(R - S < 0 \mid R = r) \qquad \ldots(5)$

---

* It is not suggested that this is an idealisation having any practical significance; it is used solely to illustrate matters of principle.

Consider now the inclusion of an additional uncertainty in analysing the behaviour of this simple idealised structure. Let the failure indicator Z become

$$Z = R - KS \qquad \ldots (6)$$

where R is the structural resistance, S is the predicted load effect, being a deterministic function of the stochastic loads W, and K is a model uncertainty relating the predicted load effect to the true load effect and which is statistically independent of R and S. Assume that R and S are continuous random variables whose statistical distributions are known in terms of long-run relative frequencies, but that K is a discrete variable that can only take the values 0.4, 0.5 and 0.6. At least two possible situations can now be envisaged:

(A) Although only one value of the model uncertainty K applies in the case of each structure, all three values are possible and the probabilities p(K) in Table 1 are observed long-run relative frequencies,

or

(B) Only one of the three values of K is in fact possible, but which is the correct value is not known (perhaps as a result of a lack of reliable data). The probabilities, p(K) in Table 1 are subjective probabilities indicating the profession's degree of belief concerning the relative likelihood of each value of K being correct.

K	0.4	0.5	0.6
p(K)	0.2	0.6	0.2

Table 1

Situations (A) and (B) are seen to be fundamentally different, but involve the same numerical probabilities p(K). In situation (A), the probabilities will not change significantly as more data becomes available. In situation (B), the probability $p(K_c)$, where K_c is the correct value of K, will tend to increase towards unity, although not necessarily monotonically, as more information becomes available.

It is possible to compute the conditional failure probability $P_F|K_i$ for each of the values K_i. From Eqn.(1)

$$P_F|K_i = \int_{-\infty}^{+\infty} F_R(K_i x) \, f_S(x) \, dx \qquad \ldots (7)$$

and hence the total failure probability P_F is

$$P_F = \sum_i p(K_i) \left[P_F|K_i \right] \qquad \ldots (8)$$

Let us now examine the meaning of the probability given by Eqn.(8). If situation (A) applies, P_F is the long-run failure frequency. If situation (B) applies, the true long run failure frequency is given by Eqn.(7) using the correct value for K; but as the correct value of K is not known, there are three possible values for this quantity. In this case, P_F given by Eqn.(8) cannot be interpreted as a relative frequency and must be treated as a subjective (or Bayesian) probability. It is a composite numerical measure of risk combining the effects of both physical variability in R and S and lack of knowledge about the true value of K.

The question now arises - Is the complement of P_F given by Eqn.(8) a useful measure of the relative reliability of different types of structure? If situation (A) applies, then it is clear that the different survival frequencies provide a useful measure of relative reliability and safety. If situation (B) applies, the question cannot be answered so simply.

Consider two different types of structures (1) and (2) which are affected by the same model uncertainty K but which show a different sensitivity to it, as given, for example, in Table 2. Judged on the basis of the total failure probability P_F,

	$P_F \mid K_i$			P_F
K_i	0.4	0.5	0.6	
$p(K_i)$	0.2	0.6	0.2	
Structure 1	10^{-5}	10^{-4}	10^{-3}	$10^{-3.58}$
Structure 2	$10^{-4.5}$	10^{-4}	$10^{-3.5}$	$10^{-3.89}$

Table 2

Structure (2) is more reliable than Structure (1), but if the true value of K is in fact 0.4 then the opposite is true. This type of reversal has an important influence on the validity of Code calibration procedures as discussed later. Assuming, however, that situation (B) applies, it is important to consider whether it is valid to use the value of P_F obtained from Eqn.(8) to assess whether a structure satisfies some externally specified target reliability level (e.g. 10^{-5} per annum). Conversely it should be asked whether it is meaningful to make such regulations. Statements of this type are often made in discussing nuclear and other industrial risks.

To answer the last question first, the view will now be advanced that probabilities considered in isolation have little meaning and that safety regulations should not be written in terms of explicit probability levels. An exception to this is if all the uncertainties entering the calculations are long-run relative frequencies when the calculated reliability will have a long-run survival frequency interpretation. This may be meaningful for mass-produced components, but has no physical interpretation for "one-off" structures. A more important issue, however, is the fact that model uncertainties cannot normally be interpreted in terms of relative frequency because they arise from engineers' inability to formulate exact mathematical descriptions for complex engineering phenomena (e.g. shear in reinforced concrete). As the behaviour of most structures is associated with at least some model uncertainty, and because in many structures this source of uncertainty may dominate, calculated reliabilities for structures can rarely have a relative frequency interpretation.

In addition, one further source of uncertainty has yet to be considered, namely parameter (statistical) uncertainty. Only in exceptional circumstances are sufficient data available to define the distributions of load and strength variables with any accuracy, and in practice the mean, variance and higher moments of these distributions are subject to increasing degrees of uncertainty. If, for the purposes of illustration, it is assumed that R and S in the earlier example are normally distributed with known variances but with uncertain means, θ_R and θ_S, and that the uncertainty in θ_R and θ_S can be described by the Bayesian probability densities f_{θ_R} and f_{θ_S} the probabilities $P_F \mid K_i$ given by Eqn.(7) become random quantities. The total failure probability P_F also becomes a random quantity with expected value (6)

$$E[P_F] = \int_0^\infty \int_0^\infty \sum_i p(K_i) \left[P_F \mid K_i\right] f_{\theta_R} f_{\theta_S} \cdot d\theta_R \cdot d\theta_S \qquad \ldots (9)$$

Because the uncertainties characterised by f_{θ_R} and f_{θ_S} stem from lack of knowledge about the true values of θ_R and θ_S and are not physical quantities, $E[P_F]$ changes as more data becomes available and cannot correspond to any long-run failure frequency that could be observed in practice. The direct comparison of such calculated reliabilities with structural survival frequencies is therefore an invalid operation, even when failures due to gross errors are eliminated. The same is true of reliabilities calculated using point estimates for the distribution parameters. However, such comparisons are frequently made in the literature.

A further complication in practice is that the model uncertainty K is generally a continuous random variable whose distribution parameters are themselves subject to considerable statistical uncertainty.

2.2 A basis for structural design decisions

As has been pointed out before (7) the calculation of values of structural reliability should be seen as an interim step in the process of making design decisions. The question should not be - "Do these values of calculated reliability have any absolute meaning?", but rather - "How can this information be used to make optimal decisions?" This applies both in the design of Codes and in the use of those Codes to design structures.

The tool for decision making in the face of uncertainty is statistical decision theory. Before examining the particular problems of applying this to the design of offshore structures let us examine the simple idealised structure considered earlier, taking into account the effect of model uncertainty but without the complication of statistical uncertainty. Assume that the statistical distributions for R, S and K are as given in Tables 1 and 3, that the initial cost of the structure is $C_I = \alpha m_R$, where m_R is both the nominal and mean value of R, and that the cost of failure, should it occur, is $C_F = \xi C_I = \xi \alpha m_R$ (i.e ξ times the initial cost).

Variable	Distribution	Mean	Coefficient of Variation
R	Normal	-	0.07
S	Normal	100	0.10
α = 10 cost units		ξ = 20	

<u>Table 3</u>

For each discrete value of K, the expected or long-run average total cost per structure is

$$E[C_T|K_i] = C_I + C_F P_F|K_i \qquad \ldots(10)$$

where $P_F|K_i$ is given by Eqn.(7).

Allowing for the possibility of different values of K, the expected value of C_T now becomes

$$E[C_T] = \sum_i [C_T|K_i] p(K_i)$$

$$= \sum_i [C_I + C_F P_F|K_i] p(K_i)$$

$$= C_I + C_F P_F \qquad \ldots(11)$$

where P_F is defined by Eqn.(8).

Let us now examine the meaning of $E[C_T]$. If situation (A) applies and p(K) in Table 1 are long-run relative frequencies, then $E[C_T]$ is the long-run average total cost per structure. If situation (B) applies and p(K) are subjective probabilities then $E[C_T]$ is the expected total cost per structure, given the information that is available. If this information changes, then $E[C_T]$ will change. However, regardless of whether (A) or (B) applies, or whether or not the distributions of R and S are associated with statistical uncertainty, the rational design objective is to minimise $E[C_T]$. It is also clear that if all the probabilities involved in this calculation are frequentist in nature with no statistical uncertainty, then minimising $E[C_T]$ is the best possible long-run design solution. However, if subjective and statistical uncertainties are also involved, then the action taken to minimise $E[C_T]$ will produce only a sub-optimal solution - but this is the most rational action that can be taken, and on average it will be the least costly.

Fig.2 shows how $E[C_T|K]$ and $E[C_T]$ vary with the nominal value of the resistance m_R chosen for the design of the structure, the overall optimum value of m_R being 78.56 units. As might be expected, the optimum value of m_R increases with increasing (deterministic) values of the model uncertainty K. However, the optimum failure probability is independent of K as shown in Fig.3 (but only when $C_F = \alpha \xi m_R$, as adopted in this example).

Two further points require discussion. Firstly, it has been assumed above that the best design objective is to minimise the expected total cost. This is true only when costs are linearly related to utility (7), but this is probably true when offshore operations are considered from the point of view of general national interest. Secondly, optimal design decisions obtained by the methods discussed above have been shown (2) to be sensitive to the ratio ξ of failure costs to initial costs. In practice, it is extremely difficult to assess the true costs and consequences C_F of the complete or partial failure of a structure, particularly an offshore jacket, and such costs are always likely to be subject to considerable subjective uncertainty. However, this additional source of uncertainty can be included in the evaluation of the expected total cost of failure. If the continuously variable quantity C_F is divided into a range of discrete values C_{F_j} with associated and subjectively assessed probabilities $p(C_{F_j})$, then the expression for $E[C_T]$ in Eqn.(11) becomes

$$E[C_T] = \sum_j \sum_i [C_I + C_{F_j} P_F|K_i] p(K_i) p(C_{F_j}) \qquad \ldots(12)$$

The minimisation of $E[C_T]$ gives the best design decision - in this case the value of the nominal resistance m_R to be used in the structure.

2.3 The effect of gross errors

The above model can be extended to provide useful insight into the relevance of reliability analysis when gross errors may occur. The criticism is often voiced that reliability analysis is of limited practical relevance because the real cause of failure is most probably a gross design or constructional error. However, the question that should be asked is - "What is the optimum nominal strength (resistance) of the structure, and how does this reflect the cost and effectiveness of gross error control?" not "What should be the target failure probability when failure is likely to be the result of a gross error?"

If it can be assumed that a gross error may reduce the resistance R by a factor G, Eqn.(6) becomes

$$Z = GR - KS \qquad \ldots(13)$$

Assuming statistical independence between the quantities G, R, K and S, and that the magnitude of the gross error factor G can be sensibly represented by a number of discrete values G_k with associated probabilities of occurrence $p(G_k)$, then the expected total cost can be expressed as

$$E[C_T] = \sum_k \sum_j \sum_i [C_I + C_{F_j} P_F|K_i,G_k] p(K_i) p(C_{F_j}) p(G_k) \qquad \ldots(14)$$

This discrete model for G is suitable for representing gross errors in quality control, such as the use of a low rather than a high grade of structural steel. In some cases, however, G would be better represented by a continuous distribution with an associated probability density function $f_G(g)$. In most cases the introduction of occasional gross errors will have a dramatic effect on the computed failure probability (and on the observed structural failure frequency). However, what is required is knowledge of the extent to which the probability of occurrence of gross errors affects the optimum value of the design nominal resistance m_R.

Before examining this problem a further refinement to the model (Eqn.14) needs to be made. Eqn.(14) represents the situation where the probabilities of occurrence of gross errors $p(G_k)$ are independent of initial costs C_I. However, the frequency of occurrence of those types of gross errors which can be classified a priori can be reduced by control and inspection. The initial costs must therefore be written as

$$C_I = C_I' + C_G \qquad \ldots(15)$$

where C_I' is the basic initial cost using "normal" construction procedures and C_G is the additional cost of gross error control.

Assuming that some appropriate function C can be found such that

$$p(G_k) = C(C_G) \qquad \ldots(16)$$

then the expected total cost becomes

$$E[C_T | C_G] = \sum_k \sum_j \sum_i \left[C_I' + C_G + C_{F_j} P_F | K_i, G \right] p(K_i) \, p(C_{F_j}) \, p(G_k | C_G) \qquad \ldots(17)$$

The best design decision is obtained by an unconstrained minimisation of $E[C_T | C_G]$ in the space of the variables m_R and C_G. This yields the optimum design values for the nominal resistance m_R and the expenditure that should be made on gross error control C_G.

This procedure is exemplified by the results given in Fig.4 for the simple structure considered earlier under the initial assumption that there is no explicit expenditure on gross error control (i.e. $C_I = \alpha m_R$). It shows the change in optimum nominal resistance $m_{R,opt}$ for specific magnitudes of gross error, G, as defined by Eqn.(13) and for different probabilities p(G) that a gross error will occur. An important result (for this particular set of assumptions) is that if control procedures are able to restrict the frequency of gross errors to less than 2%, then $m_{R,opt}$ is very insensitive to the occurrence of gross errors of any magnitude. A more important result, however, is that even if the frequency of gross errors rises to say 5%*, although $m_{R,opt}$ shows a marked increase for gross errors of moderate magnitude (0.4 < G < 0.7), the expected total cost given by

$$E[C_T | m_{R,opt}] = \sum_k \sum_i \left[C_I + C_F P_F | K_i, G_k, m_{R,opt} \right] p(K_i) \, p(G_k) \qquad \ldots(18)$$

is insensitive to the decision of whether or not to allow for the possibility of gross errors in calculating $m_{R,opt}$. (See Fig.5). At the worst, the total expected cost differs by only 15%.

---

* Note that when the probability of occurrence p(G) exceeds 2%, the optimum design resistance exhibits a snap through from a lower to a higher level at a particular value of gross error magnitude.

2.4 Methods for reliability evaluation

The central part of any method of probabilistic structural design or partial safety factor evaluation is the process of reliability analysis. This comprises two main stages. The first is the construction of an appropriate mathematical model defining the various combinations of basic variables (2) - loads, material properties, structural dimensions, model uncertainties, etc. - whose joint occurrence would cause failure. These are deterministic relationships and may be expressed as

$$Z = g(X_1, X_2, \ldots, X_n) \qquad \ldots (19)$$

which is a more general form of Eqn.(2). The function $g(\underline{X})$ may be interpreted as a hyper-surface in an n-dimensional space dividing a safe region from a failure region. It may not be possible to write Eqn.(19) explicitly. The second stage is the evaluation of the probability that any set of values of the basic variables $\underline{x}$ chosen at random will fall within the failure region. This failure probability may be written very simply as

$$P_F = \int_\Omega f_{\underline{X}}(\underline{x}) \, d\underline{x} \qquad \ldots (20)$$

where $f_{\underline{X}}(\underline{x})$ is the joint probability density function for the n basic variables. However, the accurate evaluation of Eqn.(20) may give rise to severe computational problems for practical structures. These difficulties occur in practice because of the multi-dimensional nature of the failure region, the possibility of statistical dependence between some of the basic variables, the possibility of discontinuities of slope of the failure surface arising from the existence of a number of failure modes, and for other reasons. In addition there are the effects of statistical uncertainty resulting from the use of only finite quantities of data to determine the probability distributions for the basic variables.

It would be inappropriate to try to discuss each of these items here. The aim, therefore, is to indicate which solution techniques may be used with confidence and which should be avoided. Three main procedures may be used for the evaluation of Eqn.(20), multi-dimensional numerical integration, Monte-Carlo simulation, and a variety of iterative algebraic techniques which have become to be known as Level 2 methods (2).

For all structural problems involving 5 or more basic variables, solution by direct numerical integration becomes impractical and is subject to increasing error. For complex limit-state functions (Eqn.19) involving many variables (e.g. typical offshore structure involving dynamic response calculations) solution of Eqn.(20) by such a method is never likely to be feasible and will not be discussed further.

Monte-Carlo simulation provides a useful solution tool when all other approaches fail, but the number of random trials required to obtain an accurate evaluation may be prohibitive, especially when the failure probability is small. To a certain extent this difficulty can be overcome by fitting a distribution to the Monte-Carlo output (e.g. see Ref. 9) and by "smoothing" the results when undertaking a parametric study.

By far the most useful techniques, however, are the so-called advanced Level 2 methods (reviewed in Ref.2). Although these methods were first formulated in terms of only the means and standard deviations of the basic design variables, account may now be taken of information concerning the form of each distribution when this is available (e.g. see Ref.10). The procedure may be summarised as follows.

In general the various basic variables X_i (Eqn.19) will be non-normal stochastic quantities characterised by a distribution type and a number of parameters (e.g. mean and standard deviation). If the various X_i are statistically independent, it is always possible to make the transformation

$$u_i = \Phi^{-1}\left[F_{X_i}(x_i)\right] = U_i(X_i) \qquad \ldots (21)$$

where u_i are independent standardised normal variates (i.e. having zero mean and unit variance), F_{X_i} denotes the cumulative probability of X_i at the value x_i and where Φ^{-1} is the inverse normal distribution function. By finding the inverse relationships U^{-1} giving

$$X_i = U^{-1}(u_i) \qquad \ldots (22)$$

and by substituting in Eqn. (19) we obtain

$$Z = g\left[U_1^{-1}(u_1), U_2^{-1}(u_2), \ldots, U_n^{-1}(u_n)\right] \qquad \ldots (23)$$

which may be rewritten as

$$Z = g'(u_1, u_2, \ldots u_n) \qquad \ldots (24)$$

Setting Z in Eqn. (24) equal to zero defines the failure surface in the space of the standardised normal variables $\underline{u}$. The inverse transformation functions $\underline{U}^{-1}$ provide a continuous mapping of the failure surface in the space of the real physical basic variables $\underline{X}$ to the equivalent failure surface in the space of the standardised normal variables $\underline{u}$. It should be emphasised that this is an exact transformation for all points on the failure surface. Furthermore, the probability content of the failure region in the two spaces is identical.

Working with the new variables $\underline{u}$ and the transformed failure surface it is now possible to apply a number of standard techniques to evaluate the approximate failure probability. If the transformed failure surface in u-space is in fact a hyper-plane (this will occur only when $g(\underline{X})$ is linear in $\underline{X}$ and X_i are all normally distributed), the failure probability will be given exactly by

$$P_f = \Phi(-\beta) \qquad \ldots (25)$$

where Φ is the normal distribution function and β is distance from the origin to the hyper-plane measured along a normal. In general, $g'(\underline{u})$ will not be linear in $\underline{u}$, but a close approximation to P_f will be found by taking β as the shortest distance from the origin to the transformed failure surface. An efficient algorithm for finding minimum β is given in Reference 10. Because most functions $g'(\underline{u})$ are non-linear in $\underline{u}$, this is normally an iterative procedure.

The operation $\Phi(-\beta)$ implies the linearisation of the transformed failure surface at the point $\underline{u}^*$ (the closest point on the transformed failure surface to the origin) and the evaluation of the probability content of the corresponding approximated failure region. As this linearisation occurs at the point of maximum probability density in the transformed failure region, the errors due to linearisation can in most cases be assumed to be small. This is illustrated in Fig.6 for the simple case of a cylindrical tension member for which the limit state of yielding can be expressed as

$$Z = \frac{\pi d^2 f_y}{4} - P = 0 \qquad \ldots (26)$$

The figure shows the exact failure probability (obtained from Eqn. (20) by numerical integration) of tension members designed by above Level 2 procedure for a range of target failure probabilities from 10^{-3} to 10^{-9} and covering a wide range (not necessarily typical) of coefficients of variation and distribution types. These results can be compared with the very large scatter that occurs (Fig.7) if it is assumed (as is not infrequent) that β may be approximated by:

$$\beta = \frac{g'(\underline{u}|\underline{u} = \underline{m}_u)}{\left[\sum_i \left(\frac{\partial g'}{\partial u_i}\right)^2 \bigg| \underline{u} = \underline{m}_u\right]^{\frac{1}{2}}} \qquad \ldots(27)$$

Fig.8 is a further comparison of failure probabilities obtained by the advanced Level 2 method and by numerical integration for 100 random 3-variable functions of the form

$$Z = X_1^n \oplus X_2^n \oplus X_3^n + C = 0 \qquad \ldots(28)$$

where $\oplus$ stands for + or - or x or ÷,
X_i are random variables of varying distribution types with coefficients of variation up to 50% and $1 < n_i < 6$. Whilst these functions are not necessarily representative of any structural calculation, the results show that the linearisation errors associated with the advanced Level 2 methods are generally very small.

3. APPLICATION TO THE DESIGN OF JACKET STRUCTURES

The preceding section has been concerned with some of the fundamental principles of reliability analysis and with some of the available techniques. These will now be considered in the context of offshore jacket design.

3.1 Evaluation of component reliabilities in complex structural systems

Methods of reliability analysis for complete structural systems are still in the early stages of development. One good reason for this is that the ultimate load behaviour of complex and possibly highly redundant 3-dimensional structures is itself poorly understood, particularly under dynamic loading. The existence of many alternative load paths and failure sequences makes any reliability assessment even more difficult. One line of tackling this problem has been explored by Moses (11) (see also Section 4.1).

For many jacket structures in the northern North Sea, however, (e.g. see Fig.10c) complete collapse is likely to occur at loads not greatly in excess of those causing failure (by tension, buckling or fracture) of the first leg member or brace. The probability of collapse P_F can then be approximated by the probability P_F' that any critical member in the structure will fail. This is bounded as follows:

$$\text{Max}\,[P_i] < P_F' < 1 - \prod_{i=1}^{n}(1 - P_i) \qquad \ldots(29)$$

where P_i are the individual member failure probabilities given that the other members remain intact. For jacket structures, where the loads in the members due to environmental forces are both highly variable and strongly correlated, the true value of P_F' lies close to the lower bound (namely Max $[P_i]$). This gives a good reason for spending considerable effort in undertaking an accurate assessment of the individual P_i's as discussed below.

Some details will now be given of the application of the advanced Level 2 procedures to the evaluation of component reliabilities in jacket structures. The various steps that are required for such an analysis are:

(1) Define the basic random variables for the structure and loading (e.g. see Table 4).

(2) Select the appropriate failure criterion and the associated model uncertainty for component under consideration.

(3) Develop an appropriate idealisation of the structure for the purposes of evaluating combined wave and current forces (if it is not practicable to use a full member by member representation). Such an idealisation should adequately represent:

(i) The drag and inertia loading regimes

(ii) The influences of the spatial separation of the members

and should be expressed in such a way that the parameters such as C_d, C_m and thickness of marine growth can be treated both as spatially variable and as random quantities.

(4) Develop an appropriate mathematical model relating the natural frequency of the structure in its dominant mode of vibration to the basic random variables which affect it, such as the soil and structure stiffness, superimposed deck loads, thickness of marine growth and the coefficient of added mass.

(5) Develop an efficient algorithm to determine the stochastic response of the structure under the dynamic loads.

(6) Obtain the relationship between the displaced shape of the structure and the loads and moments in the individual components of the structure by an appropriate structural analysis.

(7) Combine the mathematical models given by (2) - (6) above to obtain the g-function (Eqn. (19)) for the reliability analysis.

(8) Compute the failure probability by undertaking the appropriate failure surface transformations, as described above, and by using a suitable algorithm to find the minimum distance β from the origin to the surface.

The above represents a formidable set of calculations but a practical method. Clearly, the development of the appropriate mathematical models for a particular structure and checking them against different or more complex models is an important part of the procedure.

Some typical results from an analysis of an 8-leg jacket in 156 m of water are given in Table 4. The results are for a lower leg member failing as a strut subjected to axial load and end moment, and are part of a much larger sensitivity study (12). The variables are ranked according to the magnitude of their sensitivity factors α_i, for loading (negative α) and resistance (positive α) quantities. The absolute values of α_i show the relative sensitivity of the failure probability to each of the variables.

Variable	Distribution	Mean	Standard deviation	α_i	$x_i^* = U^{-1}(u_i^*)$
Extreme 6-hourly mean wind speed	Type I Extreme	33.9 m/sec	3.3	-0.704	43.7 m/sec
C_d	Normal	0.75	0.22	-0.575	1.15
Current speed	Normal	1.0 m/sec	0.14	-0.258	1.11 m/sec
Marine growth	Lognormal	0.16 m	0.067	-0.249	0.20 m
Deck load	Normal	24000 tonnes	720	-0.042	24100 tonnes
Steel yield	Lognormal	380 N/mm^2	18	+0.116	373 N/mm^2
Tube thickness	Lognormal	41.3 mm	0.41	+0.090	41.2 mm
Damping coeff.	Normal	0.22	0.08	+0.013	0.22
Others	-	-	-	(0.022)	-
$\beta = 3.71$				$P_F = 1.04 \times 10^{-4}$	

Table 4: Analysis of lower leg of jacket structure for 25 year reference period (Soil stiffness determinate)

In this example, the sea state is represented by the 25 year extreme 6-hourly mean wind speed and is the dominant variable, followed by the drag coefficient C_d. The set of values $\underline{x}^*$ given in the last column of the table are the values of the variables at which failure is most likely to occur.

3.2 Checks on numerical accuracy

As discussed earlier, the linearisation of the transformed failure surface implied by the use of advanced Level 2 methods introduces an uncertain degree of error in the computed failure probabilities. These errors have been shown to be quite small for simple structures (Figs.6 and 8). For complex structures involving many variables and dynamic structural response (e.g. Table 4) the complexity of the mathematical models describing the behaviour are such that an exact evaluation of P_F (conditional on the models and statistical parameters assumed) is beyond the capacity of existing computers. However, there is no reason to suppose that for practical structures the linearisation error will increase as the number of variables n increases, although the upper bound (2) on the failure probability given by

$$P_F \ngtr 1 - \chi_n^2 (\beta^2) \qquad \ldots (30)$$

increases dramatically with n. Indeed, as n becomes larger there is an increasing chance that both positive and negative errors will occur and therefore cancel. For complex structures, the only available check on the Level 2 analysis is by Monte Carlo simulation, but even this can be extremely time-consuming when spectral analysis is involved.

Fig.9 shows the transformed failure surface for a member of a jacket structure similar to that discussed above, plotted in the reduced space of the three most important variables (wind speed W, thickness of marine growth T and inertia coefficient C_m) the other variables having been treated as constants and set equal to their u^* values (see Section 2.4). (It should be noted that the sensitivity ranking here is different from that in Table 4, because of different statistical assumptions.) Re-analysis of the structure, using only the three variables gives the results shown in Table 5.

Variable	Distribution	Mean	Standard deviation	α_i	u_i^*	x_i^*
Thickness of marine growth T	Lognormal	0.16 m	0.067	-0.979	3.89	0.71 m
Extreme 6-hourly mean wind speed W	Type I Extreme	33.9 m/sec	3.3	-0.182	0.72	35.8 m/sec
Inertia coefficient C_m	Normal	2.0	0.3	-0.095	0.38	2.14
$\beta = 3.97$			$P_F = 3.3 \times 10^{-5}$			

Table 5: Analysis of lower leg of jacket using 3-variable simplification

Examination of Fig.9 shows that the failure surface is strongly convex when viewed from the failure region (i.e. the opposite side of the failure surface from the origin) and thus linearisation of the surface at the point $\underline{u}^*$ must underestimate the true failure probability. However, the P_F obtained from Level 2 analysis is 3.3×10^{-5} and can be compared with a maximum value of 9.9×10^{-5} obtained in a series of Monte-Carlo simulations. The shape of the failure surface suggests that an absolute upper bound on P_f would be given approximately by $\frac{1}{4} \chi^2 (3.97^2) = 3 \times 10^{-4}$.

Monte-Carlo simulation checks on jackets modelled using 12 random variables have shown satisfactory agreement with results obtained by Level 2 analysis, but are very expensive in computer time.

4. SOME STRUCTURAL CONSIDERATIONS

4.1 The effects of structural redundancy

It is a common feature of papers considering the reliability of offshore structures that the real system is recognised as having a considerable degree of complexity. Although from an elementary design point of view the load paths through the structure are generally clear-cut, alternative load paths generally exist in the event of removal (failure) of any member. There are, however, very wide variations between different structures in the potential importance of this effect. This is probably most notable in respect to consideration of fatigue.

Fatigue is frequently of major concern in the design of main bracing members; these are, of course, unimportant in respect to the vertical loading (apart from controlling the overall buckling of main leg members) but are subject to full stress reversal under wave loading. The welds between brace and leg are thus severely at risk. In general, the fatigue crack does not enter the main leg, but this form of failure has the very undesirable feature of low ductility.

When fixed platforms for depths exceeding 100 m were first introduced in the North Sea, discussion of fatigue was considerably influenced by ship (and to some extent, aircraft) practice (13), with emphasis on rigorous inspection and 'fail safe' design bearing in mind alternative load paths as well as the frequency and confidence of inspection, so that system failure would not occur before any crack had been detected and repaired. However, it is doubtful whether this can be achieved; any serious attempt to do so is likely to impose severe constraints on structural form and thus indirectly result in major increase in first cost. Although only a fraction of welds are currently inspected in full detail during regular in-service procedures (perhaps 10%), the cost of each inspection may reach one million pounds per structure. Considerable effort is being directed to monitoring systems to supplement inspection, but here too the difficulties are extremely great.

To illustrate the great range of availability of alternative load paths, consider the examples in Fig.10. Example (a) typifies the early steel structures in the northern North Sea reaching 140 m depth. Perpendicular to the elevation drawn there were three full planes of bracing, all fully counter-braced (i.e. X bracing with horizontals). This is clearly highly redundant, with a great multiplicity of alternative load paths. Removal of one bracing reduces the nominal ultimate load capacity of the structure by one third or less.

Example (b) has been given because it was the subject of the recent system study by Moses (11), and represents an intermediate stage between (a) and (c). It is typical of many relatively simple structures used in moderate depths. Considered as a pin-jointed plane frame, this is only one degree statically indeterminate; if (for example) member A is removed, the load path from point B passes to the top of the structure to transfer to the other bracings. Assuming ductile behaviour, so that member A continues to carry its limit load, the existence of a single parallel path of equal expected strength (uncorrelated) would reduce the coefficient of variation of structural resistance in this mode by factor $\sqrt{2}$. Moses suggests that the actual effective reduction of variability in the example is significantly greater than this as a result of member continuity, although member continuity has only a marginal effect in the conventional ultimate load calculation. It should be borne in mind that redundancy may not be without penalties; for example the X brace system raises the difficult problem of the intersection connection of two equal tubular members.

Example (c) is typical of a design approach which has achieved considerable success in the North Sea in depths 140 m to 180 m, for self-floating structures. Viewed as a plane frame resisting loads by axial member forces, there are no alternative load paths in the event of bracing failure (except, of course, for the upper panels). The complete structure comprises four faces of similar layout, with plan bracing at each level. Viewed as a tower, removal of one brace thus produces two torsionally-stiff boxes joined by a unit behaving as a channel section. The capacity of this alternative path is limited by the capacity of the plan bracing; this is individually small but may be sufficient in aggregate as alternative to braces low in the structure. The deformation in the critical panel is also large, stretching the ductility requirement if the alternative path is to be mobilised while the critical member continues to carry its limit load. It should also be pointed out that with self-floating jackets, the bending resistance of two legs will be large.

Example (d) is based on the Cognac platform in 313 m depth, and shows a dramatic reversal of the foregoing trend. The Cognac platform is founded on a base section of very complex structural layout, producing a strong box attached to well-dispersed piles; the result is an extreme example of redundancy and mutiple alternative load paths of comparable strength and stiffness.

To aid appreciation of these trends, details of a representative selection of jacket structures are shown in Table 6.

	Depth m	Barge launch	Self floating	Four legs	Many legs	X-brace	K-brace	Piles A	Piles B	Piles C
Auk	80	X			X		†	X		
Frigg DP2	103	X			X		X	X		
Maui (NZ)	110		X	X			†			X
Forties	120	X			X	X		X		
Brent A	140		X		X	X		X		
Heather	145	X			X		X			X
Thistle A	162		X	X		X				X
Ninian N	168	X			X		X	X		
Ninian S	168		X	X			X			X
Magnus	188		X		X	design in progress (early 1979)				
Cognac (USA)	313	X			X	X		X		

† N or vertical K bracing.
Piles: A, distributed; B, clusters, dispersed or combined with individual piles; C, all (or almost all) piles in clusters at corners only.

<u>Table 6</u>

4.2 Dynamic response

As jackets are constructed in progressively deeper waters the contribution of the dynamic component of the response to the total structural response is likely to become increasingly important. Fortunately, the ratio of the quasi-resonant contribution to the rms response (σ_N, say) to the quasi-static rms (σ_B, say) decreases with increasing sea-state. To a rough first approximation

$$\frac{\sigma_N}{\sigma_B} \propto \left[\frac{\hat{n}}{n_1}\right]^2$$

where $\hat{n}$ is the frequency of the peak of the wave spectrum and n_1 is the structure natural frequency. For a given structure (n_1 fixed) this ratio is roughly inversely proportional to the sea state (e.g. significant wave height). It is usual to combine σ_N and σ_B by adding variances, i.e. by root-sum-square. On this basis, if the dynamic behaviour is satisfactory at moderate sea states, the contribution of dynamic behaviour to the peak response is likely to be small. The sensitivity to the natural frequency n_1 should be noted, however, especially bearing in mind the changes in n_1 that would result from rupture of any member.

The peak values of the quasi-resonant component are not associated with large external loading. The prediction by a linearised dynamic model of a peak response expressed as an equivalent static load exceeding the ultimate limit capacity is thus not equivalent to failure; the result is an increment of inelastic deformation. Failure occurs when the total deformation exceeds the deformation capacity. Where the effect of current is small, wave loading is basically reversing, and in this situation it is widely recognised that the effective structural capacity is strongly dependent on ductility.

Methods of using stochastic process analysis to study the accumulation of small increments of deformation over a relatively long period that have been developed in relation to wind excitation (14) can be applied to this problem. However, as far as the authors are aware, no systematic study has been made taking account of shake-down to examine the effect of redundancy on the rate of accumulation of deformation in the critical element.

4.3 Member properties

It will be clear from the foregoing discussion that attention must be paid to the ductility. In this respect connections are frequently critical and consideration of overall and of local buckling of members is also required. Study is called for on the ductility and overall stiffness of joints in the presence of fatigue crack formation.

Main braces are generally sufficiently stocky that overall buckling will not have excessive adverse effect on ductility. Local buckling has received little attention. These phenomena are heavily dependent on initial imperfections, and existing design rules can be shown to have a strong lower-bound element (15); such that a member realisation close to the design bound can be expected to show a significant load plateau, while more typical realisations have substantially higher strength but low ductility if local buckling nevertheless occurs.

In these circumstances it is clear that reliability analyses should include (for example) imperfection parameters as basic variables. However, detailed analysis of each member strength is unlikely to be practical within the reliability analysis for a large structure. When preliminary studies are made to parameterise member capacities it is important that physical limits on variability are fully recognised - for example, the lower bound implicit in the affinity of imperfection mode shape to buckling mode.

5. APPLICATION TO THE DEVELOPMENT OF STRUCTURAL CODES

Because of the technical difficulties associated with the application of structural reliability analysis it is anticipated that design codes will retain a deterministic format for the foreseeable future. The main potential for reliability analysis lies in the rational development of codes - as a means of determining the combined effect of uncertainties in a wide range of design parameters, in the determination of partial factors and in the extrapolation of existing design rules for application to larger or different structures. It also has potential, through sensitivity analysis, for indicating where further research would be most beneficial by the reduction of uncertainty - particularly model uncertainty.

In developing a deterministic (Level 1) structural code the drafting committee is faced with a number of choices and has to make decisions concerning, for example:

(1) The relationship of the specified design loads (e.g. significant wave height, wind speed, etc.) to the distributions of the life-time maximum loads.

(2) The relationship of the specified material properties to the statistical distribution of those properties, taking into account the effects of quality control.

(3) Whether the formulae that are given for predicting the ultimate or working load behaviour of structural components should be mean-value correct (unbiased) or whether they should represent the 95% lower bound to the experimental data, or some other fractile.

(4) The number of partial safety factors, their magnitude and their position in the design equations (i.e. whether there are factors on loads or load effects, on material strengths or on component strengths).

(5) Whether or not the partial safety factors should be modified according to the redundancy of the structure, and the consequences of failure of different components.

In Section 2.2 it was shown how structural reliability analysis can be used to determine the optimum nominal resistance $m_{R,opt}$ for a simple structural problem and in Section 2.3 it was demonstrated how the effect of gross errors could be included. Although considerably more difficult to apply in practice, the same principles can be applied to the evaluation of partial safety factors for a code and to make optimal choices between the various possibilities mentioned above. To do this it is necessary to specify the range of structures to which the code is to apply and to undertake the numerical optimisation procedures using a representative range of those structures with appropriate weighting. If appropriate, account should also be taken of experience gained by the behaviour of actual structures under extreme conditions, through the process of Bayesian updating (16). However, this is not likely to be of much benefit unless environmental loads corresponding to the design significant wave height have been experienced by a range of structures.

An alternative approach to the evaluation of partial factors is by calibration (2) to ensure that, if appropriate, new codes have approximately the same average level of implied safety as the previous codes. However, the straight comparison of calculated reliabilities is not necessarily a good measure of relative safety, unless the calculated reliabilities have a relative frequency interpretation. As discussed earlier, this is not the case when the reliability is dominated by either model or statistical uncertainties (i.e. lack of knowledge).

ACKNOWLEDGEMENTS

Most of the work described in this paper has been undertaken as part of the programme of research of the London Centre for Marine Structures and Materials, London University, sponsored by the U.K. Science Research Council. The results shown in Table 4 were obtained during a study of the sensitivity of jacket structures undertaken jointly with Atkins Research and Development and Flint and Neill Partnership, sponsored by the E.E.C. and the U.K. Department of Energy.

The computer programs for the analysis of the jacket structure have been written by Dr. K. Ramachandran and the calculations for Figs. 6, 7 and 8 have been undertaken by Mr. J.C. Nicholls. These contributions to the work are gratefully acknowledged.

REFERENCES

1. Eight collected papers on Load and Resistance Factor Design (LRFD). Journal of the Structural Division, ASCE, Vol. 104, No. ST9, Sept. 1978.

2. Rationalisation of safety and serviceability factors in structural codes. CIRIA Report No. 63, 1977.

3. ASCE Committee on Reliability of Offshore Structures. Application of reliability methods in design and analysis of fixed offshore platforms. Tucson, Arizona, Jan. 1979, (to be published in the Journal of the Structural Division)

4. BLOCKLEY, D.I. Analysis of structural failures. Proc. ICE, Part 1, Vol. 62, Feb. 1977.

5. FREUDENTHAL, A.M., GARRETTS, J.M. and SHINOZUKA, M. The analysis of structural safety. Journal of the Structural Division, ASCE, Vol. 92, No. ST1, Feb. 1966.

6. BENJAMIN, J.R. and CORNELL, C.A. Probability, statistics and decision for civil engineers. McGraw Hill, 1970.

7. CORNELL, C.A. Bayesian statistical decision theory and reliability based design. International Conference on Structural Safety and Reliability, Washington 1969. (pub. Pergamon 1972).

8. NEUMANN, V.J. and MORGENSTERN, O. Theory of games and economic behaviour, Princeton, 1953.

9. BAKER, M.J. The reliability of reinforced concrete floor slabs in office buildings - a probabilistic study. CIRIA Report 57, March 1976.

10. BAKER, M.J. and FLINT, A.R. Safety approaches for structures subject to stochastic loads. Symposium on Integrity of Offshore Structures, IESS, Glasgow, April 1978.

11. MOSES, F. System reliability analysis of platform structures. Informal Proceedings of the ASCE Committee on Reliability of Offshore Structures, Tucson, Arizona, Jan. 1979,(to be published in the Journal of the Structural Division).

12. L.E.A. OFFSHORE MANAGEMENT. Permanently located offshore structures. Jacket sensitivity study (to be published).

13. Fatigue relevant to offshore structures. Joint meeting of the Institution of Structural Engineers, Royal Aeronautical Society, the Royal Institution of Naval Architects and the Society for Underwater Technology. Feb. 1972.

14. WYATT, T.A. and MAY, H.I. The ultimate load behaviour of structures under wind loading. Proc. Third Int. Conf. Wind Effects on Buildings and Structures, Tokyo 1971.

15. HARDING, J.E. The elasto-plastic analysis of imperfect cylinders. Proc. Inst. Civil Engineers, $\underline{65}$ pt. 2, Dec. 1978.

16. MOSES, F. Bayesian calibration of platform reliability. Informal Proceedings of the ASCE Committee on Reliability of Offshore Structures, Tucson, Arizona, Jan. 1979, (to be published in the Journal of the Structural Division).

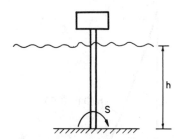

Fig 1. Simple idealisation

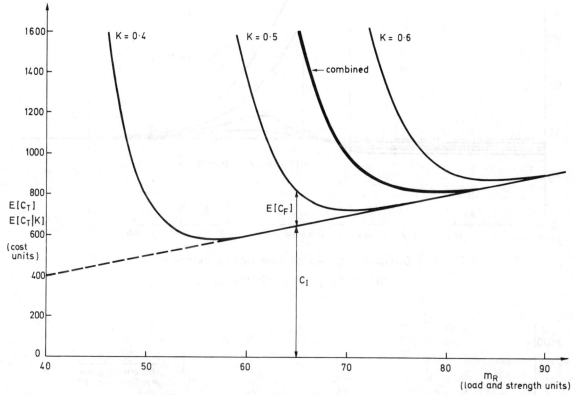

Fig 2. Variation of expected total cost with nominal resistance m_R

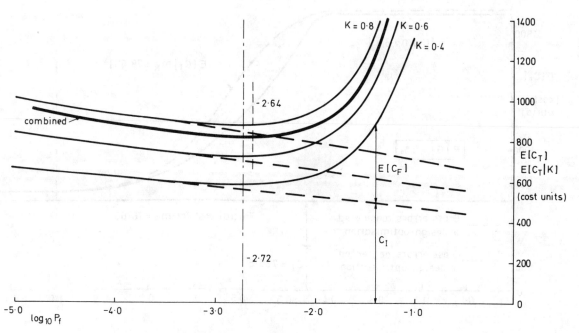

Fig 3. Variation of expected total cost with failure probability

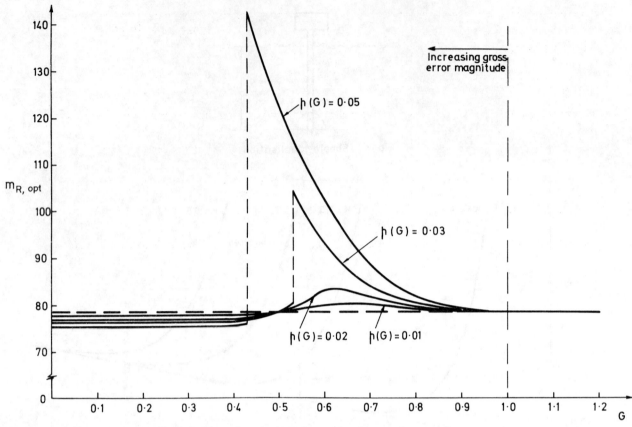

Fig 4 Optional values of nominal resistance for different gross error frequencies

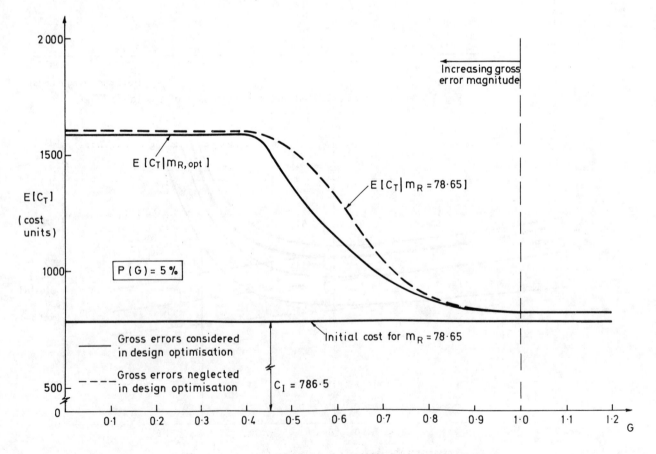

Fig 5. Expected total costs for two different design decisions

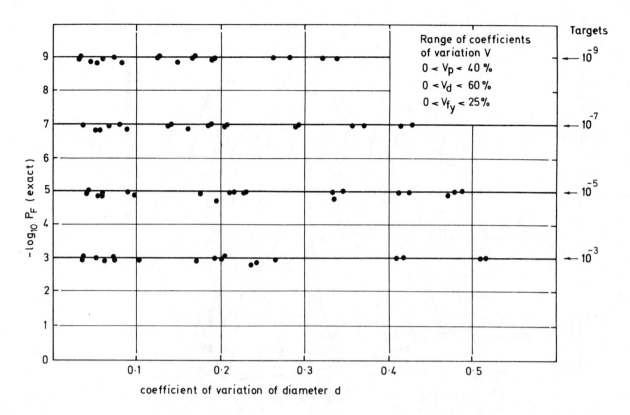

Fig 6. Exact P_F's for member designed by advanced level 2 method for targets of 10^{-3}, 10^{-5}, 10^{-7} & 10^{-9}

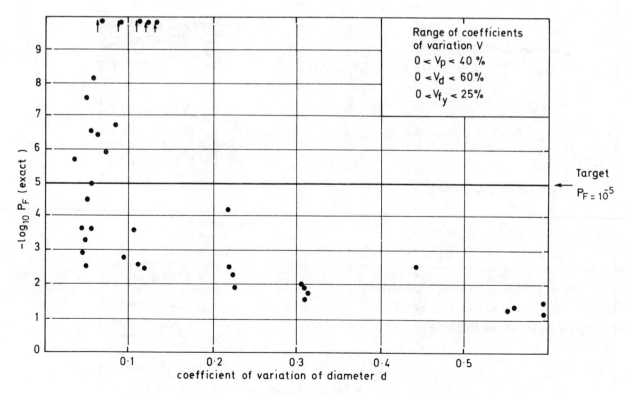

Fig. 7 Exact P_F's for member designed for target of 10^{-5} (using eqn 27)

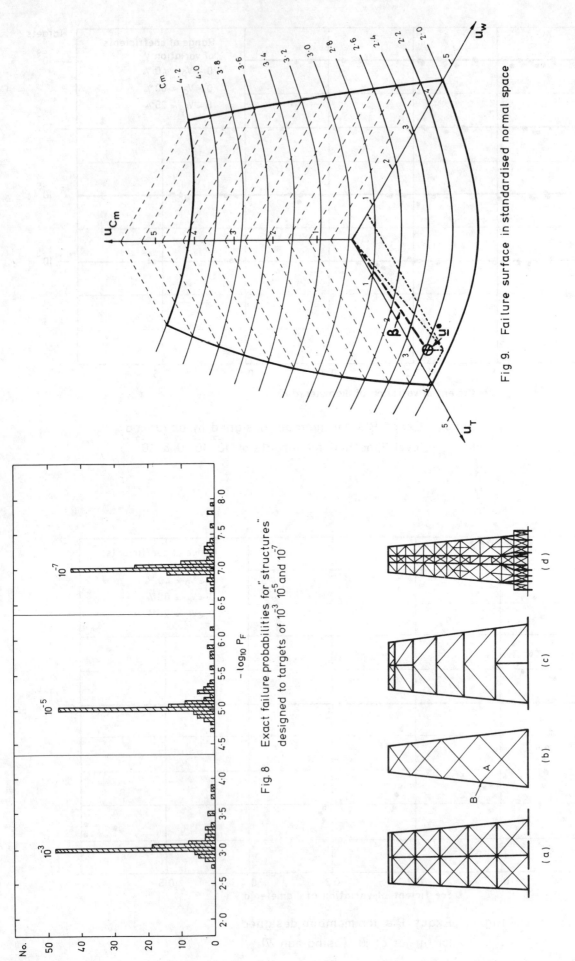

Fig 9. Failure surface in standardised normal space

Fig. 8 Exact failure probabilities for structures designed to targets of 10^{-3}, 10^{-5} and 10^{-7}

Fig.10. Typical bracing arrangements for jacket structures

BOSS'79

PAPER 85

Second International Conference
on Behaviour of Off-Shore Structures

Held at: Imperial College, London, England
28 to 31 August 1979

PROBABILISTIC RELIABILITY ANALYSIS

A.L. Bouma

University of Technology, Delft, The Netherlands

and

Th. Monnier and A. Vrouwenvelder

Institute TNO for Building Materials and Building Structures,
The Netherlands

Summary

For large structures, such as offshore structures, with high cost of construction and involving major damage in the event of failure there arises the need to assess the margin of safety with the aid of probabilistic methods. This paper reports on an introductory study of the subject. The primary aim was to test and develop probabilistic methods of analysis with reference to an example structure, i.e. providing a specific example to which the analysis could be applied. This structure has intentionally been kept simple in order to reveal the essentials of a probabilistic analysis as clearly as possible.

The probability of failure of the structure due to its maximum load capacity being exceeded and due to fatigue, and also the probability of foundation failure, have been determined, having due regard to the stochastic character of both load and structural behaviour and to the uncertainties in the mathematical models employed. The analytical techniques applied are primarily 'level II' approximate methods. For each example the mean value approximation and the approximate full distribution method are compared with each other. In addition, some 'level III' analyses have been performed.

The procedure employed have proved their usefulness. They lead to a rational and universally applicable framework of assessment, and a good deal of insight into the effect of the various factors upon the risk of failure has been obtained. The research is being continued.

Sponsored by: Delft University of Technology, The Netherlands
Massachusetts Institute of Technology, U.S.A.
The Norwegian Institute of Technology, Norway
University of London, England

Secretariat provided by: BHRA Fluid Engineering

Copyright: © BHRA Fluid Engineering
Cranfield, Bedford, England

Nomenclature

a	width foundation
A, A', B, C	constants in long-term distributions for individual, maximum and significant wave heights
C_M	inertia coefficient in Morison's equation
d	depth of water
dF	wave force on small portion of tube with height dy
D	diameter of tube
D_{tot}	accumulated damage due to fatigue (the Miner sum)
E	modulus of elasticity of steel
$F_x(\xi)$	cumulative probability distribution function
g	acceleration of gravity
H	wave height
H'	wave height corrected for directional effect
H_i	height of the ith wave
H_{max}	greatest wave height in the planned service life
H_s	significant wave height
$H_{xy}(\omega)$	transfer function from y to x
k	1) wave number; 2) fatigue parameter for steel
ℓ	length of tube
L	wave length
m_d	mass of deck structure and equipment
M	overturning moment due to wave loading
$\hat{M}$	amplitude of the overturning moment M
M_{max}	moment due to the greatest load in the service life
M_p	full plastic moment at tube cross-section
n	number of waves occurring in the service life
N	vertical load on pile foundation
$P(..)$	probability of event ' .. '
P_u	maximum load capacity of a pile group
Q_f	frictional load capacity of an individual pile
Q_p	end-bearing load capacity of an individual pile
r	ratio of actual load capacity of a pile and the design value
s_p	yield stress of steel
S_i	stress range due to i^{th} wave
S_f	fatigue parameter for steel
$S_{xx}(\omega)$	density spectrum for stochastic process $x(t)$
t	1) wall thickness; 2) time
T	wave period
T_L	planned service life
T_s	duration of a sea state

$\dot{u}_x$	horizontal acceleration of water particles
x, y, z	position co-ordinates
z	reliability function

α	1) wave steepness; 2) parameter of PM-spectrum
β	parameter of PM-spectrum
β_z	reliability index
γ	coefficient for added water mass
ζ	damping
η	vertical displacement of free water surface
μ_x, $\mu(x)$	mean of the stochastic variable x
ρ_{st}	mass density steel
ρ_w	mass density of sea water
σ_x, $\sigma(x)$	standard deviation of the stochastic variable x
φ	direction of individual wave
φ_s	predominant wave direction in a sea state
$\Phi_N(\ )$	distribution function for normal distribution
ω	circular frequency
ω_e	lowest natural frequency of the structure
ω_0	parameter of PM-spectrum

1   Introduction

In general, the design of civil engineering structures is done with the aid of deterministic or semi-probabilistic procedures. For the majority of structures such an approach is quite justifiable. However, for special structures, with, for example, high construction cost or involving major damage in the event of failure there arises the need for a more rational method of assessment. Offshore structures clearly come within this last-mentioned category. Besides, in designing an offshore structure the designer will only to a very limited extent be able to rely on experience. This is more particularly true of the deeper North Sea locations. The lack of adequate knowledge concerning the wave and sea-bed conditions existing at those locations justifies a very careful treatment of the safety problem.

The above considerations prompted the setting up of a research project on the subject of probabilistic reliability analysis for offshore structures, the results of which are briefly presented in this paper. For more detailed reports see references 1 and 2. The primary aim of the study can be formulated as the testing and, if necessary, further development of probabilistic methods of analysis with reference to a specific example structure. This example structure, as well as the modes of failure and the mathematical models considered, have intentionally been kept very simple. This has been done more particularly in order to reveal the essential characteristics of a probabilistic analysis and to demonstrate as clearly as possible the conclusions to which an approach of this kind can lead.
In view of the many simplifications and schematizations introduced in the analysis, the conclusions themselves should, however, be accepted with some caution.

2   The structure

The structure under consideration comprises a cylindrical steel tube, a deck structure and a piled foundation (fig. 1). The deck has a mass of 2000 t; its constructional details will not be considered. The tube is 120 m long and 10 m in diameter; the depth of water is 100 m. The foundation consists of a concrete footing block, supported at each of its four corners by a group of three piles.

Three main modes of failure of this structure are to be distinguished, namely:

1. failure of the tube in consequence of an extremely high wave;
2. failure of the foundation in consequence of an extremely high wave;
3. failure of the tube due to fatigue.

The failure probabilities for these respective mechanisms will successively be calculated in chapters 3, 4 and 5. The analysis will be based on a planned working life of 50 years.

Of the loads acting on the structure only the load due to wave action is taken into account, except that for the analysis of the foundation also the dead weight is considered. As for the statistical description of the wave climate the distributions given in ref. 5 for the Sevenstones location to the south of Britain are adopted. In so far as the directional effects are concerned, it is assumed that the wave climate does not have any preferred direction. This is presumably not correct for the location under consideration, but does not constitute an objection for the purpose of this study.

Before proceeding to calculate the failure probabilities, a few comments on the design of the structure are offered. In the ultimate state analysis of the tube its nominal wall thickness is 60 mm. This value has been determined on the basis of a currently used deterministic design procedure with a design wave height (100-year wave), a characteristic value of the steel strength and a factor of safety of 1.5.

The piled foundation has likewise been designed in accordance with design rules currently applied in present-day practice. On the other hand, the fatigue analysis has not been preceded by a deterministic design calculation. In chapter 5, therefore, the probability of failure has been determined for a number of wall thicknesses ranging from 150 mm to 250 mm. These larger values for the thickness are needed in order to arrive at an acceptable response of the structure. From the viewpoint of fabrication technique these dimensions are of course not realistic. However, this drawback, too, is merely of secondary importance in the present study.

3 Ultimate limit state of tube

3.1 The mathematical model

The ultimate limit state analysis of the tube is more particularly concerned with the failure of the tube in consequence of one exceptionally large wave load. In this analysis it is assumed that the lower-end cross-section of the tube is the determinative section and that the determinative load is produced by the largest wave occurring in the working life of the structure. The reliability function for this limit state can thus be expressed as follows:

$$z = M_p - M_{max} \qquad \ldots (3.1)$$

where:

M_p = full plastic moment of the tube

M_{max} = overturning moment produced by the largest wave

With the aid of elementary plasticity theory the full plastic moment M_p can be expressed in the wall thickness t, the tube diameter D and the yield stress s_p:

$$M_p = t\, D^2\, s_p \qquad \ldots (3.2)$$

The effect of the normal force has not been taken into account.

The magnitude of the applied moment M due to any particular wave with a crest-to-through height H can be calculated by means of the linear wave theory and Morison's equation, where only the inertia term need be taken into account:

$$M = \int_{-d}^{0} (d+y)\, dF \qquad \ldots (3.3)$$

$$dF = C_M\, \rho_w\, (\pi D^2/4)\, \dot{u}_x\, dy \qquad \ldots (3.4)$$

$$\dot{u}_x = gk\, \frac{H}{2}\, \frac{\cosh(ky+kd)}{\cosh(kd)}\, \cos \omega t \qquad \ldots (3.5)$$

In these expressions (see also fig. 1): d is the depth of water, y the vertical position ordinate, dF the wave force on an elementary segment dy, C_M the inertia coefficient, ρ_w the mass density of the water, D the tube diameter, $\dot{u}_x$ the horizontal acceleration of the water particles, g the acceleration of gravity, while $k = 2\pi/L$ is the wave number with L denoting the wavelength, and $\omega = 2\pi/T$ is the circular frequency with T denoting the wave period.

Substitution of (3.4) and (3.5) into (3.3) and appropriate rearrangement lead to the following expression:

$$M = \rho_w\, g\, C_M\, \left\{\frac{\pi D^2}{4}\right\}\, \frac{H}{2}\, \left\{\frac{kd \sinh(kd) - \cosh(kd) + 1}{k \cosh(kd)}\right\} \qquad \ldots (3.6)$$

The maximum moment as envisaged in (3.1) can then be found by putting $\cos \omega t = 1$ in the above expression and substituting for H the maximum wave height H_{max} occurring in the working life. In principle the reliability function z is thus completely determined. Before writing it in explicit form, we shall first introduce a new variable, namely, the wave steepness α:

$$\alpha = H/L \qquad \ldots (3.7)$$

With the aid of α the wave number k can be written as follows:

$$k = 2\pi/L = 2\pi\alpha/H \qquad \ldots (3.8)$$

For the reliability function z we then obtain:

$$z = t\, D^2\, s_p - \rho_w\, g\, C_M \left\{\frac{\pi D^2}{4}\right\} \frac{H_{max}}{2} \left[\frac{\left(\frac{2\pi\alpha d}{H_{max}}\right) \sinh\left(\frac{2\pi\alpha d}{H_{max}}\right) - \cosh\left(\frac{2\pi\alpha d}{H_{max}}\right) + 1}{\left(\frac{2\pi\alpha}{H_{max}}\right) \cosh\left(\frac{2\pi\alpha d}{H_{max}}\right)} \right] \qquad \ldots (3.9)$$

Some of the variables occurring in this reliability function will be regarded as deterministic and some as stochastic. They are fully enumerated in table 3.1. The tube diameter D, the water depth d, the mass density of the sea water ρ_w and the acceleration of gravity g are deterministic. The stochastic variables are the wall thickness t, the yield stress s_p, the wave force coefficient C_M, the wave steepness α and the maximum wave height H_{max} in the working life of the structure. The statistical properties will be discussed in the next section.

3.2 The stochastic variables (table 3.1)

<u>Wall thickness t</u>: The mean value adopted for the wall thickness t is the nominal thickness as determined from a deterministic design calculation. The relative dispersion or coefficient of variation (σ/μ) in the dimensions of steel structures may vary quite considerably, depending on the particular dimension under consideration. In the case of height and width dimensions of rolled steel sections the relative dispersion is found to have fairly low values, around 1 or 2%. For the thicknesses of flanges and plates the relative dispersion may be substantially larger. A coefficient of variation of 5% has here been rather arbitrarily chosen for the wall thickness of the tube. The statistical distribution is assumed to be a normal one.

<u>Yield stress s_p</u>: The data given in table 3.1 (mean value 280 MN/m^2, relative dispersion 7%) correspond to the usual characteristic yield stress of 240 MN/m^2 for normal structural steel. A normal statistical distribution is adopted. This assumption somewhat ignores the effect of quality control, which in all probability greatly reduces the strength deviations towards the lower values.

<u>Inertia coefficient C_M</u>: The stochastic character of the parameter C_M differs somewhat from that of the yield stress or the tube wall thickness. In the case of these two last-mentioned quantities the scatter is due to variations in the basic material and the manufacturing process; their distributions can in principle be measured. On the other hand, the stochastic character of the quantity C_M is mainly due to lack of knowledge concerning the calculation of wave force and to drastic schematization of the theoretical models. Even if refinements are introduced by considering parameters that affect C_M (such as the Reynolds number, wall roughness, Keulegan-Carpenter number, etc.) it is still impossible to state an unambiguous value for C_M, the more so as these parameters in turn are also affected by uncertainties. Basing oneself on the fairly certain knowledge that C_M must have a value between 1.2 and 2.0, C_M has been modelled as a stochastic variable with a mean value of 1.6 and a standard deviation of 0.2.

Variable	Designation	Type	Mean	σ/μ
D	Tube diameter	D	10 m	-
d	Water depth	D	100 m	-
ρ_w	Mass density of water	D	1024 kg/m^3	-
g	Acceleration of gravity	D	9.8 m/s^2	-
t	Wall thickness of tube	N	60 mm	0.050
s_p	Yield stress	N	280 MN/m^2	0.070
C_M	Inertia coefficient	N	1.6	0.125
α	Wave steepness	N	0.06	0.125
Hmax	Greatest wave height in working life	W	22.7 m	0.050

Table 3.1: Basic variables for the ultimate limit state
Type designation: D = deterministic, N = normal, W = Weibull according to (3.16)

Wave steepness α: The choice with regard to the properties of the wave steepness α was based on considerations similar to those applicable to C_M. In general, high waves are usually steep ones. The physical limit to steepness is in the region of 1:12 to 1:14. Taking account of this, a mean value of 0.06 and a relative dispersion of 12.5% have been adopted.

Maximum wave height Hmax: As already indicated in chapter 2, the wave properties have been based on measured data collected at the Sevenstones location to the south of Britain. As reported in ref. 5, the individual wave heights at that location present the following distribution:

$$F_H(h) = 1 - \exp\{-((h-A)/B)^C\} \qquad \ldots (3.10)$$

where:

h > A, A = 0.0 m, B = 1.13 m and C = 1.0

What is, however, important in connection with the ultimate limit state problem is not the probability distribution of the individual waves, but the probability distribution of the maximum wave height in the working life. If the individual waves are numbered from 1 to n, this maximum wave can be written as:

$$H_{max} = \max\{H_1, H_2 \ldots H_n\} \qquad \ldots (3.11)$$

where:

n = number of waves in the planned service life

For the sake of convenience the individual wave heights are assumed to be independent stochastic variables. In that case the probability that Hmax < h is equal to the product of n individual probabilities of the wave height falling short of the same height h:

$$P\{H_{max} < h\} = [P\{H < h\}]^n \qquad \ldots (3.12)$$

Expressed in the distribution functions, this becomes:

$$F_{H_{max}}(h) = [F_H(h)]^n \qquad \ldots (3.13)$$

For numerical computations it is advantageous to employ the following approximation:

$$F_{H_{max}}(h) = \{1 - (1 - F_H(h))\}^n \approx 1 - n(1 - F_H(h)) \qquad \ldots (3.14)$$

The exact distribution and the approximation are both presented in fig. 2. In the region which is of importance with regard to the ultimate limit state (large values of h) the error introduced by the approximation is almost nil. For lower values of h the error increases. Below a certain value of h it even becomes necessary to curtail the approximation, otherwise negative values are obtained for the distribution function. Then, on substitution of (3.10) into (3.14), the following expression is arrived at:

$$F_{H_{max}}(h) = 1 - n \exp\{-((h - A)/B)^C\} \qquad \ldots (3.15)$$

The number of waves in the working life is estimated at n = 200 million. Substitution of this value into (3.15) and some rearrangement (which is possible since C = 1.0) then give:

$$F_{H_{max}}(h) = 1 - \exp\{-((h - A')/B)^C\} \qquad \ldots (3.16)$$

where:

$h > A'$, $A' = 21.6$ m, $B = 1.13$ m, $C = 1.0$

With this result all the necessary data have been assembled and the actual failure probability analysis can be started.

3.3 Calculations and results

In order to allow a comparison between the various methods of analysis, the probability of failure has been determined in three different ways.

The <u>first</u> method is the so-called 'level II' mean value approximation, in which an estimate of the mean μ_z of the reliability function z is found by substituting the mean value for all the stochastic variables into (3.9). The variance σ_z^2 is approximated by linearization of z and evaluation of the partial derivatives by likewise substituting the mean values of the variables into them:

$$\mu_z = z\{\mu_t, \mu_{s_p}, \mu_{C_M}, \mu_\alpha, \mu_{H_{max}}\} \qquad \ldots (3.17)$$

$$\sigma_z^2 = \{\frac{\partial z}{\partial t}\}_{\underline{\mu}}^2 \sigma_t^2 + \{\frac{\partial z}{\partial s_p}\}_{\underline{\mu}}^2 \sigma_{s_p}^2 + \ldots \qquad \ldots (3.18)$$

Next, a normal distribution is assumed for z; then the probability of failure is obtained from:

$$P\{failure\} = P\{z < 0\} = \Phi_N(-\beta_z) \qquad \ldots (3.19)$$

where:

$\Phi_N(\ )$ = distribution function for normal distribution
β_z = μ_z/σ_z (reliability index)

The <u>second</u> method is the 'level II' approximate full distribution method, which likewise is based on linearization of the reliability function z, but in this case around an optimally chosen design point, while it does not employ mean values and standard deviations of the original distributions, but of substitutive normal distributions. For further information on this method see references 3 and 4.

The <u>third</u> method, finally, is an exact 'level III' analysis by means of numerical integration procedures.

	μ_z	σ_z	β_z	P {failure}
Level II mean value method	880 MNm	185 MNm	4.76	1.0×10^{-6}
Level II approximate full distribution method	975 MNm	207 MNm	4.71	1.2×10^{-6}
Level III exact			-	1.7×10^{-6}

Table 3.2: Results of ultimate limit state analysis

The principal results of the calculations are presented in table 3.2. The differences between the results of the respective methods are seen to be small. The simplest 'level II' mean value approximation and the exact 'level III' analysis differ by a factor of less than 2, which is decidedly a small difference for problems of this kind. An interesting intermediate result in the 'level II' analyses is the constitution of the variance of z, with the aid of which it is possible to draw conclusions as to the relative importance of the various stochastic variables.

This constitution is, in the case of the present problem analysed by the approximate full distribution method, exemplified in table 3.3. It is seen here that the largest contribution to the overall uncertainty as to the failure behaviour is due to the wave force coefficient C_M. If a more reliable estimate of this coefficient could be made, this could contribute greatly to reducing the variance of z. An almost equally large contribution is due to the steel yield stress s_p. The smallness of the contribution of the maximum wave height H_{max} is notable. A short analysis of this aspect is desirable.

Generally speaking, two factors determine the extent to which a variable manifests itself in the final result. One of these is the dispersion and the other is the partial derivative of the reliability function. The relative dispersion of the maximum wave height is certainly not small, and according to table 3.3 it is even greater than that of the yield stress. (Note: According to table 3.1 it is in fact just the reverse. The reason for this discrepancy is that this table gives the coefficient of variation of the original Weibull distribution, whereas table 3.3 gives the coefficient of variation of the substitutive normal distribution; it is this latter table that is relevant to the present considerations). In view of the nevertheless fairly considerable dispersion of the maximum wave height the cause of the ultimately quite minor contribution to the variance of z must be sought in a low value of the partial derivative.

x_i	$\mu(x_i)$	$\frac{\partial z}{\partial x_i}$	$\frac{\sigma(x_i)}{\mu(x_i)}$	$\{\frac{\partial z}{\partial x_i}\}^2 \sigma^2(x_i)$	$\Delta\sigma_z^2/\sigma_z^2$
t	1370 MNm	0.050		4700 (MNm)^2	11%
s_p	1540	0.070		12200	29%
C_M	- 950	0.125		14100	33%
α	690	0.125		7400	17%
H_{max}	- 650	0.105		4500	10%
				43000 (MNm)^2	100%

Table 3.3: Constitution of the variance σ_z^2 for the ultimate limit state of the tube, according to the approximate full distribution method

The relevant column in table 3.3 does indeed give a much lower value for Hmax than for the yield stress s_p. This is mainly because s_p occurs in the absolutely largest term of the reliability function (namely, the full plastic moment), whereas Hmax occurs in the smallest term (the applied moment). A subsidiary cause is, that the point of application of the wave force resultant descends lower with increasing wave height. As a result of this the applied moment becomes less sensitive to a variation in wave height. On balance, the resulting product of dispersion and derivative is greater in the case of the yield stress than in the case of the wave height. Finally the squaring in the variance calculation intensifies this effect.

The above considerations provide a technically detailed explanation for the smallness of the contribution that the wave height makes to the variance of the reliability function. In this context the following is additionally of importance. In the example the distribution (3.10) has been adopted as though it represented the actual distribution of the wave heights. A large part of the uncertainties with regard to waves, however, is in fact bound up with the reliability of this type of distribution which in many instances has been based on what actually are too short observation periods. Besides, it seldom occurs that observations are available for the precise location where a particular offshore platform is to be built. In order to obtain a correct picture of the circumstances it would therefore be necessary also to take account of those uncertainties, e.g., by considering the parameters A, B and C of the distribution (3.10) as stochastic variables with a mean value and a dispersion. This will certainly be done in the course of further study of the subject.

4 Ultimate limit state of the foundation

4.1 The mathematical model

Failure of the structure under severe wave attack may of course also occur in consequence of foundation failure. In this section of the present paper the structural safety of the foundation will be investigated. Besides the moment M due to wave loading, a vertical load N due to the dead weight of the deck, the tube and the foundation structure will be taken into account. The criterion adopted for the failure of the foundation is the simultaneous yielding of two pile groups loaded in compression. Failure in tension is ruled out, i.e. not considered as a real possibility.
In the specific case of a wave attack at an angle φ as indicated in fig. 3, failure will accordingly occur in the pile groups 1 and 2. The associated reliability function z is expressed by:

$$z = P_u - \frac{N}{4} - \frac{\hat{M} |\cos \varphi|}{2a} \text{ with } 0 < \varphi < 45^0 \qquad \ldots (4.1)$$

where

P_u is the maximum load capacity of a pile group, N is the vertical load due to dead weight, $\hat{M}$ is the amplitude of the bending moment, φ is the direction of wave attack, and a is the width of the foundation.

If wave attack comes from a different direction, other pile groups will be loaded to yielding, e.g., the groups 2 and 3 for the angle φ having a value between 45^0 and 135^0. On account of the symmetry both of the structure and of the wave climate, however, it will be sufficient to consider the problem only on the basis of (4.1).

In calculating the ultimate load capacity of a pile group P_u it is primarily assumed that all the piles under the structure are of equal strength or, in statistical terms, that their strengths are completely correlated. The formula for P_u is then:

$$P_u = 3r [Q_f + Q_p] \qquad \ldots (4.2)$$

where Q_f and Q_p respectively denote the frictional resistance and end-bearing resistance of a single pile.

In practice, a mathematical model is available for the determination of Q_f and Q_p (ref. 6). The necessary soil properties are determined on the basis of borings and penetrometer tests performed on the site. (In the study reported in this paper, soil properties relating to the northern part of the North Sea have been adopted.) Of course, both the determination of these properties and the mathematical model employed for the strength of the piles are affected by numerous uncertainties. In order to take these uncertainties into account, the stochastic variable r has been introduced into (4.2). The factor r represents the ratio between the actual load capacity and the design value thereof. Finally, in (4.2) the strength of a pile group follows from the load capacity of the single pile by application of a multiplication factor 3.

Now let us consider the loading terms in the reliability function (4.1). The vertical dead weight load N is assumed to be deterministic, with a value of 70 MN. The moment acting upon the foundation is the same as that acting at the lower end cross-section of the tube. With the aid of (3.6) the following expression is therefore obtained for the moment term in the reliability function (4.1):

$$\frac{\hat{M} |\cos \varphi|}{2a} = \rho g C_M \left\{\frac{\pi D^2}{4}\right\} \frac{H |\cos \varphi|}{4a} \left\{\frac{kd \sinh(kd) - \cosh(kd) + 1}{k \cosh(kd)}\right\} \quad \ldots (4.3)$$

where:

$$k = 2\pi\alpha/H$$

Failure of the foundation is determined by the largest value of $\hat{M} |\cos \varphi|$ in the working life of the structure. This maximum moment will not necessarily be due to the largest wave in the working life, as would be the case for the axially symmetrical tube. With regard to the foundation it may instead occur that a particular wave acting in a favourable direction (for example: $\varphi = 45°$) constitutes a less serious threat to the structure than does a smaller wave acting in an unfavourable direction (for example: $\varphi = 0°$). The complexity of expression (4.3) is something of an obstacle to an exact solution of this problem. For this reason an approximate solution has been chosen, in which it is assumed that the largest moment results from the wave having the highest value of $H' = H |\cos \varphi|$; the value of k is determined (making a small error) from $k = 2\pi\alpha/H'_{max}$. Summarizing:

$$\left[\frac{\hat{M} |\cos \varphi|}{2a}\right]_{max} = \rho g C_M \left\{\frac{\pi D^2}{4}\right\} \frac{H'_{max}}{4a} \left\{\frac{kd \sinh(kd) - \cosh(kd) + 1}{k \cosh(kd)}\right\} \quad \ldots (4.4)$$

where:

$$k = 2\pi\alpha/H'_{max}, \quad H' = H |\cos \varphi|$$

This completes the mathematical model. The data for the variables in this model are presented in table 4.1. Some short explanatory notes on the stochastic variables will now follow.

4.2 The stochastic variables (table 4.1)

Uncertainty in pile model: The mean value and the standard deviation of r are based on worldwide experimental data (ref. 7). On an average, the mathematical model is found to give estimates for the maximum load capacity which are 16% lower than reality. The dispersion r is fairly large, however: $\sigma(r) = 0.25$. It can accordingly be inferred that in many cases the actual load capacity is overestimated.

Wave data: The properties of C_M and α are identical with those considered in the ultimate limit state analysis of the tube. The distribution of the maximum wave height H_{max} differs from that in the preceding example because of the directional effect. The difference is small, however: the mean is 5% lower, while the ratio σ/μ has remained unchanged.

Variable	Designation	Type	Mean		σ/μ
d	water depth	D	100	m	-
D	tube diameter	D	10	m	-
a	foundation width	D	30	m	-
N	vertical load	D	70	MN	-
ρ	mass density of sea-water	D	1024	kg m^{-3}	-
g	acceleration of gravity	D	9.8	ms^{-2}	-
Q_f	frictional resistance of single pile	D	18.7	MN	-
Q_p	end resistance of single pile	D	2.7	MN	-
r	uncertainty of pile model	N	1.16		0.22
C_M	inertia coefficient	N	1.6		0.125
α	wave steepness	N	0.06		0.125
H'_{max}	highest value of $H' = H \|\cos \varphi\|$	W'	21.6	m	0.050

Table 4.1: Data for the analysis of the foundation

Type designation: D = deterministic, N = normal, W' = based on 3.10 and uniform distribution for φ at $[0, 2\pi]$

4.3 Calculations and results

The foundation problem has been analysed both with the 'level II' mean value approximate full distribution method. The results of the two approaches do not differ much from each other. The result of the approximate full distribution method is:

$\mu(z) = 42$ MN; $\sigma(z) = 15,5$ MN; $\beta_z = 2.7$

P {failure} = P {z < 0} = 3.6×10^{-3}

The first conclusion on the basis of these results is that the probability of failure is much higher for the foundation than for the structure itself. From the manner in which the variance is constituted, as set forth in table 4.2, it appears that this must entirely be attributed to the effect of r. In view of the considerable dispersion in r and the way in which r occurs in the reliability function (in the largest term), this is not so very surprising. The predominance of r in the end result makes it moreover interesting to investigate the effect of the assumption of complete correlation between the pile strengths.

Fig. 4 gives the results of calculations in which that assumption was not made and in which, instead, only a partial correlation was assumed to exist between the pile strengths. The correlation coefficient ρ expresses the correlation between any two piles under the structure, irrespective of whether they belong to the same group or not.

It appears from this diagram that for very low correlations the probability of failure may indeed decrease considerably, and for the case where there is no correlation at all the risk of failure is no more than 4×10^{-8}. In all probability, however, the magnitude of ρ will be in the range between 0.4 and 0.6, in which case the probability of failure is in the region of 10^{-3}, while the considerable effect of r still

x_i	$\mu(x_i)$	$\{\frac{\partial z}{\partial x_i}\}$	$\frac{\sigma(x_i)}{\mu(x_i)}$	$(\frac{\partial z}{\partial x_i})^2 \sigma^2(x_i)$	$\frac{\Delta_i \sigma^2(z)}{\sigma^2(z)}$
r	70 MN		0.220	235 (MN)^2	97 %
C_M	14		0.125	3	2 %
α	6		0.125	1	1 %
H'_{max}	7		0.050	0	0 %
				239 (MN)^2	100 %

Table 4.2: Constitution of the variance for the analysis of the foundation, according to the approximate full distribution method

remains. An increase in our knowledge concerning the prediction of the ultimate load capacity of piled foundations could therefore make a relatively large contribution to improving the safety of offshore structures.

5 Fatigue

5.1 The mathematical model

The third possible failure mechanism investigated for the structure is that of fatigue failure of the steel tube. It is assumed that this limit state can, with reasonable accuracy, be identified with the attainment of the fatigue limit at any particular point of the lower end cross-section of the tube. Let point A on the X-axis be this arbitrary point (fig. 1). It is furthermore assumed that Miner's rule in combination with S-N lines provides a fairly good fatigue model. According to Miner's rule, in consequence of each load cycle the material undergoes a degree of damage which depends only on the stress amplitude concerned. The magnitude of the damage can (according to the same principle) be calculated from the results of constant-amplitude tests to failure. The resulting formula for the total normed damage after n cycles then becomes:

$$D_{tot} = \sum_{i=1}^{n} \{\frac{S_i}{S_f}\}^k \qquad \ldots (5.1)$$

where:

S_i is the stress range for the ith stress alternation, while S_f and k are experimentally determined fatigue constants.

Failure occurs when the normed damage (or the Miner sum) attains the value 1. If the Miner sum remains below 1, no failure occurs, according to this model. Hence an obvious choice for the reliability function would be:

$$z = 1 - D_{tot} \qquad \ldots (5.2)$$

However, for numerical reasons in connection with the logarithmic normal distribution of the fatigue constant S_f (see table 5.1) the following expression is preferable:

$$z = -{}^e\log D_{tot} \qquad \ldots (5.3)$$

With the aid of (5.3) the mean value approximation produces better results, while more rapid convergence is obtained with the approximate full distribution method.

In determining the numbers and values of load cycles S_i spectral analysis techniques have been used, such as are commonly applied in fatigue analyses. The procedure adopted here is illustrated in fig. 5. Its primary conception is that the sea's wave action throughout the working life of the structure can be split up into a number of sea states in which the vertical motion of the water can be conceived as a steady-state Gaussian random process. A process of this type is further characterized by a mean value and a spectrum. In the case of the present study the mean is assumed to be zero, while the spectrum is assumed to be known if two parameters are known, namely, the significant wave height H_s and the predominant wave direction φ_s.
(Note: For complete characterization of the sea state it is also necessary to know the duration T_s; in the integration over the sea states, which is to be performed later, the duration cancels out, however). If the spectrum of the waves is known, the spectrum of the stress at point A can be determined with the aid of a transfer function. The latter distinctly splits up into two parts, namely, the transfer function from wave to load and the transfer function from load to stress. For determining the first-mentioned of these two functions, the linear wave theory and Morison's equation have in principle been employed. In the range of high frequencies where Morison's formula is no longer valid, a correction has been applied on the basis of a diffraction analysis. The transfer function from load to stress has been determined by means of schematization of the structure to a single degree of freedom system. A typical result for the spectral analysis described here is presented in fig. 6.

The object of the spectral analysis is ultimately to arrive at numbers of stress alternations and at values for the stresses concerned. The usual approach is to determine the total number of alternations on the basis of a mean frequency and to assume a Rayleigh distribution for the distribution of the stress peaks. This can, however, be justified only if the stress behaviour conforms to a so-called narrow band process. In view of the two-peaked spectrum for the median sea state (fig. 6) that requirement is certainly not fulfilled. Nevertheless, for lack of a serviceable alternative, the procedure has been applied as though this were indeed a narrow band process. It is then easy to determine a Miner sum for each sea state. Finally, the fatigue damage for the whole working life of the structure is found by summation over all the sea states.

5.2 The stochastic variables

All the data required for the analysis are enumerated in table 5.1. Three stochastic variables are to be distinguished with respect to the load (η, H_s and φ_s) and four with respect to the structure (t, γ, ζ and S_f). Some explanatory comments on these variables will be given.

<u>Vertical displacement of the water surface η</u>: It is assumed that for each sea state the vertical displacement of the surface of the water can be conceived as a steady-stated Gaussian process with a zero mean value and with a multi-directional Pearson-Moskowitz spectrum which is expressed by:

$$S_{\eta\eta}(\omega, \varphi) = \left[\frac{2}{\pi} \cos^2(\varphi - \varphi_s)\right]\left[\alpha\, g^2\, \omega^{-5} \exp\{-\beta(\omega_0/\omega)^4\}\right] \qquad \ldots (5.4)$$

where:

$\alpha = 0.008$; $\beta = 0.74$; $g = 9.8$ m/s^2 and $|\varphi - \varphi_s| < \pi/2$

For each sea state the parameter ω_0 is adjusted to the significant wave height by means of the relation:

$$H_s = 4\sigma_\eta = 4\left[\int_0^\infty \int_{-\pi}^{\pi} S_{\eta\eta}(\omega, \varphi)\, d\varphi\, d\omega\right] = 2g\,\omega_0^{-2}\,\sqrt{\alpha/\beta} \qquad \ldots (5.5)$$

Variable	Designation	Type	Mean		Dispersion
d	water depth	D	100	m	-
ℓ	length of tube	D	120	m	-
D	tube diameter	D	10	m	-
ρ_w	mass density of sea water	D	1024	kg m^{-3}	-
ρ_{st}	mass density of steel	D	7800	kg m^{-3}	-
g	acceleration of gravity	D	9.8	ms^{-2}	-
C_M	inertia coefficient	D	2.0		-
m_d	mass of the deck	D	2000	ton	-
T_L	service life	D	160×10^7	s	-
E	modulus of elasticity of steel	D	210	GN m^{-2}	-
k	fatigue parameter	D	4		-
η	water surface displacement	N	0	m	PM-spectrum
H_s	significant wave height	W	2.17	m	$\sigma/\mu = 0.60$
φ_s	predominant wave direction	U	π	rad	U on $(0, 2\pi)$
t	wall thickness	N	150-250	mm	$\sigma/\mu = 0.05$
γ	coefficient for added mass	N	0.9		$\sigma/\mu = 0.11$
ζ	damping ratio	LN	0.01		$\sigma/\mu = 0.25$
S_f	fatigue parameter	LN	5300	MN/m^2	$\sigma/\mu = 0.14$

Table 5.1: Data for fatigue analysis

Type designation: D = deterministic, N = normal, W = Weibull according to (5.6), U = uniform, LN = log-normal, PM = Pearson-Moskowitz according to (5.4)

Significant wave height H_s: The following long-term distribution for the significant wave height is adopted:

$$F_{H_s}(h) = 1 - \exp\left[-((h-A)/B)^C\right] \quad \ldots (5.6)$$

where:

$h > A$, $A = 0.60$ m, $B = 1.67$ m, $C = 1.2$

Like distribution (3.14) for the individual waves, this distribution is based on observations made at the Sevenstones location to the south of Britain (ref. 5).

Predominant wave direction φ_s: In the present paper it is assumed - quite arbitrarily - that the wave climate has no preferred direction. Hence it follows that φ_s has a uniform distribution on $[0, 2\pi]$.

Wall thickness t: As already announced in chapter 2, the fatigue analysis has been carried out for various values of the nominal wall thickness t, ranging from 150 mm to 250 mm. Furthermore, just as for the ultimate limit state, the wall thickness has been assumed to conform to a normal distribution with a mean equal to the nominal value and a relative dispersion of 5%.

Coefficient γ: In the analysis the mass of the structure is augmented with an inertia contribution due to the surrounding water. With the aid of a linear potential theory this additional mass can be determined as corresponding to a column of water with the same diameter as the tube. In order to take account of the approximations inherent in this approach, the result has been multiplied by a coefficient γ. Experts have estimated that $\mu(\gamma) = 0.9$; the uncertainty in γ has been taken care of by putting $\sigma(\gamma) = 0.1$.

Damping: By the damping ratio ζ is understood the relative damping of the structure schematized to a single-mass spring system. A low mean value (0.01) has been adopted because there is almost exclusively material damping involved, since practically no structural connections are present and the hydrodynamic damping is low on account of the large diameter of the tube. The relative dispersion has been taken as $\sigma/\mu = 0.25$. This value is not based on statistical data, but represents a subjective assessment of the uncertainty in this parameter. The log-normal distribution has been chosen more particularly in order to rule out the possibility of negative damping values.

Fatigue constants: On the basis of observed results of constant-amplitude tests it is assumed that k can be kept deterministic and that S_f has a log-normal distribution with a mean value and dispersion as indicated in table 5.1.

5.3 Calculation and results

For evaluating the mathematical model described in the foregoing, a combination of three calculation techniques has been employed. These are: a spectral analysis technique, a 'level III' treatment for the Rayleigh distribution of the stresses and the summation over the sea states, and a 'level II' treatment for the stochastic variables t, γ, ζ and S_p. The principal results are summarized in fig. 7 and table 5.2.

In fig. 7 the mean and the standard deviation of z, the reliability index and the probability of failure have been plotted as functions of the nominal wall thickness. The 'level II' calculations have again been performed both with the aid of the mean value approximation (dotted line) and the approximate full distribution method (continuous line). Thanks to the form chosen for the reliability function, the curves obtained with these two methods show considerable similarity.

A reasonable dimension to adopt for the wall thickness would appear to be 210 mm, corresponding to a failure probability of 10^{-5} in the working life. The required thickness of 210 mm can be interpreted as follows: 140 mm is needed for ensuring a mean value of zero for the reliability function. If there were no uncertainties and dispersions, this value of 140 mm would in fact suffice. Actually, however, there are dispersions to be considered, and in order to keep the probability of failure sufficiently low it is necessary to provide 70 mm additional thickness.

For the wall thickness of 210 mm the constitution of the variance is given in table 5.2. The variables with respect to the load (η, H_s and φ_s) make no contribution to the variance of z. This is so because the overall fatigue damage to the structures is due to the sum of a very large number of stress variations. Each of these variations is in itself a stochastic variable with a substantial dispersion, but the relative dispersion in their sum is negligible. Just as in the ultimate limit state analyses, however, it must be pointed out that a major source of uncertainties has not been included here, namely, the uncertainties in the statistical description of the wave climate itself. In connection with this fatigue analysis the tail portion of the Pearson-Moskowitz spectrum more particularly deserves attention, as it contains the natural frequency and is therefore very important (see fig. 6).

x_i	Designation	$\Delta \sigma_z^2$	$\Delta \sigma_z^2/\sigma_z^2$
η	water surface displacement	0	0 %
H_s	significant wave height	0	0 %
φ_s	predominant wave direction	0	0 %
t	wall thickness	0.130	21 %
γ	coefficient for added mass	0.008	1 %
ζ	damping ratio	0.156	25 %
S_f	fatigue parameter	0.325	53 %
		0.619	100 %

Table 5.2: Constitution of the variance for a nominal wall thickness of 210 mm (approximate full distribution method)

Among the variables which do contribute to the dispersion in z the fatigue parameter S_f is the most important. This parameter accounts for more than 50 % of the variance. A by no means negligible contribution is furthermore due to the damping and the wall thickness, each corresponding to more than 20 %. The uncertainty in the hydrodynamic acceleration term finally turns out to be of entirely minor importance.

6 Conclusions

In the Introduction the primary aim of the study reported in this paper has been formulated as the testing and, if necessary, further development of probabilistic analysis concepts with reference to a simple offshore structure. It can be concluded that a positive result has been achieved with regard to this aim. 'Level II' procedures have more particularly proved their usefulness. Good results are obtained with relatively little arithmetical effort. An important fact that emerged is furthermore, that, in analysing the fatigue failure probability, the 'level II' analysis was readily combinable with a spectral analysis performed partly at 'level III'.

The advantage of a probabilistic approach to the safety problem is the rational and universally applicable framework of assessment. This approach offers the possibility of arriving at an acceptable overall safety as well as achieving better interadjustment of the safety of different mechanisms and components. The first-mentioned possibility is perhaps still somewhat premature under present conditions, but the second possibility is certainly a real one. For example, the interadjustment of ultimate limit state and fatigue is conceivable in this context. Proper interadjustment is hardly possible on the basis merely of deterministic procedures. A deterministic procedure will at the very least have to be calibrated by probabilistic analysis. Another example is the interadjustment of the structure and its foundation. According to the study reported in this paper, the safety of the tubular structure is much greater than that of the foundation. Now it is not necessary for all components of the structure to have exactly the same probability of failure; indeed, the opposite is true; but in that case any differences will have to be based on rational considerations and not be the unintended consequence of a defective design procedure.

Yet another example of the probabilistic approach is the fact that it is possible to gain insight into the effect of the several variables. If a variable has a major effect on safety it should be treated with proper care. A high partial safety factor in a 'level I' procedure appears appropriate. Furthermore, if any particular variable is found to have a major effect, this may provide ground for undertaking further research

with the object of reducing the dispersion or uncertainty.
In the present study it has, with regard to this point, emerged that the long-recognized stochastic variables such as, for example, the wave heights are not even the most important sources of uncertainty. Material and structural properties and also model parameters are certainly as important. In this case the model for the pile strength and the fatigue behaviour were more particularly found to be by far the greatest sources of uncertainty.

An important part of the problems associated with the probabilistic approach is the assignment of statistical properties to the various stochastic variables. In some cases the designer has data based on observations at his disposal (e.g. for the waves, the yield stress, the fatigue behaviour). In other cases he will have to choose a distribution in an intuitive (Bayesian) manner (e.g., for the damping, the C_M value).

In either circumstance there will, anyway, be uncertainties as to the distribution adopted, and these uncertainties would in fact also have to be taken into account. For example, in the present study the suspicion has arisen that for the waves the uncertainty in the statistical characteristics could well be very important. This is indeed something that will have to be clarified by further research. In order to enhance the usefulness of the probabilistic methods it is therefore necessary to collect more and better statistical data based on observations. At the same time, however, it is important that engineers should be trained in the modelling of their uncertainties. In carrying out the present study it was found that many engineers tend to estimate 'averages' on the 'safe' side and as yet have little notion of the intrinsic meaning of a particular dispersion.

To sum up, it can be stated that the application of probabilistic methods has yielded positive experience. The research will therefore be continued towards more complex and more realistic structures.

Acknowledgements

This paper contains the results of the STUPOC V project (ref. 1), which has meanwhile been concluded, and the MATS 28 project (ref. 2), which is still in progress. For these two projects, collaborative associations were set up, in which the following persons participated:

Ir F.L. Beringen:
Fugro, Consulting Engineers, Leidschendam

Prof. Ir A.L. Bouma:
University of Technology, Delft

Prof. Dr Ir E. Bijker:
University of Technology, Delft

Ir C. Gouwens:
Institute TNO for Building Materials and Building Structures, Rijswijk

H. van Koten:
Institute TNO for Building Materials and Building Structures, Rijswijk

Ir D. van Leeuwen:
Shell Oil Company, The Hague

Ir Th. Monnier:
Institute TNO for Building Materials and Building Structures, Rijswijk

Dr Ir G. van Oortmerssen:
Netherlands Ship Model Basin, Wageningen

Ir A. Paape:
Delft Hydraulics Laboratory, Delft

Ir W.R. de Sitter:
Hollandse Beton Groep NV, Rijswijk

Ir F.P. Smits:
: Delft Soil Mechanics Laboratory, Delft

Dr Ir J. Strating:
: Protech International, Consulting Engineers, Schiedam

Ir A. Vrouwenvelder:
: Institute TNO for Building Materials and Building Structures, Rijswijk/ University of Technology, Delft

Ir C.L. van der Zwaag:
: Fugro, Consulting Engineers, Leidschendam.

Most of the funds for financing the two above-mentioned projects were provided by the Ministry for Economic Affairs of the Netherlands Government.

References

[1] STUPOC V Final Report:
'Reliability and Risk analysis for offshore structures'
(Veiligheidsbeschouwing en risico-analyse voor offshore constructies)
Published by IRO, PO-box 215, Delft, The Netherlands (1979) (in Dutch).

[2] Vrouwenvelder, A.:
'Probabilistic Fatigue Analysis for single offshore pile'
(Probabilistische Vermoeiingsberekening 'Paal in Zee')
Report B-78-442/62.6.0402, IBBC-TNO, PO-box 49, 2600 AA Delft, The Netherlands (1978) (in Dutch).

[3] Rackwitz, R.:
'Principles and methods for a practical probabilistic approach to structural safety'
Subcommittee for First Order Reliability Concepts for Design Codes of the Joint CEB-CECM-CIB-FIP-IABSE Committee on Structural Safety
Published by CEB in 1976.

[4] Hallam, M.G., Heaf, N.J., Wooton, L.R.:
'Rationalisation of safety and serviceability factors in structural codes'
Ciria report nr. 63, 1977
Ciria Underwater Engineering Group, 6, Storey's Gate, London, SW1P 3AU

[5] Battjes, J.A.:
'Long term wave height distributions at seven stations around the British Isles'
Deutsche Hydrografische Zeitschrift 25, 1972, no. 4.

[6] API-code (API-RP-2A):
'Recommended Practice for planning, designing and construction of fixed offshore platforms'
American Petroleum Institute, January, 1977.

[7] Drawing 'Numerical Comparisons on different approaches'
American Petroleum Institute, October, 1973.

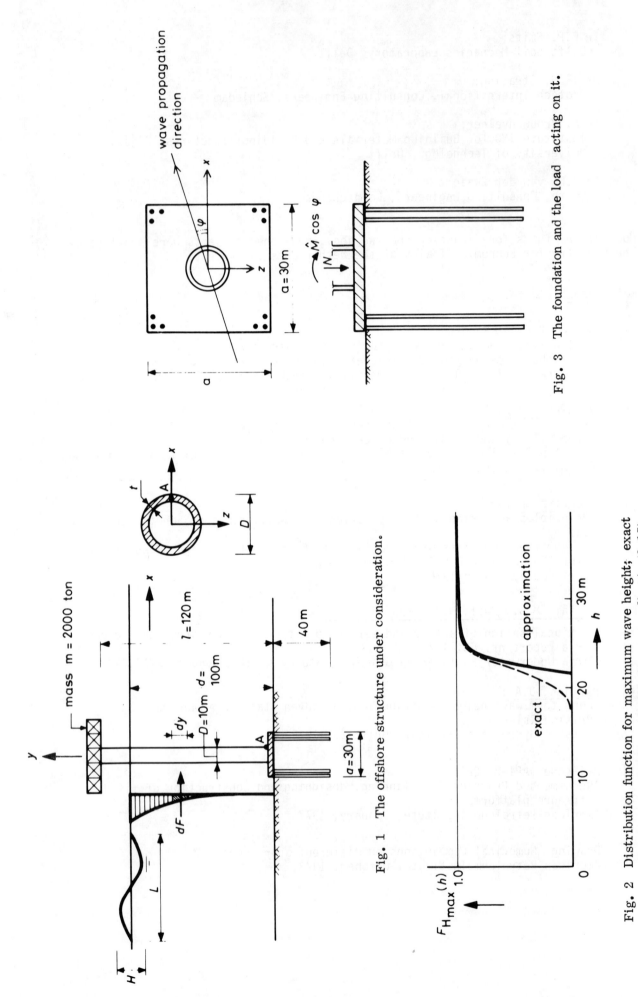

Fig. 1 The offshore structure under consideration.

Fig. 2 Distribution function for maximum wave height; exact according to (3.14) approximation according to (3.15).

Fig. 3 The foundation and the load acting on it.

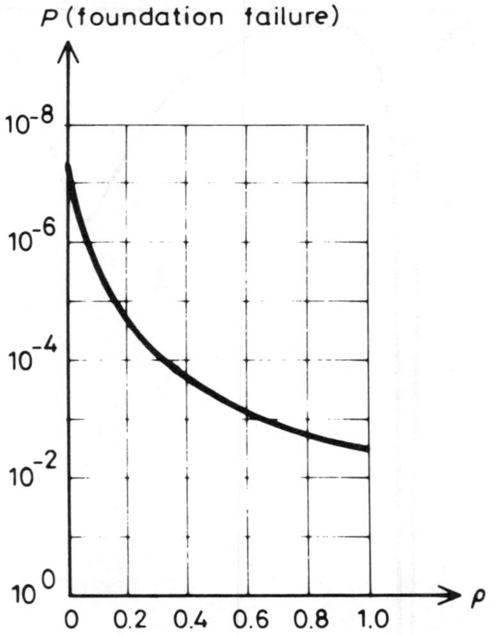

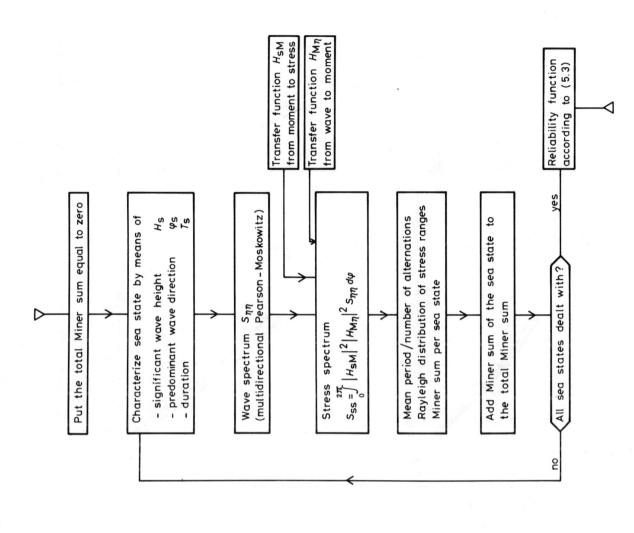

Fig. 5 Flow chart for spectral fatigue analysis.

Fig. 4 Effect of the correlation between pile strengths upon the failure probability of the foundation.

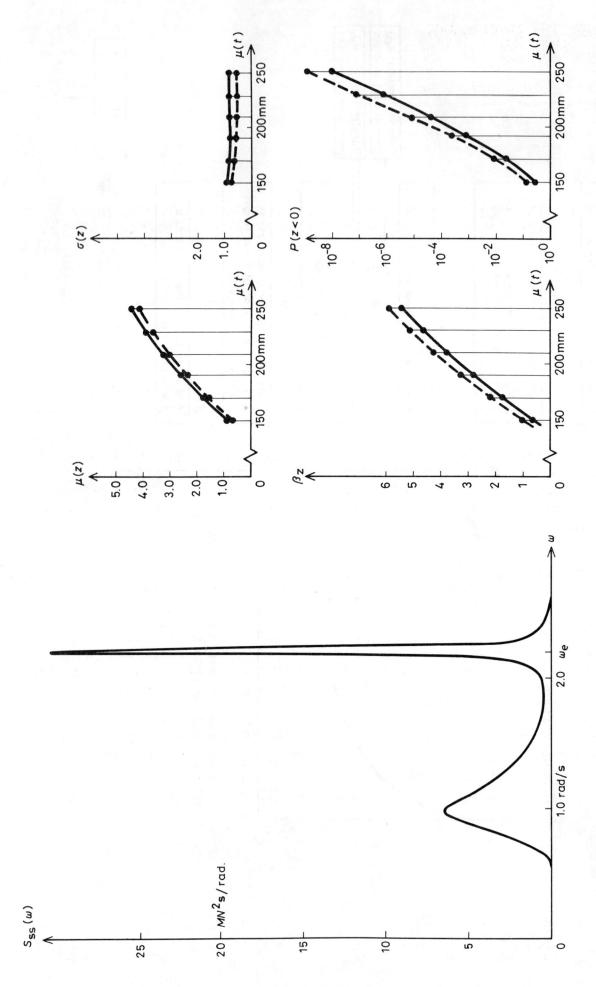

Fig. 7 Results of probabilistic fatigue analysis: continuous line - approximate full distribution method; dotted line - mean value method.

Fig. 6 Stress spectrum for point A for a sea state with $H_s = 1.83$ m (median) and $\varphi_s = 0°$.

BOSS'79

PAPER 86

Second International Conference
on Behaviour of Off-Shore Structures

Held at: Imperial College, London, England
28 to 31 August 1979

THE SAFE DESIGN AND CONSTRUCTION OF STEEL SPHERES AND END CLOSURES OF SUBMERSIBLES, HABITATS AND OTHER PRESSURISED VESSELS

D. Faulkner

University of Glasgow, U.K.

Summary

The choice of configuration and construction of end closures and the importance of avoiding local flatness is discussed. Pressure hull modes of failure are reviewed. For ring-framed elements the concept of "guaranteed strength" is acceptable but this seems less viable for unstiffened domes because of the large unexplained scatter in test results and known shape sensitivity. One purely deterministic code allows a greater margin for this. In contrast a recent semi-probabilistic code makes no specific distinction between failure margins for different pressure hull elements. But it has the merit of providing greater safety margins for slender domes (high $\sigma_y R/Eh$) because of their greater sensitivity to shape imperfection than stocky ones.

A strength formulation is derived which fits the best mean curve through the relatively sparse experimental data within 3% throughout the slenderness range. This can be used directly in design. Alternatively, by using the experimental scatter (cov = 14%) in conjunction with linear error theory to account for systematic and random variations in the four main variables, a semi-probabilistic approach to analysis and design is possible. Examples show that if a constant safety factor is adopted (as in many codes) any chosen probability or characteristic strength can be lower for slender shells than for stocky ones. This is undesirable. The examples also show that for a nominal safety factor of about 2.35 based on the mean test data, the 5% probability overall partial safety factor γ_o varies from 1.7 to 1.9. Corresponding safety indices β_f vary from 3.4 to 3.8 and these are compared with values for other structural elements and for ships.

The paper ends by indentifying the need for a greater understanding of the collapse of domes and for better shape tolerances. It stresses the need for statistical data, and outlines a semi-probabilistic basis for a design code which should provide adequate, but not excessive, safety.

Sponsored by: Delft University of Technology, The Netherlands
Massachusetts Institute of Technology, U.S.A.
The Norwegian Institute of Technology, Norway
University of London, England

Secretariat provided by: BHRA Fluid Engineering

Copyright: © BHRA Fluid Engineering
Cranfield, Bedford, England

NOMENCLATURE

R, h	nominal shell radius, thickness
p_o, p_c	safe operating pressure, collapse pressure using nominal dimensions and material properties
p_s	$2\sigma_y h/R$ membrane yield pressure for sphere
p_{cl}	classical elastic buckling pressure
p_e	elastic buckling pressure - imperfections considered
p_{em}	mean value of p_e based on test data
F	scalar safety factor p_c/p_o
Q_o	operating design load ($\doteq p_o$)
Q_d	extreme design load
R_d	design resistance of the structure
R_k	characteristic resistance ($\equiv p_c$)
E, ν	Youngs modulus, Poissons ratio
σ_y	nominal yield or proof stress - assumed to be the characteristic strength of the material
γ_m	material partial safety factor = 1.15 in elastic design, 1.3 in plastic design*
γ_f	load partial safety factor Q_d/Q_o
γ_o	overall partial safety factor $\gamma_f \gamma_m \kappa/\psi$
ψ	factor reflecting post-buckling behaviour where for shells where redistribution is not possible = 0.9*
κ	factor depending on slenderness ratio = 1 when the characteristic resistance is equal to the characteristic strength of the material; for other cases $\kappa = 1$ for $\lambda < 0.5$; $= 0.7 + 0.6\lambda$ for $0.5 \leq \lambda \leq 1.0$; $= 1.3$ for $\lambda > 1.0$*
λ, λ_m	slenderness parameters $\sqrt{p_s/p_{cl}}$, $\sqrt{p_s/p_{em}}$
x_i	random variable, e.g. σ_y, R, etc.
s_i	standard deviation of x_i
v_i	coefficient of variation of x_i
ζ_i	systematic error in x_i
v_o, v_s, v_t	objective, subjective, total uncertainty (cov)
θ, β_f	central safety factor ($\bar{p}_c/\bar{p}_o$), safety index

Note: bars are used over the parameters when mean values of the variables are used.

*Values from Ref. 3

1. CONSTRUCTION

1.1   Most end closures are domed to take advantage of membrane stressing. Their shape is often hemispherical, elliptical or torispherical as shown in Fig. 1(a). The advantages of the torispherical shape are:

  a)   it provides a form of transition between the central cap and the cylinder to which it is joined which results in a low stress concentration at this junction and a reasonable radial stiffness to match the ring-frames of the cylinder

  b)   the loss of internal volume is less than that associated with a hemispherical cap, and

  c)   it provides a large area of relatively flat structure near the axis which is suitable for penetrations such as gland, manipulators, observation windows, etc. (the stress concentration effects are less severe in a dome than they are in a cylinder).

The main disadvantage of externally convex domed closures are that the membrane stresses are in compression, and collapse pressures for domes are usually considerably less than those to cause membrane yield because of the instability mode of collapse and its sensitivity to shape imperfections. Not only do areas of local flattening introduce large bending stresses, but they may increase the membrane stresses and certainly lower the collapse pressure. This is probably the main reason why there is considerable scatter in experimental data.

1.2   In view of this it may seem tempting to turn the dome "inside out" with its concave surface subject to pressure so that its membrane stresses are in tension. This idea was conceived for the internal bulkheads of "escape compartments" in certain World War II submarines. But for external closures subjected to fluctuating pressures the risk of fatigue and unstable crack propagation makes the idea less attractive. This is discussed no further.

1.3   The effect of any possible forms of lack of sphericity upon the collapse pressure is very difficult to determine. As a result it is difficult to decide upon tolerances for manufacture and there is a tendency to choose them on the basis of what can be achieved rather than what is desirable. This is potentially dangerous in domes where every effort should be made to check sphericity as accurately as possible. BS 5500 (Ref. 1) requires spheres to be spherical to within 1% on radius, and over a local arc length of $2.4\sqrt{Rh}$ the radius of curvature must not exceed the nominal value by more than 30%. This degree of flatness only represents a departure from the true sphere of 0.166 h over a limited area. For example, with $R/h = 50$, this would be only 0.33% of the nominal radius (well within the required 1%) over an area as small as 1.4% of the area of the corresponding hemisphere (or 16° total solid angle). Clearly this needs careful detection and accurately made templates should be used.

1.4   It is likely that sensitivity to shape imperfections would be reduced if domes were stiffened. This would also allow the thickness to be reduced, thereby easing the manufacture of the dome. But research has indicated that such stiffening must be closely spaced to be effective and it would be difficult to fit either inside or outside the dome, particularly at the centre and at the position of the toroid. Although both radial-circumferential and vertical-horizontal stiffening systems have been tried, no sufficiently reliable scheme has yet been found.

1.5   For diameters up to about 4.5 m domes are usually formed by spinning. In such cases special care must be taken to check thickness as well as radius. For larger diameters, especially when combined with the use of rather thick high yield steel, the domes are usually formed by pressing petals of steel and welding them as shown in Fig. 1(b). The problem of joining the vertices of the petals is overcome by fitting a central cap. The pressing or rolling away of local areas of flattening should be undertaken with considerable care. Once the shape is right no stress-relieving is advisable.

1.6   The join to the cylinder has been mentioned. To achieve the best radial stiffness and to achieve a low stress concentration and avoid having a weld at this joint, consideration should be given to using a well designed forging at this position.

2.   DESIGN CONSIDERATIONS

2.1   The pressure hull of a submersible may fail in its cylindrical and conical elements by:

   a)   interframe collapse of the shell

   b)   general instability of the shell-frame combination between deep frames or bulkheads

and this second mode can be precipitated by premature frame tripping or by yielding of the ring frames, for example, as caused by out-of-circularity bending or bending in way of hard spots provided by internal decks and tank structure or external casing. These have been discussed reasonably fully (Ref. 2) and by and large the more recent design codes make adequate provision for safe design (Refs. 1,3) of such stiffened structures. In general the design philosophy is to ensure interframe collapse, which is less sensitive to shape imperfections, by postulating out-of-circularity frame bending criteria which virtually eliminate the possibility of general instability.

2.2   Because of the very considerable volume of test data (Ref. 4) it is thereby possible to invoke the concept of a guaranteed lower bound interframe collapse curve (Ref. 1), which would seem to imply that the implicit safety factor (of about 1.5 on this lower bound) is there solely to cater for depth or pressure excursions greater than the designed operating values. In submarines such depth "overshoots" do occur due to mal-operation of hydroplanes or depth gauges, human errors, or, in the case of military submarines, to avoid depth charge attacks. There is very little available data on such depth overshoots, but it is likely that for research submersibles there is a 10^{-2} chance that any one of them will accidentally exceed its design maximum operating depth in any given deep dive by about 1.3, whereas for fast military submarines the overshoot factor may be as high as 1.5 for the same chance in a hundred. This latter figure coincides with the ratio of the lower bound collapse pressure for cylinders and the required safe operating pressure. It is important that such data be made available by operators since it will significantly affect safety and could adversely influence the choice of codes to be used in design (see later).

2.3   Missing in the above remarks is the safety of the end closures, although this is probably best covered in BS 5500 (Ref. 1) based on the experimental data in Ref.(2). Because of the scatter in test results and the small volume of relevant test data on the collapse of fabricated steel domes, it is not quite so acceptable to invoke the concept of a guaranteed lower bound collapse curve. In time it is virtually certain, for the reasons given earlier, that further test results will yield data below the lower bound strength curve based on Fig. 9.25, Ref. (2). Moreover, the treatment of general instability, and hence out-of-circularity frame bending, for dome-ended components will in general require the compartment length to be doubled due to the relatively low radial stiffness of the dome-cylinder junction, and the advice given in Ref. (1) is most relevant.

3.   SAFETY CONCEPTS AND COMPARISONS

3.1   The only method which can combine all modes of failure for the stiffened cylinders and conical elements, and for the end closures, is a purely statistical **level-3 approach (Ref. 5)**. However, present data and knowledge does not yet permit such an approach, and so a semi-probabilistic approach would seem to commend itself (Ref. 6) and such methods are being introduced into certain offshore codes (Ref. 3). This will be compared with the best available conventional deterministic methods (Ref. 1) which are still used in many codes. In this, safety is measured by a scalar quantity known as the safety factor. The approach is simple and has served engineers well.

3.2   However, apart from not being able to combine all modes of failure, the safety factor method has two main shortcomings. Firstly, with interactive failure modes and composite loads, the approach to the failure surface is best represented by a varying vector (Refs. 6,7); and, secondly, the uncertainties which require us to apply a safety margin have strong random components (especially for domes) which are best defined statistically. Thus statistical analysis will be used as far as seems sensible in conjunction with conventional determinism to develop a semi-probabilistic approach to analysis of dome strength and to their design.

3.3   As an illustration of the level-1 safety factor approach we can note that Ref. (1) implies safety factors defined by F where

$$p_c = F p_o \qquad (1)$$

as being approximately as follows:

Structure	Based on lower bound curve	Based on mean curve
Stiffened cylinders and cones	1.5	1.75
Unstiffened domes	1.75	2.35

the greater values for domes is to cater for the greater uncertainty in their strength predictions. These safety factors are uniform throughout the slenderness range $(\sigma_y/E)(R/h)$.

3.4   As an illustration of the level-2 second moment or semi-probabilistic approach the method adopted in Ref. (3) is defined as follows (refer to the notation):

Extreme design load $\leq$ Design resistance

$$Q_d \leq R_d$$

$$\gamma_f Q_o \leq \frac{R_k \psi}{\gamma_m \kappa} \qquad (2)$$

or

$$\frac{R_k}{Q_o} = \frac{\gamma_f \gamma_m \kappa}{\psi} = \gamma_o \qquad (3)$$

For example, if $\lambda < 0.5$ as in inelastic designs

$$\gamma_o = \gamma_f \frac{1.3 \times 1}{0.9} = 1.44 \gamma_f$$

For elastic designs $\lambda > 1$ and

$$\gamma_o = \gamma_f \frac{1.15 \times 1.3}{0.9} = 1.66 \gamma_f$$

It is not clear from Ref. (3) how the load coefficient or partial safety factor γ_f is to be specified. If we take $\gamma_f = 1.3$ or 1.5 along the lines of the argument given in para. 2.2 the overall partial safety factors γ_o of eq (3) would lie in the range from inelastic to elastic designs of:

Research submersibles	1.88 to 2.17
Military submarines	2.16 to 2.49

It must be stressed that these are arbitrary assumptions open to discussion.

3.5 In view of the vagueness in certain aspects of Ref. (3) it is open to question as to whether γ_o derived above does correspond to the safety factor F of Ref. (1) as given by eq (1). In section 3.3 F for unstiffened domes was approximately 1.75 to 2.35 depending on whether the lower bound or mean collapse curves were used. There is no specific formulation given for domes in Ref. (3) but the implication of the shell design methods given there imply (insofar as they are based on an inelastic reduction of p_e) that they would be based on lower-bound information. Thus a straight comparison of γ_o with F = 1.75 from Ref. (1) suggests that Ref. (3) is somewhat more optimistic if γ_f = 1, but appreciably more pessimistic if γ_f = 1.3 or 1.5. The justification for varying γ_o with slenderness λ will be discussed later. It seems also that Ref. (3) may be appreciably more pessimistic than Ref. (1) for stiffened cylinders and cones, since the factors given above from Ref. (3) apply to all modes of failure of shell structures. In that respect it does not appear to acknowledge, as does Ref. (1), the greater uncertainty in dome strength. It would seem desirable to include a specific section on this important topic.

4. STRENGTH FORMULATION

4.1 We shall first ignore the effects of all *objective* variations in the parameters h, R, σ_y and E which influence collapse strength. That is, we shall assume we know these parameters accurately.

4.2 Although the behaviour of spherical shells under external pressure has received considerable attention in the last 60 years the prediction of their collapse pressure is still in an unsatisfactory state. This is in spite of the fact that the classical small deflection instability pressure derived by van der Neut in 1932 has received much theoretical and experimental support. This pressure is:

$$p_{cl} = \frac{2 E h^2}{\sqrt{3(1-\nu^2)}\, R^2} \qquad (4)$$

The double curvature produces a marked stiffening effect, but equally, this curvature can be greatly reduced by very small departures from the perfect spherical shape (Refs. 2,8) and this produces a marked sensitivity of collapse pressure to accuracy of manufacture and boundary conditions. For this reason spherical caps are found to collapse at pressures much less than eq (4) even when the mean stress at collapse is small compared with the yield stress. An excellent review of numerous papers attempting to explain these low buckling pressures has been written (Ref. 9), but although the theories reviewed are of interest, they do little to assist in the prediction of collapse loads. A typical one lobe collapse is illustrated in Fig. 1(c).

4.3 For this reason a semi-empirical interaction type approach is advocated. In the elastic range the equation advanced by von Karman and Tsien (Ref. 10) still provides a reasonable lower bound to the test data for very slender spheres:

$$\begin{aligned} p_e &= 0.25\, p_{cl} \\ &= 0.3\, E(h/R)^2 \end{aligned} \qquad (5)$$

However, in view of the stated preference to treat the data statistically, the author has analysed the most relevant test data from fabricated torispherical and spherical caps (Refs. 2,11,12) and the best mean curve in the elastic range is found to be:

$$\begin{aligned} p_{em} &= 0.3\, p_{cl} \\ &= 0.36\, E(h/R)^2 \end{aligned} \qquad (6)$$

stocky spheres of low $(\sigma_y/E)(R/h)$ values fail at the membrane yield pressure:

$$p_s = 2\sigma_y (h/R) \qquad (7)$$

If we postulate a circular interaction between these two extremes of failure modes we obtain the required mean collapse pressure $\bar{p}_c$ from:

$$\left(\frac{1}{\bar{p}_c}\right)^2 = \left(\frac{1}{p_{em}}\right)^2 + \left(\frac{1}{p_s}\right)^2 \tag{8}$$

or for any given applied pressure p the safety factor F is given by:

$$\frac{1}{F^2} = \left(\frac{p}{p_{em}}\right)^2 + \left(\frac{p}{p_s}\right)^2 \tag{9}$$

and the concept of a constant scalar safety factor is in this case acceptable (Ref. 7). A rearrangement of eq (8) making use of equations (6) and (7) leads to:

$$\left. \begin{array}{l} \bar{p}_c = p_s(1 + \lambda_m^4)^{-\frac{1}{2}} \\ \text{where } \lambda_m^2 = \dfrac{p_s}{p_{em}} = 5.5 \dfrac{\sigma_y R}{Eh} \end{array} \right\} \tag{10}$$

This equation is slightly conservative compared with the best mean curve drawn through the experimental data of about sixty tests (Refs. 2, 11, 12) but never more than by 3% (see Fig. 2). The experimental data had coefficients of variation in the following slenderness ranges for the torispherical tests (there is a wider variation for the hemispheres):

p_{c1}/p_s range	λ range	c.o.v. %
1.3 - 2.05	0.88 - 0.70	17
2.3 - 3.5	0.66 - 0.53	13
3.8 - 4.8	0.51 - 0.46	12
5.0 - 6.4	0.45 - 0.40	14
All tests:		
1.3 - 6.4	0.88 - 0.40	14

The consistency of the c.o.v. suggests there is little evidence from test data to support the contention in table 6.5 of Ref. (3) that the factor κ should increase from 1 to 1.3 as slenderness parameter λ increases from 0.5 to 1.0. But the sample size is small and it is natural to expect slenderer structures to be less predictable than stocky ones and therefore requiring greater safety margins. This will be examined further. The fact remains that as Ref. (9) shows, the scatter in test results is inexplicable. We shall therefore regard it as a *subjective* random uncertainty of value $v_s = 14\%$ over the entire range considered. This would be larger (about 15 to 16%) if the hemispherical tests were included in the statistical analysis.

4.4   Equation (10) can be used directly in design to arrive at a required thickness h for a given radius R and material properties and required mean collapse pressure $\bar{p}_c$. Such a direct approach is not possible using Fig. 2 (or Fig. 3.6(3) curve (b) of Ref. (1) which is based on Fig. 2 with a safety factor of about 2.35 incorporated). By substituting eq (6) and (7) in (10) and rearranging we obtain a quadratic in (R/h):

$$\left(\frac{R}{h}\right)^4 + \frac{E^2}{30.4\,\sigma_y^2}\left(\frac{R}{H}\right)^2 - \frac{E^2}{7.6\,\bar{p}_c^2} = 0 \tag{11}$$

Example (1):   Select a minimum thickness hemispherical end closure of 4 m radius made from steel having a nominal 0.2% proof stress of 540 N/mm² if it is to have a safe operating depth of 300 m.   BS 5500 standards of sphericity may be assumed.

$$p_o = 3.015 \text{ N/mm}^2, \quad \bar{p}_c = 2.35 \, p_o = 7.085 \text{ N/mm}^2$$

substituting with σ_y = 540 and E = 207,000 N/mm² in eq (11) gives:

$$R/h = 91.9, \quad h = 43.5 \text{ mm} = 44 \text{ mm selected}$$

<u>check</u> p_s eq (7) = 11.88 N/mm²

$\quad\quad\quad p_{cl}$ eq (4) = 30.31 N/mm² = 2.55 p_s

Then $\bar{p}_c/p_s$ = 0.61   from Fig. 2

$\quad\quad\quad \bar{p}_c$ = 7.20 N/mm² = 2.39 $p_o \geq$ 2.35 p_o required

So h = <u>44</u> mm is the least weight choice excluding any corrosion allowance.

5. SEMI-PROBABILISTIC STRENGTH AND SAFETY ASSESSMENTS

5.1   In addition to the *subjective* uncertainties in the test data just discussed there will be *objective* uncertainties in each of the four parameters E, σ_y, h and R which affect the mean strength as defined by eq (10). If x_i is a typical distribution of these variables, we shall assume that its *random* and *systematic* components are defined by:

$$\left. \begin{array}{l} v_i = s_i / \bar{x}_i \\ \\ \zeta_i = \bar{x}_i / \text{nom } x_i - 1 \end{array} \right\} \quad\quad\quad (12)$$

where s_i is a typical standard deviation of the variable x_i and nom x_i is the value specified in design.   If material is drawn from a wide range of stock then typical % values might be:

x_i	v_i	ζ_i
E	1	0
σ_y	5 to 10	10 to 20
h	2 to 4	-2 to 0

In making a single dome end it is of course quite likely that enough measurements would be taken from the material used to be much surer of the as-built values than these cov's imply.   But the systematic errors are likely to be realistic, and it is desirable that mean values of the variables should be used in any strength estimations, viz.:

$$\bar{x}_i = \text{nom } x_i (1 + \zeta_i) \quad\quad\quad (13)$$

5.2   As regards radius R, as mentioned earlier, local flatnesses do occur (increased curvatures are less likely) and this is recognised in some codes (Ref. 1). They are difficult to treat as systematic errors (except perhaps in specific known cases of bad flatness) because they will be randomly disposed and/or incompletely recorded. Moreover, the mean membrane stress can only be approximately assessed, unless the shape is known precisely, and the boundary conditions at the edge of the flat region will be unkown except that it is certain that bending yield will occur prematurely. It is considered best, therefore, to assume the worst flatnesses are removed and that only random errors exist, e.g. v_R = 3 to 5%, ζ_R = 0.

5.3   Applying linear error theory it will be assumed that:

    a)   all functions and their derivatives are continuous

    b)   the objective variables x_i are statistically independent - otherwise correlated errors should be allowed for (there is sometimes a weak correlation between h and σ_y, but in cases examined in this study the correlation coefficient was never worse than -0.14, which has negligible effect on the result)

    c)   the distribution of the variables is close enough to their means for second order and higher terms (derivatives) to be ignored

The total objective uncertainty in strength is then approximately given by (Ref. 5):

$$v_o^2 = \sum_{i=1}^{n} \left\{ \frac{\partial p_c}{\partial x_i} \frac{x_i}{p_c} \right\}_{x_i = \bar{x}_i} v_i^2 \qquad (14)$$

From eq (10)

$$\bar{p}_c = \frac{2 \sigma_y h/R}{(1 + \bar{\lambda}_m^4)^{1/2}} \quad \text{where} \quad \bar{\lambda}_m^4 = 30.4 \, (\sigma_y R/Eh)^2$$

Hence $\left\{ \dfrac{\partial p_c}{\partial x_i} \dfrac{x_i}{p_c} \right\}_{x_i = \bar{x}}$

$$\begin{aligned}
&= \bar{\lambda}_m^4 / (1 + \bar{\lambda}_m^4) && \text{for } x_i = E \\
&= (1 + \bar{\lambda}_m^4)^{-1} && \text{for } x_i = \sigma_y \\
&= (1 + 2\bar{\lambda}_m^4)/(1 + \bar{\lambda}_m^4) && \text{for } x_i = h \\
&= -(1 + 2\bar{\lambda}_m^4)/(1 + \bar{\lambda}_m^4) && \text{for } x_i = R
\end{aligned} \qquad (15)$$

$$\text{where} \quad \bar{\lambda}_m^4 = 30.4 \, (\bar{\sigma}_y \bar{R}/\bar{E}\bar{h})^2 \qquad (16)$$

the mean values of all the four variables must be used allowing for the systematic errors by eq (13). It will be noticed from eq (15) that for slender spheres as $\lambda_m \to \infty$ the curly bracket terms {} tend to 1, 0, 2 and -2 respectively as one would expect from eq (4). Likewise as $\lambda_m \to 0$ {} tend to 0, 1, 1 and -1 respectively as expected from eq (7) for stocky shells.

5.4   Once the objective uncertainty has been assessed as above then it is assumed the subjective uncertainty is independent, so that the total uncertainty is approximately given by:

$$v_t^2 = v_o^2 + v_s^2 \qquad (17)$$

where v_s can be taken = 14% based on the test data analysed in 4.2.

Example (2):   For the end closure of ex.(1) estimate from the nominal material properties and scantlings:

a) the expected mean strength
b) the coefficient of variation of strength
c) the 5% lower probability strength
d) the 0.1% lower probability strength

making comparisons as appropriate with the required operating pressure p_o. The random and systematic uncertainties for the four main variables are estimated to be:

	x_i	E	σ_y	h	R
v_i %		1	5	3	5
ζ_i %		0	12	-1	0

Hence $\bar{\sigma}_y$ = 1.12 x 540 = 640.8 N/mm²

$\bar{h}$ = 0.99 x 44 = 43.56 mm

$\bar{\lambda}_m^4$ = 30.4(640.8 x 4000/207,000 x 43.56)² = 2.188

a) $\bar{p}_s$ = 13.17 N/mm² from eq (7)

$\bar{p}_{cl}$ = 29.70 N/mm² from eq (4) = 2.26 $\bar{p}_s$

$\bar{p}_c$ = 7.38 N/mm² from eq (10) and (16)

= 2.45 p_o ≥ 2.35 p_o required

b) From eq (15) and (16)

$$\left\{ \frac{\partial p_c}{\partial x_i} \frac{x_i}{p_o} \right\}_{x_i = \bar{x}_i} = 0.686 \text{ for } E$$
$$= 0.314 \text{ for } \sigma_y$$
$$= 1.686 \text{ for } h$$
$$= -1.686 \text{ for } R$$

Hence from eq (14) the objective uncertainty is:

v_o^2 = (0.686 x 1)² + (0.314 x 5)² + (1.686 x 3)² + (-1.686 x 5)²

= 0.5 + 2.5 + 25.6 + 71.1 = 99.6

= 10.0%

Total uncertainty in strength from eq (17) is:

v_t^2 = 99.6 + 14²

v_t = 17.2%

c) Assume a normal strength distribution the characteristic resistance is:

R_k = $p_c(1 - k v_t)$ where k depends on the probability

5% R_k = 7.38 (1 - 1.645 x 0.172) = 7.38(0.717)

= 5.29 N/mm² = 1.76 p_o

i.e. 5% γ_o = 1.76

d) $0.1\% \ R_k$ = 7.38 (1 - 3.09 × .172) = 7.38 (0.469)

             = 3.46 N/mm² = 1.15 p_o

i.e. $0.1\% \ \gamma_o$ = <u>1.15</u>

Comment: Because of the relatively slender shell ($\bar{p}_{cl}/\bar{p}_s$ = 2.26, $\bar{\lambda}$ = 0.67, $\bar{\lambda}_m$ = 1.22) the 12% increase in yield strength causes only a minor rise in $\bar{p}_c$ and has a negligible effect on v_o. In contrast, the effects of likely uncertainties in both h and R are significant. This leads to a relatively high total uncertainty in strength and a chance in a thousand strength only 15% more than the required safe operating depth. This would be acceptable assuming the likelihood of depth overshoot to be small ($\gamma_f \simeq 1$). The characteristic strengths could be improved by quality control checks on thickness (h) and radius (R) in any given design. The design would be sensitive to "flatness" and special care would be recommended to check sphericity.

5.5   The effect of reduced slenderness can readily be seen by re-working the above example but with say 4 times the required driving depth so p_o = 12.06 N/mm². Then it can be shown that:

R/h = 34.24

$\bar{\lambda}_m^4$ = 0.304

a) $\bar{p}_s$ = 35.33 N/mm²

      $\bar{p}_{cl}$ = 213.6 N/mm² = 6.05 $\bar{p}_s$

      $\bar{p}_c$ = 30.93 N/mm² = 2.56 p_o OK

b) v_o^2 = $\{\}_E^2 \ 1^2 + \{\}_{\sigma_y}^2 \ 5^2 + \{\}_h^2 \ 3^2 + \{\}_R^2 \ 5^2$

        = $(0.233 \times 1)^2 + (0.767 \times 5)^2 + (1.233 \times 3)^2 + (-1.233 \times 5)^2$

        = 0.0 + 14.7 + 13.7 + 38.0 = 66.4

      v_o = 8.15%

      v_t = 16.2%

c) $5\% \ R_k$ = 22.69 N/mm² = 1.88 p_c

      $5\% \ \gamma_o$ = 1.88 = 7% greater than for ex (2)

d) $0.1\% \ R_k$ = 15.45 N/mm² = 1.28 p_o

      $0.1\% \ \gamma_o$ = 1.28 = 11% greater than for ex (2)

We must conclude, therefore, that as slenderness $(\sigma_y/E)(R/h)$ decreases, the overall partial safety factor γ_o increases. Thus some compensation for slenderer shells is warranted as provided by the two terms $\gamma_m \kappa$ in eq (3). However, a more explicit approach based on the above analysis is now possible, providing relevant statistical data can be obtained.

5.6   To end this section, an alternative second moment *safety index* (Ref. 6) parameter β_f will be examined which is the inverse of the cov of the margin R-Q:

$$\beta_f = \frac{\bar{p}_c - \bar{p}_o}{\sqrt{s_{p_c}^2 + s_{p_o}^2}} = \frac{\theta - 1}{\sqrt{\theta^2 v_{p_c}^2 + v_{p_o}^2}} \right\} \quad (18)$$

where $\theta = \bar{p}_c/\bar{p}_o$ central safety factor

Assume for the load there is 1 dive in every 100 to 30% more than the design operating depth p_o. Then it can be shown:

$$\bar{p}_o = 1.003\, p_o$$

$$v_{p_o} = 3\%$$

Considering ex.(2) $\theta = (\bar{p}_c/\bar{p}_o)(p_o/\bar{p}_o) = 2.45/1.003 = 2.44$, then $\beta_f = 1.44/\sqrt{2.44^2 \times .172^2 + .03^2} = \underline{3.42}$. For the same design at 4 times the operating depth $\theta = 2.56$ and $\beta_f = \underline{3.75}$. These values compare with values 0.9 to 3.4 for surface warships and values 2.9 to 6.4 for merchant ships. The relatively high values for the dome ends is because of the very low cov of the load for submersibles. For interframe shell collapse for the same designs β_f would be in the range 3.0 to 3.3.

6. DESIGN DISCUSSION AND GENERAL CONCLUSIONS

6.1   Two recent design codes (Refs. 1,3) have been advanced, each of which has considerable merit. BS 5500 is entirely deterministic using an implied safety factor somewhat greater for spheres than for ring-stiffened cylinders, but in each case the safety factor is constant over the entire slenderness range. In contrast, the recent DnV rules are based on partial safety factors, although in truth there is no application of statistics required for their use. In this sense they may also be considered to be deterministic. But, they do have the merit of acknowledging that the factors required for slender structures should be different from those required for stocky structures, no doubt to recognise:

   a)   the greater sensitivity of slender structures to shape imperfections, and

   b)   the effect of random variations in material and geometrical properties on the probability of achieving any required characteristic strength

Against this there seems to be no specific treatment of spheres, other than by implication, and a suggestion that partial safety factors would in any case be the same for both spherical and cylindrical elements.

6.2   This paper illustrates that a constant safety factor F approach, as in BS 5500, leads to designs which have varying overall partial safety factors γ_o, these being lower for slenderer shells than for stocky ones, which is against the trends of common sense. The above analysis perhaps provides a basis for unifying the two approaches to provide a better approach to safety. For example, where the designer has access to statistical data on material properties and scantlings (and this is to be encouraged) then he could arrive at a first design deterministically as was done in ex.(1) section 4.3. Then applying the statistical data as in section 5 he could arrive, for example, at the 5% R_k characteristic strength as in ex.(2) section 5.4. This can be related to the required operational diving depth Q_o (or p_o) by the overall partial safety factor γ_o which clearly might lie between 1.5 and 2.0. The required thickness h can be adjusted until the conditions are met. In this process it has been seen that as accurate an estimate of mean yield stress $\bar{\sigma}_y$ is important, so obtaining statistical data on this, h and R is clearly important.

6.3   There have been some spectacular collapses of externally loaded spherical structures, and whilst design codes vary so much it is natural for designers to select those which lead to the lightest, cheapest structure. Such structures are very shape sensitive and the detailed connection to adjoining cylinders, for example, also demands particular care. In these circumstances there is evidently a need for:

   a)   a greater understanding of the collapse of spheres, particularly in relation to the formulation of better shape tolerances

   b)   collection of relevant statistical data of the type used in this paper

c) formulation of design codes which provide adequate, but not excessive, safety margins which reflect the nature of the distribution of the main parameters

6.4 Finally, in this respect the active co-operation of designers, construction engineers, operators (for diving profile records) and researchers is required if further progress is to be made. It is unlikely, and indeed undesirable, for semi-probabilistic methods to be adopted overnight. The use of γ_o (or β_f) should emerge only after more statistical data is available and when these second moment methods have been tried alongside more conventional approaches. In this process comparison should be made with the best previous practice (not the overly safe structures) so that a sensible selection of γ_o (or β_f or other parameters) can be made and refined as our knowledge improves. A more complete study would also require a range of parametric calculations to examine the effect on safety of the various uncertainties. This could also highlight where quality control is most needed.

REFERENCES

1. Specification for Unfired Fusion Welded Pressure Vessels, British Standards Insitution, BS 5500 (1976).

2. KENDRICK, S.: Chapter 9, The Stress Analysis of Pressure Vessel and Pressure Vessel Components, ed. by S.S. Gill, Pergamon Press (1970).

3. Rules for the Design, Construction and Inspection of Offshore Structures, Det norske Veritas, Oslo (1977).

4. KENDRICK, S.B.: "Collapse of stiffened cylinders under external pressure", Proc. I.Mech.E. Conference on Vessels under Buckling Conditions, Paper C190/72, London (7 December 1972).

5. MANSOUR, A.E. and FAULKNER, D.: "On Applying the Statistical Approach to Extreme Sea Loads and Ship Hull Strength", Trans. RINA, vol. 115, pp 277-314 (1973).

6. FAULKNER, D. and SADDEN, J.A.: "Toward a Unified Approach to Ship Structural Safety", Trans. RINA, vol. 121, pp 1-28 (1979).

7. FAULKNER, D.: "Safety Factors?", Steel Plated Structures, ed by P.J. Dowling et al, Crosby Lockwood Staples, pp 764-790 (1977).

8. KENDRICK, S.B.: "Shape Imperfections in Cylinders and Spheres", Jnl. Strain Analysis, vol. 12, No. 2, pp 117-122 (1977).

9. THOMSON, J.M.T.: "Elastic Buckling of Spherical Shells", Proc. Symposium on Nuclear Reactor Containment Buildings and Pressure Vessels, Royal Coll. Sc. & Techy., Glasgow, Butterworth (1960).

10. von KARMAN, T. and TSIEN: "The Buckling of Spherical Shells by External Pressure", Jnl. Aeronautical Sciences (1939).

11. NEWLAND, C.N.: "Collapse of Domes under External Pressure", Proc. I.Mech.E. Conference on Vessels under Buckling Conditions, Paper C191/172, London (7 December 1972).

12. KIERNAN, T.J. and NISHIDA, K.: "The Buckling strength of fabricated HY80 steel spherical shells", David Taylor Model Basin, Report 1721 (July 1966).

ACKNOWLEDGEMENTS

Pergamon Press are thanked for allowing Fig. 9.25 of Ref. 2 to be used in this paper.

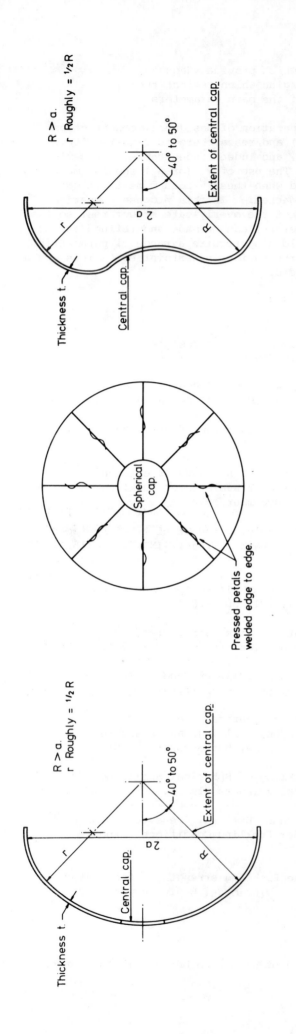

Fig. 1a Geometry of a torisphere

Fig. 1b Welded petal construction

Fig. 1c Collapse of torisphere

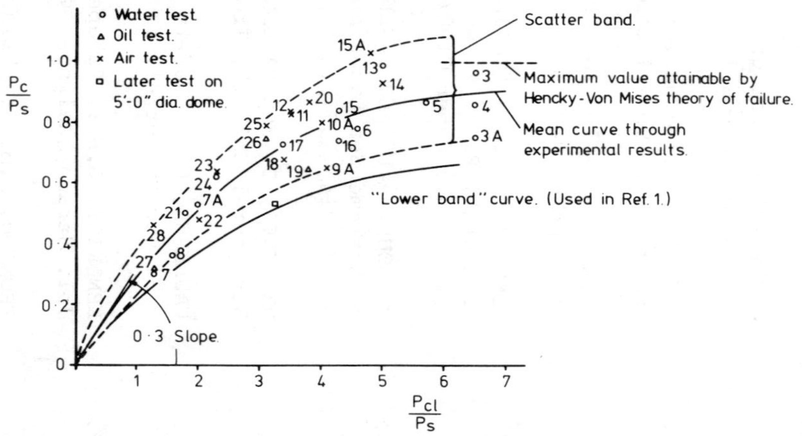

Fig. 2 Results of collapse tests on torispherical domes (Ref. 2)

BOSS'79

PAPER 87

Second International Conference
on Behaviour of Off-Shore Structures

Held at: Imperial College, London, England
28 to 31 August 1979

FUTURE SAFETY CONSIDERATIONS
A MATTER OF SCARCITY AND PROBABILISTIC APPROACH

Ch. J. Vos and B. J. G. Van der Pot

Delta Marine Consultants, Netherlands

J. K. Vrijling

Volker Stevin Group, Netherlands

Summary

Evaluating structural safety by probabilistic analysis is suggested to be common practice by research and education institutes. Designers do however meet apparently insoluble problems in their efforts to use these approaches. They are confronted with the fact that behind symbols and coefficients used in the mathematics no real values are known. Though probabilistic analysis is recommended as a tool to evaluate designs that go beyond common practice and consequentially codes. Such design work is usually required because suddenly scarcities force people to make big steps while trying to offer alternatives.

This paper gives several examples encountered by designers during the design of a gravity structure, the Dunlin Alpha platform, and a flood barrier, the Eastern Scheldt barrier. In both cases it was found that probabilistic analysis of loads and loadcases can be quite satisfactory when they are not too complex. The examples show, however, that the probabilistic analysis of material properties and of the strength of product made from this material may give unsatisfactory results in many cases. This is often because of lack of information about the types of distribution, the nature of the failure mechanism and because of time-dependent phenomena.

It is recommended that research and training institutes should concentrate research and education programmes on the observation of data required for probabilistic analysis. Such an approach will be required more and more in our world in view of works needed to meet future scarcities, such as in energy and materials, especially in offshore construction.

1. INTRODUCTION.

This paper aims to bring forward a few practical questions that can be encountered when using modern probabilistic design methods as a designer. It is essentially an anthology of recent experience of the authors in several offshore engineering design jobs. We have tried to bring this forward in a logical sequence without the pretention of being complete.

Many publications are known on probabilistic analysis and far more work is currently carried out. Although we appreciate the valuable mathematical progress being made, the exercises may usually only be valid for qualitative understanding and judgement of engineering problems. This is due to the fact that although much work on probabilistic analysis suggests accurate analysis and optimizing of results, it only uses assumed or just symbolically represented statistical distributions. It is suggested sometimes that usable statistical data do exist from phenomena that have been only very rarely or never observed or from phenomena that have happened only once or twice.

As a further introduction to the examples of probabilistic design approach in this paper it is explained why normal design practice has to go so far beyond experience consolidated in codes. It is always scarcity in existing products or materials that force man to take steps which are beyond his experience and consequentially beyond codes. Recent offshore experiencs shows such examples.

Reasonably familiar already is the probabilistic analysis using essentially stochastic loads and that is the area where the designer's experience has proven quite successful as shown in this paper. This is especially true when the number of stochastically distributed variables is small and their interdependencies are rather straight forward. Such is for instance the case where for a flood barrier the contribution of tide, waves and inside water level has to be added for the evaluation of the maximum horizontal load on the structure as will be shown in this paper. When the number of variables is larger and the relevant interactions and interrelations are more complex the analysis is already quite impossible with presently known experience. This paper will present the solution chosen for such an example where the influence of the waves including their direction had to be evaluated on the fatigue life of the deck of a gravity structure taking into account the dynamic response of the platform.

Looking into the material and product side of probabilistic design analysis the situation is quite different. Although certain construction materials are very well defined in statistical terms regarding their physical characteristics relevant for structural safety, this fact does not contribute very much to the accuracy of the probabilistic analysis. Materials including their characteristics are in fact hidden in products from which the statistical behaviour is influenced by a wide variety of phenomena such as quality of labour, inspection, tolerances, weather etc.
These factors have individually very high standard deviations and usually little or nothing is known about any interdependency. An objective way of measuring is sometimes even unknown. This paper presents an example where the quality of the material is overruling, such as is the role of the soil in foundation engineering. Here again, when the soil characteristics are influenced by compaction etc., the matter becomes more complex. Going into all factors relevant for the elastic stability of a structure does already present a rather impossible job. This is shown in an example as encountered in the concrete wall panels of a gravity structure. When however questions of more sophisticated physical behaviour of materials and products are encountered, such as long term durability, present knowledge is just insufficient for satisfactory analysis.

It is concluded that the experience and analysis of the examples shown in this paper justify certain conclusions with regard to research programmes and education.

2. SCARCITY AND PROBABILISTIC DESIGN.

Technical development is usually a process of small steps; small steps that follow codes and consequentially experience. Although worldwide research is carried out for the back up justification of codes, they cannot be regarded as a summary of scientific roots for the engineering design practice.
This becomes obvious when big steps, beyond common construction practice are taken. Lack of experience and routine cannot be compensated by codes and the design process becomes a far more conscious one than usual.

At once the designer has to question the validity of every model and algorithm, laid down in the commonly used formulae he uses. A recent example of such a case was the discovery of a printing error in most textbooks of the formula for the elastic stability of cylinders as developed by Timoshenko. The error was hidden in the section of the formula that governs only thick walled cylinders as used in the North Sea gravity structures. The designer has to check his physical model carefully with respect to possible loadings and loadcases which are usually not governing the strength. The possible liquifaction of clay in the foundation strata of the North Sea gravity structures is such as case.

In designs representing large extrapolation of experience the speed of design and construction, usually beyond any routine of most people involved, offers another source of risks, especially in the area of construction and operations. Anyone familiar with offshore construction will know this.

The largest problem however, even when all parties concerned in a job agree that all phenomena to be looked into have been satisfactory analysed, is the judgement on the level of safety to be selected. This is the main and overruling information that codes offer for standard structures and not for anything beyond that scope as it is based on experience, accepted experience in fact. This problem area makes it valid that more effort should be spend on practical questions and the collection of field data for probabilistic analysis. Such a thing should simultaneously develop engineers' capabilities in measuring risks and people's judging ability for the safety level they require.

From the above it seems rather unattractive to proceed design developments in big steps. Only one thing, scarcity, in materials, in products, in anything people need or like, will be the overruling factor in taking the challenge of big short term developments. For the examples in this paper two cases were taken, it is the Eastern Scheldt flood barrier and the Dunlin Alpha platform.

The Eastern Scheldt flood barrier was originally designed just as the other "Delta-scheme" flood protection works, being protected earth fill dams. Such a work, although huge, would have been an acceptable extrapolation of current practice. The recent increase in consciousness about our environment, stressing the scarcity of areas like the Eastern Scheldt, being of vital importance for the marine life in a wide area, forced the Dutch government to change the dam into the an open floodbarrier allowing salt water tides into the reservoir.

The Dunlin Alpha production platform, is a product of the 1973-1974 offshore concrete gallore. Here even a double scarcity effect can be spotted. The energy crisis of 1973 caused an enormous amount of construction activities around the world. They can be mainly attributed to the construction of terminals for new distribution patterns of crude and to new developments of production sources. These activities in turn caused a scarcity in production capacity of structural steel work. Although many advantages of concrete production platforms are valid, such as their cheap storage capability, their ability to cut on major offshore installation and hook-up jobs, the scarcity in structural steel production capacity in 1973-1974 really was the cause for the break-through.

All examples given in the following paragraphs relate to those two works. Loadings were treated in the Dunlin platform case statistically, mainly because conventional deterministic analyses of the platform under construction were not able to provide reliable figures for the fatigue calculation.
For the Eastern Scheldt barrier, its only function being to provide safety on a more or less defined level, a probabilistic analysis was of course the closest answer on the functional requirements. The following chapters show some of the difficulties encountered.

3. EXAMPLES OF PROBABILISTIC EVALUATION OF LOADINGS.

   Eastern Scheldt. (Fig. 1)

   In the preliminary design stage of the storm surge barrier in the Eastern Scheldt largely deterministic assumptions have been made on the combination of hydraulic conditions that will exert the maximum force on the structure.

   The design storm surge level had to be in agreement with the report of the Delta Committee which stipulates that primary sea retaining structures have to provide full protection against storm surges with an excess frequency of $2,5 \times 10^{-4}$ per annum.
   This storm surge level was combined with a maximum extrapolated single wave and a very low estimate of the inside waterlevel. Below the design condition of the most heavily loaded pier is given.

	Roompot 15
storm surge level	NAP + 5.50 m
wave height	10 m
wave period	12 s
basin level	NAP - 1.70 m
total horizontal force	173 000 kN

   In the final design stage of the storm surge barrier however a probabilistic approach to the hydraulic loading conditions of the structure was developed. At first the advice of the Delta Committee had to be refined in the sense that the "potential threat" to the storm surge barrier was in fact not a storm surge level but a force with a frequency of exceedance of $2,5 \times 10^{-4}$ per annum.
   Therefore the probability density functions of the hydraulic boundary conditions had to be found.

The distribution of storm surge levels could be derived from available statistics of a long recording period. Historical data also formed the base for the probability density function of the basin level although in both cases a correction had to be made for the closure operation of the barrier.

The most difficult problem was the probabilistic description of the various seastates as a function of the storm surge level, because only 10 years of storm wave data were recorded. By formulating a mathematical model that could hindcast the recorded combinations of wave spectra and storm surges, the conditional dimensional probability function of wave spectra and storm surge levels was extrapolated to extreme values. The distributions of the so derived hydraulic parameters and the horizontal force with an exceedance frequency of $2,5 \times 10^{-4}$ per annum are given below.

storm surge level	$\Pr(\underline{HW} > HW) = \exp(-2.3 \frac{HW - 2.94}{0.696})$
wave spectra	$\Pr(S_{\eta\eta} \mid \underline{HW} = HW)$
basin level (normal d.)	$\mu = NAP - 0.25\ m \quad \sigma = 0.65\ m$
total horizontal force	100 000 kN

Compared to initial value of 173 000 kN this means a 42% reduction.

This result led to the decision to increase the spacing of the piers from 40 to 45 m thus reducing the number of piers from 72 to 66.

Another practical use of probabilistic considerations was made in the design of the accessbridge across the Eastern Scheldt to the building site. The bridge is supported by single tubular piles of 4.8 diameter, that have to withstand a once in a 500 years design storm (Fig. 2).

An analysis of the design loading combination showed, that an improbable coincidence of the extreme gust, the extreme wave and the extreme wave impact had been assumed.

$$F_{total} = F_{mean\ wind} + F_{gust} + 2F_{sign.wave} + F_{wave\ impact} + F_{current}$$

A more realistic approach takes into account not only the correlation of the mean wind velocity, the significant wave and the current velocity during storms, but also the stochastical independance of the gust velocity the maximum wave and the wave impact.

These considerations yielded the following result:

$$F_{total} = F_{mean\ wind} + F_{sign.wave} + F_{current} + \sqrt{F^2_{gust} + F^2_{sign.wave} + F^2_{wave\ impact}}$$

The result of this very simple probabilistic approach was a 14% reduction of the design bending moment in the piles.

It should be noted that the above given procedure is of course only acceptable when the wave force is essentially linear which was the case due to the large diameter of the piles.

Dunlin A. (Fig. 3)

For the fatigue life calculation of the Dunlin A platform a spectral type of analysis has been used, because the originally chosen deterministic approach has given very unsatisfactory results. The main reasons for this is the fact that the calculated lowest natural period of the platform was about 4. sec. which, due to an unfavourable phase relation of the forces acting on the four legs of the platform showed very high resonance response.

Using a deterministic type of fatigue life calculation, the environmental data consist in general of a long term wave height distribution (Fig. 4). The total number of waves then is split into a sufficient numer of wave blocks with each a specified height and a number of occurrence. The response of the structure now is calculated to one wave in each block, assuming a "realistic" wave period and this response is thought to be representative for all the waves in the block.

In the case of the Dunlin A platform it appeared that assuming all the waves of approx. 1.0 m height to have a period which is equal to the natural period of the structure would yield an absolutely insufficient fatigue life. On the other hand, if a slightly different period was chosen there was no fatigue problem at all.

A way to overcome this type of result is a frequency domain analysis (spectral analysis). Here the environmental data are specified in terms of seastates, characterized by a wave height and a wave period. A representation of such a seastate can be given by means of a sea spectrum which described how the wave energy is distributed over the frequency band. Using the normal spectral analysis techniques one can calculate the variance density spectra of peak-stresses resulting from the seastates considered. The direct advantage of this method is that during the course of the response calculation no information about the wave description is lost and that no fixed relations between wave height and wave period have to be assumed. Using the well accepted assumptions that the stress amplitudes in a seastate are distributed according to the Rayleigh distribution, which is fully described by the zeroth moment of the stress variance spectrum, the calculation of the fatigue life using the Palmgren-Miner rule was rather easy. (Ref. 8)

A sensitivity analysis afterwards showed that although also in the spectral analysis a fair amount of assumptions had to be made concerning damping, linearity, type of distribution etc., changes in these assumptions only had minor effects on the outcome of the analysis.

It should be noted that compared to a full probabilistic analysis only one part, the description of the waves has been treated in a statistical manner. All other parameters and algorithms had been treated as if they were fully deterministic with fixed values.

An attempt had been made to incorporate also the directional variation of the waves in the analysis using directional wave spectra. Although the mathematical tools are available to perform such kind of analyses it became clear that the attempt would not yield the expected results, because of lack of data. Therefore in our analyses the directional spreading of the waves has been assumed to be fully determined, all the waves during one seastate coming from just one direction.

It is typical for the application of new techniques in design practice that the new method of analysis requires a new description of the input data. In the above case of the fatigue calculation the deterministic description of the waves during the lifetime of the structure (Fig. 4) was no point of discussion.

For the spectral analysis the description of the input data in terms of seastates (which is the basis also for the deterministic description) already required lengthy discussions with all parties involved. Finally a directional spectral analysis could not be carried out because no reliable input data were available.

4. EXAMPLES OF PROBABILISTIC EVALUATION OF MATERIALS AND CONSTRUCTIONS.

The previous chapter shows examples of probabilistic approached loadings. The strength of a construction however has also a random character.

Calculations, wherein the strength and the loading parameters were specified as stochastic quantities, have been performed to bring the reliability of the most important parts of the Eastern Scheldt storm surge barrier on target level of at most $10^{-6} - 10^{-7}$ per annum. These calculations especially provided the soil mechanical experts with a tool to compare the foundation design in an objective way with the design of other components where safety factors are rooted in experience and prescribed by national codes.

The piers are placed on a foundation bed of compacted gravel with dense sand underneath and surrounded by a sill of coarse material. The construction is subjected to cyclic wave forces and a hydrostatic force (see Fig. 5). The resistance of the foundation consists of two parts:

- the passive earth pressure at the back of the pier.
- the sliding resistance at the base of the pier along two different slide planes:
    . a slip circle crossing the various layers of the subsoil (Brinch Hansen (B.H.) 1970)
    . a straight slide plane that develops in the boundary layer between concrete base and the gravel bed (Base Sliding (B.S.))

These two failure modes have been transformed in two reliability functions $\underline{Z}$. Subsequently the distributions of the parameters have been estimated from a fair amount of triaxial tests on the different materials and on the basis of gained experience.

Basic variables X_i	Dimension $[X_i]$	Mean values $\mu(X_i)$	Standard deviation (X_i)	Contribution X_i to the variance of Z ($\sigma^2(Z)$) in %	
				B.H.	B.S.
z_o	m	11.	0.25	.7	.7
α	degr.	14.	1.5	.5	.5
γ_s	MN/m³	10.	0.7	.7	.8
V_o	MN	202	5.1	.6	.5
$H_h + H_w$	MN	103.6	5.22	76.2	78.4
ϕ_s	degr.	36	1.8	1.3	1.5
δ_s	degr.	24	1.8	0.1	.1
ϕ_b	degr.	40	2	8.1	-
$\tan \delta_b$	-	0.67	0.05	-	8.0
X_m	-	1.0	0.06	11.9	9.4

According to the state of the art the problem was solved by an approximate full distribution approach resulting in the values given below.

	B. H.	B. S.
$\bar{Z}$ (MN) :	2.41×10^4	1.95×10^4
$\sigma (Z)$ (MN) :	$.42 \times 10^4$	$.40 \times 10^4$
β :	5.73	4.83
Failure probability :	5.3×10^{-9}	6.9×10^{-7}

From these results the conclusion could be drawn that the safety used in the foundation design was sufficiently large to reach the target risk level. One should of course make sure that the variance in foundation parameters of the actual prepared foundation bed should not be more than the values in the table.

The probabilistic reliability calculations provided an insight into the relative importance of the stochastic uncertainty of the basic parameters and formed an important tool in assigning priorities in further study and quality control requirements.

An example showing the difficulties that can be encountered when a product including the materials and the tolerances in dimensions have to be evaluated is given by the survey carried out on the inner partition walls of the Dunlin A platform (see Fig. 6).

Detailed analysis showed that these walls were far more critically loaded with regard to elastic stability than was originally assumed. As not only the concrete strength is important for such a loadcase, but the complete alignment a thorough, statistically evaluated survey was carried out.

- The thickness of the walls was measured:
    number of measurements 495
    average thickness 46.7 cm (design 45 cm)
    standard deviation .85 cm
    95% limit 45.3 cm

- The deflection of the walls was measured:
    number of measurements 286
    average deflection .24 cm
    standard deviation .3 cm
    95% limit .5 cm

- The deviation of centre lines of two walls in one plane:
    number of measurements 220
    average deviation .53 cm
    standard deviation .7 cm
    95% limit 1.2 cm

- The combined deviation of all four centre lines at one crossing:
    number of measurements 33
    standard deviation .7 cm
    95% limit 1.33 cm

- Cube strength 28 days 61.7 cm:
    average N/mm²
    standard deviation 4.35 N/mm²
    95% limit 55.0 N/mm²

- Cube strength 56 days:
    average 65 N/mm²
    standard deviation 3.6 N/mm²
    95% limit 59.1 N/mm²

- Cube strength after one year:
    average 75.2 N/mm²
    standard deviation 4.52 N/mm²
    95% limit 67.8 N/mm²

- Drilled cylinder cores 100 mm Ø:
    number 24
    average 56.7 N/mm²
    standard deviation 6.12 N/mm²
    95% limit 46.7 N/mm²

Although it was not pretended to do anything more than linking this information to the semi-probabilistic safety philosophy as laid down in the code (cp 110), an analysis of the information contained in the code was not at all sufficient.

The information about tolerances could of course be used in the structural analysis, using the characteristic 95% values of the dimensions given. The information could not be used to evaluate more accurately the concrete strength to be used in the calculations, as no relevant literature was known about the relation between actual dimension and theoretical dimension assumed in the code.

About the strength of the concrete in the structure related to the sample concrete cube tests quite satisfactory information did exist (Ref. 17). Although many unknowns did remain, such as the influence of the different curing conditions from the concrete in the work compared with the test cubes and the curing conditions of the concrete in the work as investigated, it could be proven that the concrete in the work was anyhow at least equal to the value assumed in the code and further that the 1 year strength of the concrete in the work was at least 20% higher than the 28 days' strength. This information, not contained in the code, was of major importance as the design load, applied in fact during immersion, would take place when the concrete was at least one year old.

The problem, most unaccessible to probabilistic analysis encountered in practice, was the influence of the durability, especially for concrete structures with a design life to last longer than the 35 years history of prestressed concrete construction.

The problem of probabilistic durability analysis for structures with a relatively short life (25-50 years) may be solved in the near future. Current investigations in many countries of concrete structures exposed to marine and other aggressive environment show very satisfactory results. Damage due to durability is observed only in cases that can be categorized as faulty details. This does in fact mean that the contractor's headache of concreting standards for present day offshore work, requiring low water cement ratios causing unworkable mixes and cement ratios causing unpumpable concrete that nearly boils and consequentially cracks during hardening, may be relaxed.

The last objection to stop the evaluation of such investigations to be used for probabilistic analysis is the phenomena of cracks. It is only very recent that from research in the U.K. it was concluded that cracks do not influence the long term corrosion behaviour of reinforcement in concrete.

This conclusion is of enormous importance for the durability analysis, as all efforts can now be concentrated on the case where seawater and consequentially chlorides enter the concrete through the surface.

For the evaluation of the Eastern Scheldt Barrier with a 200 years design life, going far beyond the experienced characteristics, very conservative design assumptions have been made. These included a ban on any tension, even temporarily. Besides that a check was made on the behaviour of the structure when all mild steel reinforcement and prestress with only low cover (up to 100 mm) was corroded away. This status still showed a satisfactory flood barrier function.

5. CONCLUSIONS.

The examples of the previous chapters may have shown that probabilistic analysis usually fails because no data from practice are available. It seems justified to guide the efforts put into further mathematical constructions, aiming to optimize not only on loads and structure but also on construction costs, into the collection of field data.

For students it would be a very useful experience to meet, observe and record field data. This would not only train them in thinking probabilistic, but also make them feel that practice is different from numerical models as used.

Where theory of probabilistic analysis claims also cost optimisation and consequentially assumes that statistical construction cost data are available for sophisticated type of jobs such as offshore work, it should be concluded that the assumption is wrong.
This again is a reason to activate statistical observations and analysis from current construction practice.

Apart from such observations, science and research should of course concentrate on any phenomena being important for the safety of a structure, where the mechanism causing an unsafe situation is not known or accessible for analysis.

6. REFERENCES.

1. Raymond E. Davis lecture (1978).
2. John F. Breen: "But where are all the cowboys...?" Concrete International, vol. 1, no. 3, pp 11-18 (March 1979).
3. A.W. Beeby: "Concrete in the oceans - Cracking and corrosion" (1978).
4. A.R. Flint and M.J. Baker: "Rationalisation of safety and serviceability factors in structural codes". Ciria report 63 (October 1977).
5. American Concrete Institute: "Corrosion of metals in concrete" (June 1975).
6. Cur VB: "Concrete and effluent". (Beton en afvalwater) Report no. 96. (April 1979).
7. StuPOC: "Safety consideration for offshore constructions". (Veiligheidsbeschouwing voor offshore constructies)
8. D. Zijp, B. v.d. Pot, Ch. Vos and M. Otto, Andoc: "Dynamic analysis of gravity type offshore platforms experience, development and practical application" (1976).
9. Cornell, CA: "Structural safety specifications based on second-moment reliability analysis". Final report of the IABSE symposium on concepts of safety of structures and methods of design (London 1969).
10. Rackwitz, R.: "Principles and methods for a practical probabilistic approach to structural safety". Sub-committee for first order reliability concepts for design codes of the joint CEB-CECM-CIB-FIP-IABSE committee on structural safety (December 1975).

11. J.R. Morrison et al.: "The force exerted by surface waves on piles". Petroleum transactions, 189, TP 2846 (1950).
12. J. Brinch Hansen: "A revised and extended formula for bearing capacity". Danish Geotechnical Inst. Bull. 28 (1970).
13. H. Sellmeyer: "Description spring constant method". Proc. Symp. on foundation aspects of coastal structures (Delft 1978).
14. T.W. Lamb and R.V. Whitman: "Soil mechanics". Chapter 2. J. Wiley & Sons (New York 1969).
15. C.H. Buschmann: "Maximum imaginable accident". (Maximum voorstelbaar ongeval) Ingenieur no. 44 (1975).
16. "Preliminary probabilistic calculations". (Inleidende probabilistische berekeningen). Report TNO-IBBC, B-78-30 (March 1978).
17. N. Petersons: "Recommendations for estimation of quality of concrete in finished structures". Bulletin Rilem, no. 24 (November - December 1971).

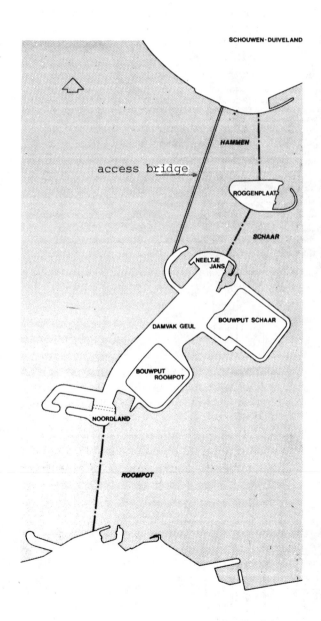

Fig. 1.

Eastern Scheldt
Storm Surge Barrier

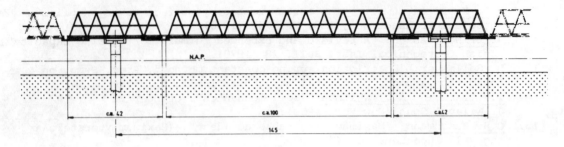

Fig. 2.   Access Bridge

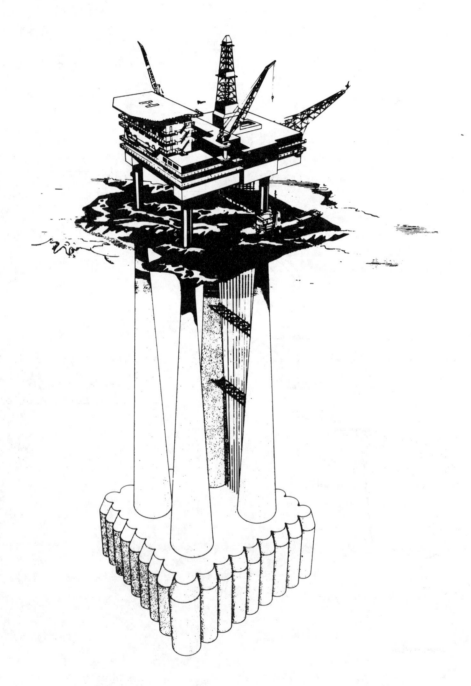

Fig. 3.   Dunlin A platform.

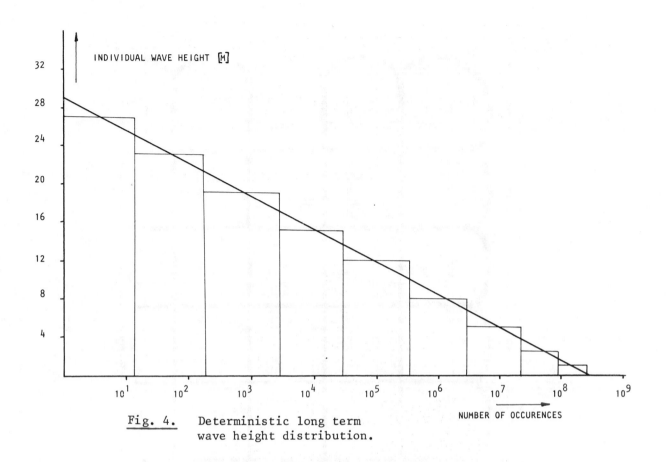

Fig. 4. Deterministic long term wave height distribution.

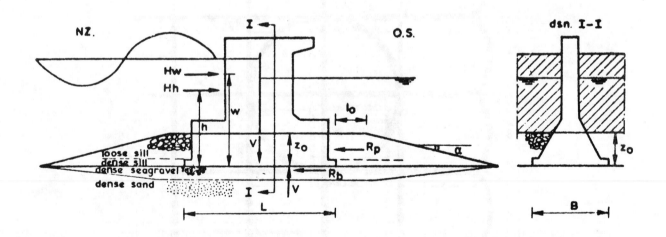

geometry: L = pier length (50 m)
B = pier width (25 m)
Z_o = embedment (11 m)
1° = sill bank (10 m)
α^o = inclination (1:4)

load: Hw, w = wave force-arm
Hh, h = headloss force-arm
V = effective weight pier
R_p = passive resistance
R_b = horizontal base resistance.

Fig. 5. Pier of storm surge barrier.

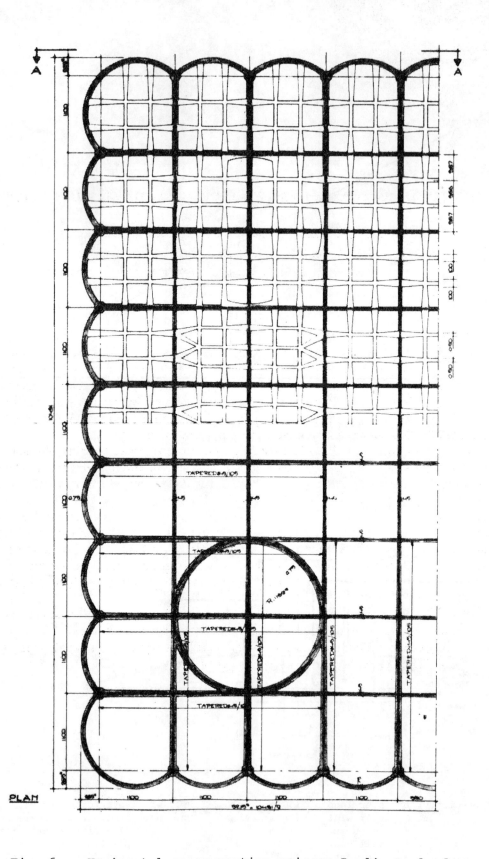

Fig. 6. Horizontal cross section caisson Dunlin A platform.

PAPER 89

Second International Conference
on Behaviour of Off-Shore Structures

Held at: Imperial College, London, England
28 to 31 August 1979

ADJUSTMENT OF STRUCTURAL CONCRETE TECHNIQUES TO OFFSHORE CONDITIONS

A.J. Harris, CBE, BSc(Eng), FICE, FIStructE, MConsE.
and B.M. Fox, MIE(Aust), MIStructE.

Harris & Sutherland, U.K.

Summary

Offshore construction in concrete has been based on building codes and regulations in which it is implicitly assumed that: (a) loads act in a single direction and (b) fatigue is not a problem. Neither of these assumptions is true in typical offshore structures; moreover, in classic engineering structures there is little economy in reducing weight whereas offshore structural weight is nearly always an expensive handicap. The consequences for reinforced concrete and prestressed concrete are examined and found to be:-

- in reinforced concrete, the reinforcement becomes the load bearing element; the concrete serves as a matrix, bonding and locating the reinforcement. A structure consisting of closely packed bars with a minimum of concrete is attractive and feasible.
- in prestressed concrete straightforward application of the codes gives about double the cross-sectional area of concrete with reversible loading than with a single direction of load. On a rupture basis, however, it is found that the same section can serve for both conditions. Given attention to the fatigue aspect, a section can thus be designed to support reversible loads of approximately half the weight of that required by the code.
- both reinforced and prestressed resist fatigue well; prestressed concrete can be so designed as to have unique characteristics of resistance to failure.
- some ignorance remains concerning behaviour under low-cycle repetitions of loads just smaller than rupture in prestressed concrete.

Sponsored by: Delft University of Technology, The Netherlands
Massachusetts Institute of Technology, U.S.A.
The Norwegian Institute of Technology, Norway
University of London, England

Secretariat provided by: BHRA Fluid Engineering

Copyright: ©   BHRA Fluid Engineering
Cranfield, Bedford, England

Introduction

The design of concrete offshore structures is for the moment governed by the phrase in "Offshore Structures" (1) "the relevant British Standard Codes of Practice and Specifications for concrete and steel structures may be used with appropriate engineering judgement".

Now there are novelties in offshore structures not present in the classic structures for which these codes and specifications were drawn up. Here are three:-

- loads are reversible in direction.

- loads are repeated so often as to make fatigue a real rather than a notional problem.

- structural weight is often an economic penalty.

This paper argues that these novelties are so revolutionary in their effect on reinforced and prestressed concrete structures that it is not enough merely to modify the codes at the promptings of engineering judgement but that new thinking is needed.

To this end:

- the novelties will be described

- their effect on classic R.C. and P.S.C. structures will be examined and changes in such structures will be suggested

- the advantages and problems of such changes will be sketched out.

Novelties in offshore structures

Reversibility of load

In classic structures, the predominant loads are caused by gravity; the idea that loads act in one direction is deeply rooted in design thinking.

In fact, of course, there are zones of continuous beams in which live load can cause reversals of bending moment but these reversals are superimposed on a permanent load and are rarely critical. Wind loads can also act in any direction, but are of primary importance only in tall slender structures such as chimneys and masts, where an exaggeration of mass can solve many problems without economic handicap. The only structures which spring readily to mind in which the major load is completely reversible are power-line pylons.

In offshore structures, however, whether fixed or floating, important loads are reversible. Tidal currents act in two directions; wind and wave can act from any direction; loads caused by hydrostatic head are often reversible according to whether e.g. compartments are empty and immersed or full and emergent or whether a hull is on a crest or in a trough. Permanent loads also act, of course, but are usually secondary compared to the reversible loads.

Repeated loads

Except in such unusual structures as railway sleepers or small span railway bridges, live load in classic structures is not often much larger than permanent load and, moreover, it is not often repeated. Failures due to fatigue are unusual.

Offshore, however, it is not unusual to design for 10^8 repetitions of a significant load; it is imperative to design for fatigue effects; failures caused by fatigue are all too common. (Structural Engineers will note that were the design vehicle to pass over a bridge once every 10 secs. day in day out it would take some 30 years to attain 10^8 loadings. In practice the design vehicle is a relative rarity).

Structural weight

The commanding factor in the design of all structures is overall cost, in considering which there is always a trade-off between rival aspects. Weight and cost are two such; a point is reached in design at which the cost of labour in manufacture needed further to reduce weight exceeds the economy obtained. In classic structures this economy seldom goes further than the actual saving of material; thus the use of very high strength concrete has seldom been found worthwhile since the saving in quantities of material has not offset the extra costs of manufacture and there have been no economies in function resulting from lighter weight. Savings in the foundations of multi-storey buildings can arise from savings in structural weight, but these savings do not go far.

Offshore the case is otherwise. Two typical cases may be quoted:

- ships, where structural weight adds to the cost of operation. (At the limit, the ship will not float).

- platforms, where structural weight adds both directly and indirectly to the cost and complication of manufacture and towing out.

It is quite usual for structural weight to be a major consideration in the design of offshore structures; in classic structures it is rare.

Consequences in Reinforced Concrete Structures

With unidirectional loading, steel is provided in the tensile flange to carry the tensile forces; if the loading is now reversed in direction, an equal amount of steel must be put in the other flange and both flanges now have enough steel to carry the tensile forces. But that steel will also be enough to carry the compressive forces so what is all the concrete for?

Codes at present require that the proportion of cross-sectional area of reinforcement to that of concrete should not exceed a certain limit, a requirement which leads to an area of concrete which is useless in tension and superfluous in compression. The weight of this concrete is a severe handicap in off-shore structures.

Now the use of lightly reinforced concrete to constitute a flange carrying compression is relatively new. The use of R.C. in structures became common in the second half of the 19th century following the famous rowing boats of LAMBOT in about 1850; these structures were typically cages of bar and wire bonded together by a fine-aggregate concrete matrix. That the steel was seen as the principal load bearing element is indicated by the fact that not until 1905, when Freyssinet built shallow arch bridges of very lightly reinforced concrete, was the phenomenon of creep revealed - hitherto, the deflections of the structure had been governed by the strains of the steel.

The concept of the concrete-bonded steel bar structure never wholly disappeared but was revivified in the form of Nervi's Ferrocement in which form it is widely used for minor structures - boat hulls, small reservoirs, shell roofs, permanent formwork. It is suggested that the idea is capable of extension to very large structures.

Characteristics of Ferrocement in large structures

The material would consist of closely packed deformed bars of large diameter bonded together by injected grout and secondary reinforcement. Such a material has much in common with fibre-reinforced plastics; the scale is such that the layout and orientation of the "fibres" can be optimised, with no more of them than is necessary at any given location. A percentage by volume of reinforcement of about 60% might well be attainable.

The fatigue strength of such a material is that of the individual bars or wires.

Cracks do not readily propagate, given that adequate secondary reinforcement be provided to prevent splitting along the line of primary reinforcement; it is not notch sensitive. Secondary reinforcement can be dimensioned to prevent buckling of bars. Behaviour at rupture is attractive, with extensive absorption of energy both in plastic strain of reinforcement and in bond slip.

Weight/strength ratios are good. Whilst two dimensional stress will generally need two fields of bars, thereby being at a disadvantage relative to steel plate, the reinforcement sections can more readily be varied. There is a particular advantage - the ability to build up very thick load bearing elements; the geometry of a ferrocement structure will have little resemblance to that of a classic steel structure. Finally, it is possible to use high tensile steel wires instead of reinforcing bars with an increase of useful tensile strength by a factor of 2 to 3.

Problems of Ferrocement in large structures

Secondary reinforcement has been mentioned; a criterion is needed for e.g. binding reinforcement in a direction perpendicular to the plane of a membrane subject to two dimensional stress. Such a criterion is available in the shear/ friction concept which may, perhaps, need confirmation unless broad safety factors are employed. Secondary reinforcement must also be provided to prevent buckling of main reinforcing bars acting in compression.

Such secondary reinforcement will clearly provide supplement to bond. Bond itself is favoured by the use of deformed bars as is the penetration of grout. Closely packed bars or wires, moreover, are known to modify the fracture behaviour of the cement matrix; one may expect plastic strain of the matrix in tension followed, presumably, by a permanent compression on removal of tensile load.

What of cover? Why have cover? A structure with exposed reinforcement is just as good as a steel structure - and steel structures appear to be satisfactory. Cover considered as a coat of paint is expensive, both in itself and in its consequences when weight costs money. The danger is less that of the corrosion of the outside layers of a thick member than of the penetration of corrosive fluids through cracks in the matrix leading to a systematic reduction of reinforcement along a given plane. Much is known of the relation between corrosion and crack width - enough, perhaps, to ensure safety. It may also be practical to use thin steel sheet as a permanent former and sacrificial protection; grouts containing fine iron powder are also known - as water penetrates, the powder corrodes and expands blocking the path to further penetration of water. This expedient is not found economically worth while in classic structures; it may well be otherwise offshore.

Consequences in Prestressed Concrete Structures

Whilst in reinforced concrete subjected to reversed load the concrete comes to play a secondary role, in prestressed concrete it continues to be a major load bearing element.

It is traditional wisdom that reversible loading is a circumstance unfavourable to the use of P.S.C. Indeed, if a section be designed on an elastic stress basis, it is the <u>range</u> of load which counts and to design for a load acting in two opposite directions instead of one is to double the section. But the major design criterion these days is the value of the rupture load and it can be so arranged that a tendon lying in the compression zone of a P.S.C. beam has its tension completely relieved prior to compressive failure of the concrete. It follows that if a member is prestressed to carry a tensile load, its rupture load in compression is not reduced; hence a symmetrical section designed to have a certain bending moment at rupture in one direction can be given an extra prestress to give as well the same strength at rupture in the other direction <u>without extra concrete.</u>

When designing for a rupture criterion, the behaviour sketched out above permits of halving both the area of a section designed for tension and the modulus of a section designed for bending under the criterion of elastic stress. There might even be some saving in prestressing steel; with a single direction of load, the eccentricity of the resultant prestressing force must lie within the kern of the section and the eccentricity of the tendons is limited, whereas with reversible load and a uniform prestress, the tendons may be placed in two groups each at maximum practical eccentricity.

Problems of concrete structures uniformly prestressed

Strain at rupture in compression

It is necessary for the strain at rupture of the concrete in compression to be large enough to relieve the prestress in the tendons in the rupture zone. This will normally require a very good bond between tendon and concrete so that the effective length over which the stress is relieved is as short as possible. It will also be desirable to provide some secondary transverse reinforcement whereby the strain at rupture of the concrete can be both substantially increased and made more reliable.

Fatigue of prestressed concrete

This has two aspects - fatigue of the steel and fatigue of the concrete.

Fatigue of the steel tendons in tension only becomes a danger with loads greater than the cracking load; at loads lower than this, the variation in stress in the steel as the load varies from positive to negative is too small to cause fatigue. After cracking, the stress in the steel increases substantially, increase of load causing corresponding decrease in fatigue life. As far as fatigue of the steel tendons is concerned, the cracking load is thus crucial; below this, there is no danger of fatigue whilst above it the danger begins to be real.

Performance in fatigue - concrete

The successive stages in the behaviour of concrete as a compressive load is increased until rupture occurs, are:

- Stable micro-cracking (0–$0.55\,f_c$, where f_c is the crushing strength). With deformation of the cementitious matrix around the more rigid particles of aggregate, microscopic cracks occur in the matrix. With the passage of time, whether the load be removed or maintained, these micro-cracks will seal themselves by re-establishment of the cementitious bond, a process encouraged by the presence of moisture. The material is stable under sustained load within this range and strains are quasi-elastic.

- Stable cracking (0.55–$0.80\,f_c$). The micro cracks link up to form larger cracks; the material is still stable under sustained load but appreciable non-linear and irreversible strains occur. Again cracks will seal themselves after some lapse of time with or without removal of load.

- Unstable cracking (0.80–$1.00\,f_c$). The large cracks link up to form planes along which sliding takes place. The material is unstable, sustained load will lead to rupture. Strains are large and dis-aggregation of particles causes increase in volume. There is still some ability for cracks to seal up with time after release of load, particularly in the presence of moisture.

In the range of stable micro-cracking (0–$0.55\,f_c$) testing cycles (rarely carried beyond 10^8) do not produce fatigue failure; one effect which has been noted, however, is an appreciable increase in strength. With loading greater than $0.55\,f_c$ fatigue failure may occur under a progressively smaller number of repetitions the higher the load. When fatigue failure occurs it is similar to static compression failure and is accompanied by large strains; fatigue failure of concrete in compression is thus not brittle but plastic and local concentrations of compressive stress under repeated loading will be relieved and redistributed prior to rupture.

This picture is modified by the following factors:

- fatigue testing of concrete is usually carried out under high frequency load but periods of rest or of rest combined with sustained moderate load between cycles of repeated load are found to increase fatigue strength, due perhaps to autogenous healing of the load-induced cracks. These conditions are closer to the conditions of service of e.g. a ship than a continuous high frequency repetition of load until rupture occurs;

- a gradient of stress across a concrete element (as is typical of the compression flange of a beam) also increases fatigue strength, due perhaps to the redistribution of stress caused by the large strains prior to rupture noted above;

- there is evidence to show that long period repetitions cause more damage than short.

It may be concluded that concrete stressed in the range 0-0.55 f_c runs no risk of fatigue failure; the strains may cause extra relaxation of the prestress and thus lowering of the cracking load.

In the range 0.55-0.80 f_c, danger of fatigue failure exists and strains will be substantial.

In the range 0.80-1.00 f_c, permanent damage is caused to the concrete and loading if repeated or maintained with cause failure. Few tests have been done on P.S.C. structures subjected to repeated loads just under the static rupture load.

A section may be postulated in which the permanent prestress is 0.3 f_c and a load adding a stress of approx. 0.4 f_c would cause cracking. A very large number of repetitions of a load causing a stress of 0.3 f_c would probably reduce the prestress enough to cause eventual cracking under this load; the section would still run no danger of fatigue failure under this load. The ratio between this load and the rupture load would depend on the geometry of the section and on the initial prestress in the steel but it could well be of the order of 0.5 x rupture load.

In an environment where fatigue risk is very real, this behaviour is attractive.

Codes and Regulations

Codes and regulations are implicitly based on uni-directional loading; the application of their provisions to R.C. and P.S.C. structures submitted to reversed loadings leads to substantial extra weight. The variants on the standard materials suggested above are sound and are worthy of the attention of engineers.

In conclusion, it is intriguing to reflect that the first industrial P.S.C. structures were Freyssinet's hollow powerline pylons of 1931 made of 110 N/mm^2 concrete with a uniform prestress of 30 N/mm^2 and successfully tested under "several million alternations" of the design load. They were installed on the coast of the Biscay and many are believed to be still in service despite a severe exposure (salt spray in winter, hot sun in summer) and a cover to the tendons of only 6 mm.

REFERENCES

1. Offshore Structures; Institution of Structural Engineers, July 1976.

PAPER 90

BOSS'79

Second International Conference
on Behaviour of Off-Shore Structures

Held at: Imperial College, London, England
28 to 31 August 1979

SEABED CONTAINMENT STRUCTURES FOR HYDROCARBON PRODUCTION

J.A. Derrington, FEng, M.J. Collard, MICE and J.M. Skillman, BSc(Eng).

Sir Robert McAlpine & Sons Limited, U.K.

Summary

There is increasing interest in the use of seabed installations for the commercial production of hydrocarbons offshore. The use of large one atmosphere dry containments, serviced by submersibles, is an attractive solution to these methods of recovery, but poses problems of installation, operation and maintenance. This paper discusses the main aspects of the structutal design of such containment vessels and concludes that the solutions using structures of high grade concrete or composite structures of steel and concrete are acceptable. No major advances of existing technology are required for these solutions.

Sponsored by: Delft University of Technology, The Netherlands
Massachusetts Institute of Technology, U.S.A.
The Norwegian Institute of Technology, Norway
University of London, England

Secretariat provided by: BHRA Fluid Engineering

Copyright: © BHRA Fluid Engineering
Cranfield, Bedford, England

1. INTRODUCTION

1.1 Background

As the oil and gas deposits in waters up to 150m depth become depleted, the oil industry seeks methods of exploiting the hydrocarbons in the deeper waters of the continental margins for which novel methods of exploration and production are necessary. Since the first offshore oil was produced in shallow water, the industry has progressed into deeper water with an advancing technology, until today exploration is possible in 1500 metres of water, and commercial production has been achieved at depths of 300 metres (Fig.1). The containment structures discussed in this paper have been developed to provide a means of producing oil in waters up to 1000m deep by locating the process equipment on the sea bed within a one atmosphere habitable environment. These structures may also serve as a basis for other deep sea scientific and technological applications.

At the present time oil and natural gas supply over 60% of the world's energy and in spite of the four-fold increase in oil prices during 1972/74 this proportion continues to increase. Many observers predict demand exceeding supply before the turn of the century and it is probable that prices will eventually reach a level at which deep water production becomes economical.

The size of the offshore hydrocarbon reserves is still uncertain. Any attempt to predict the time scale and magnitude of recovery must be treated with caution due to uncertainty both of the amount of hydrocarbons to be discovered and of the industry's ability to recover them at economic rates.

Estimates of reserves vary wildly as do the many criteria on which they are based. In 1977 the World Energy Conference attempted to quantify the future supplies of oil from conventional sources and also from deep offshore and polar regions. It is thought probable that 45% of the world's remaining recoverable oil is offshore in water less than 200 metres deep with a further 8% in deeper waters. Fig. 2 illustrates the possible distribution of reserves and is based on the assumption that it will be possible by the year 2000 to produce within a wellhead cost of 20 dollars per barrel at 1976 values.

Until further exploration drilling and survey work is completed, and the many economic and political factors are resolved, it will not be possible to predict accurately the size of the potential requirement for deepwater production hardware. Nevertheless it is certain that existing methods of production are inadequate and, consequently, new methods must be developed now.

In water depths of 300 metres and more two basic options exist : to house the production facilities on a floating structure or, alternatively, to house them at seabed level. It is this second option that is considered here and, more specifically, the problems associated with the provision of a containment structure that will maintain a dry one atmosphere environment on the seabed in which to house the production equipment. Thus, the effects of the severe surface environment are minimized, and the problem of supporting large payloads at the surface and of restricting motion to accommodate high pressure risers is eliminated.

1.2 Seabed Production Concept

Fig. 3 shows a typical seabed production complex which provides for separation, gas re-injection and water injection. Where the reserves are large enough and sufficiently close to land to justify oil export by pipeline, the tanker loading facility shown in the diagram would be unnecessary. However, it is envisaged that most deep water locations will require offshore loading facilities which can also house power generation equipment and conduct an umbilical to the seabed complex for the supply of power, air and an inert gas.

The seabed production design is based upon the use of production equipment proven on surface installations, as the development of new plant of considerably reduced size, though possible, would require a long and costly programme of development and service life. Seabed containment, therefore, must be of adequate size for installation, operation and manned intervention for maintenance and repair, depending upon their overall function.

For typical large scale production facilities a number of modules, perhaps 12m diameter and 60 metres long, would be required, whereas for single wellhead chambers, spheres about 3 metres diameter have already been used.

The fundamental requirement of the containment is to maintain a dry one atmosphere environment in which the production facilities and personnel may operate normally in warm, comfortable and well lit conditions. Mechanical and electrical plant will be selected from considerations of reliability and major replacements will be possible. Access from the surface will be by submersible which will transfer men and materials into the production complex in dry atmospheric conditions, dependent on requirements in service.

For a new system, such as that proposed, to encourage suitable personnel to work on the seabed standards of safety must be at least as high as those currently accepted in conventional industrial situations.

2. EARLY USE OF EXTERNALLY PRESSURISED VESSELS

Interest in the potentialities of the submerged pressure resistant hull developed from experiments involving wooden diving bells. In 1776 the self propelled wooden submersible Turtle became the first vessel to make a submarine attack on a man-of-war. It was powered by a hand driven propeller and was capable of diving to 7 metres for 30 minutes and of travelling at 3 knots. (Ref.1).

With the development of the combustion engine into a practical device at the end of the nineteenth century, came the first prototypes of submersibles as we know them today. The submarine Holland in the year 1900 was the first to be commissioned by the United States Navy and its steel hull was powered by a gasoline engine and batteries. It was capable of diving to a depth of 20 metres.

Steel hulls have not changed significantly over the years; present day nuclear submarine hulls are about 10 metres diameter and built to withstand external pressures of hundreds of metres of water. The demand for deeper dives has prompted the search for lighter materials and today Aluminium, Titanium, glass reinforced plastics and glass are being considered in submersible design, while descents up to 4000 metres have been achieved.

The oil industry has long used submersibles in offshore production and externally pressurised structural members are essential to the design of offshore production platforms which need buoyancy during their transport and installation phases. Both steel and concrete have been used and the new generation of concrete platforms contain pressure resistant columns which allow dry one atmosphere access to the seabed for inspection and maintenance of equipment. (Ref. 2).

Work undertaken by the Civil Engineering Laboratory Naval Construction Battalion in California has done much to establish the credibility of concrete as a material for use in very deep waters. Since 1971 the Laboratory has tested a number of spherical and cylindrical concrete vessels in water depths up to 1500 metres and proved that high quality concrete is an ideal material for deep ocean construction. (Ref. 3).

3. HYDROCARBON PRODUCTION SYSTEMS

3.1 Construction & Installation

A typical application is for a seabed containment structure for production from a medium sized oil field. It comprises separation, gas compression, water injection, controls and habitat areas. The volume of such a containment is about 40,000 cubic metres subdivided into four or five interlinked cylindrical modules fixed either within a rigid structural frame or to a common foundation, typically 80 to 100 metres square. This structure would be built within a shallow dry basin and all production equipment would be installed and tested prior to launch. The basin would then be flooded and the structure floated out, towed to site, then lowered to the seabed and fully ballasted. The submerged weight, predetermined by the design, will either be small, providing temporary stability while the foundation is piled or, alternatively, will be sufficient to ensure stability by gravity. (Fig. 4).

It is not essential to the concept for the structure to float at the construction stage since auxiliary buoyancy can be provided for the delivery stage. Also, systems have been designed in which a negatively buoyant structure is carried by flotation craft, usually a semi-submersible, which then lowers the structure to the seabed. This method is attractive for small structures but is hardly practicable for the large structures considered in this paper. Draught and space requirements are great and site availability is thereby restricted.

The design of the pressure hulls, foundation design, mode of installation, buoyancy requirements during delivery and installation, draught requirements and location of construction site are mutually interacting considerations. An optimised design depends not only upon the operating depth and location of the delivery site but also upon the stage of development and lifting capacity of specialised service vessels and equipment. Typically, the hulls will be of steel or concrete or composite construction designed to be nearly neutrally buoyant with installed payload. These rest upon a raft foundation having sufficient stiffness to limit deformations due to sea loads during tow and also when installed on the sea bed, assuming the most unfavourable mode of support.

3.2 Buoyancy

The raft will be of tubular steel or of cellular concrete construction having positive or negative buoyancy designed to control the total structure within prescribed limits. These are governed by requirements of draught at launch, freeboard during tow and submerged weight when all cells are water ballasted on the seabed. For a piled foundation only sufficient weight is required to provide stability during the relatively short period of pile driving. For a gravity design the submerged weight must be sufficient to stabilise the structure permanently against the maximum credible environmental forces on the seabed during its life. If additional weight is required, this is most economically provided by additional concrete and which must be compensated by additional air volume to maintain initial freeboard. In practice this implies deepening the raft and thereby increasing draught.

As an alternative to integral buoyancy, supplementary buoyancy may be provided by attaching flotation devices externally to the structure. Since these make no contribution to structural strength and are difficult to recover or refix in the event of retrieval of the structure, this method is less economical.

The buoyancy cells are designed to permit selective flooding so that controlled buoyancy may be achieved at all stages during the descent. There are two possible solutions to the problem of installation. The first method is to ballast the structure to a small positive buoyancy, then pull it down to the seabed by haulage lines from two or more crane barges, passing round pulley blocks fixed to anchorage points previously installed at the sea bed. In the second method the structure is made slightly negatively buoyant at which state it is then lowered to the seabed. (Fig. 5). Each method has advantages and disadvantages, though the negative buoyancy method is more suitable for the large structures under consideration. The choice of method in a particular situation will centre upon an evaluation of the critical hazards at each stage of the operation in relation to the functioning of the control system and weather considerations.

Whichever method is selected the overall response of the control system must ensure that the position and attitude of the structure is maintained within the tolerances demanded by the structural design and the installation programme. Sufficient backup and equipment redundancy must be provided to ensure reliability throughout the installation phase which, in an emergency, must be reversible.

During installation the force in each haulage line will consist of two components : a finely controllable steady force determined by ballasting and an oscillatory component due to action by waves and currents. To avoid cable snatch the steady force must be adjusted to exceed the amplitude of the oscillatory force which, at the surface, is due to relative motion between the structure and the crane vessel. The resultant cable force in relation to the capacity of the crane vessel will determine the weather tolerance of the operation. During descent the structure will be subject to lateral drift forces and oscillatory motions due to currents and a variable vertical force due to vessel motion and cable elasticity. As the hydrostatic pressure increases with depth the structure loses buoyancy due to elastic compression. It is anticipated that a period of about 8 hours will be required to place the structure on the sea bed at 500m depth.

The capacity of the buoyancy cells to withstand pressure differential is limited by the strength-density relationship of the material. Since, for any material, to increase pressure resistance is to reduce buoyancy, at extended depths it is necessary to supply compressed air to limit pressure differential.

The choice between a fully pressure resistant "hard" buoyancy solution or a "soft" buoyancy solution will depend upon a cost comparison which takes account of construction and plant costs including reliability implications. From the viewpoint of marine operations hard buoyancy is clearly preferable since the compressed air requirements required only for control purposes are much smaller. A "soft" design requires a carefully controlled compressed air supply to maintain a balanced pressure during descent and a more elaborate system of safeguards to achieve comparable reliability. An optimum design will normally lie between these extremes, the buoyancy tanks being designed to withstand a pressure differential equivalent to perhaps 2/3 of the maximum depth. This will allow a substantial operational tolerance in the supply of compressed air.

4. STRUCTURAL DESIGN OF CONTAINMENT VESSEL

4.1 Choice of Material

The material for the seabed containment structure has strength/density sensitivity without parallel in terrestrial structures and only to be found elsewhere in the aerospace field. The total structure must have a certain minimum buoyancy during delivery. This is then ballasted to provide negative buoyancy after installation. For a given material the required wall thickness and consequent weight of the pressure hulls will be dictated by the operating depth for which there is a particular strength/density relationship which optimises draught at launch, ballast requirements and seabed weight.

Choice of materials for a large structure anchored continuously on the sea bed throughout its working life differ in several respects from the case of the submarine and the submersible vehicle. The latter both require an excess buoyancy capability subject to sensitive control in order to operate. With its relatively small size and compact, generally spherical, geometry the designer of submersibles is able to take advantage of the newly developing range of non ferrous metals and lightweight composite glass and ceramic materials. (Ref. 4). While these may achieve strength/density ratios many times those of ferrous materials, present fabrication techniques and service experience are too limited for these materials to be considered seriously for the large structures required for production equipment. Steel, of progressively higher specification, continues to be the principal material for submarine hull construction.

Whereas the design philosophy of mobile submarine craft is rooted in traditional naval design practice, the need for regular dry docking and inspection raises problems for the sea bed production facility which must remain fixed to the bed of the sea throughout its working life. In all but the simplest geometrical configurations external inspection of the hull will be difficult and it may well be necessary to accept that some areas cannot be inspected. This problem is common to all marine structures and indicates the importance of durability in material selection.

4.2 Role of Steel and Concrete

The seabed habitat concept has many novel features and so the first structures should be based firmly in existing technology exploiting to the full known techniques. Initially, at least, the main structural materials will be steel and high strength concrete whose long term properties are well understood. For water depths to about 500 metres reinforced concrete is preferred not only for reasons of economy and durability, but also because the thick walls of a concrete vessel provide superior thermal insulation and resistance to impact loads from dropped objects. The judicious use of steel and concrete in composite construction, of which conventional reinforced concrete is a particular form, provides a satisfactory material for a wide range of operating depths. (Fig. 6). Since steel has a strength/density ratio about twice that of concrete, the proportion of steel within the hull wall will increase with depth tending eventually to an all-steel hull. The limiting factors at depths beyond about 1000 metres are the techniques of rolling and welding very thick high tensile steel plate.

Taking into account the advanced technology of naval submarine construction, the stiffened steel shell continues to suffer from the disadvantage of acute sensitivity to construction imperfections. It is at this point that the field of concrete technology may be able to offer solutions using techniques outside the bounds of normal practice. The in-situ strength of concrete can be increased considerably by subjecting it to a state of triaxial compression, but this would require the wall to be prestressed radially to induce a compressive stress along the minor principal axis. A possible design incorporating this principle might take the form of a sandwich hull in which the steel skins serve to distribute the radial prestress forces.

Alternatively, a high reinforcement steel ratio, intermediate between conventional reinforced concrete and traditional steel, may be used. By closely packing bars, steel areas of 50% or more may be obtained. This method, employing low cost steel, appears to offer the designer a controllable strength/density ratio and the capability of flexible geometric modelling akin to normal concrete practice. (Ref. 5).

4.3 Concrete Material Aspects

In addition to resisting the hydrostatic pressures causing implosion, the containment vessel must, above all, keep out the water. If massive prefabricated concrete components are to be used, their assembly and jointing requires special attention, and the high standards of workmanship necessary must be effectively monitored and guaranteed. Some use of precasting techniques in marine structures has already been made and the techniques for sealing joints between precast units has been established. It is normally preferable to place the concrete in-situ and its properties are therefore sensitive to a range of influences from the sources of the component materials through the mixing, transporting and placing techniques, and are dependent on the size of structural member and the ambient condition of the environment. Major advances in concrete technology developed within the construction industry during recent years and experience gained during the construction of gravity platforms enable quality variations to be reduced to a very low level. The elimination of potential leakage paths remains of major importance for the designer of concrete habitat structures.

The very high strength required for structural reasons (80 N/mm^2 or more characteristic cube strength) necessitates a high cement content which will of itself ensure high durability. The excellent performance and durability of concrete marine structures subjected to extremes of weathering, abrasion and temperature change is well documented. (Ref. 6). Those extreme environmental effects which are accentuated at the air-sea interface are virtually absent for the sea bed structure for which the ambient conditions are characterised by a constancy which increases with depth.

Water can penetrate concrete in two ways viz. inter porous percolation or through cracks. Both the cement paste and the aggregate are porous. Since the cement paste constitutes the continuous phase its permeability controls permeation through the whole. It has been demonstrated that the hardened cement paste in very high strength concrete having water/cement ratio of about 0.4 can have a coefficient of permeability of the order of 10^{-13} to 10^{-12} m/s. (Ref. 7). This tends to decrease further with time, for high strength concrete is inherently cement-rich, and the long period of hydration will steadily reduce the coefficient of permeability. Seepage due to permeation will therefore be minimal, arriving as water vapour and extracted by humidity control.

The most likely location of leakage paths are at structural discontinuities caused by penetrations. Local stresses due to penetrations can be controlled by careful attention to detailing of reinforcement aimed at matching the deformation properties of the inclusion to the concrete it replaces. Since the primary forces within the hull are compressive and the high strength concrete rich in cement, any cracks already present will tend to close and the design must ensure that local discontinuity forces and long term deformations do not induce an unstable stress state.

While permeation of concrete by water has an obvious bearing on serviceability, possible effects on material strength should also be considered. Concrete has a reduced uniaxial strength when saturated than when dry, a fact usually attributed to pore pressure. It has been suggested that this effect could lead to some reduction in the strength of saturated concrete exposed to hydrostatic pressure in deep water. (Ref. 8). While it is uncertain what degree of saturation of thick members of high grade concrete would occur in a deep sea environment, further research is needed into the effects of pore pressure on strength.

Very high strength concrete is virtually immune to chemical attack by sea water and by virtue of its very low permeability also provides protection to embedded steel reinforcement. Special attention to corrosion, which may include cathodic protection, is required in cases where the steel sections are exposed to the sea. The protection of exposed steel in marine structures is now well understood, the main difference with the subsea environment being that it is less aggressive than the surface environment both in regard to mechanical disturbance and chemical attack.

4.4 Structural Performance

The geometrical form adopted for the pressure hull is the result of a compromise between the demands of structural efficiency and internal functional layout. Curved boundaries are not usually conducive to efficient usage of space, while increasing the size of the hull favours space usage but increases structural problems. These considerations lead to the long horizontal cylinder with spherical end closures as the basic modular form for depths up to 1000m. Alternative forms having varying double curvature such as the prolate spheroid, while structurally more efficient are less efficient in space usage and are most costly to construct. Ultimately, for minimum weight, there is no alternative to the sphere, though a possible compromise solution is to enclose a cylindrical space within intersecting spheres. (Fig. 7).

While the foregoing considerations apply in general to any material, the hull form most suited to any given depth and location will to some extent, be influenced by the particular choice of material. This, as has been indicated, is likely for the first generation of subsea structures to consist of a combination of steel and concrete having properties designed to lie within a range of strength to density ratio suited to the depth of operation.

A cylindrical module to contain process equipment will have, typically, an inside diameter of 10 metres. Except at very great depths, the shell plate thickness required by a steel hull to resist the primary compression forces is less than that required for stability and stiffening rings are therefore required. The principles of designing ring stiffened cylinders are now well understood, the normal failure criterion being that interstiffener yielding occurs prior to overall collapse. A major problem with this type of structure is its sensitivity to initial imperfections and to discontinuity forces at supports and at interconnecting members. While it is possible to keep construction imperfections to a low level using the special techniques and control of workmanship employed by submarine manufacturers, the problem of discontinuities is more difficult. Considerations of space and safety require the production complex to be subdivided into several modules requiring numerous penetrations and interconnecting tunnels. These are inherent points of weakness and, in a design where the load capacity is characterised by stability, are difficult to analyse.

A reinforced concrete hull will normally be of single shell construction. For depths less than 200m the wall thickness of a cylindrical vessel of high strength concrete required to resist compression alone is thin enough for there to be some interaction with buckling. A bifurcation analysis is then carried out which takes account of the transverse restraints due to the support structure. The failure load may then be estimated by one of several methods which attempts to model the deformation properties of steel and concrete and which may also take account of the effect on load of large deflections. (Refs. 9, 10). In this situation the function of the reinforcement is as much to increase stiffness as to increase strength.

With the increased loading at depths greater than 200m walls become thick and stability is no longer a governing criterion. (Fig. 8). Reinforcement is provided to increase compressive strength and failure is characterised by plastic yield of the composite section. At depths of around 500m the proportion of steel to give the required strength/weight ratio reaches about 10% and it becomes increasingly advantageous to consider a sandwich type wall in which the space between two thin steel shells is filled with concrete. (Ref. 11).

The failure mode of an externally pressurised thick walled reinforced concrete cylinder is imperfectly understood and has received relatively little study. Tests on plain concrete cylinders and on singly reinforced concrete spheres carried out by the United States Navy Civil Engineering Laboratory (USNCEL) have led to recommendations for design in which the average stress in a cylindrical or spherical wall at collapse is related to the uniaxial compressive strength of the concrete as determined by standard tests. (Ref. 12). If the factor relating these is taken as unity, the sectional area required is very similar to that obtained assuming the section to act as a notional short column in which partial safety factors for material and for load are both taken as 1.5.

Since a concrete structure will require reinforcement to resist early thermal stresses and construction loads, the design must take account of the behaviour of the composite section under hydrostatic loading. Failure of an unreinforced concrete section, loaded in compression, initiates by splitting along planes parallel to the direction of the major principal stress, followed by buckling of the laminae so produced. Introducing a single layer of reinforcement parallel to the axis of the principal load, while increasing stiffness at small loads, also lowers the threshold for splitting, unless reinforcement is introduced to resist this effect. Premature failure due to laminar splitting of singly reinforced hollow spheres, as observed during the USNCEL tests, may have been attributable to this effect, exacerbated perhaps by pore pressure. As in a short column, obtaining full advantage of the compressive strength of steel and concrete in a composite section depends upon the effectiveness of transverse reinforcement in preventing premature splitting. Detailing is highly important since radial reinforcement can constitute a potential leak path. The amount and location of radial reinforcement will depend upon the type and density of the primary hoop reinforcement which may consist of conventional bars or structural steel members. In certain arrangements radial reinforcement may not be required. A particular case of this is the sandwich composite in which the primary reinforcement, consisting of steel plate, is placed outside the concrete, thereby imparting a small radial stress at the inner face. In addition to a knowledge of the collapse load it is also necessary to understand the effects of transfer of load from concrete to steel due to concrete creep. This may cause the steel service stress to approach the yield point.

While it is necessary to consider the pressure hull in isolation, during preliminary work, the final design must take account of the local and global effects of constraints imposed by the support structure and by penetrations. The pressure hulls will experience elastic contractions due to hydrostatic loading, whereas the support structure is effectively free of lateral loading after the structure is installed.

If these are connected rigidly large local stresses are induced and the structural modelling must be designed to control and distribute these. Similar considerations apply to interconnecting tunnels. Penetrations should, where possible, be located in the spherical rather than the cylindrical hull regions and should not be located near geometrical discontinuities. The stiffness of the included member at a penetration should match that of the lost material as closely as possible.

In applying limit state philosophy to the seabed structure, it becomes immediately apparent that partial safety factors specified in the building design codes are inappropriate. It is unfortunate that guidance included in the Rules of the Classification Societies and from Codes for marine structures is principally based on building structural codes and probably requires more careful consideration. Partial safety factors for materials and loading, are conventionally based upon lower standards of workmanship and loadings less easily defined than for seabed structures. Consideration should therefore be given to the reduction of material partial safety factors from 1.5 to say 1.15 or 1.20 for concrete commensurate with that for steel of 1.15 and of reducing the partial safety factor for loading under working conditions to 1.20. At the same time there is an argument for introducing an additional overall safety factor, where collapse can be shown to be a criterion of the design, to maintain the integrity of the structure. It is also necessary to undertake a reliability analysis to establish first order reliability criteria. (Refs. 13, 14).

5. CONCLUSIONS

The concept of building seabed containment structures to provide a one atmosphere dry environment for housing a hydrocarbon production facility and for habitat purposes is seen to be feasible. The precise form of the first prototype structures will depend on the function of the facility and the design and installation of the necessary containment is the subject of continuing study. The major problem areas have been identified and it is considered that existing technology will provide viable solutions, subject only to some extension of present principles in the detailed design.

It is evident that high strength concrete as the basic structural material has considerable advantages in the construction of suitable containments for water depths up to about 500 metres. High strength concrete is well proven for use in marine structures and provides adequate properties of durability, protection for steel reinforcement and impermeability. Due to its cement-rich nature, these properties will improve with time.

At greater depths the implications of buoyancy upon the installation procedure has increasing importance and the role of steel becomes increasingly dominant. There exists a range of potential solutions to the structural problem of the pressure hull, employing composite construction of steel and concrete. This may usefully combine the best properties of both materials. Some further research is needed to evaluate these in detail, particularly in regard to the failure mode of thick walled cylinders, to achieve an optimised solution for water depths in excess of 500 metres.

6. REFERENCES

1) Stafford, E.P. : "The Far and the Deep". Arthur Barker Ltd.

2) Derrington, J.A. : "The Construction of Gas Treatment Platform No. 1 for the Frigg Field for Elf-Norge A/S". The Structural Engineer Vol. 55, No. 2, February 1977.

3) Haynes, H.H. : "Evaluation of Pressure-Resistant Concrete Structure", in Sea-floor Construction Experiment : SEACON I, An Integrated Evaluation of Sea-floor Construction, by T.R. Kretschmer, Technical Report No. R-817, Civil Engineering Laboratory, NCBC, Port Hueneme, California, February 1975.

4) Shen-D'Ge', N.J. : "Structures/Materials Synthesis for Safety of Oceanic Deep-Submergence Bottom-Fixed Manned Habitat", J. Hydronautics, Vol. 2, No. 3, July 1968.

5) Harris, A.J. & Fox B.M. : "Adjustment of Structural Concrete Techniques to Offshore Conditions". Proceedings of Conference on Behaviour of Offshore Structures (BOSS), London 1979.

6) Somerville, G. & Taylor, H.P.J. : "Concrete Properties", Proc. FIP 8th International Congress, London, 1978.

7) Haynes, H.H. & Kahn, L.F. : "Behavior of 66-inch concrete spheres under short and long term hydrostatic loading", Technical Report R-774, NCBC, Port Hueneme, Ca. September 1972.

8) Morley, C.T. : "Theory of Pore Pressure in Reinforced Concrete Cylinders", paper presented at 8th International Congress, FIP, London, 1978.

9) Borseth, I. : "Influence of Imperfection and Reinforcement on Collapse Pressure of Concrete Cylinders", BOSS Conference 1976.

10) Holand, I. : "Ultimate Capacity of Concrete Shells", BOSS Conference, 1976.

11) Montague, P. : "The Theoretical Behaviour of Steel-Concrete Circular Cylindrical Shells subjected to External Pressure". Simon Engineering Laboratories, University of Manchester.

12) Haynes, H.H. : "Handbook for Design of Undersea, Pressure-Resistant Concrete Structures", Civil Engineering Laboratory, NCBC, Port Hueneme, Ca., September 1976.

13) CIRIA Report - 63 : Rationalisation of Safety and Serviceability Factors in Structural Codes.

14) CEB-FIP : International System of Unified Standard Codes of Practice for Structures; Vol. 1 : Common Unified Rules for Different Types of Construction and Material.

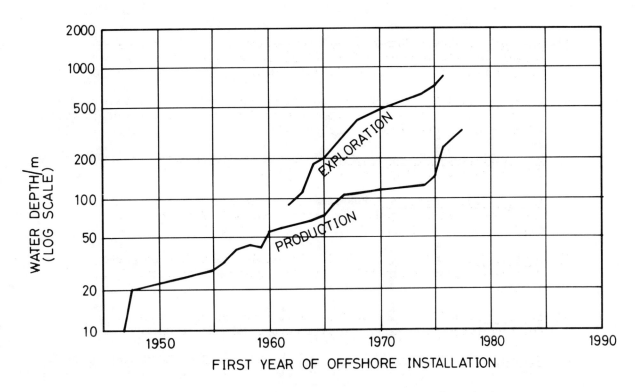

Fig. 1 Maximum water depth for exploration and production

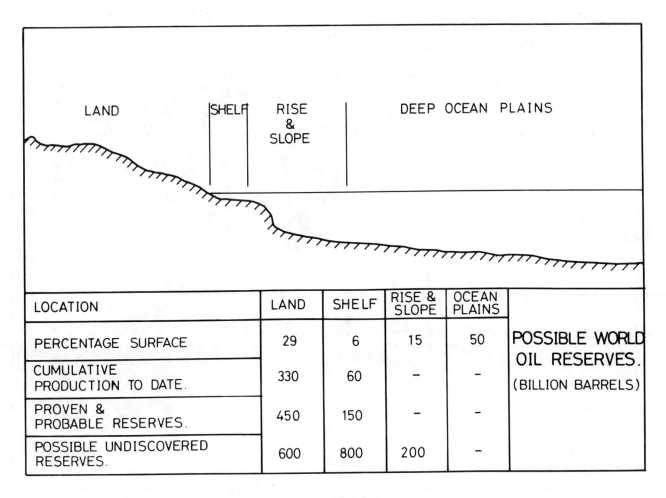

LOCATION	LAND	SHELF	RISE & SLOPE	OCEAN PLAINS	
PERCENTAGE SURFACE	29	6	15	50	POSSIBLE WORLD OIL RESERVES. (BILLION BARRELS)
CUMULATIVE PRODUCTION TO DATE.	330	60	–	–	
PROVEN & PROBABLE RESERVES.	450	150	–	–	
POSSIBLE UNDISCOVERED RESERVES.	600	800	200	–	

Fig. 2 Possible World Oil Reserves

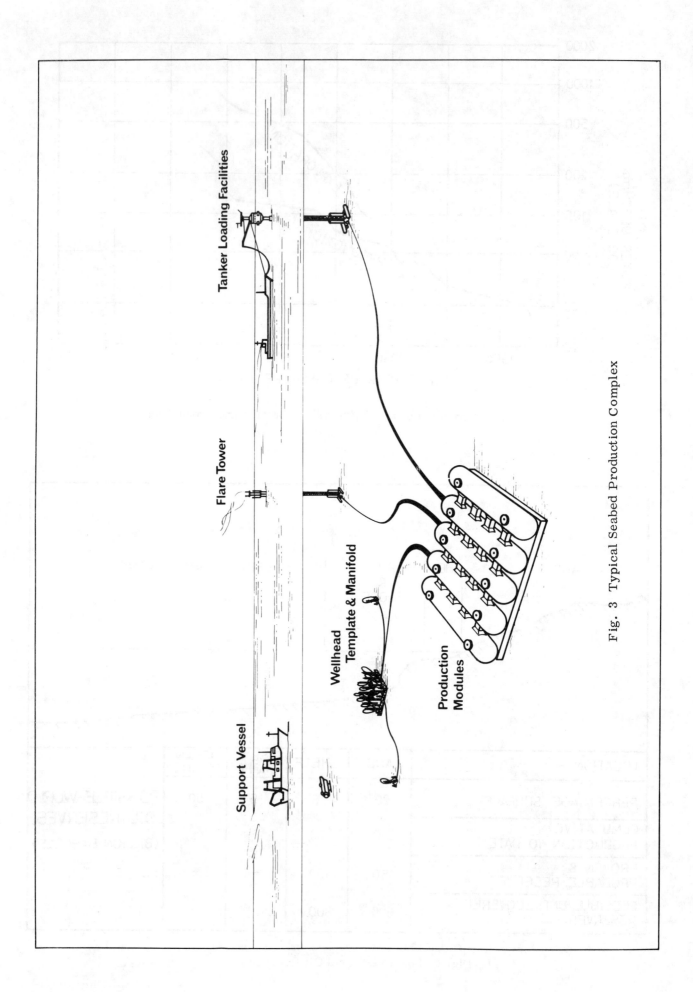

Fig. 3 Typical Seabed Production Complex

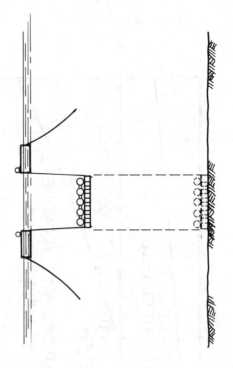

Fig. 5(a) Installation under Positive Bouyancy

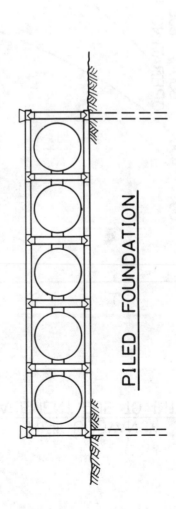

Fig. 5(b) Installation under Negative Bouyancy

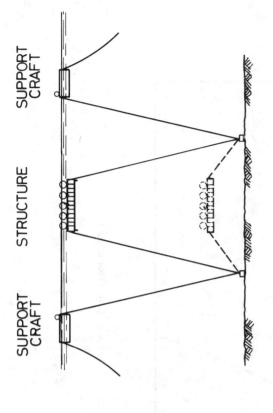

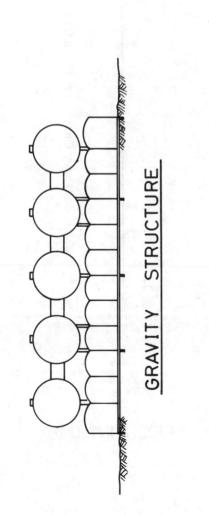

Fig. 4 Alternative Foundation Types

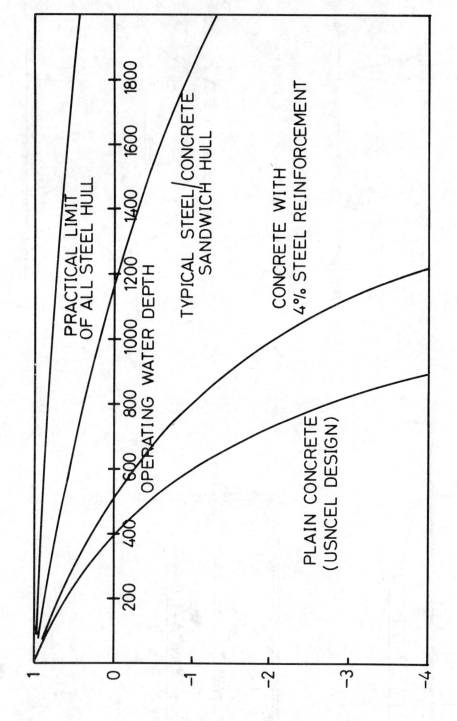

Fig. 6 Bouyancy Curves for Cylindrical Hull with Spherical End Closures

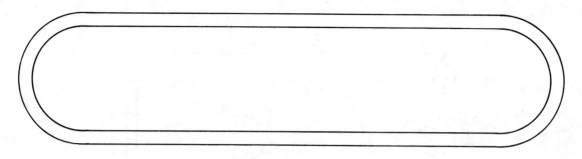

CYLINDER

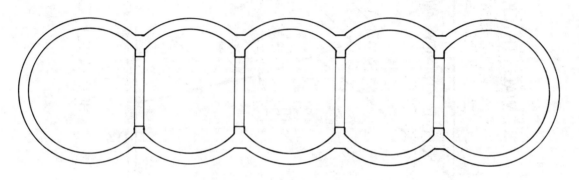

INTERSECTING SPHERES

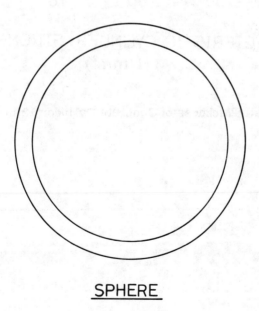

SPHERE

Fig. 7 Alternative Forms for Containment Structure

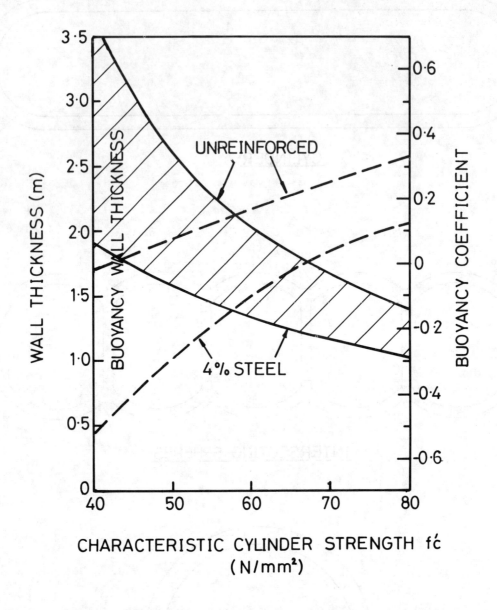

Fig. 8 Minimum Wall Thickness of Concrete Cylinder 12 m i.d. for 500m depth

BOSS'79

PAPER 91

Second International Conference
on Behaviour of Off-Shore Structures
Held at: Imperial College, London, England
28 to 31 August 1979

NETHERLANDS MARINE TECHNOLOGICAL RESEARCH

G.A. Heyning, Chairman AC.MaTS

Netherlands Industrial Council for Oceanology, The Netherlands

Summary

A review is given of ocean engineering oriented research in the Netherlands with emphasis on the operational aspects of the Netherlands Marine Technological Research organization.

Sponsored by: Delft University of Technology, The Netherlands
Massachusetts Institute of Technology, U.S.A.
The Norwegian Institute of Technology, Norway
University of London, England

Secretariat provided by: BHRA Fluid Engineering

Copyright: © BHRA Fluid Engineering
Cranfield, Bedford, England

Netherlands Marine Technological Research

The paper which I have been invited to prepare for you differs from the majority of the other BOSS 79 papers in so far as it does not deal with an interesting new engineering development or a scientific approach to solve any of the many problems confronting the users and the builders of offshore structures.
Instead, I will try to explain how marine technological research in the Netherlands has been organized and how it is operating.
There will be no pictures of impressive structures and no formulae for complex calculations in my paper, only an organization chart and appended lists of research projects undertaken.

When reviewing the ocean oriented technological research effort in the Netherlands the first impression is that this appears to have been undertaken in a rather haphazard way. Problems were tackled as and when the need arose and without any overall vision or longer term objective, at least in the initial stages.
Government entities, industrial firms, scientific bodies and sometimes even individuals often studied the problems with which they felt they were confronted on an ad hoc and almost individualistic basis. Only in recent years there are signs of a more concerted effort.

Nevertheless, the results of these predominantly application oriented research efforts have certainly served a very useful purpose. As a non-scientific senior engineer I am inclined to think that the ad hoc approach to research of the early years was probably not at all a bad proposition. In fact, even now, when we realize that research ought to be part of an overall plan for the future, the lessons learned from the individualistic and pragmatic efforts in previous years will serve as a guiding line for assigning priorities to the many proposals for continued research in this field.

Research on coastal and ocean engineering problems has, of course, been practised in the Netherlands for some considerable time already. Offshore operators, such as the oil industry and the dredging contractors, as well as government entities, among them the Royal Netherlands Navy and Rijkswaterstaat (Department of Public Works) felt the need for reliable scientifically assessed data on the behaviour of materials and structures in the ocean environment.

Their problems were dealt with by research institutes such as the Wageningen Ship Model Basin and the Delft Hydraulics Laboratory, originally established to serve the shipbuilding industry and the construction of river, coastal and harbour works respectively. These institutes had already gained international reputation in their respective fields of research when the ocean engineering problems became urgent.

In addition some of the larger companies also have their own laboratories and, of course, the three universities of technology in Delft, Eindhoven and Twente are devoting an increasing part of their research effort to ocean engineering subjects.

There are scores of associations, foundations, committees, institutes and laboratories in the Netherlands with an interest in ocean science and engineering. Some of these have been active for years already in oceanography, such as the Hydrographic Department of the Royal Netherlands Navy and the Royal Netherlands Meteorological Institute (KNMI), and have comparatively recently extended their sphere of interest to include marine technology data. Others, like the Netherlands Committee for Concrete Research (CUR) and the Netherlands Welding Institute (NIL) have right from their initiation been involved in materials research which in recent years has become more and more ocean oriented. The Foundation for Materials Research in the Sea (SMOZ), of course, devotes its activities exclusively to the behaviour in the marine environment of a wide range of materials, including metals, concrete, plastics and coatings.

Then there is the Netherlands Organization for Applied Scientific Research (TNO) with some forty specialised institutes and laboratories. Many of these go in for marine technological research and development under industry contract or government contract or joint industry/government contract.

I mention a few of them:
- Institute TNO for Building Materials and Structures (TNO-IBBC)
- Institute TNO for Mechanical Constructions (TNO-IWECO)
- Metal Research Institute TNO (TNO-MI)
- Paint Research Institute TNO (TNO-VI)
- Institute of Applied Physics TNO-TH (TPD).

In general the associations, foundations and committees decide on the research programs, arrange the financing and supervise the progress of research projects, whereas the institutes and laboratories carry out the projects, including calculations, model testing, etc., and prepare the reports.

With so many interested and competent parties in the field there is, of course, some overlap and at times even a tendency towards competition and rivalry. As a rule, however, the experts know each other personally well enough to be able to come to terms. Often they serve together on more than one committee or studygroup. And a certain amount of competition is of course healthy, even in research.

Still, there clearly was a need for more coordination in the national research effort.

For the shipbuilding and shipping industry the Netherlands Maritime Institute (NMI) was established in 1974. Co-sponsors are the Netherlands Association of Shipbuilders (CEBOSINE), the Royal Netherlands Shipowners Association (KNRV) and the Department of Shipping of the Ministry of Transport. Dredging research is being coordinated by the Association of Dutch Dredging Contractors (CB), established in 1935 initially with the object of promoting the economic and social interests of its members.

For the offshore industry it might of course have been possible to incorporate their research effort in one of the existing bodies, for instance NMI, and this has been contemplated more than once. However, there were good reasons for setting up a separate coordinating organization.

Technological development in the offshore industry has a considerable higher impetus than in most other industrial activities and concentrated research effort in this field was therefore thought to be preferable to a combination with other, albeit closely related, interests. Moreover, the Netherlands Industrial Council for Oceanology (IRO), a foundation established in 1971 as a body representing ocean oriented industry and commerce in the Netherlands, had meanwhile been entrusted with the responsibility to supervise and administer two series of marine technological research projects.

In 1974 IRO formed a Steering Group on Offshore Structures' Problems (StuPOC) supervising a number of research projects financed by the Ministry of Economic Affairs (EZ). These projects had been proposed by the Division for Underwater Technology of the Royal Institution of Engineers in the Netherlands (KIvI), but EZ preferred to entrust IRO rather than KIvI with the supervision. The reports on these projects have been published in 1978. See appendix 2.

When the Netherlands Welding Institute (NIL) applied for government support for their Neptunus program, a series of underwater welding research projects, IRO was again requested to nominate a Steering Group on Underwater Welding (StuLas) to supervise the program. These projects (see appendix 3) are receiving 70% financing from EZ, the balance being funded by NIL.

An additional and probably the most important reason for establishing a separate organization for marine technological research was the following.

In 1975 the government nominated a Planning Group for Longterm Oceanographic Research (POOL), reporting to the Ministry of Science, on which government, science and industry were represented. One of the POOL Working Groups with a strong industry participating and dealing with Marine Technology, proved to be particularly active and by the end of 1976 had formulated a proposal for a series of technological research projects, which they deemed to have a high priority.

Unfortunately the Ministry of Science was not in a position to guarantee the necessary financing of these proposals. The Netherlands Industrial Council for Oceanology (IRO) was then found willing to approach the Ministry of Economic Affairs (EZ) - with whom they already had such excellent contacts - for financial assistance. Largely due to the favourable impression StuPOC had made on EZ this Ministry agreed in principle to continue their support for marine technological research.

In contrast to the StuPOC research program, which was wholly government-sponsored, EZ now however made financial support conditional on adequate financial participation by industry. Moreover, a new supervisory body, replacing StuPOC with its limited mandate, was to be nominated by IRO for approval by EZ.

Early 1978 the Netherlands Marine Technological Research Organization (Marien Technologisch Speurwerk, MaTS) was established. It consists of an Advisory Committee (Advies Commissie, AC.MaTS) and a number of Steering Groups (Stuurgroepen, StuMaTS) and Project Groups (Projektgroepen, ProMaTS). See appendix 1.

The four members of AC.MaTS are appointed by IRO. They in turn appoint the StuMaTS chairmen. The StuMaTS chairmen, in consultation with AC.MaTS, invite four additional members to join their StuMaTS team, who, upon accepting, are then appointed by AC.MaTS. EZ appoints a sixth member in every StuMaTS and IRO provides an administrative assistant for each StuMaTS. Per project (sometimes more than one) a ProMaTS is appointed by the relevant StuMaTS, usually consisting of two to four experts.

The most important task assigned to AC.MaTS is to develop a longterm program for marine technological research. Clearly, this is a very difficult assignment and in the first operational year of MaTS hardly any headway has been made in this respect. AC.MaTS spent most of its time and effort on manning the StuMaTS teams and getting them started on the various research projects proposed by the POOL Working Group on Marine Technology.

As these proposals had been quite well prepared and in many cases even included tenders from appropriate institutes or laboratories, the StuMaTS teams were asked to say either yes or no to the project proposals assigned to them, rather than altering or improving their content. This procedure was adopted for this so-called first phase of the MaTS program only, the firm intention being that StuMaTS teams will in future be required to fully prepare the research project proposals, whether generated by themselves or initiated by outsiders, i.e. industry or scientific bodies.

In so doing they will have to weigh the proposals in accordance with certain criteria. For each proposal an evaluation will have to be given as to its expected importance for the Netherlands in regard of:
    a) promoting industrial/commercial capability
    b) improving safety aspects
    c) improving environmental aspects
    d) enhancing fundamental know-how.
Further, each proposal must contain a cost estimate, preferably based on a preliminary tender, and an estimate of the time required for carrying out the project.
An indication of industry participation is also required.

The proposals are then sent to AC.MaTS, who on the basis of the same four criteria and taking into account available financial and manpower resources list the proposals in order of priority. This list is then submitted via IRO to EZ.

At various stages of the procedure there is, of course, close consultation between EZ and MaTS, primarily through the EZ representative on each of the StuMaTS teams. As a result EZ are well informed on the various proposals at an early stage and therefore in a good position to make a final decision regarding financial support from government funds for the marine technological research program.

When a proposal has been accepted AC.MaTS assign the project to the appropriate StuMaTS, who in turn appoint a ProMaTS for close supervision of progress.

A formal contract agreement with an institute or a laboratory is drwan up, signed on behalf of MaTS by the responsible StuMaTS chairman and the AC.MaTS chairman and also by the chairman of IRO.

The contractor in consultation with the relevant ProMaTS submits quarterly progress reports to StuMaTS. ProMaTS chairmen also attend the StuMaTS meetings to report in person on the progress of their project or projects. AC.MaTS are being kept informed by copy of the minutes of the StuMaTS meetings.

Final reports on completed projects are prepared by the contractor in consultation with ProMaTS and StuMaTS. As a rule the full report will be in dutch and printed in limited numbers. In addition summaries in dutch and in english are available for wider distribution.

AC.MaTS give guidance as to uniformity of format and regarding lists of recipients. On this latter aspect AC.MaTS do, of course, consult the sponsors - EZ in particular - because of the possible commercial advantage these sponsors might wish to reserve for themselves for a period of time. An embargo of up to two years may sometimes result from these consultations.

For the same reason MaTS do not feel free to conclude formal agreements with corresponding organizations abroad. We do value, of course, the contacts we already have with for instance the Science Research Council and CIRIA in the UK and with Det Norske Veritas in Norway and we are in favour of exchanging general information on the subjects being studied by each one of us.
The fact that such contacts may from time to time prevent costly duplication of effort is in itself an advantage. However, as regards closer cooperation we feel that this should be restricted to ad hoc instances on the level of a few specific projects where international cooperation will clearly be to the advantage of all parties concerned.

Appendix 4 lists the MaTS projects initiated in 1978. Total funds available from EZ for this so-called first phase amount to ƒ 5.425.000. As mentioned before, the majority of projects stem from the POOL proposals, with only a few later additions proposed by others.

These proposals were prepared and formulated during long and interesting deliberations by scientists from institutes, laboratories and universities and experts from industry. Incidentally, the individuals taking part in these discussions felt that the exchange of views with their colleagues was in itself a valuable exercise and well worth the time and effort spent. It might well have been of interest to you to hear how some of these discussion groups arrived at their proposals, but time does not permit. Moreover, I am certainly neither a scientist nor an expert and I would not be the right person to inform you properly on this subject.

As a general observation I would like to say that the resultant MaTS framework as shown in appendix 1 may again give you the impression of a rather haphazard approach. It may, for instance, seem odd to you that research on concrete and on plastics has been lumped together under one StuMaTS. You may also query the fact that there is no StuMaTS for, say, soilmechanics or for instruments or electronics/acoustics, just to mention a few spheres of undoubted interest in marine technological research. And should ProMaTS S-1 for design criteria not be a separate StuMaTS, taking into account the behaviour of concrete as well as that of steel and of the soil on which the structures will stand?

The answer is, of course, that the POOL proposals handed down to us for action led to this subdivision for the very practical reason of trying to spread the workload as evenly as possible.

Naturally, these points have been discussed on several occasions but fortunately every one concerned agreed that it was better to get going with a somewhat lopsided framework than to sit and wait for the perfect organization. Still, it may well be that by the time this paper is introduced to you orally I shall have to show you a revised organigram.

And by that time I also expect to be able to report on progress made on the various research projects now in hand and even on some of the new projects due to be initiated in early 1979.

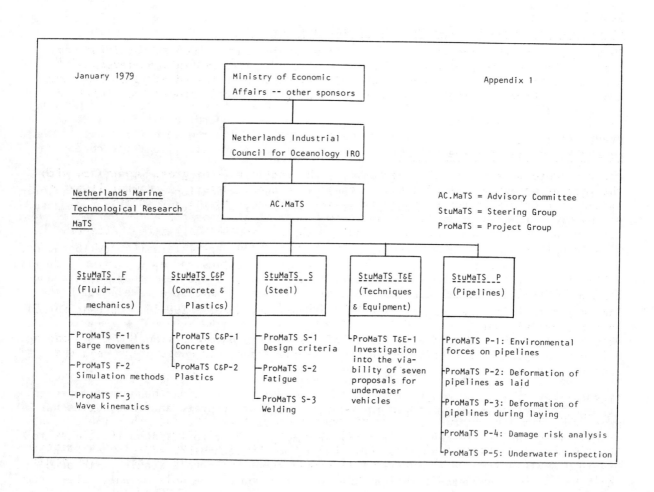

Appendix 2

StuPOC Projects

In 1974 the Netherlands Industrial Council for Oceanology, the Division for Underwater Technology of the Royal Institution of Engineers and the Ministry of Economic Affairs set up the Netherlands Steering Committee on Offshore Structures' Problems (StuPOC).

Under the auspices of StuPOC several studies were performed in the field of the design, the construction and the installation of jacket-type structures and gravity structures at sea. The list of subjects is as follows:

StuPOC-I-1-1
Repeated loading on clay

The main purpose of this project was to investigate the fundamental behaviour of clay subjected to cyclic loading. The report is primarily based on a study of results from laboratory testing on one particular soil, the plastic clay.

StuPOC-I-7
Axial bearing capacity of open-ended steel pipe piles

This report contains the results of an introductory study, aimed at obtaining some basic insight in the phenomenon of cone resistances, measured in sand layers at shallow depths below the seabed. The study has been concentrated on North Sea subsoil conditions, in which extensive experience with offshore cone penetration tests has been gained.

StuPOC-II-2
Measuring methods for the unevenness of the seafloor

Accurate determination of the unevenness of the seafloor is of great importance with the increasing activities of the oil industry in deeper water. In general it can be stated that errors in depth recording at sea increase with the waterdepth.
An inventory is made of available and promising optical, acoustic and electronic techniques for the depth recordings and for the horizontal positioning. Attention is also paid to the processing of results and some recommendations are drawn up.

StuPOC-II-6
Wave- and current forces on jackets

The report contains the results of a literature study about "wave- and current forces on circular cylinders".
Attention is also paid to the influence of the various parameters on the wave loading and to some specific subjects such as an analysis method for measurements and the spectral and probabilistic approach to wave loading.

StuPOC-II-8
Aqua-planing of gravity structures in the last stage of sinking

A theoretical consideration of aqua-planing has been set up from which it is plausible that in the last stage of sinking and already with small heeling angles an asymmetric pressure distribution on the bottom will occur. The accompanying destabilising moment will tend to move the structure side ways.

StuPOC-III-3
Fatigue of concrete

To obtain a better understanding of the behaviour of plain concrete in compression under varying repeated stresses, constant-amplitude tests and programmed-stress tests were done to verify the Palmgren-Miner rule.
Applicability of this rule for concrete gives the possibility to make a reasonably reliable expectation of the lifetime of concrete loaded with irregular varying compressive stresses.

StuPOC-III-3-1
Fatigue of concrete, safety considerations

The main intention of the safety considerations is to indicate a line along which the investigation in this field has to be done in the future to set up a reliable probabilistic calculation of fatigue.
These considerations are based on the experience and insight gained from the investigation of fatigue of concrete (StuPOC-III-3).

StuPOC-IV-4
Fatigue properties of connections in jackets

This literature study focuses on gathering as much information as possible about the influence of some factors on the fatigue properties of connections in jacket structures.
Those factors are: material, environment, loading and design.

StuPOC-V-5-3-(1)
Inventory of accidents with offshore structures

Risk analysis is a useful instrument for the evaluation of a great number of aspects and for the purpose of the analysis an inventory was made of accidents with offshore structures during the period 1966 - 1975.

StuPOC-V-5-3-(2)
Inventory of accidents with fixed platforms in the North Sea

As a sequence to the report "Inventory of accidents with offshore structures" a separate report was made of the accidents with fixed platforms in the North Sea during the period 1966 - 1975.

StuPOC-X-2
Pile driving tests and loading tests in overconsolidated sand strata

This is a summary of the results of a study, which was coordinated by Fugro-Cesco and in which StuPOC participated.
The test program is a sequel to StuPOC-1-7 "Axial bearing capacity of open-ended steel pipe piles".

These studies have led to the following reports:
- Repeated loading on clay
- Axial bearing capacity of open-ended steel pipe piles
- Measuring methods for the unevenness of the seabed
- Wave- and current forces on jackets
- Aqua-planing of gravity structures in the last stage of sinking
- Fatigue of concrete
- Fatigue of concrete, safety considerations
- Fatigue properties of connections in jackets
- Inventory of accidents with offshore structures during the period 1966 - 1975
- Inventory of accidents with fixed platforms in the North Sea during the period 1966 - 1975
- Pile driving tests and loading tests in overconsolidated sand strata
- Foundations of gravity structures
- Deep sea attack on concrete
- Safety considerations and risk analysis for offshore structures
- Netherlands associations involved directly or indirectly in offshore activities.

These reports can be ordered at the costs of printing from the Netherlands Industrial Council for Oceanology (IRO), P.O. Box 215, 2600 AE Delft.

Appendix 3

StuLas Projects

Welding in the dry under increased pressure

The behaviour of the electric arc and the melt when welding under increased pressure will be studied with the object of developing suitable equipment for such conditions.

Remote controlled underwater welding

An investigation into the possibility of developing an automatic welding procedure with remote control for underwater welding.

Underwater stud-welding

Stud-welding is already practised in the dry. Investigation into the influence of the wet environment and increased pressure on the welding parameters and the power supply, when welding underwater, will permit developing procedures for application in the marine environment.

MIG-welding

Existing MIG-welding techniques will be investigated with a view to develop the process for underwater application.

Underwater welding with mantle-electrodes

Optimisation of this welding procedure by investigating the characteristics of the electrodes, the steel and the power supply and the resultant weld properties.

Explosive clamping and welding

Based on some recent experience with explosive clamping and welding, under conditions where other processes could not be applied, the procedure will be developed for wider application.

Bonding of steel/steel and concrete/concrete connections

The aim is to develop reliable bonding procedures for underwater repairs on steel and/or concrete structures.

Appendix 4

MaTS Projects

Ageing of fibre-reinforced plastics in a marine environment

A literature review and a limited market investigation is expected to result in a proposal for exposition and simulation tests.

Collapse behaviour of fibre-reinforced plastics in a marine environment

As a rule fibre-reinforced plastic structures are over-dimensioned with regard to the stresses to which they are submitted. Insight in collapse behaviour may therefore lead to cost reductions. Based on a literature review a test program will be proposed.

Wider application of fibre-reinforced plastics in offshore engineering

An investigation into production processes and construction procedures and methods to improve the cost-effectiveness thereof will, hopefully, lead to wider offshore application of these interesting materials.

Fatigue of concrete structures

To obtain a better insight in the deterioration of reinforced concrete in a marine environment preparations are made for research into
- stationary and non-stationary random stresses under compression
- stationary and non-stationary random stresses under tension and bending forces.

Inspection and monitoring methods

Methods will be investigated by which it will be possible to detect cracks in concrete and corrosion or fracture of steel reinforcement bars at unaccessible parts of offshore structures.

Repair and protection of concrete

A state of the art report on underwater repair and protection of concrete will be prepared indicating those techniques which merit further research and development.

Influence of oil on concrete

Based on a literature review and interviews with industry a proposal will be prepared for further investigation into possible harmful effects of oil and oil/seawater on reinforced concrete.

Influence of temperature changes on concrete

An inventory of available research in this field will be prepared. There is some doubt regarding the conclusions of french and british research on the magnitude of observed cracks.

New materials for concrete

A state of the art report will indicate the applicability of new materials in concrete structures.

Concrete in the deep sea environment

Application of reinforced and prestressed concrete at depths greater than 100 m requires investigation of the behaviour of concrete at such depths. A proposal for further research will be prepared.

Stress- corrosion of prestressing steel wires

There is insufficient knowledge on this phenomenon and a proposal for further research will be prepared.

Underwater inspection of pipelines

New and/or better methods to obtain information on the condition of pipelines on/in the seabed will be investigated.

Forces acting on pipelines and the behaviour of the seabed

A literature review will cover the following subjects:
- Forces acting on slender cylindrical bodies
- Behaviour of the seabed, in particular vertical movements and the behaviour of underwater dunes
- Behaviour of the seabed around pipelines due to currents and waves
- Penetration of ships anchors with a view to required burial depth of pipelines
- Behaviour of the soil cover over pipelines.

Deformation of pipelines as laid

Investigation of buried pipelines behaviour (onshore, at roadcrossings, etc.) and model tests in a pressure tank.

Deformation of pipelines during laying

A computer program, based on the finite element method, for defining the static equilibrium of the pipeline will be developed, taking into account large displacements but only small deformations.

Damage risk of pipelines

Available information on the impact of anchors, fishing gear, etc., on steel pipes and their concrete coating will be listed. Secondly, an inventory of accidents to pipelines in the North Sea will be prepared.

Dynamic behaviour of offshore structures

Based on a review of available information a proposal will be prepared for further research into:
- Defining actual loads on offshore structures
- Relative importance of loads caused by different phenomena
- Influence of damping effects on the dynamic behaviour of offshore structures.

Guidelines for deciding on the optimum configuration of a structure under predetermined conditions

Based on existing configurations and their environmental conditions an optimalisation procedure for configurations at future locations with known environmental conditions will be developed.

Probabilistic safety analysis for offshore structures

Developing and testing of a probabilistic safety analysis procedure for gravity and jacket structures, taking into account wave loads, own weight and deck loading and foundation characteristics.

Kink and collapse calculations on floating, fixed and gravity structures

State of the art reports will be prepared on kink and collapse research into sheet steel and concrete shell structures. As a result proposals for further research may be expected.

Life expectancy of offshore structures

A theoretical study sequent to the EEC research on fatigue and corrosion fatigue behaviour of offshore steel structures.

Full-scale fatigue tests in seawater

This forms part of the extensive EEC research on fatigue.

Evaluation of defects

The aim is to provide a calculation method for the life expectancy of structures based on observed defects, assuming that the critical defect magnitude is known for the material concerned.

Protective coatings of steel structures in the marine environment

A literature review investigating the possibility to:
- Develop products which will reduce the detrimental effects of the subsoil
- Develop methods to clean the subsoil in order to improve the protective effect of existing coating materials.

Optimalisation of cathodic protection of offshore structures

A review of available information, indicating areas of inadequate know-how. The market potential for Netherlands industry will also be reviewed.

In situ fatigue tests

Under EEC auspices a large number of laboratory tests have been performed. To check whether the results are representative for actual offshore conditions fatigue tests will be carried out in the real marine environment.

Effect of stress-relieving

COD-tests will be carried out on small and large stress-relieved testpieces.

Guidelines for heavy-wall welding

The aim is to define the elements of the welding procedure which influence crackresistance in heavy-wall welded structures and to formulate recommendations.

Welding information courses

This project consists of arranging information and demonstration courses for industry personnel, the course attendants being "vertically integrated", i.e. welders, draftsmen designers, managers, etc.

Non-destructive inspection of offshore structures

Development of NDT techniques for inspection during construction and monitoring when installed.

Current and wave loads on cylindrical bodies

Measured forces on a vertical pile due to constant currents and regular waves will be analysed as a first phase.

Mathematical simulation methods for floating structures

The various phenomena will be listed and the possibility of describing their effects in mathematical terms will be analysed. Treatment of these aspects in existing simulation techniques will be reviewed, including their restrictions.

Wave kinematics

Results of measurements will be compared with those of spectral and deterministic calculation methods.

Movement of barges

The aim of this study is to evaluate the applicability of existing calculation methods, to indicate their restrictions and to propose ways to improve their reliability.

Techniques and Equipment

Investigation into the viability of seven proposals for underwater vehicles.

BOSS'79

Second International Conference on Behaviour of Off-Shore Structures

Held at: Imperial College, London, England
28 to 31 August 1979

PAPER 92

MARINE FOULING ON PLATFORMS IN THE NORTHERN NORTH SEA

R. Ralph, BSc, PhD
Aberdeen University, U.K.

Summary

The pattern of marine growth on platforms in the northern North Sea is described. At levels near the surface mussels or the soft hydroid **Tubularia** are dominant but at deeper levels the growth becomes more complex. There are some deep water species that may be significant in the operation of subsea completions. A unit has been established in Aberdeen to investigate some of the problems associated with marine growth.

Sponsored by: Delft University of Technology, The Netherlands
Massachusetts Institute of Technology, U.S.A.
The Norwegian Institute of Technology, Norway
University of London, England

Secretariat provided by: BHRA Fluid Engineering

Copyright: © BHRA Fluid Engineering
Cranfield, Bedford, England

The presence of marine growth on offshore oil and gas platforms has a number of engineering implications. Some of these are obvious, for example, the increase in structural loading and interference with inspection procedures, while others like the relationship between marine growth and corrosion, and the effect of growth on different corrosion protection systems, are not clear. Marine biologists have a role to play in identifying the different species present on existing platforms, predicting the rates of growth of different fouling communities, and in giving general biological advice on the methods of attachment of different organisms, their ease of removal, rate of recolonisation etc.

Gas platforms in the southern North Sea experience levels of marine growth, particularly of mussels (Mytilus edulis), to the extent that regular cleaning is carried out to relieve structural loading. It is not clear yet whether such cleaning will become a regular requirement in the more northerly fields but some cleaning will have to be done to allow inspection of important areas. Although there is an enormous amount of information available on the general subject of marine fouling, this is almost all experience that has been gathered from ships and shallow water installations. The practical difficulties faced by biologists in sampling hard surfaces in depths of more than a few metres mean that there is almost no prior information on what organisms might be expected on offshore structures. This is especially true of the rates at which fouling might appear and changes in the fouling communities with time. The oil industry itself is the most important source of information and although many aspects of the industry's operations are necessarily confidential, an interchange of data on marine growth is desirable.

From the Zoology Department in Aberdeen we have had opportunities of examining the marine growth on platforms and have produced a number of marine growth inspection reports. Some trends in the pattern of growth are appearing. In the upper 30 m one of two organisms are dominant, either mussels or the soft colonial hydroid Tubularia. The presence of one or the other depends upon the position of a platform in relation to currents that may carry mussels larvae from coastal populations. A platform out of the path of such a current will have very few, if any, mussels and Tubularia becomes dominant. Tubularia is present all over the North Sea living on shells, stones and boulders while mussels are restricted to rocky shores. The presence of one or the other on a platform is an important point because mussels can be a potential cause of loading problems while the soft Tubularia will have a negligible effect. Below about 30 m the picture becomes more complicated. Platforms are colonised initially by a community of species, all of which have hard calcareous skeletons. These include tubeworms, barnacles and saddle-oysters. These animals are very quick to colonise new surfaces and may completely cover an area in one season. The community is a relatively shortlived one and after three to four years is overgrown by a more complex community of hydroids, sponges, anemones, seasquirts and bryozoans, almost all of which are softbodied. The tubeworms and others are smothered and die but their skeletons remain firmly cemented to the surface and may be there for many years. Corrosion is known to take place beneath barnacles and other organisms and its development under the skeletal remains of these initial colonisers may be important. The ultimate thickness of this secondary community is not yet known but is unlikely to be more than an average level of 10 to 15 cm. Below 80 to 100 m marine growth is sparser generally, slower to appear and may not reach its full extent for perhaps a decade. However there are some deeper water species that may interfere with the operation of subsea completion units. One hydroid Obelia longissima can form very dense mats of material that will impair inspection by unmanned submersibles while the barnacle Balanus hameri may impede remotely controlled operations. This barnacle can grow to a basal diameter of 5 to 6 cm and a similar height. The body of the animal can be easily removed but the baseplate is very firmly attached to the substrate. Both of these organisms settle on structures very quickly and have rapid growth rates.

As yet there is not much information on the fouling of concrete platforms. The surface fouling is likely to be similar to that appearing on steel platforms but a number of organisms can burrow into concrete, including the boring sponges and some